IB	IIB	IIIA	IVA	VA	VIA	VIIA	noble gases
							He 2
		B 5	C 6	N 7	O 8	F 9	Ne 10
		Al 13	Si 14	P 15	S 16	Cl 17	Ar 18
Cu 29	Zn 30	Ga 31	Ge 32	As 33	Se 34	Br 35	Kr 36
Ag 47	Cd 48	In 49	Sn 50	Sb 51	Te 52	I 53	Xe 54
Au 79	Hg 80	Tl 81	Pb 82	Bi 83	Po 84	At 85	Rn 86
111	112	113	114	115	116	117	118

Tb 65	Dy 66	Ho 67	Er 68	Tm 69	Yb 70	Lu 71
Bk 97	Cf 98	Es 99	Fm 100	Md 101	No 102	Lw 103

INTRODUCTORY CHEMISTRY

SAMIA SARKAR

INTRODUCTORY CHEMISTRY

Robert J. Ouellette
THE OHIO STATE UNIVERSITY

HARPER & ROW, PUBLISHERS
New York, Evanston, and London

INTRODUCTORY CHEMISTRY

 For information address Harper & Row, Publishers, Inc., 49 East 33rd Street, New York, N.Y. 10016.

LIBRARY OF CONGRESS CATALOG CARD NUMBER: 73-96234

Cover photomicrograph by Mortimer Abramowitz. Oxalic acid recrystallized from a melt was photographed in polarized light using a Nikon S-Ke Polarizing Microscope and a Nikon EFM Microflex camera. See "The Amateur Scientist," *Scientific American,* April 1968, p. 125, for details of the method.

CONTENTS

PART FOUR: ORGANIC CHEMISTRY *327*

PREFACE

At the Ohio State University four introductory chemistry courses are offered for groups designated as honors, science majors, engineers, and nonscience majors. It has been both a joy and a demanding experience for the author to have taught the nonscience majors for seven years. In that time the heterogeneity of the class, in terms of both background and interest, has provided a teaching challenge that has led ultimately, through a long process of experimentation, revision, and refinement, to the preparation of this text. Students with little or no chemistry background are intermixed with those who have at least been exposed to all the terms but possess varying degrees of understanding. The instructor is confronted with nursing, agriculture, and home economic students as well as a substantial number of arts and science students, all of whom have chosen chemistry to fill some specific need in their education.

The interests, requirements, and abilities of this group dictate an approach quite different from those generally used in majors' or honors' courses. This book has been written with the assumption that most students have had no previous course in chemistry and that their primary interests lie outside the physical sciences. There is, however, no reason that these students cannot master the basic principles of chemistry, and, with allowances made for their limited mathematical background, they can be exposed to almost as rigorous a treatment of the fundamentals as a chemistry major. For those students who have had a high school chemistry course, the author has attempted to alter the approach in order to hold their interest while reviewing familiar material and to prevent their being lulled into a false sense of mastery.

In addition to the principles of chemistry traditionally covered in an introductory course, the author has chosen to emphasize organic and biochemistry. To this generation of students the products of research in organic and biochemistry are topics in the popular press and affect their daily life almost immediately. The "relevance" of organic and biochemistry to them needs little elaboration. In addition, organic and biochemistry serve as excellent vehicles for teaching the fundamental concepts. The physical properties of organic compounds can be explained in terms of intermolecular attractive forces and other concepts of bonding theory. The mechanisms of organic reactions serve to illustrate the principles of kinetics and can be interpreted in terms of the structures of reactants and products. The vital relation of kinetics and thermodynamics to life processes is shown through its impact on metabolism in Chapter 28.

Although descriptive inorganic chemistry is necessarily treated less comprehensively, it has not been totally omitted. Chapters on metals, nonmetals, and transition elements emphasize the trends in chemical properties of inorganic materials rather than their nomenclature and variety of compounds. Some topics often taught via inorganic chemistry, such as molecular geometry and stereochemistry, are found instead in the organic or biochemistry sections, where the author believes they can be most logically and economically developed.

One premise that has guided the pedagogy of this text is that for nonmajors an understanding of the method of the scientist and a grasp of fundamental principles are more important that manipulative skill in problem solving or an array of descriptive facts. Accordingly, the author has deemphasized the mechanics of equation balancing and stoichiometric calculations. Although these subjects form part of the basis for atomic mass determination, the mole concept, and much of the rest of chemistry, the nonscience major need not develop a proficiency in setting up and solving all the conceivable stoichiometric problems. It is more important to illustrate that there is a way to approach some simple examples of these problems and that their solution rests on certain basic principles. Beyond that point chemical calculations are exercises in math. For this reason, the problems chosen stress method rather than computation. If the solution to a problem is properly set up, obtaining the answer requires only the simplest arithmetic.

In this book the meaning of the term *model* and its importance to chemistry are key themes. Throughout the text, the student is reminded that our model for a particular phenomenon is just that—a model with certain inherent limitations and therefore subject to change. The intensity of this warning decreases as the book develops, in the hope that the student progresses and interprets all statements in the light of earlier reading. In short, an attempt is made to show the student that the scientist actually "knows" little and "believes" much but is open to changing his mind about his beliefs as additional information becomes available.

A semihistorical approach to the development of models is taken in the early chapters. Selected facts and ideas are interwoven to suggest a model to the student. In some cases the actual model may have arisen from another combination of facts. The main goal in the approach is to delay the ultimate statement of the model until the student is prepared to accept it, rather than to confront him with a model and then point out how the model is consistent with reality. The semihistorical approach allows the student to enjoy vicariously the accumulation of experimental observations without restricting his understanding by an introductory statement that gives him the conclusion but denies him the insight of seeing its evolution.

The course at Ohio State for which this text was written is two quarters long and covers the material of Parts I–IV, the fundamentals, inorganic chemistry, and organic chemistry. The inclusion of a biochemistry section makes the book appropriate for a full-year course and provides greater flexibility to instructors in two-quarter or one-semester courses. For example, for a short course that must include biochemistry, the instructor may wish to delete Part Three on inorganic chemistry; this can be done without loss of any material vital to the discussion of organic or biochemistry. Other selections of material and corresponding teaching schedules are detailed in an instructor's manual available on request from the publisher.

R. J. O.

PART ONE

THE PROPERTIES OF MATTER

The portion of knowledge called chemistry can be summarized and discussed in terms of a collection of fundamental principles. These principles are consistent with all known experimental facts of the physical and chemical properties of matter and provide a framework that helps organize and explain a vast body of information.

Historically, the development of chemistry rested primarily upon the observation of chemical change. However, the physical properties of matter also provide criteria for evaluating the validity of hypotheses and theories. In particular, the physical properties of matter are the crucial evidence for any theory of the structure of matter. In this part the emphasis is on physical properties, because they can be examined without extensive knowledge of the constitution of matter. Chemical properties and changes are treated in increasing detail in subsequent parts of this text.

1

CHEMISTRY AND MEASUREMENT

1.1 EARLY HISTORY OF CHEMISTRY

From the time of man's first appearance on earth he has been concerned with and fascinated by his surroundings. Initially man had to adapt to his environment and hope for the best in his attempt to survive. By observing the processes that occurred about him he was able to predict future events, thereby improving his adaptability. Basically this is still the chief concern of man, because he can survive only if he can adapt to future events. Although early man may have considered and examined nature in an attempt to control it, he probably was not concerned with the reasons behind occurrences.

Basic chemical industry as it appeared between 4000 and 3000 B.C. involved the very practical processes of production of metal, glass, pottery, pigments, dyes, and perfume. Gold, which occurs free in nature, was used by the ancients for ornamental purposes, much as it is today. The metallurgy of copper, lead, tin, iron, and silver developed before 3000 B.C. Ornamental glass and glazed pottery were produced in China, India, and Egypt. Early chemistry, as utilized and developed by these societies, was more a practical art than a science.

The beginnings of philosophy, which occurred in Greece about the time of Socrates (500 B.C.), are generally associated with the beginnings of science. Science is the knowledge of facts, phenomena, and proximate causes obtained and verified by observation, organized experiment, and ordered thinking. Although other civilizations had recorded and utilized natural processes, they did not develop the habit of asking penetrating questions about nature or of attempting to explain natural phenomena. Unlike the practical orientation of chemistry in other civilizations, the Greeks approached chemistry as *rationalists:* They asked questions and sought answers based on reason alone. While the human mind is extremely powerful and capable, it depends on continuing observations in order to progress. The Greeks were unwilling to become *empiricists* because they did not rely on knowledge derived from the various sensory experiences and to some extent were unwilling and unable to control the variables associated with natural phenomena in order to make additional observations. There are similarities between what certain Greek philosophers thought about nature and what is now believed. However, their theories were products of contemplative philosophy rather than of an experimental science.

It was not until the sixteenth century that Europeans began to realize that Greek rationalism was an imperfect way of studying nature. The experimental method was emphasized in Galileo's study of mechanics. In chemistry the alchemical laboratory of the ninth to sixteenth centuries was the center of experimentation. Although the experimental approach was directed toward discovery of a philosopher's stone to change common metal into gold and the preparation of the elixir of life, the net result was the development of many methods that would form the bases of the field of chemistry.

The *iatrochemists* of the sixteenth and seventeenth centuries sought to merge the fields of medicine and chemistry under one discipline. They believed that man was a composite of chemicals and that health depended upon the proper proportions of the elements. This belief, although somewhat simplistic, is essentially correct. The experimental method continued to flourish in the laboratory of the physician-chemist, and the standards of the developing field of chemistry were increased.

Modern chemistry can be traced to the era of Robert Boyle, whose study of the properties of gases is a classic example of the experimental methods of science. His book, *The Skeptical Chymist,* published in 1661, established chemistry as an open field of study relying on experiment and seeking explanations for natural phenomena that are consistent with facts gathered in the laboratory.

In the past two centuries the growth of science has been rapid and

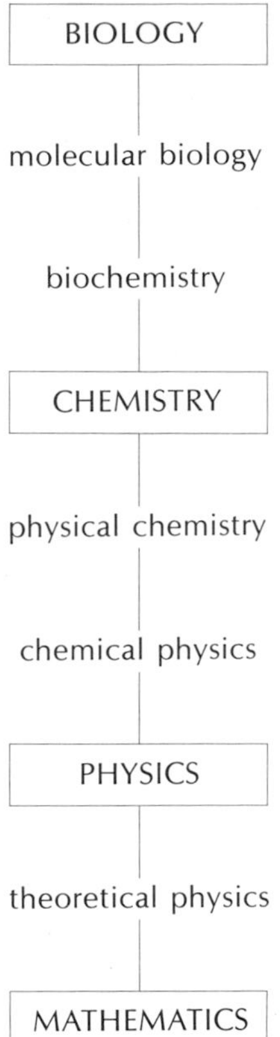

Figure 1.1
Interrelation among certain scientific disciplines.

diverse. Although mathematics is fundamental to all sciences, chemistry and physics have provided the base from which the other sciences have evolved (Figure 1.1). Today chemistry is a broad discipline with many subdivisions. However, chemists essentially are interested in determining the composition, properties, and behavior of matter. Chemists, like other scientists, seek to understand nature and to discover general principles to account for their observations of natural phenomena. To the scientist understanding implies the correct prediction of events by a law that has accurately predicted many similar occurrences. If events are not predictable, the principles behind them are not understood completely.

1.2 CLASSIFICATION OF MATTER

PROPERTIES OF MATTER

Any given sample of matter has unique properties that distinguish it from other substances. The list of properties that can be used in characterizing matter is long. Among the common qualitative observations that are made are color, odor, taste, hardness, and fluidity. Quantitative measurements, such as mass and volume, also can be employed in describing a sample of matter. These two characteristics are common to all matter and serve as a definition: All matter occupies space and possesses mass. In order to study matter it is necessary to classify it and to have a means of determining the mass and space that it occupies.

In describing matter both *extrinsic* and *intrinsic* properties are used. Extrinsic properties are qualities that are not characteristic only of the type of matter contained in a sample but are also functions of the particular sample. The extrinsic properties of a given sample of water are its volume, mass, shape, and temperature. These properties do not serve to define the substance water from the chemist's viewpoint. Intrinsic properties are qualities that are characteristic of any sample of a particular substance regardless of the size, shape, or mass of the sample. Examples of the intrinsic properties of water are transparency, fluidity, boiling point, and freezing point.

MIXTURES

Classification of matter was a difficult problem for early scientists (Figure 1.2). Only occasionally did they find matter in a pure form devoid of other substances. Most substances existed in what must have seemed an infinite array of mixtures. It soon became obvious that two types of mixtures, *homogeneous* and *heterogeneous,* were possible. Heterogeneous mixtures are more easily identified, because they can be seen to contain two or more substances and their properties are not uniform throughout. An example of a heterogeneous mixture is salt and sand. The differences between the two substances in the mixture can be seen, and it is possible

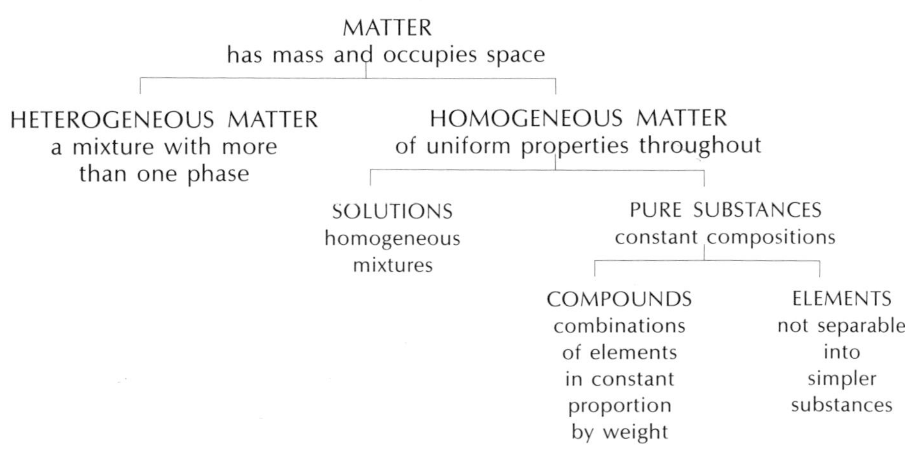

Figure 1.2
Classification of matter.

to separate the two-phase system manually into its components. A *phase* is a part of a system throughout which the properties are uniform. In the salt-sand mixture the salt constitutes one phase and the sand another. Oil and water form a heterogeneous mixture in which individual droplets of each component can be seen. Homogeneous mixtures are single-phase systems and are more difficult to identify properly since they may be mistaken for pure substances, which also consist of one phase. Examples of homogeneous mixtures are salt-water, air, and brass, a mixture of copper and zinc.

The distinction between heterogeneous and homogeneous mixtures is one of uniformity. A homogeneous mixture and any fraction of it have the same intrinsic properties; a heterogeneous mixture is nonuniform, consisting of two or more phases. The properties of the heterogeneous mixture therefore can vary in different fractions. Variability of composition is common to both types of mixtures. Salt and sand can be mixed in any proportions to form an infinite number of heterogeneous mixtures. Similarly, an infinite number of homogeneous salt-water mixtures are possible.

A single phase or a homogeneous system may be either a pure substance or a homogeneous mixture. The distinction between a pure substance and a homogeneous mixture is that the former cannot be separated into components by physical means, whereas the mixture can be separated. Physical separations include any process by which mixtures are separated without altering the identity of the individual components. Historically, difficulties were encountered in distinguishing between pure substances and homogeneous mixtures because even failure to find methods of separating a sample into components did not necessarily mean that the substance was pure. Some substances that once were thought to be pure later proved to be mixtures. At the present time, with the large number of physical techniques available, pure substances can be distinguished from homogeneous mixtures with relative ease.

PURE SUBSTANCES

Pure substances can be divided into two classes, *elements* and *compounds,* with elements forming the smaller class. The concept of elements was prevalent from the time of the early Greek civilization. Aristotle thought that all matter consisted of four elements: earth, air, fire, and water. However, elements were not formally defined until Robert Boyle did so in 1661. His definition represents the evolution of the idea that the universe is composed of relatively few simple substances. Boyle's classic definition can be stated concisely: An element is a substance that cannot be constructed from or decomposed into simpler substances. This definition has stood the test of time until relatively recently, when it was recognized that elements are composed of even more elementary substances, which are discussed later. For the time being the stated definition will serve our purposes. Indeed, chemistry made admirable progress for more than 200 years using this simple definition.

At present there are 103 known elements that constitute a unique group of substances. The isolation of chemical elements has occurred for the most part since 1700. The elements gold, silver, sulfur, and copper, which occur in the elemental state in nature, were among the first elements to be identified. Other elements were discovered as techniques were developed to decompose compounds into their constituent elements. Of the known elements only about 90 are found in nature, the remainder being formed from nuclear reactions in the laboratory. Of the elements found in nature, only about 30 occur in significant quantities in the earth and atmosphere, as indicated in Table 1.1.

The elements are symbolized by abbreviations of their names, as illustrated in Table 1.2. Thirteen of the elements are written with just a capital letter. When two letters are used, the first is written as a capital and the second as lowercase. The choice of the second letter in the chemical symbol varies considerably. Often it is the second letter in the name of the element, although it may be any letter. Occasionally the symbol bears no formal resemblance to the English name of the elements but is derived from another language. The symbol Au for gold is derived from the Latin name *aurum,* and the symbol W for tungsten comes from the German name *Wolfram.*

Compounds are composed of elements combined in such a way as to be inseparable by physical means. Compounds can be broken down into their constituent elements by ordinary *chemical reactions.* The term *chemical reaction* is difficult to define, since the concept requires a basic knowledge of chemistry. Chemical reactions that are employed in separating compounds into their constituent elements involve fundamental changes in the constitution of matter and definite quantities of matter. For example, wood will burn in air. If insufficient air is available, the wood will not burn completely; if excess air is available, only a definite amount will be used in the combustion process, which is a chemical reaction. The fact that compounds are of definite composition distinguishes them from homogeneous mixtures, which are of variable composition. For example, the elements hydrogen and oxygen may be mixed in all proportions to form many homogeneous gaseous mixtures, but water, which is a com-

Table 1.1

Percent composition of the earth, oceans, and atmosphere

oxygen	49.5	phosphorus	0.1
silicon	25.6	carbon	0.09
aluminum	7.5	manganese	0.08
iron	4.8	sulfur	0.06
calcium	3.4	barium	0.05
sodium	2.6	chromium	0.05
potassium	2.4	nitrogen	0.03
magnesium	1.9	fluorine	0.03
hydrogen	0.9	nickel	0.02
titanium	0.6	strontium	0.02
chlorine	0.2	all others	0.07

Table 1.2

Chemical symbols and elements

Ac	actinium	Hf	hafnium	Pr	praseodymium
Al	aluminum	He	helium	Pm	promethium
Am	americium	Ho	holmium	Pa	protactinium
Sb	antimony	H	hydrogen	Ra	radium
Ar	argon	In	indium	Rn	radon
As	arsenic	I	iodine	Re	rhenium
At	astatine	Ir	iridium	Rh	rhodium
Ba	barium	Fe	iron	Rb	rubidium
Bk	berkelium	Kr	krypton	Ru	ruthenium
Be	beryllium	La	lanthanum	Sm	samarium
Bi	bismuth	Lw	lawrencium	Sc	scandium
B	boron	Pb	lead	Se	selenium
Br	bromine	Li	lithium	Si	silicon
Cd	cadmium	Lu	lutetium	Ag	silver
Ca	calcium	Mg	magnesium	Na	sodium
Cf	californium	Mn	manganese	Sr	strontium
C	carbon	Md	mendelevium	S	sulfur
Ce	cerium	Hg	mercury	Ta	tantalum
Cs	cesium	Mo	molybdenum	Tc	technetium
Cl	chlorine	Nd	neodymium	Te	tellurium
Cr	chromium	Ne	neon	Tb	terbium
Co	cobalt	Np	neptunium	Tl	thallium
Cu	copper	Ni	nickel	Th	thorium
Cm	curium	Nb	niobium	Tm	thulium
Dy	dysprosium	N	nitrogen	Sn	tin
Es	einsteinium	No	nobelium	Ti	titanium
Er	erbium	Os	osmium	W	tungsten
Eu	europium	O	oxygen	U	uranium
Fm	fermium	Pd	palladium	V	vanadium
F	fluorine	P	phosphorus	Xe	xenon
Fr	francium	Pt	platinum	Yb	ytterbium
Gd	gadolinium	Pu	plutonium	Y	yttrium
Ga	gallium	Po	polonium	Zn	zinc
Ge	germanium	K	potassium	Zr	zirconium
Au	gold				

pound of hydrogen and oxygen, is always composed of 11.19 percent hydrogen and 88.81 percent oxygen. The physical properties of water are different from those of any of the homogeneous gaseous mixtures of hydrogen and oxygen, including the one consisting of 11.19 percent hydrogen and 88.81 percent oxygen. A general definition for all compounds is as follows: A homogeneous substance composed of two or more elements in fixed proportions by mass is called a *compound.*

From the relatively small number of known elements approximately 4 million compounds have been identified. Many millions more are possible. That only 4 million are known is a reflection of the fact that many elements are unreactive and form few if any compounds. In addition, many elements occur in relatively low abundance and have not been examined extensively. Compounds containing more than four elements are rarities. If this were not the case, the number of possible compounds would soar beyond comprehension.

The compounds that contain carbon and hydrogen are far more numerous than all of the other compounds of the other elements. Because they occur naturally in living material, carbon-containing compounds are called *organic compounds* and are discussed in later chapters, separate from the consideration of inorganic compounds.

Under the proper conditions most matter can exist in each of three *states:* solid, liquid, and gaseous. The three states are recognized easily and will not be defined at this point. Changing the state of an element or compound does not alter the chemical identity of the substance. It is therefore a physical change, not a chemical one. Oxygen is still oxygen in the liquid state at −183°C. Alcohol is still alcohol as a solid at −117°C and as a gas at 78°C. At certain temperatures elements and compounds exist in two or more different states simultaneously. Water can exist as a liquid and as a solid at 0°C and as a liquid and as a gas at 100°C. The significance of these special temperatures is discussed in Chapter 3.

1.3 THE METRIC SYSTEM

Standards of measurement have always been defined in an arbitrary manner. In England the inch was once defined as the length of four barleycorns, and in many locales the foot was defined as the length of a particular man's foot. Scientific units of measurement are just as arbitrary, but an effort has been made to choose standards that can be precisely measured and reproduced.

In scientific work the units of length, volume, and mass are expressed in metric terms. The *metric system,* while not common to most Americans, is used throughout most of the rest of the world. At present some American industries as well as the U.S. Army use metric units. The English system of weights and measures lacks the advantages of the metric system, in which all the units are based on a decimal system. Standard units of mass, length, and volume have been established by definition, using Latin prefixes to designate subunits and Greek prefixes to designate multiple units. The prefixes deci-, centi-, and milli- refer to $\frac{1}{10}$, $\frac{1}{100}$ and $\frac{1}{1000}$ of the basic unit. The prefixes deka-, hecto-, and kilo- refer to 10, 100, and 1000 times the basic unit.

The standard of length in the metric system is the *meter,* which was originally thought to be equal to 1/10,000,000 of the distance from the North Pole to the equator. More accurate surveys have indicated that the earth is not a perfect sphere and that the original value for the circumference of the earth is not correct. Nevertheless, the arbitrary defined length is still maintained as a standard and is equal to the distance between two marks on a particular platinum–iridium bar that is preserved under controlled conditions in the International Bureau of Weights and Measures near Paris. Secondary standards are maintained in many other countries, and the unit of length is accurately known and is reproducible worldwide. The meter was redefined in 1960 as 1,650,763.73 wavelengths of the orange light emitted by gaseous krypton-86. This definition, adopted by the Eleventh General Conference on Weights and Measures, does not change the length of the meter but makes it a more convenient known standard for the scientist. The orange light can be generated in any part of the world because it is based on a natural, universally measurable and reproducible phenomenon. In Chapter 5 the significance of the orange light and the term *krypton-86* becomes obvious.

Most chemical measurements require units of length that are smaller than the meter, such as the centimeter or millimeter. The interrelationships of the metric units of length are given in Table 1.3 along with their common abbreviations.

The term *volume* refers to the amount of space occupied by matter or available for occupancy. Since the meter is the standard for length, the cubic meter (m^3) could have been chosen as the standard unit of volume, but it is an unwieldy standard. The cubic centimeter (cm^3 or cc) offers an alternative standard that is often used in laboratory work. The *liter* has been chosen as a volume standard intermediate between cubic meters and cubic centimeters. A liter was designated originally as the volume of 1 kilogram (kg, a standard unit of mass) of water at 4°C. This volume was thought to equal 1000 cm^3. Thus one milliliter (1 ml $= \frac{1}{1000}$ liter) should be equivalent to 1 cm^3. Precise measurement indicates that a liter is actually

Table 1.3
Metric units

units of length				
millimicron	(mμ)	=	0.001	μ
micron	(μ)	=	0.001	mm
millimeter	(mm)	=	0.001	m
centimeter	(cm)	=	0.01	m
decimeter	(dm)	=	0.1	m
kilometer	(km)	=	1000	m
units of mass				
microgram	(μg)	=	0.001	mg
milligram	(mg)	=	0.001	g
centigram	(cg)	=	0.01	g
kilogram	(kg)	=	1000	g
units of volume				
microliter	(μl)	=	0.001	ml
milliliter	(ml)	=	0.001	liter

equal to 1000.028 cm^3. For practical purposes the liter can be considered to be equal to 1000 cm^3, and the cubic centimeter to 1 ml. The difference is unfortunate but insignificant in all but the most accurate computations and measurements. A current suggestion is that 1 liter be defined as 1 cubic decimeter (dm^3). The metric units of volume are given in Table 1.3.

The most practical unit of mass in chemistry is the *gram* (g), and it is equivalent to an arbitrarily chosen standard amount of matter. The mass of a standard platinum–iridium cylinder, kept by the International Bureau of Standards, is defined as 1000 g or 1 kilogram (kg). Smaller quantitites of matter may be described conveniently in terms of the centigram (cg) or the milligram (mg).

The equivalents of the metric and English systems relating mass and weight as given in Table 1.4 should be qualified. Weight refers to the force exerted on a quantity of matter (which by definition possesses mass) by the gravitational attraction of the earth. Since weight is actually a force exerted on a mass, weight and mass are not identical. The mass of a specific object is a constant independent of the gravitational field in which it is located. While the weight of a substance may vary as a function of its geographical location on earth, its mass is unchanged. Although the terms *mass* and *weight* often are used interchangeably, mass is the real measure of the quantity of matter and is used exclusively in this text. There is no English verb that can be used to denote the act of determining the mass of an object. Therefore, it will often be necessary to use somewhat complex grammatical construction to describe this process.

A specific object may be described by the extrinsic properties of mass and volume. These extrinsic properties can be expressed as a quotient to produce the intrinsic property *density*. The density of a type of matter is a useful property because it is independent of sample size. Since a variety of units can be used in measuring the volume and mass of a substance, its density can be expressed in various units. Usually the density of liquids and solids is expressed in grams per milliliter (g/ml), because the resultant values avoid small fractions and extremely large numbers. For the same reason the densities of gases are expressed in grams per liter. A list of the densities of some common substances at 25°C is given in Table 1.5.

Table 1.4
Approximate English and metric equivalents

1 in.	=	2.54 cm
1 ft	=	30.5 cm
1 yd	=	0.9 m
1 grain	=	0.07 g
1 avoirdupois oz	=	28.4 g
1 lb	=	453.6 g
1 fluid oz	=	29.6 cm^3
1 qt	=	946.0 cm^3
1 tsp	=	4.0 cm^3

Table 1.5
Densities of substances at 25°C and atmospheric pressure

liquids (g/ml)		*solids (g/cm³)*		*gases (g/liter)*	
alcohol	0.79	iron	7.86	carbon dioxide	1.80
bromine	3.12	gold	19.3	carbon monoxide	1.25
ether	0.74	lead	11.3	hydrogen	0.08
olive oil	0.92	rock salt	2.2	helium	0.16
turpentine	0.87	sugar	1.59	methane	0.66
water	1.00	uranium	19.0	oxygen	1.43
mercury	13.53	wood, balsa	0.12	nitrogen	1.25

1.4 TEMPERATURE AND ENERGY

TEMPERATURE

Many properties of matter are functions of temperature. The volumes of solids, liquids, and especially gases are affected by temperature. For this reason the standards of mass, length, and volume are all defined at a particular temperature, and it is necessary to be able to specify the temperature of a substance before discussing any of its physical properties. In a qualitative sense an individual can determine whether a substance is hot or cold by touching it. When the substance is in contact with the skin, a burning, freezing, or neutral sensation is transmitted to the brain. This sensation is produced by heat flow or the lack of it between the skin and the object. Heat energy flows spontaneously from a substance of high temperature to a substance of low temperature. When we touch a cold object heat flows from us to the object, whereas with a hot object the reverse process occurs. This principle allows us to make qualitative statements about temperature.

Thermometers are devices to measure (meter) the thermal energy (temperature) of substances. A thermometer must come in contact with the substance being examined and must undergo some physical change that reflects heat transfer. With few exceptions, liquids and solids expand when heated and contract when cooled, and the thermal expansion of a substance often is used to define the temperature scale of a thermometer. In the mercury thermometer, which consists of a glass bulb sealed to a capillary tube, a small change in the volume of the mercury in the glass bulb leads to a large change in the height of the mercury column in the capillary tube. Graduations on the capillary tube indicate the degree of mercury expansion in terms of defined degrees of temperature.

Because water is one of the most important substances in the maintenance of life, it is not surprising that water was chosen to provide scales for the measurement of temperature as well as of volume. The Fahrenheit scale is most commonly used in American nonscientific communities. Gabriel Fahrenheit took as the zero for the scale the lowest temperature he could obtain in a mixture of salt and ice. For the 100° standard he chose what he believed to be normal body temperature. On this scale the freezing point of water (or the melting point of ice) is 32°F and the boiling point is 212°F. Therefore, there are 180 Fahrenheit units between these two arbitrary reference points. Placement of 180 uniform divisions between the two reference points makes up the Fahrenheit scale. Temperatures above and below the reference points are defined by extension of the uniform degree division.

The Celsius scale is used most frequently in scientific work. On the Celsius scale the freezing point and boiling point of water are defined as 0 and 100°C, respectively. There are 100 uniform degree intervals between these two points. At one time this scale was called centigrade, but it is now referred to as the Celsius scale after the Swedish astronomer Ander Celsius, who developed it in 1742. The Fahrenheit and Celsius temperature scales are compared in Figure 1.3.

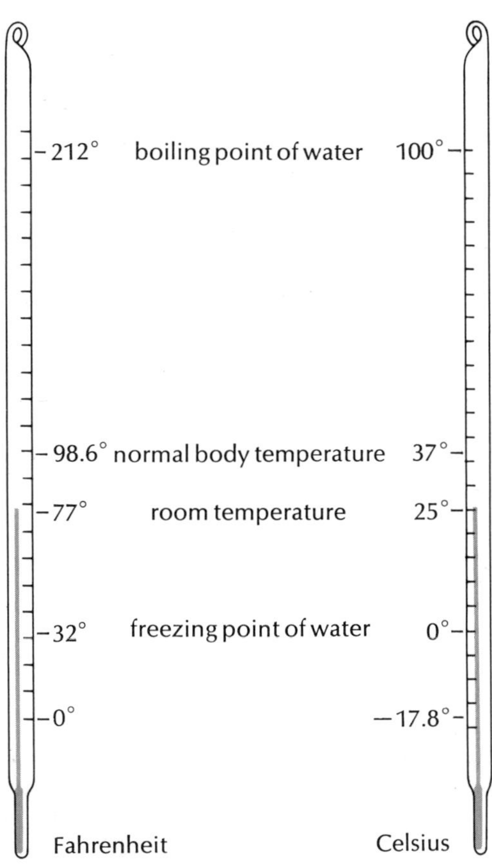

Figure 1.3
The Fahrenheit and Celsius scales.

The temperature of an object can be expressed on both the Fahrenheit and the Celsius scales because the position of the mercury column is independent of the scales placed on the thermometer. A given temperature on one scale can be converted to the other scale simply by remembering the basis of their origin. There are 180 Fahrenheit units for every 100 Celsius units, because these are the differences between the two reference points. Therefore, the Celsius degree is $\frac{180}{100} = \frac{9}{5}$ as large as the Fahrenheit degree. This conversion factor, properly applied along with the knowledge of the temperature equivalents of the two reference points, allows conversion from one scale to the other. The temperature 122°F is 90 Fahrenheit units above the reference freezing point of 32°F. There are fewer Celsius units per Fahrenheit unit and, therefore, the temperature is $\frac{5}{9} \times 90 = 50$ Celsius units above the reference point of 0°C for the Celsius scale, or 50°C. The temperature −15°C is 15 Celsius units below the reference point of 0°C. Since there are more Fahrenheit units per Celsius unit, the temperature is $15 \times \frac{9}{5} = 27$ Fahrenheit units below the Fahrenheit reference of 32°F, or 5°F.

EXAMPLE 1.1

What is the Fahrenheit equivalent of 150°C?

The temperature 150°C is 50° above the 100°C reference point. Since there are more Fahrenheit units per Celsius unit over any range on the temperature scales, the Fahrenheit units above the same standard reference temperature must be equal to $\frac{9}{5}(50) = 90$ Fahrenheit units. Therefore, the temperature is 212°F plus 90 Fahrenheit units, or 302°F.

ENERGY

Energy is defined as the ability to do work. Although this definition is supportable only in mechanical terms, it is nevertheless the broadest possible definition. There is a variety of types of energy that are usually encountered in chemistry. These are potential energy, kinetic energy, heat energy, electrical energy, and chemical energy.

Potential energy refers to the energy that a substance possesses because of its position relative to a second possible position. A wound watch spring possesses potential energy and does work as it unwinds. Similarly, water behind a dam possesses potential energy that can be obtained if the water is made to turn a turbine while going through the dam. The energy released by both systems described is transformed into kinetic energy of some object (the gears of a watch or blades of a turbine) that is moving. Thus the energy possessed by a moving object is called *kinetic energy.*

Chemical energy is the potential energy stored in elements or compounds that can be released during a chemical reaction or a physical transformation. As the subject of chemistry is developed in this text it is shown that the chemical energy stored in elements and compounds determines the reactions that these substances will undergo. In the latter chapters it is shown that most life processes can be discussed in terms of the storage and release of chemical energy.

The chemical energy required or generated may be in the form of either heat or electrical energy. A reaction that occurs with the evolution of heat is *exothermic,* while a reaction in which energy is required from the surroundings is *endothermic.* The electrical equivalent of an exothermic reaction is one in which an electric current is generated. Conversely, the electric equivalent of an endothermic reaction is one in which an external electric current is required for the reaction to occur.

The most common unit of heat energy is the *calorie* (cal). It is defined as the amount of energy required to raise the temperature of 1 g of water 1°C from 14.5 to 15.5°C. All other measures of energy can be converted to their equivalents in calories by appropriate conversion factors (Table 1.6). A kilocalorie (kcal), equivalent to 1000 cal, is the unit used in dietary tables, although it is referred to as a Calorie.

Table 1.6
Energy units

1 kilocalorie (kcal or Calorie)	=	1000 cal
1 calorie (cal)	=	4.184×10^{7} ergs
1 electron volt (eV)	=	1.602×10^{-12} ergs
1 eV	=	3.80×10^{-20} cal
1 eV/molecule (or atom)	=	23.06 kcal/mole
1 liter atmosphere (liter-atm)	=	24.22 cal

EXAMPLE 1.2

How many calories are required to heat 500 g of water from 25 to 50°C?

In order to calculate the desired quantity it is necessary to assume that the definition of the calorie applies to any temperature range for the water sample. In other words, the calorie should heat 1 g of water from 25 to 26°C or any other 1° temperature range. This assumption is reasonably valid. To heat 1 g of water from 25 to 50°C requires 25 times the number of calories required to heat the sample up 1°C. Therefore,

25 cal are required for the 1-g sample. Since the given sample has a mass of 500 g, the number of calories required will be 500 times that of the 1-g sample. Therefore, 500(25) cal = 12,500 cal are required.

1.5 THE SCIENTIFIC METHOD

OBSERVATION AND LAWS

All natural sciences are dependent upon observations for their growth. The method of making observations concerning a problem of interest, of cataloging observations, and of the ultimate use of these observations in the solution of the problem is called the *scientific method.*

In the development of science, qualitative observations were made very early. Man probably had been making general observations about his surroundings since his appearance on earth. Conveyance of these observations required communication, and written records gradually accumulated. Quantitative observation is the major stepping stone from speculation, exaggeration, and omission to the orderliness required for an understanding of one's surroundings. Once quantitative observations have been made, recorded, and delineated with respect to the conditions under which they were obtained, it is then possible for other individuals to reproduce the observations. The ability to reproduce an observation is a requirement in the development of science. Reproducibility ensures that an observation is a fact and not the creation of the observer's prejudices or desires.

Sufficient numbers of quantitative observations of the regularities in the behavior of the universe led to the so-called laws of nature by a process of deductive reasoning. A law is simply the conclusion of a deductive process, which in turn is merely an explicit statement of fact that is already inherent in the information obtained by observations. Nothing new in understanding nature is obtained by stating a law; the law merely summarizes what has been observed.

MODELS

Scientists find it convenient to use models for natural phenomena. The term *model* refers to an idea that corresponds to what is responsible for the natural phenomena. A model may be a mental picture of a phenomenon; but to communicate with others, verbal or written descriptions of the model are necessary. For precise communication and in order to remove ambiguities of language, geometrical or mathematical models commonly are used.

Models are conceived or postulated by an inductive process and are used to make predictions. There are no simple formulas for arriving at models; they simply evolve according to the variables of time, place, observations, and man. If a model enables the scientist to make accurate predictions, it is said to be validated. If it does not allow accurate predictions, alteration or complete replacement of the model may be required.

Scientists employ a general pattern for testing their models, which is called the *scientific method.* All available experimental measurements and observations are utilized to create a model, which is then checked against further observations and measurements that are predicted by the model. Finally, the model may be modified or replaced as necessitated by the new observations.

The terms *hypothesis* and *theory* are often used in science. A hypothesis refers to a tentative model, whereas a theory describes a model that has been tested and validated many times. The dividing line between the two is arbitrary and obviously cannot be precisely defined.

A scientist must be ready to remove himself from his previous training, ideas, and personal prejudices so that faulty models are not promulgated beyond their usefulness (see Section 1.6). Although every scientist should realize this difficulty, it is easy to forget. New models often sound untenable to an established scientist because they are in opposition to the very basis of what he believes about the world. He often cannot abandon old ideas and accept radically new ones. Therefore, it is not too surprising to find that many important creative ideas are conceived by young scientists. Sir Isaac Newton suggested the concept of gravity at age 24; Évariste Galois developed the theory of groups in mathematics before he was 21; Albert Einstein's most original work was done while he was still in his twenties.

In this book some of the systematized knowledge of chemistry, which represents what is now believed to be correct, is presented. An attempt has been made to present more than just facts. It is hoped that an understanding of the evolutionary process that science has undergone will be gained and that a boundary knowledge of the field of chemistry will be attained.

1.6 LAWS OF CHEMICAL CHANGE

LAW OF CONSERVATION OF MASS

It has been shown repeatedly during the last two centuries that the products of a chemical reaction have the same total mass as the sum of the masses of the starting materials. These observations have led to the general acceptance of the *law of conservation of mass:* There is no experimentally detectable gain or loss of mass during an ordinary chemical reaction. The general concept involved in this law was stated first in 1756 by the Russian scientist M. V. Lomonosov. However, it was the French chemist Antoine Lavoisier who gradually persuaded the scientific community to accept the concept of conservation of mass in 1774 by his careful research on the mass relationships between oxygen and metals in compounds such as tin and mercury oxide. Usually Lavoisier is given exclusive credit for the law of conservation of mass, a fact that probably reflects the high regard shown for his unique contribution in developing an understanding of combustion.

Prior to the experiments of Lavoisier it was theorized that a substance called *phlogiston* was lost whenever a material was burned. The phlogiston theory accounted for the apparent decrease in the mass of the material (ash) left after burning wood. All combustible substances were thought to contain phlogiston, which escaped when the material burned. Noncombustible substances were thought to lack phlogiston. The theory was based on inadequate quantitative observations; a gaseous reactant (oxygen) and gaseous products (carbon dioxide and water vapor) were not identified. The demise of the phlogiston theory began when quantitative data indicated that the calces (oxides) of metal that were obtained after combustion (that is, after loss of phlogiston) were of higher mass than the metal sample. At this point the phlogiston model for combustion was patched up and severely strained by assuming that the phlogiston contained in metals had negative mass. There are many reasons why the phlogiston theory eventually was discarded. In general, it became easier to interpret the chemical facts by the modern theory of combustion than to place unlikely restrictions on the nature of phlogiston. However, the phlogiston theory served a useful function in providing stimulus for further experiments.

In the last half century the law of conservation of mass has had to be qualified. There is an experimentally detectable loss of mass in nuclear reactions. The Einstein equation, which states that mass m and energy E are interconvertible according to the formula $E = mc^2$, where c is the velocity of light, is found to account for the loss of mass. The conversion of mass to energy in nuclear reactions accounts for the tremendous quantities of energy released.

LAW OF DEFINITE PROPORTIONS

After the pioneering research of Lavoisier additional careful analyses of chemical changes were reported by Joseph Proust and Jeremiah Richter. They observed that not only was matter conserved in chemical changes but that the quantities of each element involved also remained unchanged. In addition they found that the composition of a specific compound is independent of its source, providing it has been rigorously purified. The composition of pure water, for example, is always 88.81 percent oxygen and 11.19 percent hydrogen, whether it is obtained by the purification of seawater or rainwater. The composition can be determined by breaking the compound down into its constituent elements. Alternatively, the constituent elements can be recombined in the proper amounts by appropriate chemical reactions to produce the desired compound. Proust and

Richter's observations can be summarized in the *law of definite proportions:* When elements combine to form compounds, they do so in definite proportions by mass. The law is really equivalent to the operational definition of a compound. Like the law of conservation of mass, the law of definite proportions is subject to some limitations, which are presented in Chapter 4.

LAW OF MULTIPLE PROPORTIONS

Two or more elements often combine under different experimental conditions to produce different compounds. Experimentally, it has been found that there is a simple relationship between the masses of the elements involved in forming the compounds. For example, the combustion of carbon in the presence of sufficient oxygen produces a gas that is nonpoisonous and noncombustible. If an insufficient amount of oxygen is available, a poisonous, combustible gas is produced. Analysis of the noncombustible gas, carbon dioxide, reveals that for every 1.000 g of carbon there is combined with it 2.670 g of oxygen. In the case of the combustible gas, carbon monoxide, every 1.000 g of carbon is combined with 1.335 g of oxygen:

carbon dioxide: 1.000 g **C** + 2.670 g **O**
carbon monoxide: 1.000 g **C** + 1.335 g **O**

The ratio of the mass of oxygen combined with the same mass of carbon in the two compounds is 2.670/1.335, or 2/1. A statement of the experimental facts described above is known as the *law of multiple proportions:* When two or more elements combine to form more than one compound, the masses of one element that combine with a fixed mass of another element are in the ratio of small whole numbers.

EXAMPLE 1.3

A compound of carbon and oxygen consists of 0.0890 g of oxygen combined with 0.1000 g of carbon. Show that this compound and the poisonous combustible gas described above are consistent with the law of multiple proportions.

In order to compare the compound described with carbon monoxide gas it is first necessary to treat the data such that a fixed mass of one element in each compound may be compared. Consistent with the law of definite proportions, the poisonous gas consists of 0.1335 g of oxygen combined with 0.1000 g of carbon. The mass of oxygen combined with the same mass of carbon are therefore in the ratio of 0.1335/0.0890 = 3/2.

1.7 UNITS AND CONVERSION FACTORS

In describing the behavior of matter it often will be necessary or convenient to employ mathematical equations. These equations are not difficult, but the student is discouraged from simply "plugging in" values and solving the equations. The numbers always have units associated with them. The number without the units has no meaning in an equation describing a chemical situation. The units provide clues to solving physical problems and serve as an effective check on the method of solution.

Units can be treated as numbers in calculation. Any mathematical manipulation can be performed on them. Therefore, if the units of both sides of an equation are equal, the correct answer is assured, providing computational errors do not occur. Units are helpful in deciding which of two complementary mathematical operations must be employed. For example, consider the calculation of the density of a substance that occupies 10 ml and has a mass of 20 g. The two numbers could be multiplied to obtain 200 g-ml or divided in two possible ways to obtain 2 g/ml and 0.5 ml/g. Which of these numerical values is correct? Since density is expressed in grams per milliliter the correct answer must be 2 g/ml, because the correct units can be obtained only by dividing 20 g by 10 ml.

Often it is necessary to convert a number and its corresponding unit into the equivalent in another unit. This is accomplished by the use of conversion factors. Conversion factors are numbers having two or more unit systems associated with them (Table 1.7). They are obtained from a knowledge of equivalent values on two scales of units. For example, from the definition, 1 kg is equivalent to 1000 g. The two possible conversion factors can be obtained by dividing the numbers and the units in two ways. The conversion factors relating grams and kilograms are 1000 g/1 kg = 1000 g/kg and 1 kg/1000 g = 0.001 kg/g. If the equivalent of 456 g in terms of kilograms is required, the correct answer can be obtained by

$$456\ \not{g} \times 0.001\ \frac{\text{kg}}{\not{g}} = 0.456\ \text{kg}$$

multiplying 456 g by 0.001 kg/g: The correct answer is assured because the

Table 1.7
Conversion factors

2.54 cm/in.	0.394 in./cm
453.6 g/lb	0.0022 lb/g
0.454 kg/lb	2.2 lb/kg
0.946 liter/qt	1.057 qt/liter
0.00254 m/in.	39.37 in./m
100 cm/m	0.01 m/cm
1000 m/km	0.001 km/m
1000 ml/liter	0.001 liter/ml
1000 g/kg	0.001 kg/g

units of gram will cancel from the numerator and denominator, leaving only the unit kilogram.

Of course, the student should not rely solely on units any more than he should "plug in" numbers. It is important to develop a sense of the magnitude of the correct answer. In the conversion of 456 g into 0.456 kg, it should be realized in advance that the numerical answer will be less than 1. The mass of 456 g is less than 1000 g, and its equivalent should be a fraction of a kilogram. Of the two factors derived, only one will produce the desired result. Thus both units and a sense of the magnitude of the answer will aid in the solution of mathematical problems.

EXAMPLE 1.4

Calculate the speed in feet per second (ft/sec) equivalent to 60 miles per hour (mi/hr).

In order to convert the speed from miles per hour into feet per second, it is necessary to change both units, the hours and the miles. A factor analysis suggests that the miles can be converted into feet by multiplying by the factor 5280 ft/mi:

$$60 \text{ mi/hr} \times 5280 \text{ ft/mi} = 60(5280) \text{ ft/hr}$$

The mile-foot conversion yields a reasonable number because the number of feet traversed in an hour must be larger than the number of miles. In order to convert speed in feet per hour into feet per second, a factor having the units hour per second(hr/sec) must be applied:

$$60(5280) \text{ ft/hr} \times 1 \text{ hr/3600 sec} = 88 \text{ ft/sec}$$

Suggested further readings

Beveridge, W. I. B., *The Art of Scientific Investigation.* New York: Random House, 1961.

Brown, J. C., *History of Chemistry.* London: J. A. Churchill, 1920.

Farber, E., *The Evolution of Chemistry.* New York: Ronald Press, 1952.

Garrett, A. B., "The Discovery Process and the Creative Mind," *J. Chem. Educ.,* **41,** 479 (1964).

Howlett, L. E., "International Basis for Uniform Measurement," *Science,* **158,** 72 (1967).

Ihde, A. J., *The Development of Modern Chemistry.* New York: Harper & Row, 1964.

Kesselman, B., "The Skeptical Chemist (Robert Boyle)," *Chemistry,* **39,** 9 (1966).

Partington, J. R., *A Short History of Chemistry*. New York: Harper & Row, 1960.
Pauling, L., "Chemistry," *Sci. Amer.*, 32 (September 1950).
Schurr, S. H. "Energy," *Sci. Amer.*, 92 (September 1963).
Smith, H. M., *Torchbearers of Chemistry*. New York: Academic Press, 1949.
Stimson, H. F., "Celsius versus Centigrade: The Nomenclature of the Temperature Scale of Science," *Science*, **136,** 254 (1962).
Terman, L. M., "Are Scientists Different?" *Sci. Amer.*, 25 (January 1955).
Waddington, C. H., *The Scientific Attitude*. Baltimore: Penguin Books, 1948.
Weaver, W., "Fundamental Questions in Science," *Sci. Amer.*, 47 (September 1953).
Weeks, M. E., *Discovery of the Elements*. Easton, Pennsylvania: Chemical Education, 1945.

Terms and concepts

alchemistry
Boyle
calorie
compound
Celsius
centigrade
chemical energy
chemical reaction
conservation of mass
conversion factor
definite proportions
density
element
empiricist
endothermic
energy
exothermic
extrinsic
Fahrenheit
gram
heterogeneous
homogeneous
hypothesis
iatrochemist
intrinsic
kinetic energy
Lavoisier
law
liter
mass
matter
meter
metric system
models
multiple proportions
phase
phlogiston
physical properties
potential energy
rationalist
science
scientific method
states
symbols
theory
thermometer

Questions and problems

1. What physical properties serve to distinguish a mixture that contains 11.19 percent hydrogen and 88.81 percent oxygen from the compound water?
2. How could two samples of liquids, one a solution and the other a pure compound, be distinguished?
3. How could each of the following mixtures be separated?
a. salt and water
b. iron filings and sand
c. oil and water
d. alcohol and water

4. Make the following conversions using the appropriate conversion factors:

a. 25 mm into centimeters
b. 100 mg into kilograms
c. 250 ml into liters
d. 2 kg into milligrams
e. 5 km into centimeters
f. 3 liters into milliliters

5. A cube is 10 cm on each side. What is its volume in liters? What is the volume of the cube in cubic meters (m^3)?

6. A snail named Sammy runs at 1 cm/min. What is his velocity in furlongs per fortnight? (A furlong is 1320 ft, and a fortnight is 14 days.)

7. What is the equivalent of 167°F on the centigrade scale?

8. If a new temperature scale were created on which the freezing point of water was defined as −100°U and the boiling point of water as 300°U, what would be the equivalent of 25°C on this scale?

9. A cube of metal that is 2 cm on each side has a mass of 24 g. What is its density?

10. Two different compounds of nitrogen and hydrogen have been found to have the following mass compositions: one compound contains 41.62 g of nitrogen combined with 1.00 g of hydrogen; the second compound contains 4.64 g of nitrogen combined with 1.00 g of hydrogen. Show from these data that the two compounds are in agreement with the law of multiple proportions.

11. How many calories are required to heat the following systems to the indicated temperature (assume that the definition of a calorie applies to all temperature changes)?

a. 1.0 g of water from 10 to 75°C
b. 25 g of water from 14 to 15°C
c. 25 g of water from 10 to 75°C

12. Approximately 0.05 cal is required to heat 1 g of tin from 0 to 1°C. How many calories are required to heat 1 kg of tin from 0 to 100°C?

13. Silver melts at 961°C. What is the melting point of silver in degrees Fahrenheit?

14. Thermometers containing alcohol and a red pigment are used to measure very low temperatures. Alcohol freezes at −117°C. What is the coldest temperature in degrees Fahrenheit that such a thermometer could record?

15. The speed of light is 186,000 mi/sec. What is the corresponding value in centimeters per second?

16. A professional athlete may earn his weight in gold in one year if he is talented. Calculate the dollar equivalent for a 200-lb athlete if the price of gold is $35/oz.

2

THE GASEOUS STATE

Although gases are elusive and more difficult to handle experimentally than liquids or solids, the gaseous state was historically the first to be studied in detail. Gases represent the simplest state of matter, with behavior that can be predicted quantitatively on the basis of a remarkably simple model. The first strides in investigating the nature of matter were made with the development of techniques for capturing and measuring gas samples.

Gases are easily recognized and distinguished from liquids and solids. They have no characteristic shape and can be contained in any size or shape vessel. They are transparent, compressible, diffuse rapidly, and show little resistance to flow.

In discussing the properties of gases we shall use a model gas, the *ideal gas,* that reflects the properties of most real gases—whether they are elements or compounds. The utilization of an abstract idea, such as an ideal gas, allows a study to be made of the characteristics that are generally common to all gases. When the behavior of a given real gas deviates significantly from the ideal gas behavior, then the specific reason for its individual properties can be sought.

At this point it is important to develop general broad principles to describe the gaseous state at the macroscopic level (that which is visible to the eye). With such a description it is possible to devise a model for the submicroscopic (invisible even with the aid of a microscope) behavior of atoms and molecules. A critical evaluation of the simple quantitative mathematical relationships developed 200 years ago then will be used to provide information on the sizes of atoms and molecules and the forces they exert on each other. The four properties of gases that define their behavior are volume, temperature, mass, and pressure. Each of these is discussed here as a background for understanding the gaseous state.

2.1 PROPERTIES OF GASES

VOLUME

The volume of a gas is different from the volume occupied by liquids and solids. Solids and liquids have clearly visible boundaries: The amount of space occupied by a particular sample can be seen and defined. A gas, on the other hand, fills completely any container in which it is placed and is bounded only by the walls of the container. If a colored gas such as bromine is placed in a container with a window, the red gas can be seen to be distributed evenly throughout the container. If the same quantity of gas is transferred to a larger or differently shaped container with a window, the distribution is still uniform.

Gases are *completely miscible,* that is, they can be mixed in all proportions. If two or more gases are mixed together in a common container of fixed dimensions, the volume of the gaseous mixture is still equal to the volume of the container. The volume of a mixture of gases is independent of the original volumes of the individual gases.

TEMPERATURE

The gaseous state exhibits the same properties of heat transfer as the liquid and solid states. A thermometer placed in a stream of hot gases, such as those emanating from a hot air heating system, will register the temperature of the gases. Matter in the gaseous phase must come in contact with the thermometer in order to produce an observed temperature rise. It is important to determine the temperature of any gas sample under scientific investigation because, as is illustrated shortly, the physical properties of a gas are temperature dependent.

If two or more gases at different temperatures are placed together in a common container, the resultant temperature of the mixture is intermediate between those of the separate gases. This intermediate temperature of the mixture is rapidly achieved.

The mass of a gas contained in a given unit volume is usually much smaller than the mass of a liquid or solid of the same volume. Therefore, the amount of matter actually present in the gaseous state is smaller than in the liquid or solid state. Accordingly, it is most convenient to express the density of a gas in units of grams per liter instead of grams per milliliter, as is used for liquids and solids.

Whereas the mass of a gas sample is independent of variables such as temperature, volume, and pressure, the density is a function of the temperature, volume, and pressure. If 1 g of a gas is placed in a 1-liter container, its density is 1 g/liter. However, if the container is diminished in size to 500 ml, the density is increased to 2 g/liter. The quantity of matter has not increased, but the volume that contains that quantity has been halved. Therefore, the density is increased by a factor of two.

PRESSURE

The concept of *pressure,* which is equal to force per unit area, is somewhat more difficult to deal with than that of volume, temperature, and mass. It can be illustrated by some commonly encountered examples.

The pressure an individual exerts on the soles of his shoes is a function of his mass and the area of his soles. Consider two individuals of the same mass who wear different size shoes; the individual with the larger shoes exerts less force per unit area on the soles of his shoes although the total force involved is the same in both cases. In the case of certain shoes designed for women, the force per unit area on the spike heels is very high, and it is for this reason that they can cause substantial damage to floors. On the other extreme, the low pressure accounts for the effectiveness of snowshoes which serve to distribute the force of the individual wearing them over a wider area.

In an automobile tire the enclosed gas exerts a pressure to keep the tire extended against the external forces tending to push it in. The mass of the car forces the tire to flatten somewhat against the road surface. The area of the tire that is flattened is a function of the pressure inside the tire. The higher the tire pressure, the less the tire will flatten. The flat surface of the tire exerts a force per unit area on the ground. If the pressure is multiplied by the total area flattened, the force exerted against the ground is obtained. This total force is equal to that exerted by the car on the tires.

EXAMPLE 2.1

Calculate the flattened area per tire caused by the force of a 3600-lb motor vehicle supported by four tires whose total pressure is 30 lb/in.^{2}.

The total area flattened, which is necessary to provide support for the car, is equal to the weight (a reflection of gravitational attraction) of the car divided by the pressure of the tires.

$$\text{total area flattened} = \frac{3600 \text{ lb}}{30 \text{ lb/in.}^2} = 120 \text{ in.}^2$$

It should be noted that the above mathematical operation is dictated by the units involved and sought in the solution of the problem. The area flattened per tire of course is one-quarter of the total area flattened, or 30 in.^{2}.

In determining temperature, account is taken of the direction of heat flow from a hot body to a cold body. In a similar manner, pressure is determined by the flow of mass that normally moves from regions of high pressure to low pressure. This principle can be used to define a scale of pressure just as heat flow is used to define temperature.

One additional facet of pressure must be considered before we can define this concept quantitatively. Pressure in a fluid medium (gases or liquids) is independent of direction at a given point. If an object is submerged to a given depth under water, the pressure exerted upon it is independent of its orientation. An individual swimming at a depth of 10 ft will feel the pressure of the water all about him, and it will be the same regardless of whether he swims on his back, side, or front.

THE BAROMETER

In measuring pressure we can make use of our observations on fluid states. The air that surrounds the earth exerts a pressure on us just as water does on a submerged swimmer. We are accustomed to this pressure and usually do not consider it, but it affects our lives in many ways. The weather patterns and air movements are controlled by the variance in pressure at different points on the earth. The cooking time of food in boiling water is a function of the pressure exerted by the atmosphere (Chapter 3).

The *barometer,* which is the standard device for measuring the pressure of the atmosphere, was invented by Evangelista Torricelli in the seventeenth century. It consists of a long tube closed at one end, which is filled with mercury and inverted in a vessel of mercury. If the tube is more than 76 centimeters (cm) long, part of the mercury will run out of the tube when it is inverted, but a column of mercury approximately 76 cm high will remain in the tube (Figure 2.1). The height of the column can be found experimentally to be independent of the length of the tubing and of the size or shape of the vessel containing the mercury.

The reason that a column of mercury remains in the tube of the barometer is that the atmosphere exerts a pressure on the surface of the mercury. This pressure transmitted through the mercury to the base of the column supports the mercury in the tube. In the barometer above the level of the mercury there exists a near vacuum. Therefore, the pressure at the base of the mercury column is due to the mass of the mercury in the column. The mercury falls until the pressure exerted by its weight is equal to the atmospheric pressure. Two barometers of different cross sections show the same column height since the pressure at the base of each is the same. The mass of liquid in each column is directly proportional to the cross section of the tube. Hence the force per unit area (pressure) is the same. The dimensions of the vessel are unimportant as long as atmospheric pressure has access to some area of the surface.

Pressure should be expressed in units of force per unit area. In the barometer the downward force exerted by the mercury column is proportional to the mass of the liquid supported in the column, which in turn is proportional to the height of the column. Therefore, the pressure can be expressed in centimeters of mercury. It is understood that pressure

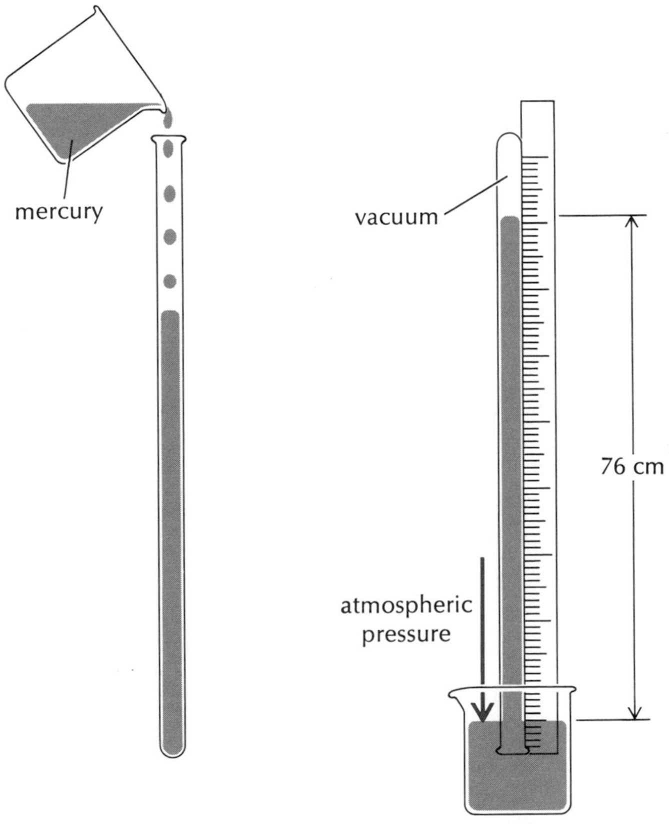

Figure 2.1
The construction of a simple barometer: A glass tube, filled with mercury, is closed and inverted in a beaker of mercury. When the end of the column is opened, the column drops to about 76 cm.

is not the same thing as length, but by convention the term *centimeter of mercury* (Hg) indicates the pressure or force per unit area exerted by the column of mercury.

Atmospheric pressure is a function of altitude and the local weather, and it fluctuates from day to day. At 0°C at sea level the average atmospheric pressure supports a column of mercury 76 cm high. This pressure is called a *standard atmosphere* and is referred to as 1 atm. The unit of pressure, 1 mm of Hg, is also called 1 torr. Therefore, the standard pressure (1 atm) is 760 torr.

Because the density of mercury is 13.6 g/ml, the mass of a column of mercury in a tube with a cross-sectional area of 1 cm^2 is (76 cm)(1 cm^2)(13.6 g/cm^3) = 1033.6 g. One atmosphere will support the 1033.6 g of mercury in a column 1 cm^2, or in terms of its equivalent in pounds per square inch (psi) the standard atmosphere is 14.7 psi. If a different liquid were used in place of mercury in a barometer, the height of the column would be different from 76 cm. For a liquid less dense than mercury the column is longer, but in every case the pressure at the base of the column is equal to atmospheric pressure.

EXAMPLE 2.2

Calculate the height of a column of water in a water barometer that can be supported by 1 atm.

The density of water is approximately 1 g/ml. The height of the column of water should be 13.6 times as high as a mercury column because the density of water is 1/13.6 that of mercury. One atmosphere will support 76 cm (13.6) = 1033.6 cm of water. Note that this column of water in a tube of cross section 1 cm^2 contains a mass of water equal to 1033.6 g. Regardless of the density of a liquid, the mass that can be supported by 1 atm must be the same.

THE MANOMETER

A manometer (Fig. 2.2) uses a variation of the barometer principle to measure the pressure of a gas sample. Mercury is placed in the U tube at the open end. A gas sample trapped in the closed end of the tube by the mercury exerts a pressure on the mercury on that side of the tube. Atmospheric pressure pushes on the open end, and any difference in the levels of the two mercury columns must be equal to the difference between atmospheric pressure and the gas pressure. If the levels are identical the pressure of the gas sample is equal to atmospheric pressure. When the height of the column open to the atmosphere is lower than

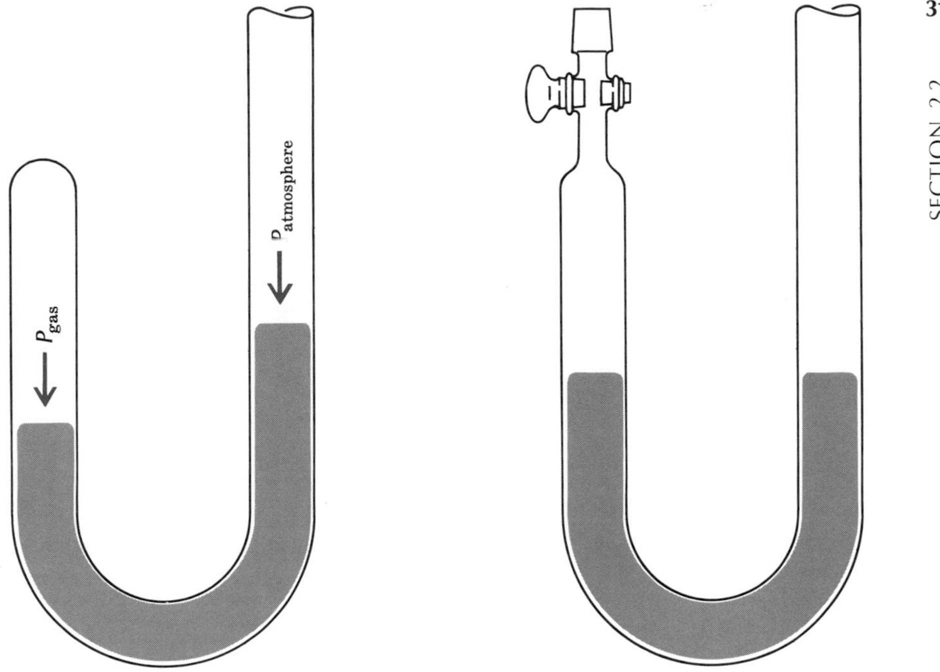

Figure 2.2
Examples of manometers.

that on the gas side of the tube, the pressure of the gas sample is less than atmospheric pressure. If the closed end contains mercury at a height 10 cm higher than the open end, the pressure of the gas sample is 66 cm of mercury. The manometer can be modified by altering the top of the closed end to attach a stopcock. Connection of such a manometer via the stopcock to another gaseous system allows the determination of the pressure of the system.

2.2
GAS LAWS

BOYLE'S LAW

In the course of experimenting with vacuums and vacuum pumps, Robert Boyle in 1662 made the first systematic study of the relationship between volume and pressure in gases. He measured what he termed "the spring of air," the pressure with which a gas sample pushes back when it is compressed, using what seems today to be remarkably simple devices.

The classic experiments of Boyle can be illustrated by using the modified manometer shown in Fig. 2.3. If mercury is added to the manometer with the stopcock open, the mercury level in the two columns is

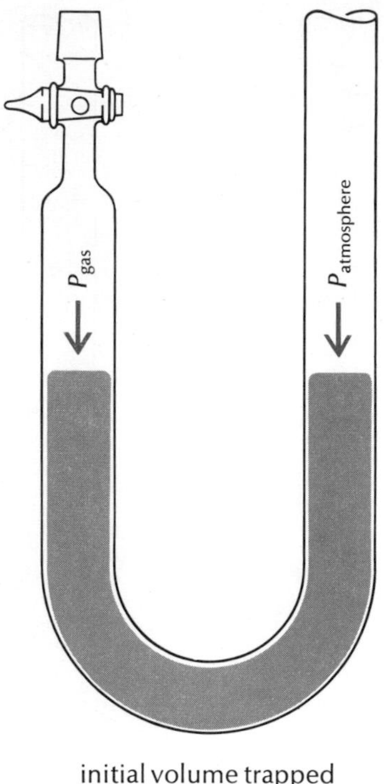

initial volume trapped
by closing stopcock

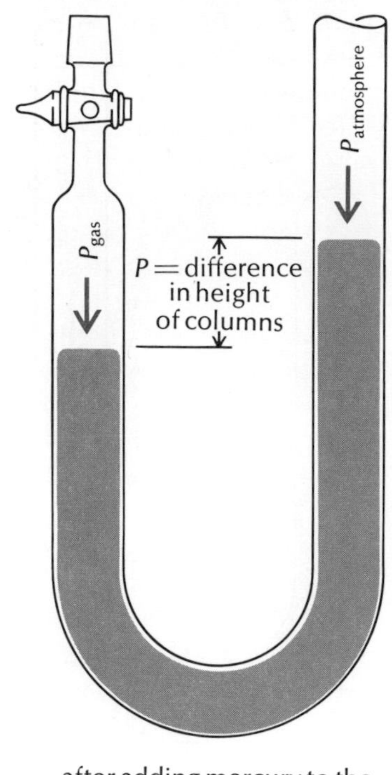

after adding mercury to the
right arm of manometer device

Figure 2.3
Boyle's law experiment.

equal. However, when the stopcock is closed, a volume of air is trapped under atmospheric pressure, and further addition of mercury to the open end of the tube increases the pressure exerted on the gas sample. The gas pressure is equal to the sum of atmospheric pressure and the difference in the level of the two mercury columns. The volume of the trapped gas sample is known from the dimensions of the tube. In such a series of measurements the pressure is gradually increased by adding the mercury, and the volume occupied by the gas sample decreases. An idealized plot of the observed pressure-volume dependence at constant temperature for three points *a, b,* and *c* is shown in Fig. 2.4. The graph of P vs. V illustrates the inverse pressure-volume relationship. Its hyperbolic form is typical of equations of the general type $PV = k_1$ (Appendix 1), where P and V represent the variables pressure and volume, respectively, and k is some constant number. The value of one variable increases as the other decreases so that their product is always the same. An alternate way of representing the inverse proportionality between pressure and volume, $P \propto 1/V$, is shown in the plot of P vs $1/V$. The equation for the resultant straight line is $P = k_1(1/V)$, and the value of k_1 is the slope of the line.

The series of measurements can be repeated using the same initial volume at the same initial pressure, and the same value of k_1 will be obtained. If the experimental conditions are varied, the general inverse pressure–volume relationship is still observed, but the value for k_1 will differ. When a large volume of gas is trapped at atmospheric pressure, the value for k_1 is larger by a factor equal to the ratio of the initial volumes. The temperature also affects the value for k_1. As the temperature of the gas sample increases, the value for k_1 also increases.

If the variation of k_1 as a function of the initial volume of the gas sample and temperature were noted by a modern scientist, he could proceed to investigate the relationships still further. However, Boyle noted that the value of PV for a fixed amount of gas was only approximately constant. He also observed that when a gas was heated its volume was increased even if the pressure was kept constant. Although Boyle did not investigate this phenomenon further, his contributions to the theory of gases are highly regarded.

Boyle's observations and those of subsequent workers on gas volume as a function of pressure now can be stated: At constant temperature the volume of a given quantity of an ideal gas varies inversely as the pressure exerted on it. This statement has been tested many times with a variety of gases. The same value of k_1 is obtained for different gases whenever the same initial volume at a given pressure and temperature is used. This observation is important and interesting because the same volume of different gas samples contains different amounts of matter. It therefore follows that for the same mass of two or more different gases, the values for k_1 will vary. This behavior will be examined later in this chapter.

There are gases that do not obey Boyle's law, and they are said to be nonideal. The deviations from Boyle's law are most serious at low temperatures and high pressures, but at room temperature and normal atmospheric pressure most gases behave ideally. Nonideality of gases is

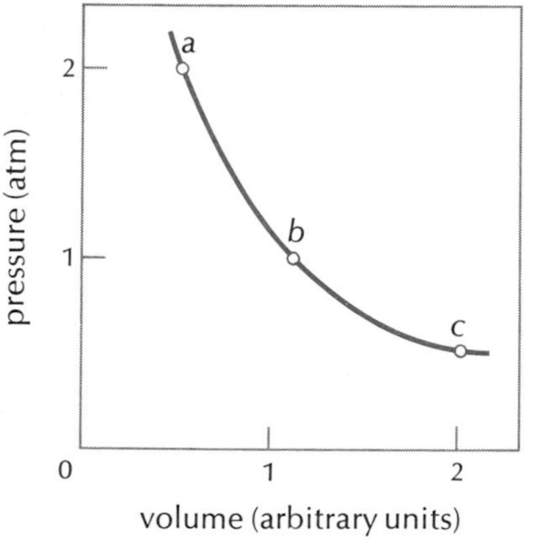

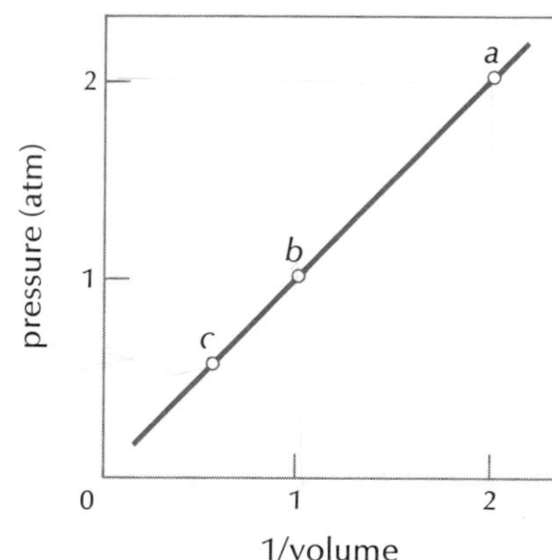

Figure 2.4
Boyle's law relationship.

discussed in Chapter 7, where the nature of the structure of matter is developed.

Boyle's law can be used to predict the volume or pressure of a mass of gas at a given temperature if its volume and pressure are defined under some other conditions. For example, if a sample of gas is placed inside a chamber containing a movable piston of negligible mass, the pressure on the system is that of atmospheric pressure (Fig. 2.5). The pressure on the system can be increased by placing a weight on the piston. If the weight exerts a force on the piston equal to 14.7 psi, then the pressure on the system is doubled. The piston will descend to a point such that the internal pressure of the gas equals the external pressure (2 atm). If a second weight is added the piston will descend still further until the internal pressure is equal to the external pressure. When the pressure on the gas is doubled, the volume is decreased by one-half; and when the pressure is tripled, the volume is decreased to one third of its initial volume.

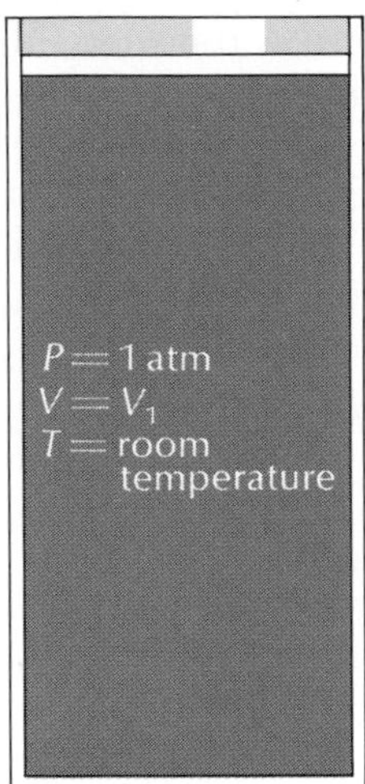

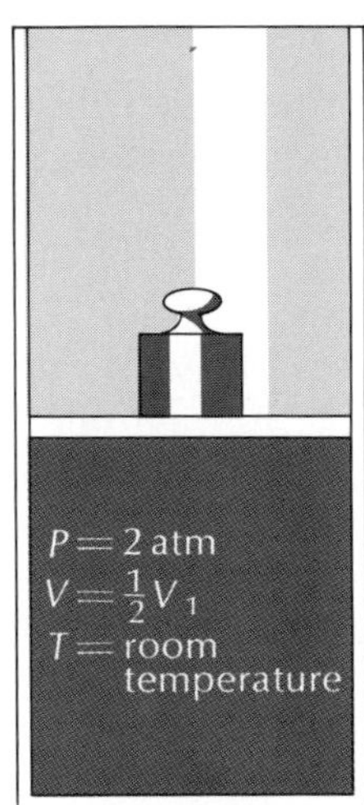

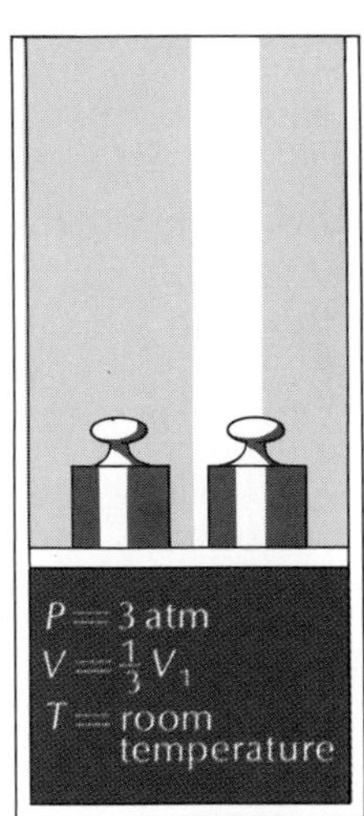

Figure 2.5
Behavior of gases predicted by Boyle's law.

EXAMPLE 2.3

A 20-liter sample of a gas at 0°C at 10 atm is allowed to expand until its pressure is 3 atm. What will be the volume under these new conditions if the temperature is not changed?

Since the volume and pressure of a gas are inversely proportional, either term can be evaluated from the knowledge of the factor by which the other term varies. The initial pressure P_i is decreased by the factor of $\frac{3}{10} = P_f/P_i$.

$$P_i\left(\frac{P_f}{P_i}\right) = P_f$$

$$10 \text{ atm}\left(\frac{3}{10}\right) = 3 \text{ atm}$$

The factor by which the pressure changes is a dimensionless number as the units of the pressure terms cancel. According to Boyle's law, the volume should vary inversely with the pressure. Therefore the final volume V_f should be $\frac{10}{3}$ that of the initial volume V_i, or 66.7 liters.

$$V_f = V_i\left(\frac{P_i}{P_f}\right)$$

$$V_f = 20 \text{ liters}\left(\frac{10}{3}\right) = 66.7 \text{ liters}$$

It should be noted at this point that calculations can be made by applying Boyle's law to a macroscopic sample of a gas about which little is known in terms of its submicroscopic structure. The application of a law to a problem does not advance our understanding of the phenomenon. In order to understand phenomena such as those observed for the gaseous state, it is necessary to devise models to represent gases. Before we can postulate such a model, we must consider additional experimental observations and derived laws.

CHARLES' LAW

Gases expand when heated under constant pressure and contract when cooled. One of the unfortunate examples of this behavior is witnessed by small children who have their balloons filled in an air-conditioned store and then leave to go into the hot summer sunlight. The balloon may burst because of the increase in its volume. Another example of the relationship between the volume and temperature of a gas is encountered by individuals who let air out of their tires to release the pressure buildup caused by high-speed driving. This practice is inadvisable for several reasons. One obvious problem is the result of a decrease in the volume of the tire when the temperature decreases. The driver who neglects to add air to his tires at the end of his trip will have flat tires the next day.

The French physicist J. A. C. Charles observed in 1787 the relationship between volume and temperature at constant pressure. His quantitative observations can be repeated using a glass tube of small diameter closed at one end. The other end is filled with a mercury "piston" made by warming the tube slightly and then placing it in a sample of mercury. As the tube cools mercury is drawn into it by the decreasing pressure in the tube.

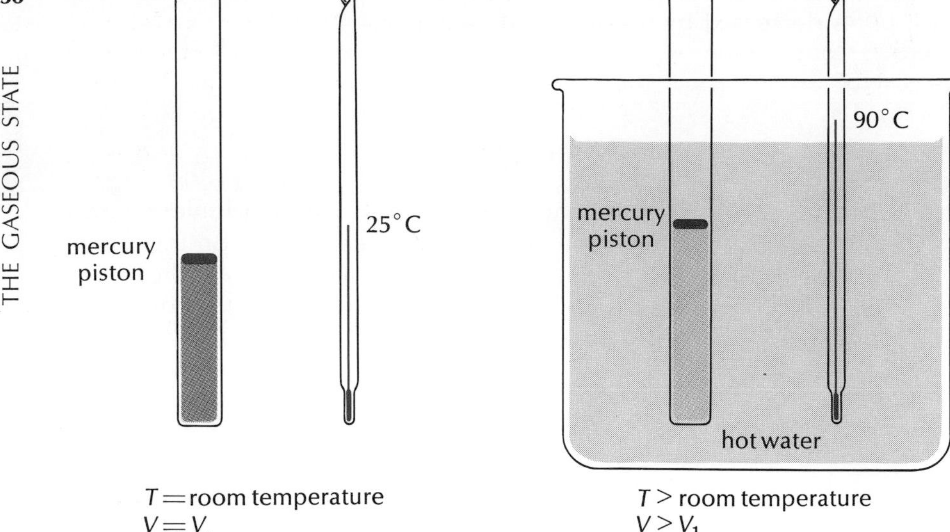

Figure 2.6
Charles' law experiment.

Rapid inversion of the tube yields the necessary mercury piston, which continues to settle in the tube until the temperature of the tube is equal to that of its surroundings (Fig. 2.6). The pressure on the trapped gas sample is equal to the sum of the atmospheric pressure and the length of the mercury column.

If the tube is placed in a beaker containing warm water, the mercury "piston" rises to a new stationary position. Under these conditions the gas pressure is unchanged: The gas pressure equals atmospheric pressure plus the length of the piston. The volume of the gas is known from the dimensions of the tube. A plot of the volume as a function of temperature yields the straight line (Fig. 2.7) characteristic of two variables that are directly

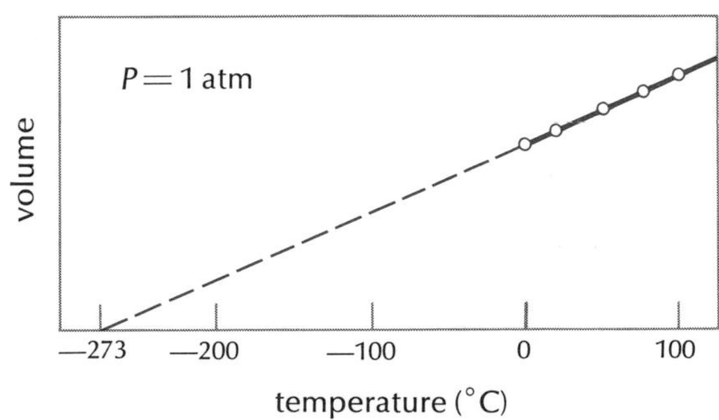

Figure 2.7
Dependence of volume on temperature: Charles' law.

proportional. When the plot is extrapolated to lower temperatures it leads to a point of zero volume. The temperature at which the extended line predicts zero volume is −273.15°C. This temperature is called *absolute zero.* A temperature scale using absolute zero as the lowest possible temperature is called the Kelvin scale, after Lord Kelvin who proposed the scale in 1848. The temperature interval for one degree on the Celsius scale is equal to that on the Kelvin scale. Therefore, temperature on the Celsius scale can be converted to temperature on the Kelvin scale by the addition of 273.15° to the Celsius value:

$$°K = °C + 273.15°$$

Using the Kelvin scale, the volume and temperature of the gas sample in the tube in Fig. 2.8 are directly proportional. Their relationship can be mathematically expressed as

$$V \propto T$$
$$V = k_2 T,$$

(Appendix I) where T is in degrees Kelvin. The value for k_2, the slope of the line, is a function of the experimental conditions. If the pressure on the gas sample is increased as a result of higher atmospheric pressure or an increase in the mass of the mercury piston, the value for k_2 decreases. When a smaller sample of gas is trapped, the value for k_2 also decreases. Accordingly, the slope of the line in Fig. 2.8 becomes smaller in both cases. However, the extrapolated value of −273.15° remains unchanged. The value of k_2 is independent of the gas chosen for investigation when the same volume is under the same conditions of initial pressure and temperature for each gas. Again, as was observed in the Boyle's law discussion, it fol-

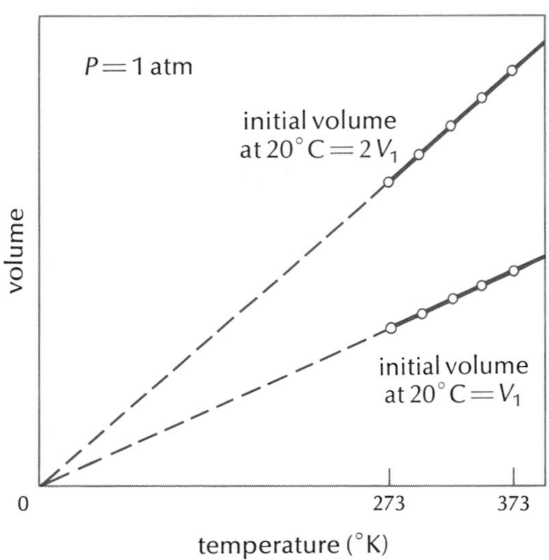

Figure 2.8
Dependence of Charles' law constant on the initial volume.

lows that the constant will vary for the same mass of different gases since their densities are different.

The experimental results of Charles can be summarized as Charles' law: For a given mass of an ideal gas at constant pressure, the volume is directly proportional to the absolute temperature. While many real gases do conform approximately to Charles' law, most show deviations from expected behavior at high pressures and low temperatures. The most common deviation occurs at low temperatures. The observed volume of a gas near its liquefaction temperature is less than predicted by Charles' law.

As was the case in Boyle's law, Charles' law can be utilized as an empirical result without understanding why the law works. A cylinder with a movable piston that keeps the gas sample under constant pressure is shown in Fig. 2.9. If the temperature of the gas is increased from 0 to 91°C, the piston moves upward. Since the temperature has increased from 273 to 364°K, or by a factor of $\frac{364}{273}$, the volume increases by the same factor of $\frac{364}{273}$ or $\frac{4}{3}$. If the volume at 273°K is 3 liters, the volume at 364°K will be 4 liters.

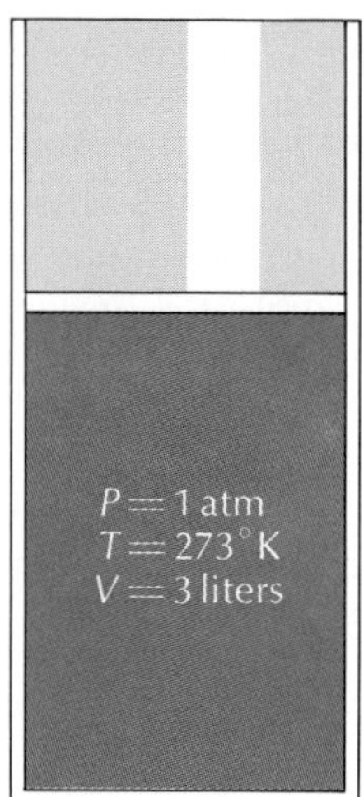

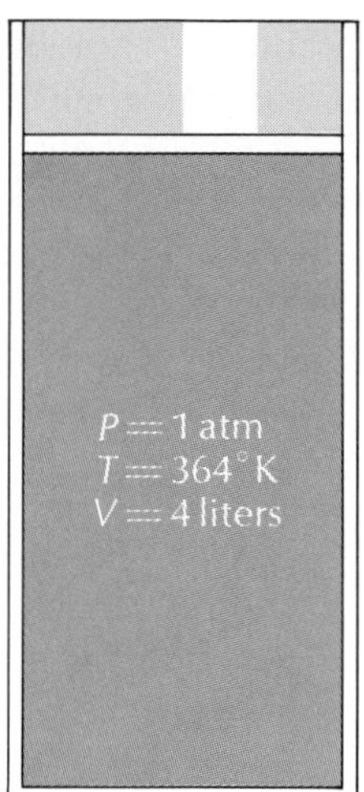

Figure 2.9
Behavior of a gas predicted by Charles' law.

EXAMPLE 2.4

A 2-liter sample of air is cooled from 127 to −73°C at constant pressure. What will be the volume of the gas at the lower temperature?

In order to solve this problem it is first necessary to convert the given temperatures into °K. The gas is cooled from 400 to

200°K. Therefore the temperature has decreased by a factor of two.

$$T_f = \left(\frac{T_f}{T_i}\right) T_i$$

$$200°\text{K} = \left(\frac{200}{400}\right) 400°\text{K}$$

According to Charles' law, the volume should decrease by the same factor. The final volume will be 1 liter.

$$V_f = \left(\frac{T_f}{T_i}\right) V_i$$

$$V_f = \left(\frac{200}{400}\right) 2 \text{ liters} = 1 \text{ liter}$$

GAY-LUSSAC'S LAW

Of the three variables controlling the gaseous state, two of the possible pairs of variables have been related by keeping the third constant. Once Boyle's and Charles' laws are known, the relationship between pressure and temperature follows easily. The pressure of a given mass of gas under conditions of constant volume is directly proportional to absolute temperature. This is called Gay-Lussac's law, although it is also referred to as Amonton's law. Guillaume (William) Amonton first made use of the relation between pressure and temperature in 1703 by constructing a gas thermometer that recorded temperature changes in terms of the pressure of a fixed volume of gas. The behavior of a gas under constant volume can be observed by use of a rigid container attached to a manometer. Upon heating, the gas can expand only by an insignificant amount into the manometer, and an increase in pressure is recorded by the change in the heights of the mercury columns in the manometer (Fig. 2.10).

The mathematical statement relating the two variables of pressure and temperature is $P = k_3T$, where the temperature is in degrees Kelvin. The value of k_3 is a function of the amount of the gas sample initially trapped for an experiment and also the volume of the container. If the same initial volume at the same initial temperature and pressure is considered, the value of k_3 is independent of the type of gas. Of course the mass of each of several gases under such identical conditions will not be equal because their densities are not equal. Thus for the third time it is noted that gases behave in a manner dictated by the equivalence of volumes rather than of masses.

COMBINATION OF GAS LAWS

The three variables of pressure, volume, and temperature can be interrelated in one expression. Since volume is inversely proportional to pressure

$$V \propto \frac{1}{P}$$

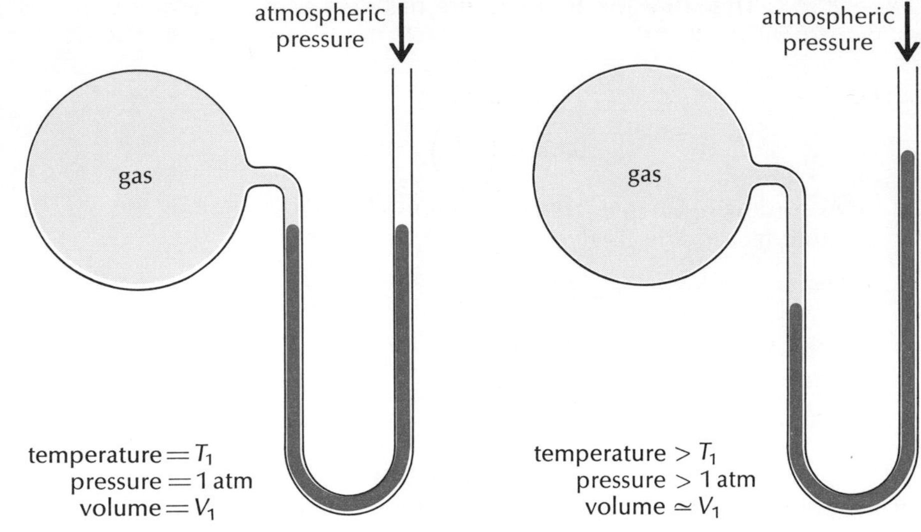

Figure 2.10
Gay-Lussac's law: Heating the gas sample with almost no increase in volume produces a pressure increase.

and directly proportional to temperature

$$V \propto T$$

it follows that volume is proportional to the quotient of the temperature divided by the pressure.

$$V \propto \frac{T}{P}$$

The proportionality symbol ($\propto$) may be replaced by a constant k_4, and the equality

$$V = k_4 \frac{T}{P}$$

is established. The value for k_4 is a function of the amount of gas under consideration. If equal volumes of different gases at the same temperature and pressure are compared, the value of k_4 is independent of the type of gas. Again, as noted in each of the preceeding sections relating pairs of variables, if samples of the same mass of different gases are compared the value of k_4 is a function of the gas.

The general expression relating all variables could be used to calculate any unknown variable if the others and k_4 were known. However, it need not be used as expressed, and the value of k_4 need not be calculated. A correct answer to a problem involving gases can be obtained by the use of factors. For example, if a given mass of a certain gas at 1 atm and 273°K occupies 500 ml, what will its volume be at 5 atm and 273°C? The answer can be obtained by considering separately the change in each variable. An increase in pressure from 1 to 5 atm decreases the volume by a factor of $\frac{1}{5}$. Increasing the temperature from 273 to 546°K increases the volume by a factor of $\frac{546}{273}$. Therefore, the factor by which both changes alter the volume is $(\frac{546}{273})(\frac{1}{5})$, or $\frac{2}{5}$. The final volume is 200 ml.

EXAMPLE 2.5

A 5-liter sample of a gas at 273°C and 2 atm is cooled to 0°C and its pressure changed to 1 atm. What will be the new volume under these conditions?

The temperature decreases by a factor of two (that is, from 546 to 273°K). Therefore, the volume will decrease by this same factor (see Example 2.4 for a previous example of this reasoning). The pressure decreases by a factor of two and, therefore, the volume will increase by a factor of two. The two factors exactly balance, and the resultant volume will be the same as the initial volume!

$$V_f = V_i\left(\frac{T_f}{T_i}\right)\left(\frac{P_i}{P_f}\right)$$

$$V_f = 5 \text{ liters}\left(\frac{273°\text{K}}{546°\text{K}}\right)\left(\frac{2 \text{ atm}}{1 \text{ atm}}\right) = 5 \text{ liters}$$

DALTON'S LAW OF PARTIAL PRESSURE

In previous sections we dealt with the behavior of a single gas, but mixtures can be treated with equal mathematical simplicity. Dalton observed the behavior of mixtures of gases in 1801 and found that the total pressure exerted by a mixture of gases is equal to the sum of the partial pressures of the gases. The term *partial pressure* of a gas in a mixture of gases is the pressure that gas would exert if it were the only gas present under those conditions of volume and temperature.

Dalton's law can be demonstrated by placing samples of hydrogen, oxygen, and nitrogen in three separate containers at different pressures but at the same temperature (Fig. 2.11). The pressure could be determined by the use of manometers, which have been eliminated for simplicity in the diagram. When the stopcocks are opened the gases mix, and a uniform pressure throughout the system results. The final observed pressure is 2 atm. This pressure can be calculated by use of Dalton's law. Hydrogen, originally at a pressure of 3 atm in a volume of 1 liter, has a final partial pressure of $\frac{1}{2}$ atm; since it expands by a factor of 6 its pressure must decrease by $\frac{1}{6}$.

$$P_{\text{hydrogen}} = P_i\left(\frac{V_i}{V_f}\right) = 3 \text{ atm}\left(\frac{1 \text{ liter}}{6 \text{ liters}}\right) = \frac{1}{2} \text{ atm}$$

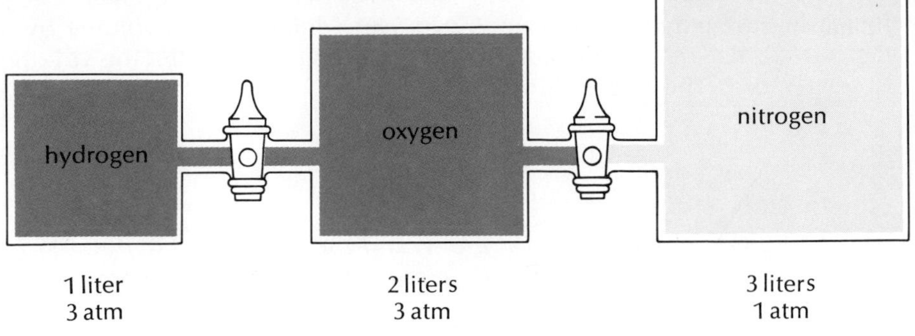

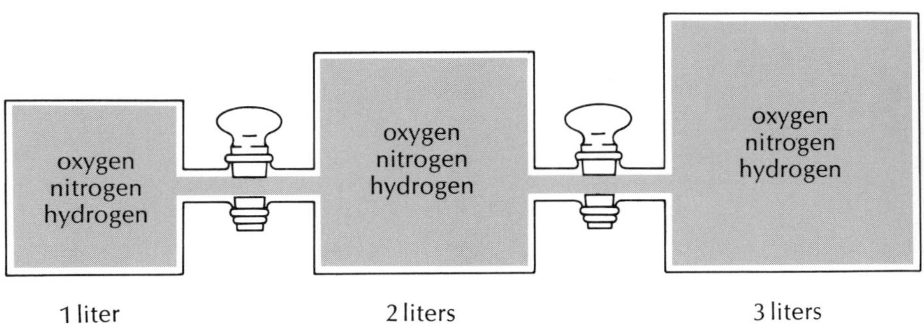

Figure 2.11
Dalton's law of partial pressures:

$$\begin{aligned} \text{Pressure} &= P_{\text{hydrogen}} + P_{\text{oxygen}} + P_{\text{nitrogen}} \\ &= \tfrac{1}{2} \text{ atm} + 1 \text{ atm} + \tfrac{1}{2} \text{ atm} \end{aligned}$$

The partial pressure of oxygen is 1 atm because the oxygen expands by a factor of 3:

$$P_{\text{oxygen}} = P_i \left(\frac{V_i}{V_f}\right) = 3 \text{ atm} \left(\frac{2 \text{ liters}}{6 \text{ liters}}\right) = 1 \text{ atm}$$

The partial pressure of nitrogen is $\frac{1}{2}$ atm because it expands by a factor of 2:

$$P_{\text{nitrogen}} = P_i \left(\frac{V_i}{V_f}\right) = 1 \text{ atm} \left(\frac{3 \text{ liters}}{6 \text{ liters}}\right) = \frac{1}{2} \text{ atm}$$

Therefore, the sum of the partial pressures is 2 atm:

$$\begin{aligned} P_{\text{total}} &= P_{\text{hydrogen}} + P_{\text{oxygen}} + P_{\text{nitrogen}} \\ &= \tfrac{1}{2} \text{ atm} + 1 \text{ atm} + \tfrac{1}{2} \text{ atm} = 2 \text{ atm} \end{aligned}$$

Dalton's law is subject to the same limitations as are Boyle's and Charles' laws. It is assumed that the gases behave ideally at all temperatures and pressures. In addition, it is also necessary that the gases maintain their chemical identity and do not react with each other. If chemical reactions occur, Dalton's law does not apply.

GAY-LUSSAC'S LAW OF COMBINING VOLUMES

Some gases when mixed react with each other to form compounds or elements. A mixture of hydrogen and oxygen can exist indefinitely at normal temperatures and pressures, but when a spark is introduced water is formed in an explosive reaction. At constant pressure and at a constant temperature above the boiling point of water, 2 liters of hydrogen and 1 liter of oxygen produce 2 liters of water vapor. In a similar manner a mixture of hydrogen and chlorine can exist until exposed to ultraviolet light. Irradiation produces the compound hydrogen chloride. In this case 1 liter of hydrogen and 1 liter of chlorine react to produce 2 liters of hydrogen chloride under conditions of constant pressure and temperature.

Observations of several gaseous reactions led Gay-Lussac to a generalized statement commonly called the *law of combining volumes:* At a given temperature and pressure gases combine in simple proportions by volume. The volumes of the gaseous products are in simple whole-number ratios to the volumes of the gaseous reactants.

2.3 GASES AND ATOMIC THEORY

AVOGADRO'S HYPOTHESIS

In earlier sections of this chapter it was observed that Boyle's and Charles' laws could be applied to a variety of gases and that a single constant serves to relate all of the variables of pressure, volume and temperature if the same initial volume is compared. This suggests that equal volumes of different gases under the same conditions are similar in a way which is not indicated by their masses. The law of multiple proportions described in Chapter 1 and the law of combining volumes described in this chapter also suggest that the determining feature of matter and its reactions is not its mass alone.

Amadeo Avogadro in 1811 suggested that equal volumes of gases under the same conditions of temperature and pressure contain the same number of submicroscopic particles. For example, equal volumes of helium and argon at 1 atm pressure and 0°C (called *standard temperature and pressure* or STP) contain the same number of atoms. These atoms are tiny particles that cannot be subdivided and still maintain the characteristic physical and chemical properties of the element. The atomic concept had been considered by many individuals since the time of Democritus in ancient Greece. However, the derivation of the concept was philosophical and was not based on any experimental data. Avogadro's hypothesis utilized the atomic concept in a quantitative manner and provided a basis for further experimental tests. It provided a rationale for all the then-known gas laws and was in agreement with the laws of chemical combination rationalized by John Dalton (Chapter 4) in his atomic hypothesis.

If, as Avogadro hypothesized, equal volumes of different gases contain the same number of submicroscopic particles (atoms in the case of some elements and molecules in the case of gaseous compounds), then the behavior of the gaseous state is independent of the mass of the particles. Only the number of particles is important in understanding the ideal gas laws.

The concept of molecules is a valid model with which to explain the observed behavior of gaseous reactants expressed by the laws of multiple proportions and combining volumes. The reaction of hydrogen and oxygen cannot be explained if these elements are monatomic (Fig. 2.12). The combination of 2 liters of monatomic hydrogen with 1 liter of monatomic oxygen dictates the combination of two hydrogen atoms per atom of oxygen. The resultant molecules, H_2O, containing two hydrogen atoms and one oxygen atom, must be produced in a number equivalent to the number of oxygen atoms available. Since only 1 liter of oxygen was used, only 1 liter of water vapor should be produced according to Avogadro's hypothesis. Because 2 liters are produced, the monatomic assumption must be incorrect. Other assumptions regarding the nature of the elements hydrogen and oxygen are necessary. The assumption that both hydrogen and oxygen are diatomic adequately explains observed combining volumes. The volume of oxygen contains a certain as yet unknown number of diatomic oxygen molecules, O_2. Twice that number of oxygen atoms are potentially available for a given reaction. In the hydrogen sample containing diatomic hydrogen molecules, H_2, there are the same number of molecules per unit volume as in the oxygen sample. The observed combination of two volumes of hydrogen per volume of oxygen requires the interaction of two hydrogen molecules per molecule of oxygen. If the water molecule consisted of four hydrogen atoms and two oxygen atoms, only one volume of water vapor would be produced. The two volumes produced require the union of two hydrogen atoms per oxygen atom to form water.

A balanced chemical equation that gives the chemical makeup of hydrogen, oxygen, and water and that is in agreement with the experimentally observed volume change is the following:

$$2H_2 + O_2 \longrightarrow 2H_2O$$

The subscripts represent the number of atoms involved in each molecule. The prefix numbers indicate the relative number of molecules required for reaction.

RELATIVE ATOMIC AND MOLECULAR MASS

Avogadro's principle can be used to determine the relative mass of atoms and molecules. Since equal volumes of gases under the same conditions

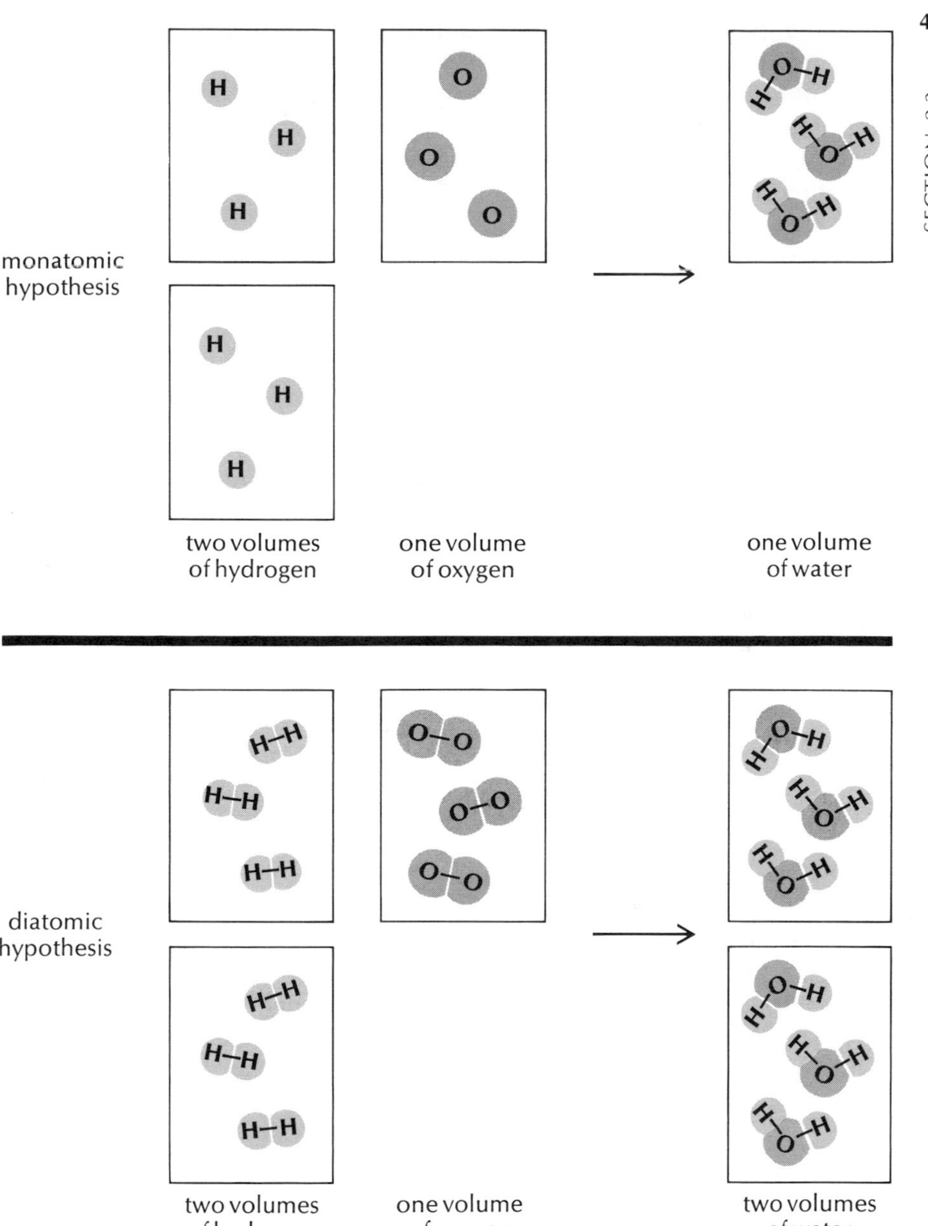

Figure 2.12
The combination of hydrogen and oxygen to yield water.

of temperature and pressure contain the same number of submicroscopic particles, the masses of the gas particles must be in the same ratio as the gas densities. The densities of the monatomic gases, helium and argon, at STP are 0.179 and 1.79 g/liter, respectively, and indicate that one argon atom is 10 times as massive as one helium atom. If the absolute mass of one atom of one element were known, the absolute mass of an atom of the other could be calculated. However, even without a knowledge of absolute atomic masses, the relative atomic masses of all gaseous elements can be established from density measurements and the knowledge of whether the element is monatomic or diatomic.

Hydrogen, the lightest element known, has a density of 0.090 g/liter at STP. Comparison with the density of helium, 0.179 g/liter, indicates that the helium atom is twice as massive as the hydrogen molecule. Since hydrogen is diatomic, the mass of the hydrogen atom is one-fourth that of the helium atom. Oxygen, which is diatomic, has a density of 1.43 g/liter at STP and, therefore, the oxygen atom is 4 times as massive as helium and 16 times as massive as the hydrogen atom.

The relative masses of the atoms were set at one time on a scale in which the mass of the oxygen atom was 16.000 *atomic mass units* (amu). On this scale the lightest element, hydrogen, had a mass of 1.008 amu. The reason for this arbitrary choice was one of convenience, as the atomic mass units of most elements were nearly integers on this basis. Recently the basis of the atomic mass unit system has been changed to designate the atomic mass of an isotope (Chapter 5) of carbon as 12 amu. The change to a different standard of reference makes only small changes in most atomic masses. The reason for the change becomes evident in Chapter 5.

EXAMPLE 2.6

The density of fluorine gas (which consists of diatomic molecules F_2) is 1.70 g/liter at STP. What is the relative atomic mass of a fluorine atom?

The density of fluorine gas is larger than that of hydrogen gas by the factor 1.70/0.090. Therefore, the fluorine molecule is more massive than the hydrogen molecule by the same factor. The approximate molecular mass of the hydrogen molecule on the relative mass scale is 2 amu. The approximate molecular mass of the fluorine molecule is

$$2 \text{ amu}\left(\frac{1.70}{0.090}\right) = 38 \text{ amu}$$

and the atomic mass of a fluorine atom is one-half this quantity, or 19 amu.

The relative mass of compounds can be established on the same scale as that for atoms. The density of carbon dioxide (CO_2) is 1.97 g/liter at

STP, which is 1.37 times the density of diatomic oxygen. The relative molecular mass of CO_2 is therefore 44 amu:

$$\text{molecular mass of } \mathbf{CO_2} = 1.37 \times \text{molecular mass of } \mathbf{O_2}$$
$$= 1.37 \times 32 \text{ amu} = 44 \text{ amu}$$

It should be noted at this point that a knowledge of the number and type of atoms contained in a molecule yields the relative molecular mass directly without density measurements. Carbon dioxide consists of one atom of carbon and two atoms of oxygen per molecule. Since the relative atomic masses of carbon and oxygen are 12 and 16 amu, respectively, the relative molecular mass of carbon dioxide is 12 amu + 2(16 amu) = 44 amu.

EXAMPLE 2.7

Calculate the density of carbon monoxide (CO) at STP.

The relative molecular mass of carbon monoxide is the sum of the atomic mass of carbon and oxygen, or 12 amu + 16 amu = 28 amu. The molecular mass of carbon monoxide is $\frac{28}{2} = 14$ times the mass of the hydrogen molecule whose density is 0.090 g/liter at STP. The density of carbon monoxide at STP is larger than that of hydrogen by the factor $\frac{28}{2}$ or

$$0.090 \text{ g/liter } (28/2) = 1.26 \text{ g/liter}$$

A table of relative atomic mass values for the elements is located on the inside back cover of this text. The values are the most correct currently known and are given to more significant figures than will be used in the problems given in this text.

THE MOLE

Since atoms and molecules are postulated to be extremely small, any experiment involving observable amounts of elements or compounds must involve tremendous numbers of atoms or molecules. It has been found convenient to define a quantity of atoms or molecules as a basic working unit in chemistry. This unit is the *mole* and is no different from any other arbitrarily defined unit such as the dozen, pound, or acre. A *mole of atoms* is a collection of atoms in sufficient quantities such that their combined mass in grams is equal in magnitude to the relative atomic mass of the atom in atomic mass units. A mole of helium atoms contains enough atoms for its mass to be 4 g. Similarly, the mole concept can be applied to molecules. A mole of carbon dioxide molecules contains 44 g of CO_2 molecules.

Since the mass of a mole of any element or compound is equal in magnitude to the mass of any atom or molecule on the atomic mass scale, it follows that 1 mole of any element or compound contains the same number of basic particles. A mole of helium atoms contains the same number of atoms as a mole of argon atoms. A mole of carbon dioxide contains the same number of molecules as a mole of water molecules. Therefore, the discussion of chemical reactions can be transformed from the interactions of atoms and molecules to a measurable unit, the mole. In the case of the reaction of hydrogen and oxygen to produce water, 2 moles of hydrogen molecules react with 1 mole of oxygen molecules to produce 2 moles of water molecules. Or, in terms of mass, 4 g of hydrogen combine with 32 g of oxygen to produce 36 g of water:

$$\begin{aligned} \mathbf{2H_2} + \mathbf{O_2} &\longrightarrow \mathbf{2H_2O} \\ \text{two molecules of } \mathbf{H_2} + 1 \text{ molecule of } \mathbf{O_2} &\longrightarrow \text{two molecules of } \mathbf{H_2O} \\ 2 \text{ moles of } \mathbf{H_2} + 1 \text{ mole of } \mathbf{O_2} &\longrightarrow 2 \text{ moles of } \mathbf{H_2O} \\ 4 \text{ g of } \mathbf{H_2} + 32 \text{ g of } \mathbf{O_2} &\longrightarrow 36 \text{ g of } \mathbf{H_2O} \end{aligned}$$

AVOGADRO'S NUMBER AND ABSOLUTE ATOMIC MASS

The concept of a mole was firmly established in chemistry before the exact number of atoms or molecules contained in the unit was known. It is now possible to determine the number of particles making up 1 mole by many modern experimental techniques, the details of which will not be discussed in this text. The value of this number is 6.0225×10^{23} and is referred to as *Avogadro's number.*

Accepting the value of Avogadro's number, it becomes possible to determine the actual mass of any atom or molecule. The mass of a mole of atoms or molecules can be calculated from the table of relative atomic mass values, and Avogadro's number indicates the number of atoms or molecules contained in a mole. Division of the mass of 1 mole by Avogadro's number yields the mass per basic particle.

EXAMPLE 2.8

Calculate the mass of a helium atom.

The relative atomic mass of the helium atom is 4 amu. According to the definition of a mole, the mass of 1 mole of helium atoms is 4 g. This quantity of helium atoms contains Avogadro's number of atoms. The mass of the helium atom is

$$\frac{4 \text{ g/mole}}{6.02 \times 10^{23} \text{ atom/mole}} = 6.64 \times 10^{-24} \text{ g/atom}$$

The concepts of the mole, Avogadro's hypothesis, and Avogadro's number allow the determination of the molecular mass of an unknown compound. The volume occupied by 1 mole (32 g) of oxygen molecules at STP has been determined to be 22.4 liters. Since equal volumes of gases contain the same number of basic particles, 22.4 liters must be the volume occupied by 1 mole of any gaseous substance at STP.

The density of methane CH_4 is 0.714 g/liter at STP; therefore, the mass of 22.4 liters is 16 g. On the atomic mass scale (amu), the mass of a methane molecule is 16 amu, as the mass of a mole is 16 g. This is in agreement with the formula CH_4, since the atomic masses of carbon and hydrogen are 12 and 1, respectively.

EXAMPLE 2.9

The density of a gas at 91 °C and $\frac{1}{3}$ atm is 0.65 g/liter. What is the molecular mass of the gas?

The molecular mass can be found from the known molar volume and the gas density if we first find the gas density at standard temperature and pressure. At STP the volume of a gas sample will be smaller owing to both the change in temperature and pressure. Accordingly, the density will increase although the mass of the gas sample is unaltered. Only the volume change is responsible for the change in density.

$$\text{density at STP} = 0.65 \text{ g/liter} \left(\frac{364}{273}\right)\left(\frac{1}{1/3}\right) = 2.60 \text{ g/liter}$$

$$\text{mass of 22.4 liters at STP} = 2.60 \text{ g/liter} \times 22.4 \text{ liters} = 58.2 \text{ g}$$

$$\text{molecular mass} = 58.2 \text{ g/mole}$$

GRAHAM'S LAW OF DIFFUSION

When the volume available to a gas is suddenly increased, the gas atoms or molecules gradually spread out through the entire volume accessible to them. This process, which eventually results in the even distribution of the gas, is called diffusion. Graham examined the rate of this process in 1829 and observed that the relative rates of diffusion of two gases at the same temperature are inversely proportional to the square root of their densities under the same conditions. Since the densities of gases are directly proportional to their atomic or molecular masses, *Graham's law* may be stated as the rate of diffusion of a gas is inversely proportional to the square root of its molecular or atomic mass. The mathematically equivalent statement of Graham's law is

$$\frac{r_1}{r_2} = \frac{\sqrt{m_2}}{\sqrt{m_1}} = \sqrt{\frac{m_2}{m_1}}$$

where r_1 and r_2 represent the rates of diffusion of two gases and m_1 and m_2 are the corresponding atomic or molecular masses. Since a ratio of two masses is utilized, it is not necessary to use the absolute masses of atoms or molecules. The relative masses (amu) can be used because they are directly proportional to the absolute masses of the particles.

The compound sulfur dioxide (SO_2) diffuses at a rate that is one-quarter that of helium. Therefore, sulfur dioxide must be $4^2 = 16$ times more massive than helium.

$$\frac{r_{He}}{r_{SO_2}} = 4 = \sqrt{\frac{m_{SO_2}}{m_{He}}} \qquad m_{SO_2} = 16m_{He} = 16 \times 4 \text{ amu} = 64 \text{ amu}$$

This value is in agreement with the formula for sulfur dioxide. The relative atomic masses of sulfur are 32 and 16 amu, respectively, and therefore the mass of SO_2 must be 32 amu + 2(16) amu = 64 amu.

IDEAL GAS LAW

The combination of Boyle's and Charles' laws to give a general equation for a specified volume of an ideal gas is of the form:

$$V = k_4\frac{T}{P}$$

Now that Avogadro's principle has been discussed, the significance of a constant value of k_4 for the same initial volumes of various gases under the same conditions becomes evident. The same initial volumes, according to Avogadro's principle, contain the same number of atoms or molecules.

The combined gas laws can be written as a more general expression by using a term n to represent the number of moles of a gas under consideration and R as a new proportionality constant. Making these substitutions yields the *ideal gas law* in terms of P, V, T, and n. According to Avogadro's principle, volume is directly proportional to the number of moles, so that the general equation relating all four variables is $PV = nRT$. The proportionality constant R is called the *universal gas constant.* It can be evaluated numerically from the knowledge that 1 mole of a gas at STP occupies 22.4 liters.

$$R = \frac{PV}{nT} = \frac{(1 \text{ atm})(22.4 \text{ liters})}{(1 \text{ mole})(273°\text{K})} = 0.082\frac{\text{liter-atm}}{\text{deg-mole}}$$

The value for R is 0.082, and the units are liter-atmosphere per degree per mole (liter-atm/deg-mole). The term R could be evaluated for any units of P and V desired, but atmospheres and liters are most commonly used.

The molecular mass of an unknown gas that behaves ideally can be calculated from the volume occupied by a known mass of gas at any specified pressure and temperature.

EXAMPLE 2.10

A mass of 1.24 g of a gas occupies 2 liters at 91 °C and 0.5 atm. What is the molecular mass of the gas?

The number of moles in this sample is

$$n = \frac{PV}{RT}$$

$$n = \frac{(0.5 \text{ atm})(2 \text{ liters})}{(0.082 \text{ liter-atm/mole-deg})(364°\text{K})} = 0.031 \text{ mole}$$

Since 0.031 mole of the gas has a mass of 1.24 g, the mass of one mole is

$$\frac{1.24 \text{ g}}{0.031 \text{ mole}} = 40 \text{ g/mole}$$

2.4 KINETIC THEORY OF GASES

MODEL OF GASES

The diffusion of gases and their spontaneous expansion from regions of high pressure to regions of low pressure suggest that gases consist of particles in a state of motion. Another phenomenon that also suggests that the units of matter are constantly moving is called *Brownian motion* (Fig. 2.13). Robert Brown, a Scotch botanist, observed in 1827 that small

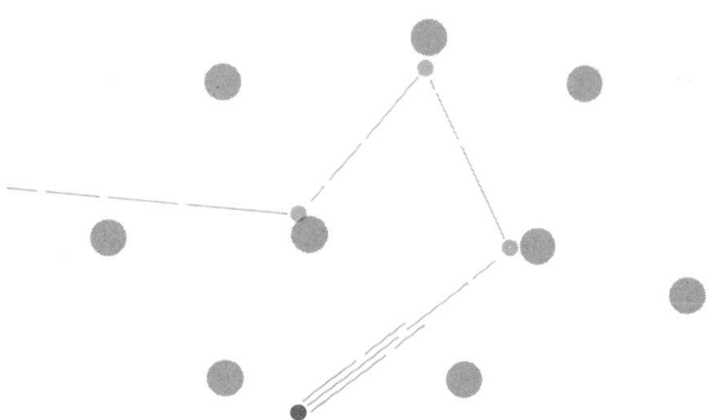

Figure 2.13
Brownian motion.

particles suspended in either a liquid or a gas tended to move constantly in a zigzag manner. This movement can be observed for dust particles on the surface of still water and smoke in a room in which there are no air currents. In either system the motion of the suspended particle does not cease but continues to move in an irregular path. Collision of moving submicroscopic particles such as atoms or molecules with the suspended macroscopic particles can account for Brown's observation.

The concept of moving atoms and molecules is known as the *kinetic theory* of gases, which is a model proposed to explain the observed facts of the behavior of the gaseous state. By extension, it also applies to the liquid and solid states (Chapter 3). In this discussion we shall first list the basic postulates and then summarize the justification for the assumptions. Finally, the experimentally established gas laws will be interpreted using the kinetic theory of matter.

The assumptions made in the kinetic theory of matter can be summarized as follows:

1. Gases are composed of atoms or molecules that are widely separated from one another. The space occupied by the atoms or molecules is extremely small compared with the space accessible to them.
2. The atoms or molecules are moving rapidly and randomly in straight lines. Their direction is maintained until they collide with a second atom or molecule or with the walls of the container.
3. There are no attractive forces between molecules or atoms of a gas.
4. Collisions of molecules or atoms are elastic. That is, there is no net energy loss upon collision, although transfer of energy between molecules or atoms may occur in the collision.
5. In a gas sample individual atoms or molecules move at different speeds and possess different energies of motion—kinetic energies. The kinetic energy of a particle is given by the expression K. E. $= \frac{1}{2}mv^2$, where m is the mass of the particle and v is its velocity. For a given temperature the average kinetic energy is constant. As the temperature increases, the average kinetic energy increases and, therefore, the average velocity also increases. The average kinetic energy is directly proportional to the absolute temperature.

The first assumption is in agreement with the observed ease of compression of gases. In addition, the density of gases indicates that the amount of matter per unit volume is extremely low. At STP the actual volume occupied by the atoms or molecules is approximately 0.04 percent of the total observed volume. This approximation can be supported by information about atomic and molecular dimensions, given in Chapter 5. While the observed volume of a gas is essentially empty space, it is occupied by particles that move through all regions with time.

The second assumption is supported by Brownian motion observations. The suspended particles reflect the motion of the atoms or molecules. In addition, it is our experience that moving objects of macroscopic dimensions travel in straight lines unless acted on by some force.

The third assumption is suggested by the spontaneous expansion of gases from regions of high pressure into all of the volume accessible to them at low pressure. Even in highly compressed gas samples, where atoms and molecules are close enough that intermolecular (between molecules) forces could be operative, the gas will spontaneously expand if the pressure is released. This assumption is related to the second assumption in that if attractive forces did exist, atoms and molecules would not travel in straight line motion but would be affected by neighboring particles.

The fourth assumption is in agreement with many observations of closed gaseous systems. If the collisions were nonelastic the particles of gas should eventually lose kinetic energy and velocity and settle to the bottom of the container. Such behavior would mean that a gas sample in an insulated container would gradually decrease in temperature. This never has been observed. Gases in insulated containers maintain their pressure and temperature and exhibit Brownian motion. Therefore, they do not lose kinetic energy.

The fifth assumption is largely intuitive. A range of velocities and kinetic energies must result from collisions between particles, because it seems inconceivable that all particles travel at the same velocity and continue at that speed after collision. Some particles must speed up and some must be slowed as kinetic energy is transferred between particles without net loss in energy. When heat is added to a gaseous system, thermal energy is converted into kinetic energy as the observed temperature of the gas increases. Brownian motion also supports the assumption of increasing kinetic energy with increasing temperature. Dust particles suspended in a gas move more rapidly at higher temperatures and reflect the higher average velocity of the atoms or molecules.

THE MAXWELL-BOLTZMANN DISTRIBUTION

In the kinetic theory of matter a distribution of velocities for gaseous samples was presented as being intuitively reasonable. An analogy between the motion of billiard balls on a billiard table and the motion of atoms or molecules can be made. In the case of billiard balls a distribution of velocities can be observed experimentally. However, the analogy is not an entirely valid one because the motion of billiard balls eventually ceases owing to the inelasticity of their collisions. Nevertheless, the average velocity or average kinetic energy of a collection of billiard balls could be calculated at a given instant by recording the individual velocities and kinetic energies of each billiard ball. In the case of submicroscopic matter, such a bookkeeping procedure is not possible. The problem of tabulating the velocities of individual atoms or molecules contained in a mole of a gas at a given instant would be too much even for computers. And if made, the tabulation would be valid for less than 10^{-9} sec because atomic and molecular collisions occur several billion times a second at room temperature.

Since there is a large number of atoms or molecules in a gas sample,

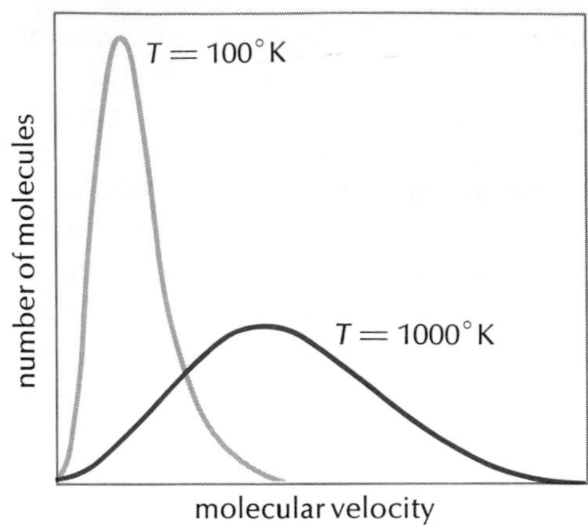

Figure 2.14
Maxwell–Boltzmann distribution.

it is possible to use statistical methods to describe the velocities and kinetic energies of the particles. Although there is a constant exchange of energies, the fraction of particles in a gas sample that have a given kinetic energy remains constant at a specified temperature. It is not necessary to specify the velocity or kinetic energy of any given particle at a given instant. The mathematical equation describing the speed distribution of atoms and molecules was derived by Clerk Maxwell and Ludwig Boltzmann in 1860. The actual equation and its derivation will not be given, but a graphical representation is shown in Fig. 2.14, where the relative number of molecules with specified speeds is plotted on the ordinate (vertical axis) and the corresponding speeds on the abscissa (horizontal axis). The curve shows that at any temperature a wide range of molecular velocities exists but that the largest fraction have some intermediate velocity. A much smaller number of particles have very high or very low kinetic energy. Experimental determinations of the statistical distribution predicted by Maxwell and Boltzmann have verified their equation.

The effect of temperature on the distribution of speeds also is shown in Fig. 2.14. At high temperatures the average speed of particles is higher than at a lower temperature. The maximum of the curve that represents the most probable speed is shifted to a higher velocity, and the distribution curve is broadened to show a much larger number of particles at high and low velocities. The Maxwell–Boltzmann distribution is discussed later in this book in the examination of other chemical phenomena such as the rates of reactions.

THE GAS LAWS AND THE KINETIC THEORY

Boyle's law expresses the relation between volume and pressure under conditions of constant mass and temperature. Pressure, which is defined as force per unit area, reflects the collisions of the gas with the walls of the container. The pressure, then, is controlled by the force exerted by a single collision and the number of collisions per unit area per unit time. Since the temperature remains constant, the average kinetic energy and impact per collision does not vary under the conditions of Boyle's law. In addition, the requirement of constant mass means that the number of particles is a constant and that the pressure is not affected by this potential variable. The explanation of Boyle's law in terms of the kinetic theory is that molecules or atoms in a reduced volume must collide with the walls more frequently. That Boyle's law applies at ordinary pressures is understandable, as very little of the available volume is actually occupied in a given instant. Therefore, compression simply involves a diminution in free space accessible to the atoms or molecules.

Gay-Lussac's law involving the relation between pressure and temperature is a reflection of the direct proportionality between kinetic energy and temperature. As the kinetic energy increases, the average impact per collision must increase as well. In addition, since the average velocity increases with temperature and the distance between container walls is unchanged, the number of collisions per unit time must increase also.

Charles' law can be interpreted in the same manner as Gay-Lussac's law, except that in the case of Charles' law temperature and volume are related at constant pressure. Since the external pressure in Charles' law is a constant, the increased velocity and increased impact per collision must push the walls of a container or a piston outward, thus increasing the volume occupied.

Dalton's law follows from the assumption of no attractive forces between atoms or molecules. Since the particles are independent of each other, all the particles of a given gas in a mixture will collide with the walls of the container with the same average frequency as they would if there were no particles of another gas present.

The fifth postulate of the kinetic theory can be used to derive Graham's law. At a given temperature the average kinetic energies of all gases are equal. However, since the masses of two different gases are not equal, it follows that their average velocities must be nonidentical. Kinetic energy is defined as $\frac{1}{2}mv^2$, where m is the mass of a particle and v is the velocity. Since a gas consists of a collection of particles traveling at a variety of speeds, v represents the average velocity. For two gases, 1 and 2, the kinetic energies are equal, and the following is an equality:

$$\text{K. E.}_1 = \text{K. E.}_2$$

$$\tfrac{1}{2}m_1v_1^2 = \tfrac{1}{2}m_2v_2^2$$

Recognizing that the rate at which a gas diffuses is directly proportional to its average velocity, we can convert this equality to a form like Graham's law. By rearranging the equation and solving for v_1/v_2 and setting this ratio equal to the ratio r_1/r_2, Graham's law is obtained:

$$\frac{v_1^2}{v_2^2} = \frac{m_2}{m_1}$$

$$\frac{v_1}{v_2} = \sqrt{\frac{m_2}{m_1}} = \frac{r_1}{r_2}$$

The ideal gas law relates *P*, *V*, *T*, and *n*. The variables *P*, *V*, and *T* already have been discussed in terms of the kinetic theory. The variable *n* is directly proportional to the number of impacts per unit time per unit area. It is possible to relate all four variables with mathematical exactness and derive the ideal gas law. However, the derivation involves some principles of physics. Therefore, the student is asked to accept the fact that the law can be derived and to refer to other texts for its derivation.

DEVIATIONS FROM IDEALITY

Most real gases deviate from ideal gas behavior at high pressures and low temperatures. The reasons for these deviations can be found in the assumptions of the kinetic theory. At high pressure the volume of free space is comparable to the actual volume of the gas particles (Fig. 2.15). The available free space is such that additional pressure will involve actual compression of the gas particles. This further compression is resisted by the impenetrability of atoms and molecules. While Boyle's law is obeyed in compressions from 1 to 2 atm, it is unlikely that it will be obeyed in compressions from 500 to 1000 atm. In the first case, a diminution in volume by a factor of 2 will result; the second case will lead to a change by a factor other than 2.

The deviations at low temperatures reflect the fact that contrary to the assumptions of the kinetic theory there actually are attractive forces between molecules and atoms. Indeed, if there were no attractive forces, the liquid and solid states could not exist and all matter would be in the gaseous state. The assumption of no attractive forces is approximately cor-

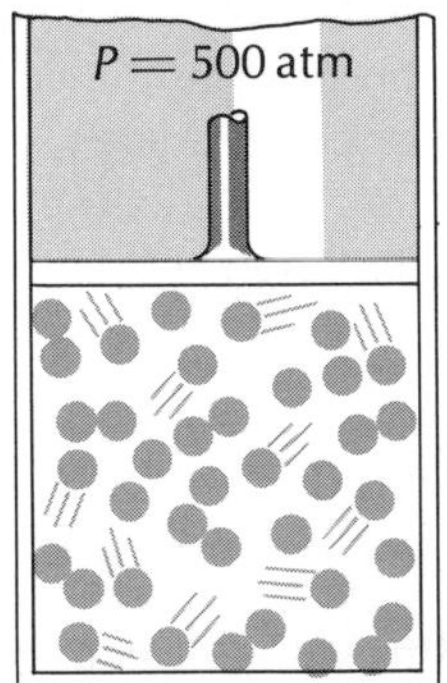

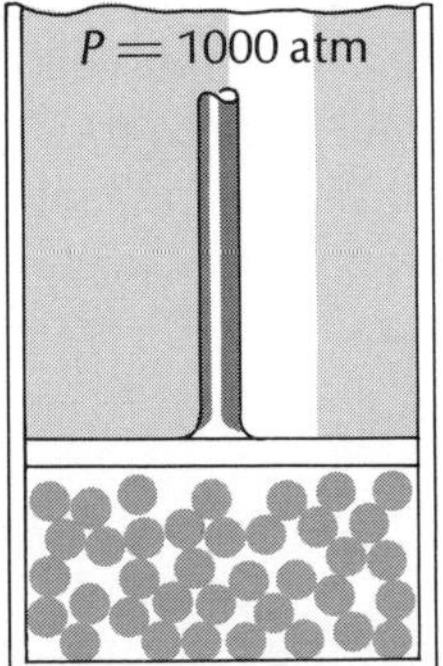

Figure 2.15
Deviation from Boyle's law at high pressure.

rect at high temperatures where they are small with respect to the average kinetic energy of the gas particles. As the temperature is lowered, the average kinetic energy decreases until it becomes of comparable magnitude to the energy of attraction between the gas particles. The forces of attraction become relatively more important as the velocity of the particles decreases.

In Chapter 7 these deviations are treated quantitatively. However, this cannot be accomplished until the actual volume of atoms and molecules and the nature of the attractive forces have been discussed in Chapters 5 and 6.

Suggested further readings

Avogadro, A., "Essay on the Manner of Determining the Relative Masses of Elementary Molecules and the Proportions in Which They Enter into These Compounds," in W. C. Dampier and M. Dampier (eds.), *Readings in the Literature of Science.* New York: Harper & Row, 1959.

Hall, M. B., "Robert Boyle," *Sci. Amer.,* 97 (August 1967).

Kieffer, W. F., *The Mole Concept in Chemistry.* New York: Reinhold, 1962.

Labbauf, A., "The Carbon-12 Scale of Atomic Masses," *J. Chem. Educ.,* **39,** 282 (1962).

Meldrum, W. B., "Gram-Equivalent Weights," *J. Chem. Educ.,* **32,** 48 (1955).

Neville, R. G., "The Discovery of Boyle's Law, 1661–62," *J. Chem. Educ.,* **39,** 356 (1962).

Reilly, D., "Robert Boyle and His Background," *J. Chem. Educ.,* **28,** 178 (1951).

Ruckstuhl, A., "Thomas Graham's Study of the Diffusion of Gases," *J. Chem. Educ.,* **28,** 594 (1951).

Terms and concepts

absolute zero
atmosphere
atom
atomic mass
Avogadro's hypothesis
Avogadro's number
barometer
Boyle's law
Brownian motion
Charles' law
chemical equation
combining volumes
compressible
Dalton's law
diffusion
force
Gay–Lussac's law
Graham's law
ideal gas law
Kelvin
kinetic theory
macroscopic
manometer
Maxwell–Boltzmann
mole
molecular mass
molecules
pressure
standard temperature and pressure
submicroscopic
torr
universal gas constant

1. Calculate the pressure exerted on a floor at the point of contact by a 120-lb woman balancing on the spike heel of one shoe. The area of the surface of the heel is 0.75 $in.^2$.
2. Explain why the level of a mercury barometer drops when the atmospheric pressure decreases.
3. Why is atmospheric pressure less in the mountains than at sea level? Explain why it is more difficult to work at high altitudes than at sea level.
4. If a liquid of density 2 g/ml were used in a barometer, how high a column of the liquid could be supported by 1 atm?
5. A 500-ml sample of a gas at 5 atm is allowed to expand until it reaches a final pressure of 2 atm at constant temperature. What is the final volume of the gas?
6. A 500-ml sample of a gas at −91°C is heated to 91°C at constant pressure. What is the final volume of the gas?
7. A sample of a gas in a rigid container is heated from 273°K to 273°C. If the gas was initially at 1 atm pressure, what is the final pressure?
8. A 1-liter sample of a gas at −23°C and 38 cm of Hg is heated to 127°C, and the pressure is increased to 1 atm. What is the volume of the gas under the new conditions?
9. How many moles of a gas are present under each of the following conditions?

- **a.** 1.12 liters at 2 atm and 0°C
- **b.** 560 ml at 5 atm and 182°C
- **c.** 22.4 liters at 273°C and 2 atm

10. How many molecules of sulfur dioxide (SO_2) are contained in a 2-mole sample of the gas?
11. A 1.12-liter sample of a compound has a mass of 2.2 g at standard temperature and pressure. What is the mass of a molecule of this compound on the atomic mass scale?
12. A 5.6-liter sample of a gas at 182°C and 38 cm of mercury has a mass of 4.8 g. What is the mass of a mole of this gas?
13. The rate of diffusion of ammonia (NH_3) is 1.414 times faster than that of a noxious gas at the same temperature. What is the molecular mass of the noxious gas? The atomic mass of nitrogen is 14 amu.
14. A 200-ml sample of oxygen at 2 atm and 0°C is transferred to a rigid 1-liter vessel containing hydrogen at 3 atm and 0°C. What will be the partial pressure of the oxygen in the mixture? How will the pressure of hydrogen be affected by the introduction of the oxygen? What will be the total pressure of the mixture?
15. For a volume of an ideal gas at a given pressure and temperature, describe how the average number of collisions per unit area per unit time will be affected by each of the following changes in the experimental variables:

- **a.** decreasing the volume at constant temperature

b. increasing the temperature at constant volume
c. adding a mass of a gas at the same temperature and at constant volume

16. Consider two rigid vessels of the same capacity at the same temperature, one containing a given mass of H_2 and the other an equal mass of O_2:
a. Which vessel contains more molecules?
b. In which vessel is the pressure greater?
c. In which vessel are the molecules moving faster on the average?
d. In which vessel do the molecules have the greater average kinetic energy?

17. Consider that samples of H_2 and O_2 are placed in separate containers of equal size. The pressures of the samples are equal but the temperatures of the H_2 and O_2 samples are 200 and 800°K, respectively:
a. Which sample contains more molecules?
b. Which sample contains molecules of the higher average kinetic energy?
c. In which sample are the molecules moving faster?

3

LIQUID AND SOLID STATES

The solid state of matter represents an extreme contrast to the gaseous state. The ease of compression and high rate of diffusion of gases are characteristic macroscopic properties of this state, while the extreme rigidity, resistance to compression, and fixed geometry of solids are their complete antitheses. The liquid state is an intermediate state. The semblance of regularity of shape and resistance to compression are macroscopic properties that liquids have in common with solids; the fluidity and rate of diffusion of liquids are more akin to the properties of gases.

In this chapter the liquid and solid states, the condensed phases, are discussed and compared with the gaseous state by presenting their general properties and examining them in terms of the kinetic theory developed for gases. It will be necessary to modify the kinetic theory to include the effect of attractive forces between submicroscopic particles in order to present a consistent picture for the macroscopic properties of liquids and solids. The states will be examined in general as gases were. Only a few specific examples are presented here. A detailed discussion of the macroscopic properties of liquids and solids in terms of submicroscopic structure is given in Chapter 7 after an examination of atomic structure (Chapter 5) and bonding (Chapter 6).

3.1
LIQUIDS AND SOLIDS

LIQUIDS

Unlike gases, liquids maintain a constant volume but do not have a characteristic shape. They will fill a container from the bottom, not using all of the volume accessible to them. They are practically incompressible, and their volume is little changed when subjected to even high pressure.

Diffusion of liquids is very much slower than the diffusion of gases. If a colored dye is placed in a liquid, the color slowly becomes dispersed throughout the liquid. By comparison, diffusion of one gas throughout a volume of another gas is much more rapid at the same temperature.

Matter in the liquid state slowly evaporates from open containers. While the rate of evaporation is a function of the type of matter in the liquid state, it is also a function of the surface area and temperature of the liquid and air currents in the vicinity of the liquid. Liquids evaporate fastest (1) from vessels with large surface areas, (2) at higher temperatures, and (3) in the presence of strong local air currents.

SOLIDS

Solids are characterized by definite volume and shape, which are independent of the size or shape of the container in which they are placed. The shape of a solid is fixed, and it does not flow to fit the shape of a container. Solids, for all practical purposes, are incompressible. For example, the volume of a given piece of coal beneath the enormous pressure of the earth is virtually the same as it is on the surface of the earth.

Diffusion of solids is extremely slow and requires eons to proceed a few inches. This fact enables geologists to determine the ages of rock deposits in the earth and to define various periods of the development of earth from the boundaries between rock layers.

Solids form crystal structures of definite geometric shapes, indicating that matter must be arranged in an orderly fashion in the solid state. Many crystals, when subjected to pressure or split mechanically, cleave into well-ordered smaller crystals.

3.2
KINETIC THEORY OF LIQUIDS AND SOLIDS

The kinetic theory of matter as applied to gases can be extended to the liquid and solid states with only minor changes. All gases can be liquefied if sufficiently high pressures and low temperatures can be achieved. Since matter in the gaseous phase is moving, there is little reason to expect it to stop in the liquid phase. Indeed, Brownian motion and the diffusibility of liquids both indicate that the particles of the liquid state do move. In the liquid state atomic and molecular motion must be more restricted than in the gaseous state as a result of attractive forces between neighboring atoms and molecules. The balance between these attractive forces and the kinetic energy of the particles produces a fluid with a semblance of cohesion but not a rigid structure.

The incompressibility of a liquid is a reflection of the nearness of neighboring particles. Compression of a liquid would require a squeezing of the molecules or atoms, and matter tends to resist such deformations.

Although diffusion in the liquid state is slow, it does take place at a measurable rate. Since particles in the liquid state are closer than in the gaseous state, it is reasonable that the progress of an individual particle in a given direction would be slowed. Repeated collisions with neighboring particles hinder the progress of a particle in a given direction and tend to randomize its motion. In the gas phase a particle can travel relatively long distances in a straight line before collision.

Evaporation of a liquid involves transfer of matter from the liquid phase to the gaseous phase. In a liquid the individual particles are traveling at different rates of speed, and those particles of high velocity may possess sufficient kinetic energy to break away from the attractive forces of their neighbors and become independent in the gas phase. This process is the reverse of the liquefaction of a gas.

Most liquids when cooled sufficiently solidify to form crystals. Cooling decreases the kinetic energy of the atoms or molecules, allowing the attractive forces between neighboring particles to exert a greater influence. In crystals the motion of the particles has been largely overcome so that

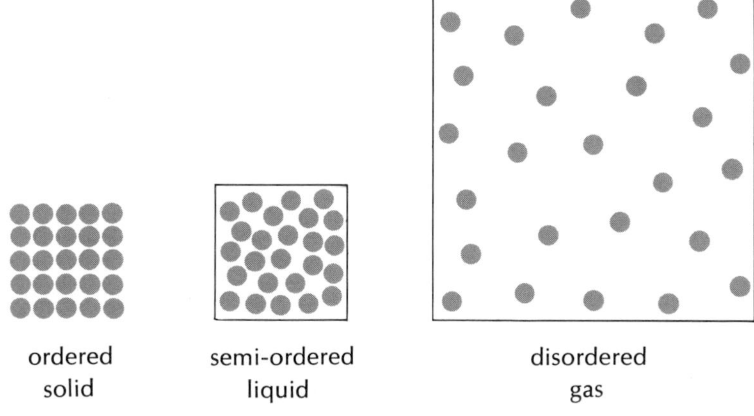

Figure 3.1
The degree of order of the states of matter.

they can vibrate only about fixed positions. Because of the diminished motion of matter in the solid state as compared to the liquid and gaseous states, diffusion occurs at a slow rate. In addition, the rigidity and decreased motion of the solid structure decrease the probability of matter entering the liquid or gaseous states. Only the most energetic particles on the surface of the solid may leave.

A representation of the order of the three states of matter is given in Figure 3.1.

3.3 PHYSICAL PROPERTIES OF LIQUIDS

HEAT CAPACITY

When energy is added to a liquid at a temperature below its boiling point its temperature increases. As in gases, the temperature of a liquid is closely related to the movement of matter. Increasing the temperature increases the average kinetic energy and the average velocity of particles in the gaseous state. In a similar manner, an increase in the temperature of the liquid state also indicates an increase in the average kinetic energy of matter.

The number of calories required to increase the temperature of 1 g of a substance 1°C is called the *heat capacity* of that material. Heat capacities have been shown experimentally to be a function of the type of liquid. In other words, the same number of calories supplied to two liquids of the same mass will not necessarily produce the same increase in temperature. Therefore, the average kinetic energies of the two liquids are not increased by the same amount. Unlike gases, liquids cannot be treated in terms of a simple model in which attractive forces between particles are neglected. Moreover, equal masses of two or more different types of matter do not contain the same number of atoms or molecules. Therefore, the heat capacity expressed in calories per degree per gram would be expected to be a function of the type of matter even if attractive forces could be neglected. A quantity that takes into account the different molecular

Table 3.1
Heat capacities of liquids at 20°C

substance	*heat capacity (cal/g-deg)*	*molar heat capacity (cal/mole-deg)*
alcohol (C_2H_6O)	0.581	26.7
bromine	0.107	17.1
chloroform ($CHCl_3$)	0.234	26.9
ether ($C_4H_{10}O$)	0.545	40.4
mercury	0.033	6.8
water	1.00	18.0

masses of liquids is the *molar heat capacity*—the number of calories required to increase the temperature of 1 mole of a substance 1°C.

The heat capacities and molar heat capacities of some common substances are given in Table 3.1. Heat capacities in terms of grams are convenient when dealing with substances whose molecular or atomic mass is unknown. In addition, heat capacities can be measured for homogeneous mixtures.

EVAPORATION

The energy transfer associated with evaporation has many familiar results. Any individual who has exercised and perspired or who has stood in a breeze immediately after stepping out of the swimming pool is well acquainted with the cooling effect of evaporation. In the liquid phase, as in the gaseous phase, there is a distribution of particles of different kinetic energies. The most energetic particles leave the liquid phase most readily, causing a decrease in the average kinetic energy of the remaining particles; this is felt as a lowering in the temperature of the liquid. The evaporation process continues at the same rate if a heat source is available to maintain the temperature of the liquid. A glass of water in a closed room without air currents will evaporate without any noticeable cooling. As the most energetic particles leave the liquid phase, heat is transferred from the surroundings to the liquid, maintaining the temperature. A redistribution of particles with various kinetic energies yields the same Maxwell–Boltzmann distribution as existed before the most energetic particles left.

Evaporation of matter from the liquid phase is less likely as the temperature decreases, since the average kinetic energy of the particles and hence their ability to escape into the gaseous phase decreases. This is verifiable by experience. Housewives know that it is more difficult to dry clothes on a cool day than on a hot day under identical wind conditions.

Even at the same temperature, different liquids evaporate at different rates. For example, gasoline evaporates faster than water at room temperature, and water evaporates faster than lubricating oil. The ease of evaporation in the liquid phase is a function of the liquid. Since the average kinetic energies of particles in two different liquids at the same temperature are identical, the escaping tendency must be controlled by the attractive forces between neighbors. If attractive forces between neighboring particles are large, the escaping ability of a given particle is retarded by its neighbors. In later chapters, when the structure of matter has been detailed, we shall see why the evaporation rate depends on intermolecular forces.

VAPOR PRESSURE

Evaporation from an open vessel indicates that the liquid particles have an escaping tendency. This tendency also should be manifested in a closed vessel. However, when a liquid is placed in a closed vessel there is only a small decrease in the volume of the liquid with time. This should not be too unexpected because the densities of matter in the liquid and gaseous phases are drastically different. Evaporation of a small quantity of liquid results in a large volume of gas. Whenever a liquid is placed in a closed vessel, particles leave the liquid phase and enter the gaseous phase. As the number of particles in the gaseous phase increases, it becomes more likely that they will collide with the liquid surface and return to it. Eventually a balance is achieved: The rates at which the particles leave and return to the liquid phase become equal (Figure 3.2). When such a balance occurs, the system is in *equilibrium*. In a system at equilibrium no net macroscopic change occurs. Various submicroscopic processes, such as evaporation and condensation, occur continuously, but in such a manner that they balance each other.

The particles in the gas phase exert a pressure like those of any gas. This pressure of a gas in equilibrium with its liquid phase is called the *vapor pressure* of the liquid. The vapor pressure is a measurable quantity, just as is the pressure of a gas in the absence of a second phase. It indicates the escaping tendency of the liquid, which in turn is characteristic of the individual liquid and its temperature.

Vapor pressure of liquids can be measured using the manometer and barometer described in Chapter 2. In Figure 3.3 the sample of liquid is sealed in a vial. The mercury levels are identical, indicating that the pressure inside and outside the bottle are equal. Plunging down the rod breaks the vial, and the liquid is liberated. The increase in pressure noted on the manometer results from the partial pressure of the vapor. Such an experiment is meaningful only if liquid remains in the bottle in equilibrium with the vapor. If this condition is not satisfied, the observed pressure will be

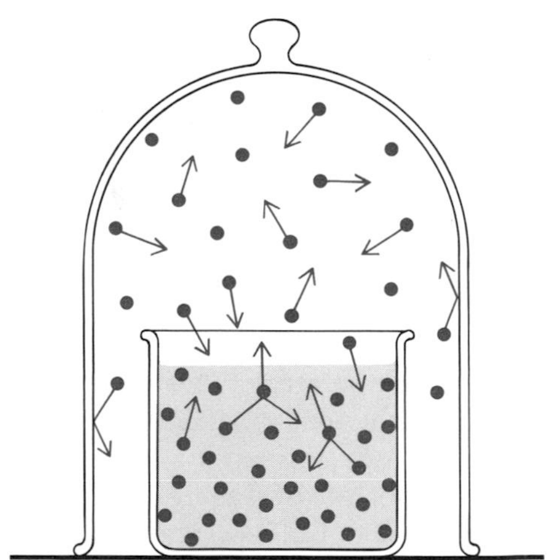

Figure 3.2
A model of the equilibrium vapor pressure of a liquid.

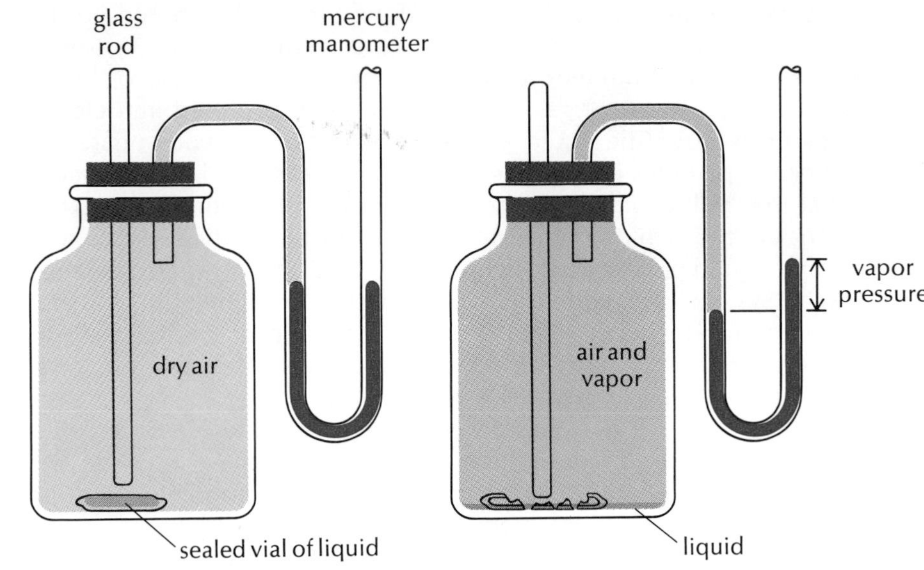

Figure 3.3
The determination of the vapor pressure of a liquid.

less than the vapor pressure of the liquid. Equilibrium between phases is necessary to determine the vapor pressure.

Another means of measuring vapor pressure involves the use of a manometer in which a vacuum exists in the space above the mercury column. If a dropper containing a liquid is placed beneath the column of mercury and the contents are ejected, the liquid will rise to the top of the column since most liquids are less dense than mercury. The results of such an experiment with water, chloroform, and ether are illustrated in Figure 3.4. A portion of each liquid evaporates, and the resultant pressure pushes the mercury column downward. The decrease in the length of the column is equal to the vapor pressure of the liquid. Again, some liquid must remain in the tube in order to measure the vapor pressure of the liquid.

The experiments illustrated in Figures 3.3 and 3.4 can be performed at various temperatures. As would be anticipated from our observations on the phenomenon of evaporation, the vapor pressure of liquids increases with temperature. Table 3.2 lists the vapor pressures of some common liquids as a function of temperature. The molecular masses of ether, chloroform, alcohol, and water are 74, 119.5, 46, and 18 amu, respectively. Clearly, molecular mass is not the only determining feature of vapor pressures. If it were the order of increasing vapor pressure on this basis would be chloroform, ether, alcohol, and water. Experimentally the vapor pressure increases in the order of water, alcohol, chloroform, and ether. Therefore, attractive forces between particles must be an important factor in determining the ability of a particle to escape from the liquid phase.

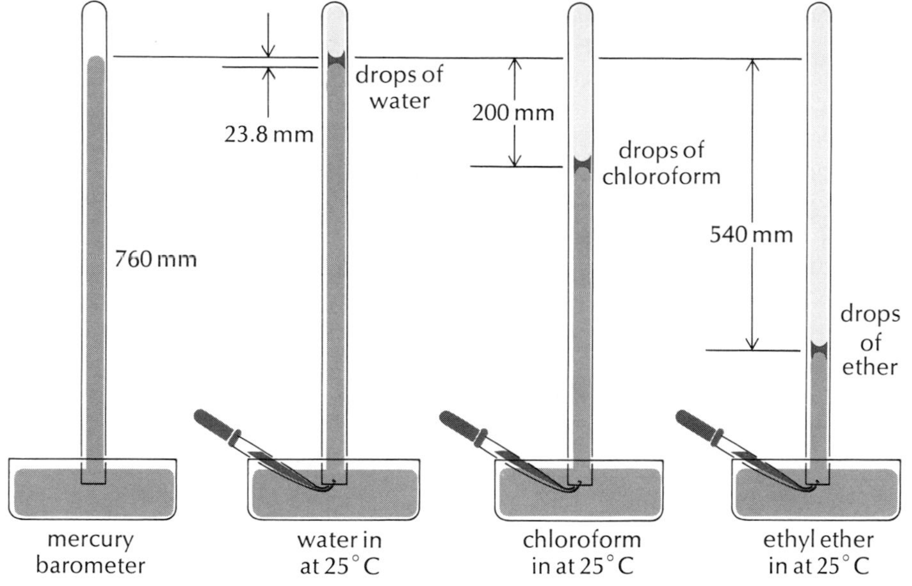

Figure 3.4
A comparison of the vapor pressures of different liquids. Adapted from Figure 7-9 of General College Chemistry, *3rd ed., by C. W. Keenan and J. H. Wood (New York, Harper & Row, 1966).*

BOILING POINT

Evaporation as a physical process is rather unspectacular, but upon heating to a specific temperature any liquid eventually will undergo a very pronounced transformation. Bubbles are formed throughout the liquid, and they rise rapidly to the surface, bursting and releasing vapor in large quantities. This process is called boiling, and the temperature at which it occurs is called the *boiling point* of the liquid. At the boiling point it can be shown experimentally that the vapor pressure of the liquid is equal to that

Table 3.2
Vapor pressures of liquids (in cm of Hg)

°C	ether (74 amu)	chloroform (119.5 amu)	alcohol (46 amu)	water (18 amu)
0	18.5	6.2	1.2	0.5
20	44.2	14.5	4.3	1.8
40	92.0	34.7	13.2	5.5
60	173.0	72.5	34.7	14.9
80	300.0	138.0	81.4	35.5
100	486.5	246.0	178.0	76.0
120	749.5	417.5	353.5	148.9

of atmospheric pressure. Owing to the large escaping tendency of the liquid at its boiling point, bubbles of vapor may be formed within the volume of the liquid. This is in contrast to the process of evaporation that is largely a surface phenomenon.

The vapor pressure of water at 80°C is 35.5 cm of mercury. If we lived on a planet with a normal atmospheric pressure of 35.5 cm of mercury, water would boil at 80°C. At this temperature the vapor pressure of the liquid would equal that of atmospheric pressure. The boiling point of water or any other liquid can be made to occur at any temperature if the external pressure is increased or decreased appropriately. To avoid ambiguity, the *standard* or *normal boiling point* is usually referred to and is the boiling point at one standard atmosphere.

The fact that liquids boil lower at reduced pressures can be confirmed by anyone who has attempted to cook food in boiling water at high altitude where the atmospheric pressure is low. In Figure 3.5 the effect of altitude on the boiling point of water is illustrated. Since water at the boiling point maintains a constant temperature, no amount of fuel can raise the temperature above the low boiling point. Food requires longer to cook at the lower temperature, and it is not uncommon to have to use cooking times twice as long as normal at altitudes of 7000 ft above sea level. Food can be cooked faster at the high pressures obtained in a pressure cooker. When the pressure gauge is set for 5 psi, the boiling point

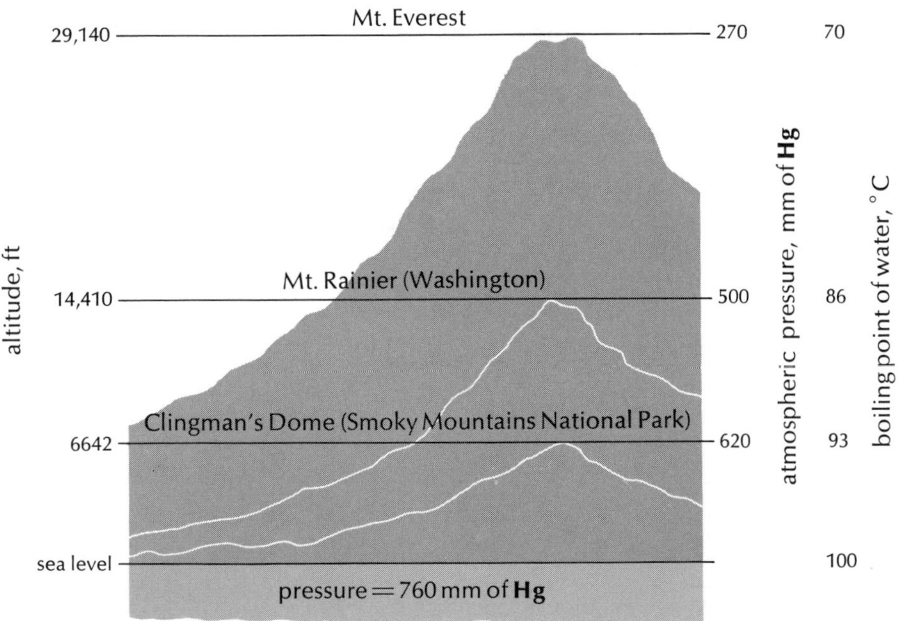

Figure 3.5
The effect of altitude on the boiling point of water.

of the liquid in the pressure cooker is about 108°C. Under these conditions food cooks twice as fast as at 100°C.

HEAT OF VAPORIZATION

The average kinetic energies of particles in gaseous and liquid phases at the boiling point are equal. However, energy is required in order to maintain boiling with the resultant transfer of matter from the liquid to the gaseous phase. The heat added does not increase the temperature of the liquid at the boiling point but provides the energy necessary for the most energetic particles to continue to escape. The quantity of heat energy required to transform 1 g of a substance at its boiling point from a liquid into a gas is called its *heat of vaporization.* The heat of vaporization of water is 540 cal/g, a value that is rather large compared to other liquids. Its size is a reflection of the strong attractive forces between neighboring water molecules in the liquid phase.

While the heats of vaporization of substances are usually listed in calories per gram, it should be pointed out again that mass per se is not of primary importance in understanding matter. The number of atoms involved in a phenomenon is more basic. The heats of vaporization along with the boiling points of several common substances are listed in Table 3.3 in terms of calories per mole of compound. If the heat of vaporization per mole of a substance is divided by its boiling point on the Kelvin scale, an average value of 21 cal/mole-deg is obtained for many liquids.

The empirical observation that for most liquids the quotient of the molar heat of vaporization and the boiling point is nearly always 21 cal/mole-deg is called Trouton's rule. Why this ratio is constant is an intriguing question. At the boiling point, the relatively unified liquid phase is transformed into a random gaseous phase with no change in the average kinetic energy of particles. However, energy has been required for the process and has been utilized in randomizing the system. At the boiling point vapors can be liquefied with the recovery of energy equal to the

Table 3.3
Heats of vaporization

substance	*molar heat of vaporization (cal/mole)*	*boiling point (°K)*	*Entropy change (ΔS) (cal/deg-mole)*
alcohol	9,220	351	26.2
carbon tetrachloride	7,170	350	20.5
chloroform	7,020	334	21.0
ether	6,500	308	21.1
hydrogen sulfide	4,480	212	21.2
mercury	14,100	630	22.4
water	9,720	373	26.0

heat of vaporization. In the liquefaction process matter becomes less random and more ordered. The constant value of 21 cal/mole-deg indicates the constant relation between heat energy and temperature for the randomization process.

The term *entropy* (S) is used to indicate the degree of randomness of a system. As the molecular chaos of matter increases, its entropy is said to increase. The entropy of matter in the liquid phase is less than that in the gaseous phase, and the value 21 cal/mole-deg is a measure of the change in entropy (ΔS) for the transformation. Thus, for the general process of converting matter from the liquid to the gaseous phase, the increase in the randomness of the system is a constant.

In the case of water and alcohol, the Trouton's constant is approximately 26 cal/mole-deg, a higher-than-average value. Since the value represents a change in entropy for the process of vaporization, it must be concluded that there is something about the molecular structure of these compounds that leads to higher than average ordering in the liquid state. If this is the case, then the change in the degree of molecular chaos would be larger for the vaporization process that leads to the very random gaseous state.

EXAMPLE 3.1

The heat of vaporization of carbon bisulfide is 6400 cal/mole. Estimate its normal boiling point.

If carbon bisulfide is a liquid with approximately average ordering, Trouton's rule can be used to estimate the normal boiling point. The problem can be solved by dividing the heat of vaporization by the value 21 cal/mole-deg.

$$\frac{\text{heat of vaporization}}{T_b} = 21 \text{ cal/mole-deg}$$

$$T_b = \frac{\text{heat of vaporization}}{21 \text{ cal/mole-deg}}$$

$$= \frac{6400 \text{ cal/mole}}{21 \text{ cal/mole-deg}} = 305°\text{K}$$

The calculated boiling point is approximately 32°C. The actual value is 46°C.

3.4 PHYSICAL PROPERTIES OF SOLIDS

HEAT CAPACITY

The *heat capacity* of a solid is the number of calories required to raise the temperature of 1 g of solid matter 1°C. Heat capacities vary with the type

Table 3.4
Heat capacity of solid metals

element	*heat capacity* (cal/g-deg)	*atomic mass*	*molar heat capacity* (cal/mole-deg)
lithium	0.92	6.9	6.3
magnesium	0.235	24.3	5.7
aluminum	0.216	27.0	5.8
potassium	0.178	39.1	7.0
calcium	0.157	40.1	6.3
iron	0.118	55.8	6.0
nickel	0.105	58.7	6.2
zinc	0.093	65.4	6.1
silver	0.0565	107.9	6.1
tin	0.053	118.7	6.3
lead	0.031	207.2	6.4

of solid material and are related to the magnitude of attractive forces between neighbors. In addition, the heat capacity is a function of atomic or molecular mass because the number of atoms or molecules per gram of matter is determined by the mass per atom or molecule. The energy required to increase the temperature of 1 mole of a substance 1°C is called the *molar heat capacity;* this value is also a function of attractive forces in the solid.

The French physicists Pierre Dulong and Alexis Petit proposed in 1818 that heat capacities could be useful in estimating the atomic mass of metallic elements. On the basis of the heat capacities of the metals known at that time, they suggested that the average molar heat capacity is 6.3 cal/mole-deg. The molar heat capacity is the product of the heat capacity and the atomic mass of the metal:

$$\text{molar heat capacity (cal/mole-deg)} = \text{heat capacity (cal/g-deg)} \times \text{atomic mass (g/mole)}$$

Examples of the molar heat capacities of metals are given in Table 3.4. Because most solid metals do fit the approximation of Dulong and Petit within 10 percent, it provides a method of obtaining a rough value for atomic mass. Although the exact atomic mass is not obtainable from the approximate relationship, some uncertainties in atomic masses were resolved in the early years of chemistry by the use of this method. The fact that the approximation of Dulong and Petit works as well as it does indicates some basic similarity in the submicroscopic structure of metals.

EXAMPLE 3.2

Estimate the atomic mass of silver from the heat capacity of 0.0565 cal/g-deg.

The molar heat capacity divided by the heat capacity in grams is equal to the atomic mass. Using the approximation of Dulong and Petit, the atomic mass of silver is 118 g/mole. The actual value is 107.87 g/mole.

$$\frac{6.1 \text{ cal/mole-deg}}{0.0565 \text{ cal/g-deg}} = 118 \text{ g/mole}$$

VAPOR PRESSURE

Solids, like liquids, have a vapor pressure. However, the idea that a solid can evaporate is not as common among the nonscientific community as the concept of evaporation of liquids. The evaporation of ice is the reason that even frozen clothes will dry outdoors in subzero weather. The vaporization of solid carbon dioxide (Dry Ice) and mothballs are other common examples.

The kinetic theory applies to solids in much the same way as to liquids. Particles in the solid state have a range of energies and are distributed among the energies according to the Boltzmann curve. Those of the more energetic particles that lie at the surface of the solid can escape into the gas phase. This is in contrast to the liquid state, in which the mobility of the particles may result in the transfer of a particle from deep within the liquid to the surface where it may escape.

The vapor pressure of a solid can be measured by the same apparatus described for the determination of the vapor pressure of a liquid. A list of vapor pressures of solids is given in Table 3.5. As is the case with liquids, the vapor pressure is a measure of an equilibrium process. If a solid such as ice is placed in a closed container, the most energetic surface particles escape. However, eventually the gaseous particles return to

Table 3.5

Heats of fusion

substance	*formula*	*melting point (°K)*	*cal/g*	*cal/mole*
water	H_2O	273	80	1440
carbon tetrachloride	CCl_4	249	4.2	641
alcohol	C_2H_6O	159	24.9	1150
acetone	C_3H_6O	178	23.4	1360
silver	**Ag**	1234	24.9	2690
lead	**Pb**	600	5.5	1140
zinc	**Zn**	692	26.5	1740
gold	**Au**	1336	15.4	3050
hydrogen	H_2	14	7.0	14
fluorine	F_2	53	4.9	186
neon	**Ne**	24	4.0	80

the solid. The pressure of the vapor gradually increases to a maximum value; then, the rate of escape is equal to the rate of return to the solid.

The larger the attractive forces between molecules or atoms, the smaller will be the vapor pressure of the solid. With increasing temperature the vapor pressure of the solid increases, because the average kinetic energy of the particles increases. However, the vapor pressure of most solids rarely gets very large since the solid usually melts at high temperatures. For most solids the tendency to enter the liquid phase is larger than that for entering the gaseous phase. Dry Ice, solid carbon dioxide, is an exception in that upon heating it is converted directly to the gaseous state; the liquid state is not observed at atmospheric pressure. Direct vaporization of a solid without melting is called *sublimation*.

MELTING POINT

When heat energy is added to a solid, the temperature increases until the solid starts to melt. That point at which the added heat energy is used only to melt the solid without raising the temperature of the solid or liquid is called the *melting point*. At the melting point the solid and liquid states exist in equilibrium. Particles from the solid, which consists of ordered arrays, escape and enter the more random liquid state at the same rate particles from the liquid are deposited on the surface of the solid.

The amount of heat energy required to transform 1 g of a solid into a liquid at the melting point is called the *heat of fusion*. The heat of fusion of water is 80 cal/g. Like the heat of vaporization for water, this value is higher than that of many solids (Table 3.5)—yet another indication that water has strong attractive forces between neighboring molecules. The melting point of solids can be considered an approximate indication of the intermolecular attractive forces. For substances of similar molecular mass, those with the higher melting points have the stronger intermolecular forces. However, there are many more variables that affect the melting point of a solid. Foremost among these is the packing arrangement of the particles or the geometrical arrangement of one particle with respect to its neighbors (Chapter 7).

Although the effect of pressure on the melting point is not as dramatic as its effect on the boiling point, the pressure at which a melting point is determined should be stated. For most solids the melting point of a substance increases with pressure. Water is atypical; its melting point decreases with increasing pressure. The melting point decrease is approximately 0.01°C/atm. This fact is part of the basis for the ease with which skaters can travel on the surface of the ice. Under the pressure exerted by the narrow edge of a hollow-ground skate blade, the ice melts to provide water as a lubricant.

3.5
LE CHÂTELIER'S PRINCIPLE

Le Châtelier in 1884 proposed a principle that predicts the behavior of an equilibrium when it is subjected to an external force: If an external force is applied to a system at equilibrium, the system will readjust to reduce the stress imposed upon it if possible. This principle, which is intuitively reasonable, explains the effect of pressure on the boiling point of a liquid and the melting point of a solid. In the first case, evaporation of a liquid always involves a tremendous increase in volume. Application of pressure will cause the equilibrium system of a liquid and its vapor at its normal boiling point to shift toward the liquid state, because the resultant decrease in volume tends to diminish the pressure on the equilibrium system. Therefore, in order to boil a liquid at high pressures, higher temperatures are necessary. In the case of the melting point of a solid, most substances decrease in volume in going from the liquid to the solid state, and application of pressure will cause an equilibrium system of liquid and solid to shift toward the solid state. This shift decreases the volume of the system and counteracts the applied pressure. Therefore, in order to melt most solids under pressure, higher temperatures are necessary. That only small increases in melting points are usually observed is a reflection of the small change in volume between solids and liquids. The abnormal behavior of water is the result of the increase in volume of this compound in going from water to ice. Thus, an increase in pressure shifts ice-water systems toward water.

3.6
HEATING CURVES

The relationship between the heat energy added to a substance and the temperature change is an interesting one. Heating a solid, liquid, or gas produces an increase in the temperature of that particular state. The temperature increase is a reflection of higher average energy of the particles making up the substance. However, when a change of state occurs, such as when a solid melts or a liquid vaporizes, the heat energy is employed in disordering matter. Under these conditions no temperature increases are observed.

Changes of state and temperature variations can be graphically illustrated by means of a heating curve. Consider the changes that occur when a 1-g sample of a solid is subjected to a constant input of energy per unit time. In the graph in Figure 3.6 the horizontal axis represents an arbitrary temperature scale. The vertical axis indicates the number of calories added. At the beginning of the experiment the temperature of the solid is T_0, which may be any temperature below the melting point. As heat is added the temperature of the solid increases, reflecting increased motion of particles in the solid state. The ratio of the number of added calories to the temperature increase for the 1-g sample is given by the slope of the line, which by definition is the heat capacity of the solid.

Eventually a temperature is reached at which the addition of heat energy no longer leads to a temperature increase. This temperature is the melting point, and its attainment is reflected by the first vertical line shown in Figure 3.6. The line continues as long as the solid is present in equilibrium with the liquid. When the last of the solid disappears, an increase in temperature occurs. Therefore, the length of the vertical line represents the number of calories necessary to melt the solid and is equal to the heat of fusion. The heat of fusion indicates the energy required to increase the randomness of the substance in going from the solid to the liquid state.

After complete conversion of the solid into liquid, the slope of the heating curve represents the number of calories necessary to cause a change in temperature of the liquid state. By definition, this ratio is the heat capacity of the liquid state, which reflects the increase in the average kinetic energy of the particles in the liquid state.

The next temperature at which a vertical line is obtained is the boiling point of the liquid. At this temperature additional heat energy again is used to increase the randomness of the substance. The number of calories

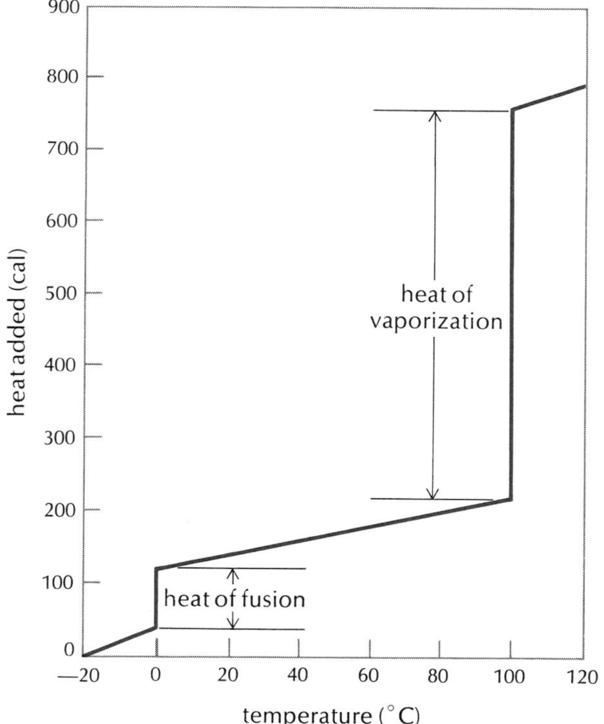

Figure 3.6
Heating curve for 1 g of ice at −20°C.

represented by the length of the second vertical portion of the heating curve is the heat of vaporization. Finally, the liquid is completely converted to gas, and a temperature increase again results. The heat capacity of the gas is equal to the slope of the final portion of the graph and represents the amount of heat energy required to raise the average kinetic energy in the gaseous phase.

The heating curve drawn in Figure 3.6 is precisely the one found experimentally for a 1-g sample of ice. The two vertical portions of the curve occur at 0°C, the melting point of ice, and at 100°C, the boiling point of water. The lengths of these two straight lines correspond to the heat of fusion and the heat of vaporization for the sample.

$$\text{heat of fusion} = 80 \text{ cal/g}$$
$$\text{heat of vaporization} = 540 \text{ cal/g}$$

The slopes of the three rising regions of the curve give the heat capacity of water as a solid, liquid, and gas.

Suggested further readings

Azaroff, L. V., *Introduction to Solids.* New York: McGraw-Hill, 1960.
Bernal, J. D., "The Structure of Liquids," *Sci. Amer.,* 124 (August 1960).
Cullity, B. D., "Diffusion in Metals," *Sci. Amer.,* 103 (May 1957).
Frenkel, L., *Kinetic Theory of Liquids.* New York: Dover, 1955.
Holden, A., *The Nature of Solids.* New York: Columbia University Press, 1965.
Wannier, G. H., "The Nature of Solids," *Am. Sci.,* 39 (December 1952).

Terms and concepts

attractive forces
boiling point
Dulong and Petit
entropy
equilibrium
evaporation
heat capacity
heat of vaporization
heating curve
incompressible
Le Châtelier's principle
liquefaction
melting point
molar heat capacity
normal boiling point
sublimation
Trouton's rule
vapor pressure

Questions and problems

1. The space above the column of mercury in a barometer is considered to be a vacuum. Is this idea strictly correct?

2. What factors control the drying of clothes at temperatures below the freezing point of water?

3. How many calories must be removed from 500 g of a substance to cool it from 0 to −10°C if the heat capacity of the substance is 0.5 cal/g-deg?
4. What can be stated about two liquids that have identical boiling points but differ in molecular mass?
5. The rate of evaporation of a beaker of water inside of a larger sealed container at constant temperature decreases with time. Explain this phenomenon and contrast it with the rate of evaporation of the same beaker in a room at constant temperature.
6. A beaker of water in a closed room at constant temperature does not decrease in volume on a given day. Why does the water not evaporate?
7. The vapor pressure of liquid *A* at 25°C is equal to that of liquid *B* at 50°C. If the molecular masses of *A* and *B* are identical, which of the two liquids has the stronger intermolecular attractive forces?
8. What is the highest temperature at which water vapor will condense to yield liquid water under 1 atm pressure? At what temperature will water boil under pressure of 148.9 cm of mercury?
9. Is the vapor pressure of water at 25°C at an altitude of 10,000 ft in the mountains less than, equal to, or more than the vapor pressure of water at sea level at 25°C?
10. Before the time of temperature control for rooms, a tube of water was often placed in fruit cellars to prevent fruit from freezing in the winters and to slow the spoiling process in the summer. Why did this work?
11. Diagram apparatus suitable for measuring the vapor pressure of ice at 0°C.
12. How many calories are necessary to heat 1 g of ice at −20°C to steam at 100°C? The heat capacities of ice and water are 0.5 and 1.0 cal/g-deg, respectively.
13. State how the heats of vaporization of two liquids can be compared from their respective heating curves.
14. Steam at 100°C causes more severe burns than water at 100°C. Why does this happen?
15. How will the slope of a heating curve for the solid state vary if the mass of the substance under investigation is increased from 1 to 2 g?
16. The heat capacities of solid gold and aluminum are 0.03 and 0.21 cal/g-deg, respectively. Will it take more, the same, or less energy to heat 1 kg of gold from 10 to 80°C as compared to heating 500 g of aluminum from 20 to 25°C?
17. The heat capacity of uranium is 0.028 cal/g-deg. Estimate its atomic mass. The approximate atomic mass was thought to be 120 in the mid-nineteenth century. What relationship does this number bear to your calculated value?
18. The boiling point of octane is 126°C. Estimate its heat of vaporization.
19. Ice floats on water. However, most solids sink to the bottom of the liquid phase of the same substance. Why does this difference occur? What would happen to the life cycles of aquatic species in a small pond during a prolonged cold spell if ice was like the majority of solid substances?

20. The heat of vaporization of toluene is 87.0 cal/g at its normal boiling point, 110.6°C. Estimate the molecular mass of toluene.

21. The molar volumes of gases at the same temperature and pressure are approximately equal, while the molar volumes of liquids are significantly different from each other under the same conditions of pressure and temperature. Why?

22. The boiling points of methane (CH_4), ammonia (NH_3), and water (H_2O) are −161.5°C, −33.5°C, and +100°C, respectively. The molecular masses are approximately the same. What statement can be made about the intermolecular forces that exist between the molecules. (See Chapter 7 for a discussion of these molecules and their properties.)

23. Calculate the entropy change for the conversion of 1 mole of ice at 0°C to water at 0°C. Compare this value with the entropy change for the vaporization of water.

4

ATOMIC THEORY

In the preceding chapters the principal emphasis was on the physical aspects of matter and on providing a consistent rationale for these macroscopic properties in terms of the atomic theory of matter. However, the atomic theory was originally conceived in response to the evidence of both physical and chemical phenomena. Indeed, the atomic theory is vital to understanding the relationships between elements and compounds and their chemical reactions. As illustrated throughout this book, the atomic theory has allowed chemists and other scientists to organize, correlate, and understand vast quantities of information concerning the chemical behavior of matter.

In this chapter the contributions of Dalton and Cannizzaro in the interpretation of chemical phenomena are interwoven with the contributions of Avogadro and Gay-Lussac that were described in Chapter 2. The discussion of the contributions of several individuals to the development of the atomic theory illustrates that the evolution of these ideas was not a simple process that occurred in a short period. Rather, over a period of 50 years chemists grappled with a problem that at times seemed to be essentially resolved but, with the addition of more experimental data, again appeared to be almost insoluble.

4.1 JOHN DALTON

BASIC ATOMIC THEORY

The French chemist Joseph Proust showed in 1799 that elements combine to form compounds in a manner such that the ratio of their masses is always a constant. This generalization is now called the *law of definite proportions.* For example, the masses of oxygen and hydrogen that combine completely to form water are always in the ratio of 1:8.

EXAMPLE 4.1

A 3.024-g sample of hydrogen is oxidized to yield 27.024 g of water. What is the mass percent composition of hydrogen in water? How much oxygen is needed to react with 1.008 g of hydrogen?

The mass percent composition of hydrogen in water is equal to the fraction of the compound that is hydrogen times 100:

$$\frac{3.024 \text{ g of hydrogen}}{27.024 \text{ g of water}} \times 100 = 11.19 \text{ percent} \frac{\text{hydrogen}}{\text{water}}$$

The ratio of the mass of hydrogen and the mass of oxygen that react to form water must always be equal to a constant. In the described experiment the ratio is

$$\frac{3.024 \text{ g of hydrogen}}{24.000 \text{ g of oxygen}}$$

Therefore, the amount of oxygen needed to react with 1.008 g of hydrogen is

$$1.008 \text{ g of hydrogen}\left(\frac{24.000 \text{ g of oxygen}}{3.024 \text{ g of hydrogen}}\right) = 8.000 \text{ g of oxygen}$$

In 1805 John Dalton suggested a set of astonishingly simple generalizations that account for the law. The self-consistency of his assumptions and their agreement with the experimental data available at that time led to a general acceptance of his theory of the atomic structure of matter.

Dalton proposed that (1) elements consist of indivisible atoms, (2) atoms of any specific element are the same (however, see Chapter 5), (3) the atoms of one element differ from those of another element in their mass and in their ability to react with other substances, and (4) in chemical reactions the atoms combine to form molecules. The law of definite proportions is consistent with the atomic theory of Dalton because if atoms exist and have characteristic masses and if compounds are formed by a specific interaction between atoms, then the mass percent composition of a substance must be a constant. However, consistency does not constitute proof of a theory or designate the accuracy of it. The only valid statement that can be made about Dalton's theory is that the law of definite proportions is a reasonable consequence of the model. Since no theory more reasonable than Dalton's has been proposed, scientists accept the concept of atoms and molecules as a convenient model by which to depict matter.

ATOMIC MASS AND MOLECULAR FORMULAS

Although the postulation of atoms and molecules was conceptually easy to deal with in a qualitative sense, it was quite another matter to evolve the quantitative aspects of the early atomic theory. First, if atoms exist and have different masses, then what are the relative atomic masses of the elemental particles? Second, when atoms combine to form molecules, they do so in masses corresponding to a definite ratio, but these ratios do not yield directly the number of atoms of each element making up a molecule of the compound. These two problems are interrelated. If the atomic masses of the atoms and the masses of a molecule were known, then the number of atoms involved could be established. If the molecular formula, which gives the number and type of atoms per molecule, were known, then this information could yield the atomic masses of the constituent elements.

When Dalton proposed the atomic theory he was faced with two unknown quantities, atomic masses and molecular formulas. One could not be determined without information about the other. Dalton made the most reasonable assumption possible at that time and attempted to rationalize the known data in terms of that assumption. He assumed that when two or more elements combine to form a compound they do so in the ratio of one atom to one atom. Therefore, he assumed that the formula for water was HO. Since water consists of 8 g of oxygen for every 1 g of hydrogen, it follows that the relative atomic masses of oxygen and hydrogen are 8:1. Unfortunately, Dalton's assumption was incorrect, and it follows that the atomic mass scale derived on this basis was not correct. As we saw in Chapter 2, information from the gas laws and Avogadro's hypothesis indicates that two atoms of hydrogen combine with one atom of oxygen to produce a molecule of water with the molecular formula H_2O.

EXAMPLE 4.2

The percent composition of ammonia is 82.3 percent nitrogen and 17.7 percent hydrogen. Assuming that the formula of ammonia is NH, calculate the atomic mass of nitrogen on the basis that the atomic mass of hydrogen is 1 atomic mass unit (amu).

If one atom of nitrogen combines with one atom of hydrogen, the ratio of the masses of the atoms must equal the ratio of the masses of the elements contained in the compound. Using an arbitrary sample of 100 g of ammonia, the ratio of the atomic masses can be calculated:

$$\frac{\text{mass of } \mathbf{N} \text{ in ammonia}}{\text{mass of } \mathbf{H} \text{ in ammonia}} = \frac{\text{mass of } \mathbf{N} \text{ atom}}{\text{mass of } \mathbf{H} \text{ atom}}$$

$$\frac{82.3 \text{ g of } \mathbf{N}}{17.7 \text{ g of } \mathbf{H}} = 4.65 \frac{\text{g of } \mathbf{N}}{\text{g of } \mathbf{H}}$$

$$\text{mass of } \mathbf{N} \text{ atom} = 4.65 \text{ (mass of } \mathbf{H} \text{ atom)}$$

$$\text{mass of } \mathbf{N} \text{ atom} = 4.65 \text{ (1 amu)}$$

$$\text{mass of } \mathbf{N} \text{ atom} = 4.65 \text{ amu}$$

EXAMPLE 4.3

The formula for ammonia is actually NH_3. Using the atomic mass of H equal to 1 amu, calculate the atomic mass of nitrogen.

If three atoms of hydrogen combine with one atom of nitrogen, then the ratio of the masses of the elements contained in the compound must equal the ratio of the mass of one nitrogen atom and three hydrogen atoms:

$$\frac{82.3 \text{ g of } \mathbf{N}}{17.7 \text{ g of } \mathbf{H}} = 4.65 \frac{\text{g of } \mathbf{N}}{\text{g of } \mathbf{H}}$$

$$\frac{\text{mass of one } \mathbf{N} \text{ atom}}{\text{mass of three } \mathbf{H} \text{ atoms}} = 4.65$$

$$\text{mass of one } \mathbf{N} \text{ atom} = 3(4.65)(1 \text{ amu})$$

$$\text{mass of one } \mathbf{N} \text{ atom} = 14.0 \text{ amu}$$

4.2 JOSEPH LOUIS GAY-LUSSAC

Three years after the proposal of the atomic theory, Joseph Louis Gay-Lussac developed a method that eventually became the basis for deducing molecular formulas and atomic masses. The law of combining volumes (Section 2.2) summarized his observation that the combining volumes of gaseous reactants under the same conditions of temperature and pressure are in the ratio of small whole numbers. The compounds nitrous oxide, nitric oxide, and nitrogen dioxide (N_2O, NO, and NO_2) provide a good example of the general observations made by Gay-Lussac. In the formation of these compounds, two, one, and one-half volumes of nitrogen, respectively, react completely with each unit volume of oxygen:

$2N_2$	$+ O_2$	$\longrightarrow 2N_2O$
two volumes of nitrogen	one volume of oxygen	two volumes of nitrous oxide
N_2	$+ O_2$	$\longrightarrow 2NO$
one volume of nitrogen	one volume of oxygen	two volumes of nitric oxide
$\frac{1}{2}N_2$	$+ O_2$	$\longrightarrow NO_2$
one-half volume of nitrogen	one volume of oxygen	one volume of nitrogen dioxide

At the present time the law of combining volumes can be readily interpreted in terms of the concept of atoms and molecules. However, Dalton raised two objections to Gay-Lussac's observations because of the implication that the number of atoms contained in equal volumes of gases were either equal to or integral multiples of each other. The first objection was based on the fact that the density of the water vapor produced from the reaction of oxygen with hydrogen is less than the density of oxygen. Dalton considered the chemical reaction of hydrogen and oxygen to involve the addition of one atom of hydrogen to one atom of oxygen, according to the equation $H + O = HO$.

Because the individual HO molecules must be heavier than oxygen O atoms alone, fewer molecules of water than of oxygen must be contained in the same volume in order to be consistent with the density data. At the present time, with the molecular formula of water known to be H_2O, the density data can be readily explained, once it is also realized that hydrogen and oxygen actually exist as the diatomic molecules H_2 and O_2: $2H_2 + O_2 = 2H_2O$. The mass of an oxygen molecule is greater than the mass of a water molecule. Therefore, if equal volumes of these two substances contain the same number of molecules, the density of water vapor should be less than the density of oxygen.

Dalton also pointed out that Gay-Lussac's observations concerning the combination of one volume of oxygen and one volume of nitrogen to produce two volumes of nitric oxide were inconsistent with the concept of indivisible atoms and with the hypothesis of equal volumes containing equal numbers of elemental particles. If the equal volume–equal numbers of particles assumption is correct, then it must be concluded that the x atoms of nitrogen and x atoms of oxygen contained in equal volumes must produce $2x$ particles of nitric oxide to account for the two volumes of product produced. This is clearly impossible, because the combination of x atoms of an element with another element to produce more than x molecules of a compound would require that individual atoms be split. If the atoms are not split, a sample of nitric oxide would have to contain one-half the number of molecules in a unit volume as compared to nitrogen or oxygen. The whole problem, with its apparent inconsistencies, could have been resolved if Dalton had examined his assumption that the particles of gaseous elements must be indivisible atoms. Instead Dalton chose to reject Gay-Lussac's equal volume–equal number of particles hypothesis.

4.3 AMADEO AVOGADRO

In 1811 Amadeo Avogadro suggested that some elements might consist of polyatomic molecules rather than atoms. His suggestion provided the missing key to understanding the available data on the combining volumes of gases. Although his rationale added yet another assumption to the growing list, it did lead to an internally consistant picture that is still considered valid.

The reaction of equal volumes of nitrogen and oxygen to produce two volumes of nitric oxide can be explained by assuming the equal volume–equal number of particles hypothesis to be correct. Thus x molecules of the diatomic element nitrogen N_2 and x molecules of the diatomic element oxygen O_2 produce $2x$ molecules of nitric oxide NO. Cleavage of the two diatomic molecules to produce two molecules of nitric oxide upon recombination is consistent with the experimental facts:

$$\begin{array}{rcl}\text{nitrogen} + \text{oxygen} & \longrightarrow & \text{nitric oxide} \\ \mathbf{N_2} + \mathbf{O_2} & \longrightarrow & \mathbf{2NO} \\ x \text{ molecules of } \mathbf{N_2} + x \text{ molecules of } \mathbf{O_2} & \longrightarrow & 2x \text{ molecules of } \mathbf{NO} \\ \text{one volume of } \mathbf{N_2} + \text{one volume of } \mathbf{O_2} & \longrightarrow & \text{two volumes of } \mathbf{NO}\end{array}$$

However, the experimental data is also consistent when tetratomic molecules such as N_4 and O_4 are considered. In order for the volume changes to be observed, nitric oxide must be N_2O_2:

$$\begin{array}{rcl}\mathbf{N_4} + \mathbf{O_4} & \longrightarrow & \mathbf{2N_2O_2} \\ x \text{ molecules of } \mathbf{N_4} + x \text{ molecules of } \mathbf{O_4} & \longrightarrow & 2x \text{ molecules of } \mathbf{N_2O_2} \\ \text{one volume of } \mathbf{N_4} + \text{one volume of } \mathbf{O_4} & \longrightarrow & \text{two volumes of } \mathbf{N_2O_2}\end{array}$$

Indeed, any molecule of nitrogen or oxygen consisting of an even number of atoms is consistent with the data.

In order to distinguish between the various possible molecular formulas that can be written and to bring some order to the data that were accumulating in the early nineteenth century, it was necessary to find a way to determine atomic masses and molecular formulas. However, since these two quantities are interrelated, it was felt by many chemists that the puzzle was insoluble.

One volume of O_2 combines with two volumes of chlorine gas to give two volumes of a gaseous compound. Does this data indicate whether chlorine is monatomic or diatomic?

The oxygen molecule must cleave into two atoms in order to form two volumes of the compound. Since the chlorine atoms must combine with oxygen atoms on at least a 1:1 basis, the compound could be OCl. If OCl is formed, then the volume data is consistent only if chlorine is monatomic.

O_2	$+ 2Cl$	$\longrightarrow 2OCl$
one molecule of oxygen	two atoms of chlorine	two molecules of **OCl**
one volume of O_2	two volumes of **Cl**	two volumes of **OCl**

However, the compound formed could be OCl_2 (which it is). If it is, then chlorine must be diatomic in order to be consistent with the volume data.

O_2	$+ 2Cl_2$	$\longrightarrow 2OCl_2$
one molecule of oxygen	two molecules of chlorine	two molecules of OCl_2
one volume of oxygen	two volumes of chlorine	two volumes of OCl_2

Without the molecular formula, the question of the nature of chlorine gas cannot be answered from the data given. It also should be noted that the formula of the compound cannot be determined unless it is known whether chlorine is monatomic or diatomic.

4.4 STANISLAO CANNIZZARO

Stanislao Cannizzaro resolved the atomic mass–molecular formula dilemma in 1858 by following up the line of Avogadro's thinking. He made the reasonable assumption that a molecule must contain a whole number of atoms of each of the constituent elements. If this were not the case, the hypothesis of indivisible atoms would have to be abandoned. Therefore, it follows that there must be at least one atom of a constituent element in each molecule of a compound although, of course, there could be more than one atom per molecule.

If a series of compounds containing a common element is examined, the mass of the common element contained in 1 mole of each compound must be an integral multiple of its atomic mass. Thus it would be possible in theory to deduce the atomic mass of the common element by finding that common factor. While the concept is simple, it remained for Cannizzaro to determine the molecular mass of the series of compounds or to provide some means of assuring that the same number of moles of compound are considered. Cannizzaro accepted Avogadro's assumption that equal volumes of gases contain equal numbers of molecules. Therefore, the masses of two different gaseous compounds contained in equal volumes must be in the same ratio as their molecular masses. It remained to define the atomic mass of one element; then the whole problem of atomic masses would be resolved. Cannizzaro set the relative molecular mass of the diatomic molecule of hydrogen at 2. Only the development of accurate methods of quantitative analysis were necessary.

Cannizzaro's method of analysis involved the determination of the density of gaseous compounds containing oxygen in relation to the density of hydrogen gas at the same temperature and pressure. The molecular mass of each compound is simply the molecular mass of hydrogen times a factor equal to the ratio of the densities of that compound to hydrogen. The molecular masses of several compounds of oxygen are given in Table 4.1.

Cannizzaro also determined the mass of oxygen in a mole of each compound. He accomplished this by determining the percent by mass of oxygen contained in each compound. The mass of oxygen contained in 1 mole of a compound is a product of the mass percentage of oxygen in the compound and the molecular mass of the compound. These data also are given in Table 4.1.

The smallest mass of oxygen contained in a mole of any compound was 16 g. All other values were multiples of 16. Therefore, Cannizzaro assumed the atomic mass of oxygen is 16. The number 16 is also a multiple of 8, 4, and 2. It could be argued that the atomic mass of oxygen is 8 and that there just happens to be even numbers of oxygen atoms in the compounds examined by Cannizzaro. However, if this were the case, there should be found compounds in which odd multiples of 8, such as

Table 4.1
Cannizzaro's analysis

molecule	*molecular mass*	*percent oxygen*	*mass of oxygen*
hydrogen	2		
water	18	88.9	16
sulfur dioxide	64	50.0	32
carbon dioxide	44	72.7	32
nitric oxide	30	53.3	16
nitrous oxide	44	36.3	16
nitrogen dioxide	46	69.6	32

24 and 40 g of oxygen, are present. This has never been observed, and the atomic mass of oxygen is accepted as 16.

By using procedures similar to those outlined above it is theoretically possible to determine the atomic mass of any element that will form a series of gaseous compounds. Cannizzaro's contribution to resolving the problems that were presented by the combination of the hypotheses of Dalton, Gay-Lussac, and Avogadro was a landmark in the science of chemistry. Once a means of determining the atomic mass scale was provided, molecular formulas could be established easily since the mass of the proposed molecular formula must be equal to the experimentally determined molecular mass.

EXAMPLE 4.5

The density of carbon suboxide is 34 times that of H_2 at the same pressure and temperature. The mass percent composition is 47 percent oxygen. Calculate the relative mass of oxygen combined in a molecule of carbon suboxide relative to $H_2 = 2$ amu.

The molecular mass of carbon suboxide must be 34 times that of H_2 if the densities differ by that factor. Therefore, the molecular mass is 68 amu:

$$34 = \frac{\text{density of carbon suboxide}}{\text{density of } \mathbf{H_2}}$$

$$34 = \frac{\text{mass of molecule of carbon suboxide}}{\text{mass of molecule of } \mathbf{H_2}}$$

$$68 \text{ amu} = \text{relative mass of molecule of carbon suboxide}$$

The relative mass of oxygen contained in carbon suboxide is

$$(68 \text{ amu})(0.47) = 32 \text{ amu}$$

The atomic mass of an element that does not form gaseous compounds can be determined by a simple extension of the Cannizzaro method. If an element of unknown mass is combined with an element of known atomic mass, the ratio of the masses of the two elements combined in the compound is either an integral multiple or rational fraction of the ratio of the atomic masses. This basic procedure of determining atomic masses was used far into the twentieth century as an accurate analytical method. An example of the chemical determination of atomic mass is provided by the combination of bromine with silver to form silver bromide.

EXAMPLE 4.6

The atomic mass of silver is 107.87. By analysis it can be shown that 1.000 g of silver combines with 0.7419 g of bromine. What is the atomic mass of bromine?

The ratio of the reacting masses of silver to bromine is

$$1.000/0.7419 = 1.350$$

The value 1.350 must be equal to the ratio of the atomic mass of silver to bromine or some multiple or rational fraction of the atomic mass ratio. If one atom of silver per one atom of bromine is involved in the formation of silver bromide, the formula for the compound would be AgBr. If this were the case, the atomic mass of bromine would be

$$\text{atomic mass } \mathbf{Br} = 107.87 \text{ amu}/1.350 = 79.91 \text{ amu}$$

If the formula was $AgBr_2$, then the atomic mass of bromine would be

$$\tfrac{1}{2}(79.91 \text{ amu}) = 39.95 \text{ amu}$$

The correct formula is AgBr, and the atomic mass of bromine is 79.91.

EXAMPLE 4.7

A 1.60-g sample of calcium is converted to 2.24 g of an oxide of calcium. If the formula for the oxide is CaO, calculate the atomic mass of calcium relative to the atomic mass 16.0 amu for oxygen.

The formula of the oxide indicates that every atom of calcium contained in the compound is combined with one atom of oxygen. The mass of oxygen contained in the compound is 0.64 g:

$$\begin{aligned} \text{mass of oxide} &= \text{mass of calcium} + \text{mass of oxygen} \\ 2.24 \text{ g} &= 1.60 \text{ g} + \text{mass of oxygen} \\ 0.64 \text{ g} &= \text{mass of oxygen} \end{aligned}$$

The atomic mass of calcium relative to the atomic mass of oxygen must stand in the same ratio of the masses of the two elements contained in the compound.

$$\frac{\text{atomic mass of } \mathbf{Ca}}{\text{atomic mass of } \mathbf{O}} = \frac{\text{mass of } \mathbf{Ca} \text{ in oxide}}{\text{mass of } \mathbf{O} \text{ in oxide}}$$

$$\frac{\text{atomic mass of } \mathbf{Ca}}{16 \text{ amu}} = \frac{1.60 \text{ g}}{0.64 \text{ g}}$$

$$\text{atomic mass of } \mathbf{Ca} = 40 \text{ amu}$$

4.5
CHEMICAL EQUATIONS AND STOICHIOMETRY

A chemical equation is a symbolic way of summarizing the reactants and products and the quantities of each involved in a chemical reaction. Hydrogen reacts with chlorine to produce hydrogen chloride, as described by the following equation:

$$H_2 + Cl_2 \longrightarrow 2HCl$$

The equation represents a mathematical equality that contains all the necessary quantitative information to describe the reaction of any amounts of hydrogen and chlorine. On a molecular scale the coefficients associated with each substance indicate the relative number of molecules involved in the reaction. Thus one molecule of hydrogen reacts with one molecule of chlorine to produce two molecules of hydrogen chloride. Of course, any multiple of these quantities also would react.

As additional chemical equations appear in this book, they will be seen to be mathematically correct and chemically meaningful. The law of conservation of matter requires that there may never be creation or destruction of matter in total during a chemical reaction. The products always must contain the same number and types of atoms as the reactants contained, but the atoms may be arranged differently.

Although the atoms and molecules are the fundamental units of matter involved in describing chemical reactions, it is more convenient to deal with larger quantities of these units. The equation indicating the reaction of hydrogen peroxide to produce water and oxygen is written as follows:

$$\begin{aligned} \mathbf{2H_2O_2} &\longrightarrow \mathbf{2H_2O + O_2} \\ \text{two molecules of } \mathbf{H_2O_2} &\longrightarrow \text{two molecules of } \mathbf{H_2O} + \text{one molecule of } \mathbf{O_2} \\ 2x \text{ molecules of } \mathbf{H_2O_2} &\longrightarrow 2x \text{ molecules of } \mathbf{H_2O} + x \text{ molecules of } \mathbf{O_2} \\ 2 \text{ moles of } \mathbf{H_2O_2} &\longrightarrow 2 \text{ moles of } \mathbf{H_2O} + 1 \text{ mole of } \mathbf{O_2} \\ 68 \text{ g of } \mathbf{H_2O_2} &\longrightarrow 36 \text{ g of } \mathbf{H_2O} + 32 \text{ g of } \mathbf{O_2} \end{aligned}$$

The equation states that $2x$ molecules of hydrogen peroxide produce $2x$ molecules of water and x molecules of oxygen. The value for x may be any integer, but if it is equal to Avogadro's number (6.023×10^{23}), then the equation represents the conversion of 2 moles of hydrogen peroxide to yield 2 moles of water and 1 mole of oxygen. Such a statement in terms of moles provides a basis for describing in manageable units the quantities of reactants and products involved in a reaction.

The number of moles of a substance involved in a reaction can be converted to its equivalent in terms of grams. For example, 2 moles of hydrogen peroxide have a mass of 68 g. Therefore, the chemical equation can be read as follows: 68 g of hydrogen peroxide produce 2 moles of water or 36 g of water and 1 mole of oxygen or 32 g of oxygen. Such a statement in terms of grams of the substances allows the calculation of the quantities of matter involved on any scale. The equation tells us, for example, that 6.8 g of hydrogen peroxide produce 3.6 g of water and 3.2 g of oxygen or that 6.8 lb of hydrogen peroxide produce 3.6 lb of water and 3.2 lb of oxygen.

EXAMPLE 4.8

Ferric oxide reacts with carbon in a blast furnace to produce iron and carbon monoxide according to the equation

$$\mathbf{Fe_2O_3 + 3C \longrightarrow 2Fe + 3CO}$$

How much iron will be produced from 31.94 g of ferric oxide? (The atomic masses of iron, carbon, and oxygen are 55.85, 12.0 and 16.0 g/mole, respectively.)

The equation states that 1 mole of Fe_2O_3 produces 2 moles of Fe. One mole of Fe_2O_3 has a mass of 2(55.85) + 3(16.00) = 159.70 g. The quantity of ferric oxide involved is a fraction of a mole:

$$\frac{31.94 \text{ g } \mathbf{Fe_2O_3}}{159.70 \text{ g } \mathbf{Fe_2O_3}/\text{mole } \mathbf{Fe_2O_3}} = 0.2 \text{ mole } \mathbf{Fe_2O_3}$$

Therefore, 2(0.2) moles or 0.4 moles of Fe will be produced. The mass of 0.4 mole of Fe is

$$(0.4 \text{ mole } \mathbf{Fe})(55.85 \text{ g } \mathbf{Fe}/\text{mole } \mathbf{Fe}) = 22.34 \text{ g } \mathbf{Fe}$$

EXAMPLE 4.9

How many liters of oxygen at STP will be produced by the decomposition of 340 g of hydrogen peroxide? (The atomic masses of hydrogen and oxygen are 1.0 and 16.0 g/mole, respectively.)

$$\mathbf{2H_2O_2 \longrightarrow 2H_2O + O_2}$$

The equation states that 2 moles of H_2O_2 produce 1 mole of

O_2. The quantity 340 g of H_2O_2 is more than a mole of H_2O_2, as its molecular mass is 34.0 g.

$$\frac{340\text{ g }\mathbf{H_2O_2}}{34.0\text{ g }\mathbf{H_2O_2}/\text{mole }\mathbf{H_2O_2}} = 10\text{ moles }\mathbf{H_2O_2}$$

From the equation it can be seen that 10 moles of H_2O_2 will produce 5 moles of O_2. The volume of 1 mole of O_2 at STP is 22.4 liters and, therefore, the quantity of O_2 produced is

$$(5\text{ mole }\mathbf{O_2})(22.4\text{ liters }\mathbf{O_2}/\text{mole }\mathbf{O_2}) = 112\text{ liters }\mathbf{O_2}$$

Although this presentation of chemical equations, the mole concept, and chemical *stoichiometry* is very brief, these concepts are fundamental. The chemist must be able to deal with the quantitative aspects of chemical transformations with great facility. By using the basic logic involved in solving Examples 4.8 and 4.9, it is possible to deal with the most complex chemical reactions. Any amounts of solids, liquids, or gases can be dealt with in terms of densities, volumes, molecules, atoms, and moles at any temperature and pressure.

It is not the purpose of this text to teach the methods of solving complex computational problems. The purpose is to provide an introduction to the logic involved in their analysis without introducing unnecessary stumbling blocks.

Suggested further readings

Meldrum, W. B., "Gram Equivalent Weights," *J. Chem. Educ.*, **32,** 48 (1955).
Neville-Polley, L. J., *John Dalton*. New York: Macmillan, 1920.
Tilden, W., *Famous Chemists*. New York: Dutton, 1921.

Terms and concepts

Avogadro	molecular formula
Cannizzaro	Proust
Dalton	stoichiometry
Gay-Lussac	

Questions and problems

1. A 1.40-g sample of silicon can be converted to 3.00 g of SiO_2. What is the atomic mass of silicon if the atomic mass of oxygen is 16.0?

2. When an oxide of iron of mass 1.00 g is heated with hydrogen gas, 0.70 g of metallic iron is obtained. What is the formula for the iron oxide?

3. A 1.000-g sample of a metal oxide can be converted to 0.927 g of pure metal. The heat capacity of the metal is 0.0332 cal/g. What is the atomic mass of the metal and the formula for the oxide?

4. The heat capacity of arsenic is 0.081 cal/g. Hydrogen and arsenic combine to form a compound that is 96.15 percent arsenic by mass. A 3.48-g sample of the compound occupies 1 liter at standard temperature and pressure. Calculate the atomic mass of arsenic. What is the formula for the compound?

5. One volume of arsenic vapor combines with six volumes of hydrogen to produce four volumes of a gaseous product. Assuming that each molecule of product contains one arsenic atom, determine the number of atoms of arsenic in the arsenic molecule. What is the formula for the compound?

6. A compound AY_3 contains 25 percent Y by mass. What are the relative atomic masses of A and Y?

7. A 0.1000-g sample of a gaseous compound consisting of only carbon and hydrogen is burned in oxygen to give 0.3300 g of carbon dioxide and 0.0899 g of water. What is the simplest ratio of whole numbers that expresses the relative number of atoms of each element in the compound? What are the possible molecular formulas? What information is needed to determine the molecular formula?

8. Two volumes of a gaseous compound consisting of carbon, hydrogen, and nitrogen give upon burning in oxygen four volumes of CO_2, seven volumes of water, and one volume of nitrogen (nitrogen is diatomic). What is the formula of a compound that would yield these products? Is this formula the molecular formula?

9. A 0.1156-g sample of tin is reacted with fluorine, and the resultant compound has a mass of 0.1889 g. What is the formula of the compound formed?

5

THE PERIODIC TABLE AND ATOMIC STRUCTURE

By the mid-nineteenth century, when Cannizzaro carried out his analysis of compounds to determine the atomic masses of constituent elements, there were approximately 60 known elements. With a list of elements and their atomic masses it was only natural that man, a creature of intellectual curiosity, began to probe the nature of atoms in greater detail. The questions were many: Why do the masses of atoms of various elements differ? Are there any common features among the elements that can be discovered? What are the tiny particles called atoms really like?

In this chapter the nature of the atom is unveiled from a semihistorical review of the developments during two exciting periods of chemistry. First the discovery of a method of interrelating elements in terms of their behavior is presented (1865–1870) and then, using this as a foundation, the nature of the subatomic particles present in atoms (1897–1932) is considered.

5.1 CLASSIFICATION OF THE ELEMENTS

EARLY CLASSIFICATION SCHEMES

Early in chemical history it was recognized that many elements have similar sets of properties, and these were used for the purpose of classification. The earliest division entailed only two groups of elements designated as metals and nonmetals. Metals are elements that are malleable and ductile and have a lustrous appearance. In addition, they are conductors of heat and react with certain uniform chemical characteristics that need not be described here. Elements that do not fit into this category, nonmetals, also have certain common physical and chemical characteristics.

Further attempts were made to relate elements to each other by the German chemist Johann Dobereiner in 1829. He observed that certain elements have similar physical properties and chemical reactivities, such as exhibited by the "triad" chlorine, bromine, and iodine. In light of present-day knowledge, the members of the triads of elements appear to be related in terms of atomic mass in that the atomic mass of one element of the triad is approximately the average of the atomic masses of the other two. The approximate atomic masses of chlorine and iodine are 35.5 and 126.6 amu, respectively. The average of these two numbers is 81 amu, a value close to the atomic mass of bromine, which is 79.9 amu. Therefore, it appears likely that there are families of related elements. Actually the grouping of only three elements together is not significant because there were only a limited number of elements to compare when Dobereiner made his observations. Nevertheless, such observations provided the background and incentive to seek further classification schemes for the elements.

It was not until 1866, when the English chemist John Newlands examined the then known elements, that an order of great significance was realized. Newlands arranged the elements in an increasing order of their atomic mass. He noted that chemically similar elements occur at regular intervals in the list (Figure 5.1). Every eighth element has similar properties. Lithium, the second in the list, is comparable to sodium, the ninth in the list, and to potassium, the sixteenth in the list. In like manner, beryllium, magnesium, and calcium are third, tenth, and seventeenth in the list. The similarity of this periodic reoccurrence of the properties of elements to the octave on the musical scale prompted Newlands to postulate a *law of octaves.* However, Newlands' comparison with the octave of the musical scale was not long lasting. In later years the discovery of helium, which is of an intermediate atomic mass between hydrogen and lithium, invalidated the law of octaves. In addition, gases similar to helium, such as neon (which is heavier than fluorine) and argon (which is heavier than chlorine), have been discovered. While the law of octaves is no longer valid, if Newlands had known of these additional elements he might have proposed a law of nonaves.

	No.		No.		No.		No.		No.		No.		No.
H	1	Li	2	Be	3	B	4	C	5	N	6	O	7
F	8	Na	9	Mg	10	Al	11	Si	12	P	13	S	14
Cl	15	K	16	Ca	17	Cr	18	Ti	19	Mn	20	Fe	21

Figure 5.1
Newlands' arrangement of elements in octaves.

There are some valid criticisms of Newlands' octaves that are independent of historical development. Several elements do not fit well in his scheme. Chromium fits poorly under aluminum although both are metals. The metals manganese and iron do not resemble the nonmetals phosphorus and sulfur. Unfortunately, Newlands did not leave places for elements that might yet be discovered, a process that would have eliminated some of the obvious difficulties and suggested other arrangements of elements. In addition, Newlands did not critically consider the accuracy of the then known atomic masses. For instance, titanium is actually of lower atomic mass than chromium.

The most important conclusion that can be derived from Newlands' work is that the regularity of similar properties of elements appears to be more than a coincidence. An ordering of the elements more extensive than the triads of Dobereiner is indicated. At this point the idea emerges that there might be a regularity of atomic structure indicated by the atomic mass. If the elements are composed of more basic particles and if their assemblage follows a repeating order, then the regularity of the properties of elements could be rationalized.

THE PERIODIC TABLE

In 1869 two chemists working independently proposed similar classification schemes for the elements. The Russian chemist Dimitri Mendeleev, who had no knowledge of Newlands' law of octaves, published a classification in which he arranged all of the elements in a table according to their atomic masses. Mendeleev's periodic table (Figure 5.2) was a considerable improvement over that of Newlands, because its slightly greater complexity allowed for proper placement of the elements both in terms of atomic masses and physical and chemical properties. The German chemist Lothar Meyer independently proposed a similar table at the same time, but Mendeleev is mainly credited with the concept. Mendeleev did more than just arrange the elements—he also contributed valuable experimental data. For example, the atomic mass of chromium was revised from 43.4 to 52.0 amu. Although there was a place in which to fit chromium (between calcium and titanium) with its faulty value of 43.4 amu, the properties of chromium did not seem proper for such a placement. The incorrect value for the atomic mass of chromium was derived from an

row	group I	group II	group III	group IV	group V	group VI	group VII	group VIII
1	H=1							
2	Li=7	Be = 9.4	B=11	C=12	N=14	O=16	F=19	
3	Na=23	Mg=24	Al=27.3	Si=28	P=31	S=32	Cl=35.5	
4	K=39	Ca=40	?=44	Ti=48	V=51	Cr=52	Mn=55	Fe=56, Co=59, Ni=59
5	Cu=63	Zn=63	? = 68	? = 72	As=75	Se=78	Br=80	
6	Rb=85	Sr=87	Y = 88	Zr=90	Nb=94	Mo=96	? = 100	Ru=104, Rh=104, Pd=106
7	Ag=108	Cd=112	In=113	Sn=118	Sb=122	Te=125	T=127	
8	Cs=133	Ba=137	Dy = 138	Ce = 140				
9								
10			Er = 178	La = 180	Ta=182	W=184		Os=195, Ir=197, Pt=198
11	Au=199	Hg=200	Tl=204	Pb=207	Bi=208			
12				Th=231		U=240		

Figure 5.2
Mendeleev's periodic table in 1869.

incorrect molecular formula for the compound of chromium and oxygen. Once the molecular formula was established by Mendeleev, the correct atomic mass was obtained. As another example, the atomic mass of indium was taken as 77 amu before the time of Mendeleev's development of a periodic table of elements. This value required that it be placed between arsenic and selenium where there was no vacant position in the table. The properties of indium suggested that it was related to other elements in Group III of the table. A reevaluation of the molecular formulas of compounds of indium revealed that the atomic mass should be 114.8 amu. It should be noted that in the case of both chromium and indium the correct atomic masses are in simple whole number ratios to the incorrect atomic masses. The ratio is 6:5 in the case of chromium and 3:2 in the case of indium.

It was necessary for Mendeleev to leave gaps in his table because there were no known elements to fill all the spaces. His classification of elements attracted the immediate attention of all chemists; an order of nature had been unveiled, and the several gaps in the table suggested that new elements could be found to fill them. In addition, the location of these gaps enabled Mendeleev and other chemists to make predictions of the chemical and physical properties of the unknown elements. These predictions guided the search for new elements, because it is invariably easier to find something that can be defined as compared to a completely unknown quantity. In 1871 Mendeleev suggested that elements similar to aluminum and silicon should exist. Gallium (similar to aluminum) was discovered in 1875, and germanium (similar to silicon) was discovered in 1886. The atomic mass and density of the element similar to silicon were predicted to be 72 amu and 5.5 g/cm^3, respectively. Germanium has an atomic mass of 72.6 amu and a density of 5.37 g/cm^3.

Mendeleev's table did include some exceptions to the relationship between properties and atomic mass. Today, in the modern periodic table shown in Figure 5.3, cobalt (58.93 amu) is placed before nickel (58.71 amu) and tellurium (127.60 amu) is placed before iodine (126.90 amu). In addition, argon, which was not discovered until many years later, is placed before potassium in spite of its larger atomic mass. More exceptions existed in Mendeleev's time, but redetermination of the atomic masses removed these apparent discrepancies. In spite of the three exceptions of nickel, iodine, and potassium, which are not arranged according to atomic mass in the periodic table, the arrangement of the other elements does allow the general statement of a *periodic law:* The chemical and physical properties of the elements are periodic functions of their atomic mass.

The development of the periodic table helped to systematize chemistry. It was no longer necessary to consider each element as a unique entity with no relation to any of the other elements. All of the available physical and chemical properties could be discussed in a general manner and interrelated. The problem of committing large quantities of data to memory was diminished somewhat as the concept of periodicity was accepted. As new data were accumulated at an ever-increasing rate, the information was organized in terms of the periodic table. It is a tribute to the genius of Mendeleev that an essentially equivalent form of his table (Figure 5.3) is used today, almost a century after its conception. The elements are arranged in the periodic table in order of increasing atomic mass (excluding the three exceptions previously noted), with elements of similar properties placed under each other in a vertical column called a *group.* There are eight main groups designated I, II, III, IV, V, VI, VII, and 0. The elements intervening between Groups II and III are called the transition metals (Chapter 14). Each vertical column of the transition metals is called a subgroup. The horizontal sequences of the periodic table are called periods and are numbered from the top down.

5.2 FUNDAMENTAL PARTICLES OF THE ATOM

ELECTRICAL NATURE OF MATTER

Even in early recorded history the electrical nature of matter was observed. The Greeks noted that an amber (elektron) rod rubbed with wool acquired an electric charge as did a glass rod if rubbed with silk. Either would attract bits of paper. When the charged glass and amber rods were brought together a spark jumped between them. There was evidence that

period	IA	IIA	IIIB	IVB	VB	VIB	VIIB	VIII	VIII	VIII
						transition elements				
1	1 **H** 1.008									
2	3 **Li** 6.939	4 **Be** 9.012								
3	11 **Na** 22.990	12 **Mg** 24.31								
4	19 **K** 39.102	20 **Ca** 40.08	21 **Sc** 44.956	22 **Ti** 47.90	23 **V** 50.94	24 **Cr** 51.996	25 **Mn** 54.94	26 **Fe** 55.85	27 **Co** 58.93	28 **Ni** 58.71
5	37 **Rb** 85.47	38 **Sr** 87.62	39 **Y** 88.91	40 **Zr** 91.22	41 **Nb** 92.91	42 **Mo** 95.94	43 **Tc** 99	44 **Ru** 101.07	45 **Rh** 102.91	46 **Pd** 106.4
6	55 **Cs** 132.91	56 **Ba** 137.34	* rare earths	72 **Hf** 178.49	73 **Ta** 180.95	74 **W** 183.85	75 **Re** 186.2	76 **Os** 190.2	77 **Ir** 192.2	78 **Pt** 195.09
7	87 **Fr** 223	88 **Ra** 226.05	† actinides							

* *rare earths*	57 **La** 138.91	58 **Ce** 140.12	59 **Pr** 140.91	60 **Nd** 144.24	61 **Pm** 145	62 **Sm** 150.35	63 **Eu** 151.96	64 **Gd** 157.25
† *actinides*	89 **Ac** 227	90 **Th** 232.04	91 **Pa** 231	92 **U** 238.03	93 **Np** 237	94 **Pu** 242	95 **Am** 243	96 **Cm** 247

Figure 5.3
The modern periodic table.

two types of charge existed. Benjamin Franklin arbitrarily suggested that the charge on a glass rod was negative and that the charge on the amber rod was positive. Electricity as defined by Franklin was thought to flow from positive to negative regions of matter. His choice was unfortunate, since we now know that electricity is the result of negative particles, *electrons,* flowing from negatively charged regions to positively charged regions.

Many basic properties of electricity were known by the beginning of the twentieth century: Oppositely charged objects are attracted to each other while objects of like charge are repelled. Thus the direction of a moving charged particle can be altered by placing charged objects along its path. This behavior is represented in Figure 5.4 by a beam of negatively

IB	IIB	IIIA	IVA	VA	VIA	VIIA	O inert gases
						1 **H** 1.008	2 **He** 4.003
		5 **B** 10.81	6 **C** 12.011	7 **N** 14.007	8 **O** 15.999	9 **F** 19.00	10 **Ne** 20.183
		13 **Al** 26.98	14 **Si** 28.086	15 **P** 30.974	16 **S** 32.064	17 **Cl** 35.453	18 **Ar** 39.948
29 **Cu** 63.54	30 **Zn** 65.37	31 **Ga** 69.72	32 **Ge** 72.59	33 **As** 74.92	34 **Se** 78.96	35 **Br** 79.909	36 **Kr** 83.80
47 **Ag** 107.870	48 **Cd** 112.40	49 **In** 114.82	50 **Sn** 118.69	51 **Sb** 121.75	52 **Te** 127.60	53 **I** 126.90	54 **Xe** 131.30
79 **Au** 196.97	80 **Hg** 200.59	81 **Tl** 204.37	82 **Pb** 207.19	83 **Bi** 208.98	84 **Po** 210	85 **At** 210	86 **Rn** 222

65 **Tb** 158.92	66 **Dy** 162.50	67 **Ho** 164.93	68 **Er** 167.26	69 **Tm** 168.93	70 **Yb** 173.04	71 **Lu** 174.97
97 **Bk** 249	98 **Cf** 251	99 **Es** 254	100 **Fm** 253	101 **Md** 256	102 **No** 253	103 **Lw** 257

charged particles that are attracted by the positive electrode (*anode*) and repelled by the negative electrode (*cathode*). Positively charged particles would be deflected downward. The extent of deflection of the beam is directly proportional to both the charge on the plates and the charge on the particles and is inversely proportional to the mass of the particles. In a similar manner the beam of charged particles is deflected by a magnetic field. The direction of the deflection, as shown in Figure 5.5 for negatively charged particles, is at right angles to the magnetic field. Positively charged particles would be deflected in the downward direction. The deflection is directly proportional to the charge on the particle and to the strength of the magnetic field; it is inversely proportional to the mass of the particle.

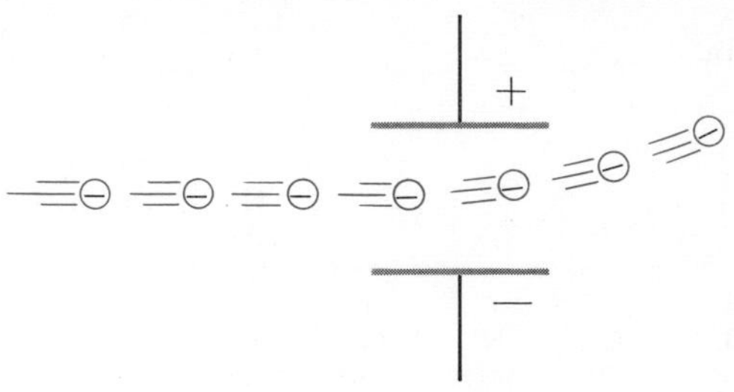

Figure 5.4
Effect of an electric field on negative particles.

THE ELECTRON

Chemists were examining the reactions of elements and compounds, which were postulated to contain atoms and molecules, respectively, for nearly a century prior to the discovery of subatomic particles. At the same time physicists were studying the phenomena of electricity and magnetism. Near the turn of this century these two areas of the natural sciences were joined to produce great advances in our understanding of the nature of matter.

In 1897 Sir J. J. Thomson at Cambridge University proposed that the rays of the *cathode-ray* tube consisted of basic, negatively charged particles called electrons. The phenomenon of the cathode-ray tube had been known since J. Plucker's work in 1858, not until almost 40 years later was the concept of an electron evolved to account for it. Two electrodes sealed in a closed, partially evacuated glass tube and connected to a source of electrical potential constitute a cathode-ray tube (Figure 5.6). If the tube contains helium gas it exhibits a pink glow; a red glow is observed when neon is used (neon signs). Electricity flows in both cases independent of the gas, and upon removing the source of electric potential the glow ceases. Since the glows are different, the substances responsible for them cannot be a common, basic electric particle.

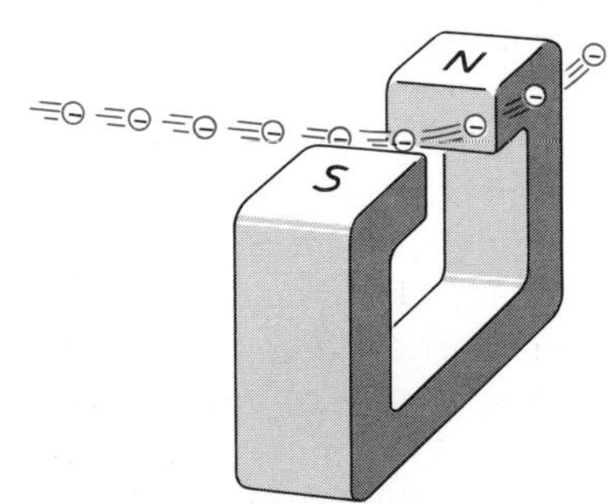

Figure 5.5
Effect of a magnetic field on negative particles.

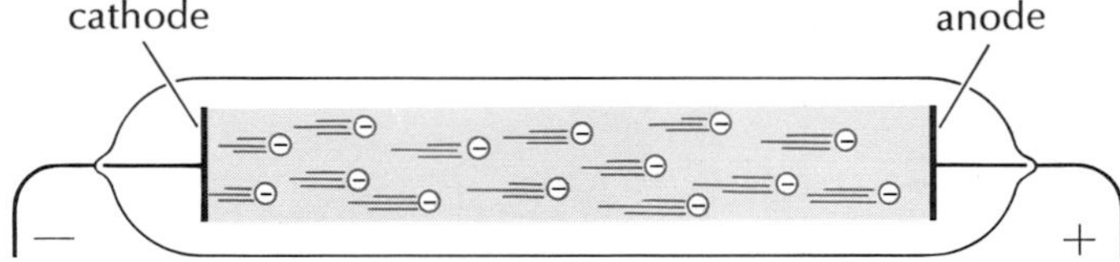

Figure 5.6
Cathode-ray tube.

When a plate with a slit is placed near the cathode and a zinc sulfide screen is positioned at a slight angle to the longitudinal axis of the tube, an interesting effect is noted. A line fluoresces across the zinc sulfide screen. When a magnet is placed near the tube, as shown in Figure 5.7, the line on the zinc sulfide screen is deflected upward. According to the most satisfactory interpretation of the cathode-ray tube experiment, invisible negatively charged particles (cathode rays) emanate from the cathode and move toward the anode. When they collide with the zinc sulfide screen they produce light. The direction of deflection by a magnet is indicative of the assigned negative character of the particles.

Thomson's experiments showed that cathode rays were common to all of the gases that were placed in the tube. Using an apparatus similar to that shown in Figure 5.8, he measured the deflection of the cathode rays produced by electric and magnetic fields of varying strength. The anode is hollow to allow the passage of some negative particles past the anode. The presence of the particles can be detected by the fluorescence caused by their impact with the zinc sulfide screen at the end of the tube. By varying the charge on the electric plates and the strength of the magnetic field, Thomson was able to balance the respective deflections so that no net deflection occurred. It was then possible to calculate the ratio of the charge e to the mass m of the particles. Since both e and m were unknown quantities, only the value for their ratio e/m resulted from Thomson's calculations. The value for e/m was determined to be -1.76×10^8 coulombs/g and is independent of the gas used. Thus it was proposed that

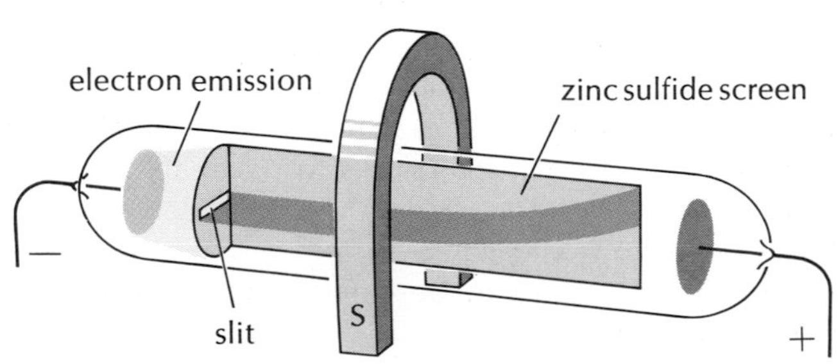

Figure 5.7
A magnetic field deflects cathode rays.

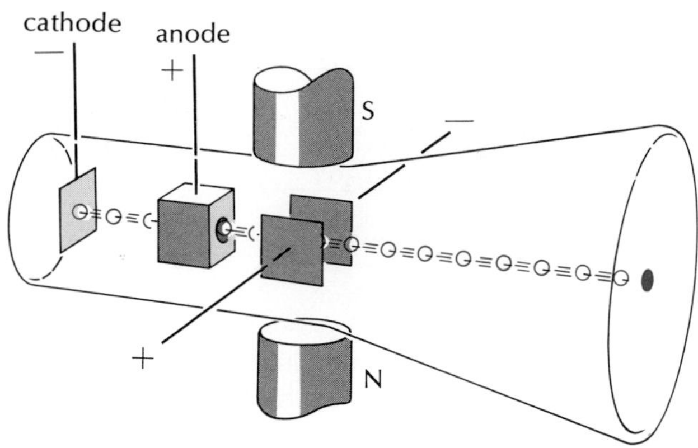

Figure 5.8
The Thomson experiment.

all matter contains a negatively charged subatomic particle called an *electron.*

Once the charge-to-mass ratio of the electron was determined, determination of either the charge or the mass of the electron would yield a calculated value for the other. The ingenious oil droplet experiment of R. A. Millikan in 1909 allowed the determination of the charge of an electron. The basic features of the experiment are illustrated in Figure 5.9. When oil droplets are sprayed between two charged plates they drop because of gravity. If the oil droplets are made to acquire a negative charge by x-ray irradiation, the attraction of the positive plate acts against gravity and either slows the descent of the oil droplet or actually causes it to rise. By adjusting the plate charge an oil droplet can be made to remain stationary, allowing the e/m ratio for the oil droplet to be calculated. The ingenuity of Millikan is reflected by the fact that the mass of the electron is unimportant in this experiment. The mass in the e/m value is essentially the mass of the oil droplet, as the mass of an electron is much smaller than the mass of the droplet. The mass of the droplet can be determined by more conventional physical means, and e can be calculated. Millikan found many values for e in his experiments, but they were all simple, whole number multiples of one value, -1.60×10^{-19} coulombs, which is the accepted charge of an electron. His observations can be explained by the assumption that several electrons create the charge on the oil droplets. In his experiments Millikan never observed an oil droplet containing fewer than four electrons.

The mass of the electron can be calculated by dividing the charge of an electron by its e/m ratio.

$$\frac{-1.60 \times 10^{-19} \text{ coulombs}}{-1.76 \times 10^{8} \text{ coulombs/g}} = 9.1 \times 10^{-28} \text{ g}$$

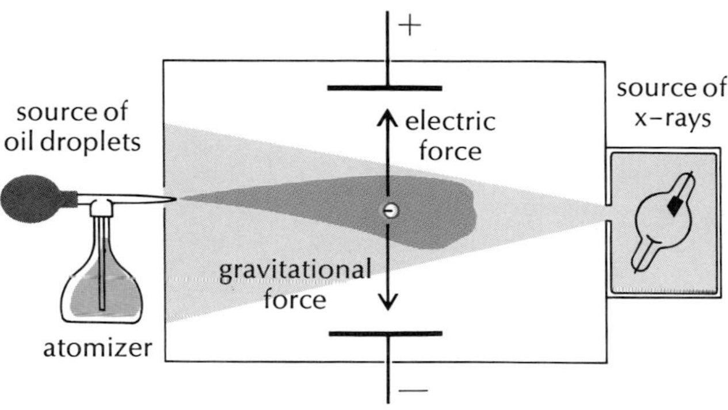

Figure 5.9
The Millikan experiment.

The mass of an electron is much smaller (1/1840) than that of the simplest atom, hydrogen. Division of the mass of 1 mole of hydrogen atoms by Avogadro's number yields a value of 1.66×10^{-24} g for the mass of the hydrogen atom:

$$\frac{1 \text{ g/mole}}{6.02 \times 10^{-23} \text{ atom/mole}} = 1.67 \times 10^{-24} \text{ g/atom}$$

It is evident that the majority of the mass of an atom resides in the positively charged species that remain after removal of an electron in the cathode-ray experiments.

THE PROTON

If electrons are separated from atoms in the cathode-ray tube, then a positively charged residue must be formed. In 1886 the German physicist Eugen Goldstein showed that positive particles (ions) are present in the cathode-ray tube. His modified tube (Figure 5.10) contained a cathode with a hole through which any positively charged particles could pass if they had the necessary velocity and direction. Most ions would be expected to collide with the cathode, but in the region beyond the cathode Gold-

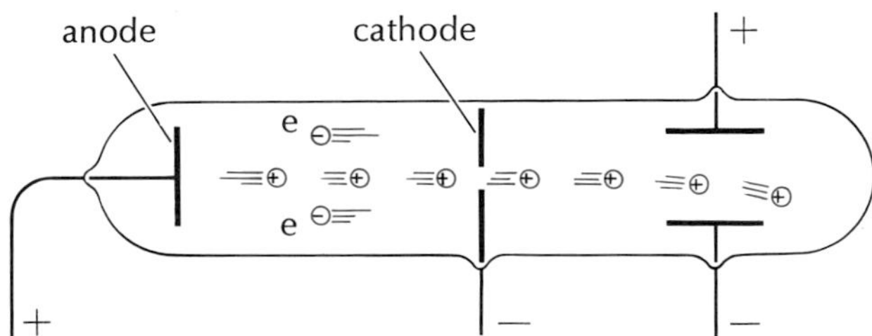

Figure 5.10
The Goldstein tube.

stein observed a glow resulting from the presence of positively charged particles. That the ions are positive can be shown with a magnet or charged plates since the ions are deflected in the opposite direction from electrons. If the electrodes in Thomson's tube are reversed so that the positive particles enter the end of the tube containing the detecting screen, the value of the charge-to-mass ratio q/m for the ions could be measured, and this ratio could be shown to be dependent on the gas used. The charge-to-mass ratios can be more accurately determined by means of a mass spectrograph, described in the next section. However, if the experiment with the cathode-ray tube could be done for the hydrogen atom, the q/m ratio would be 9.60×10^4 coulombs/g. If it is assumed that the hydrogen atom has only one electron, then the hydrogen atom minus its electron (hydrogen ion) must have a charge of $+1.60 \times 10^{-19}$ coulomb. Therefore, the mass of the hydrogen ion is 1.67×10^{-24} g:

$$\frac{1.60 \times 10^{-19} \text{ coulomb}}{9.60 \times 10^4 \text{ coulombs/g}} = 1.67 \times 10^{-24} \text{ g}$$

This value is equal to the mass of the hydrogen atom. This evidence coupled with the results of many other experiments led to the postulation that the hydrogen atom minus its electron is a basic subatomic particle with a charge of $+1.60 \times 10^{-19}$ coulombs, which is considered the basic unit charge on the atomic level. This particle is called the *proton.* The mass of the hydrogen atom is largely due to the proton.

The hydrogen atom consists of one electron and one proton, which balance each other electrically. Either the addition or removal of an electron to a neutral atom produces a charged particle called an *ion.* A proton is produced by removing one electron from a hydrogen atom and is called a *positive hydrogen ion.*

Other elements have a lower charge-to-mass ratio for the ions produced by loss of electrons. The lower value cannot be due to a smaller value for the charge, as it cannot be less than the basic unit positive charge that results from the loss of one electron. Therefore, m must be larger for the ions of other elements. It is reasonable to postulate at this time that other elements contain more electrons and protons than hydrogen.

MASS SPECTROMETER AND ISOTOPES

In the years prior to World War I, J. J. Thomson and F. W. Aston developed a technique of deflecting beams of gaseous ions by means of magnetic and electrical fields. Their apparatus, called a *mass spectrograph,* is shown in Figure 5.11 in its simplest form. In the ionization chamber, which is basically a cathode-ray tube, electrons are knocked off atoms and ions result. As we shall see later it is a relatively simple matter to form a singly charged ion because the removal of additional electrons requires higher electrical potential. Therefore, ions of unit positive charge can be produced and passed through the hole in the cathode and enter the curved

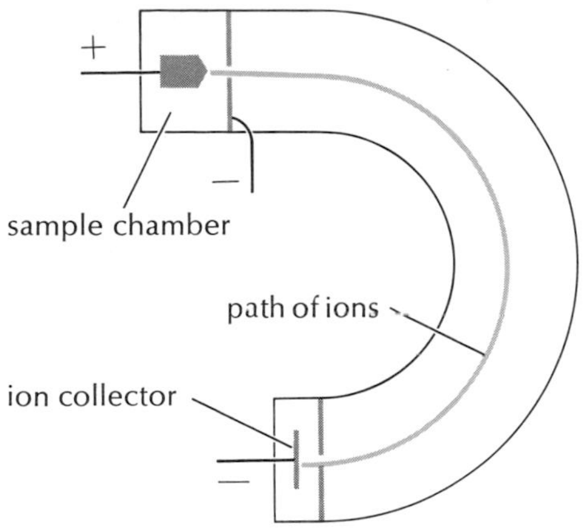

Figure 5.11
The mass spectrograph.

chamber. By means of a magnet, the paths of the particles are curved so that they eventually reach the detecting chamber. Because an ion of high mass is not deflected as much as an identically charged ion of low mass at the same magnetic field strength, the q/m ratio of ions can be calculated with great accuracy from the extent of the deflection by use of a mass spectrograph.

When helium, a monatomic (single atom) element, is placed in a mass spectrograph and one electron per atom is removed by application of the minimum electrical potential, an ion of $q/m = 2.42 \times 10^4$ coulombs/g is obtained. This value is one-fourth that of the proton. Application of a higher electrical potential results in a particle whose q/m ratio is 4.84×10^4 coulombs/g. Obviously, since the mass of the electron is small, the mass of the ion in each case is essentially constant and, therefore, the difference between this second ion and the first is its higher positive charge. Because application of higher electrical potential does not produce any ions with charge-to-mass ratios higher than 4.84×10^4 coulombs/g, it may be concluded that helium contains only two electrons. In order to be electrically neutral, helium must then contain two protons whose mass should be $2(1.67 \times 10^{-24}\text{ g}) = 3.34 \times 10^{-24}$ g. However, the mass of the helium atom is 6.68×10^{-24} g:

$$\frac{4.00\text{ g/mole}}{6.023 \times 10^{23}\text{ atom/mole}} = 6.68 \times 10^{-24}\text{ g/atom}$$

The entire mass of the helium atom cannot be accounted for in terms of just electrons and protons. Apparently there are additional subatomic particles to be discovered.

It is not possible to account for the entire mass of any atom except hydrogen by using the mass of electrons and protons. In order not to upset the electrical neutrality of atoms, a basic subatomic particle without a charge has to be sought. Since it has no charge, the conventional means of electric and magnetic fields are not applicable. Not until 1932 was the *neutron* identified as a subatomic particle that has no charge and has a mass of 1.67×10^{-24} g. After its discovery by J. Chadwick, chemists and physicists were able to picture the atom as an assemblage of three types of subatomic particles: the electron, the proton, and the neutron, with the protons and neutrons accounting for essentially the entire mass of the atom. For example, helium consists of two electrons, two protons, and two neutrons. The presence of two neutrons is deduced from the mass of 3.34×10^{-24} g, which remains unaccounted for after the subtraction of the mass of two protons from the mass of the helium atom.

The mass spectrograph has been used to investigate a basic problem in chemistry that is illustrated by the element neon. Neon ions, produced by the loss of one electron, show two q/m ratios, 4.80×10^3 and 4.33×10^3 coulombs/g. Since the value for the charge is identical for both types of ions, they must have different masses: the calculated masses are 3.34×10^{-23} g and 3.68×10^{-23} g, respectively. The difference of 3.4×10^{-24} g corresponds to the mass of two neutrons. The neon atoms that produce these ions are called *isotopes.* Isotopes of an element contain the same number of protons and electrons but differ from each other in the number of neutrons they contain.

Most elements consist of mixtures of isotopes. However, the presence of a variable number of neutrons does not significantly alter their chemical reactions. It is now possible to see why discrepancies arise in the correlation of properties of elements and their atomic mass. If the elements are arranged according to the number of protons in an atom, which is the *atomic number* of an element, the correlation of physical and chemical properties is excellent. In general the atomic number and atomic mass increase simultaneously, but the addition or subtraction of a neutron from an element may reverse the order dictated by atomic number and upset the correlation. The determination of the mass of a mole of an element reflects the relative amounts of the various isotopes that occur in nature. The atomic mass recorded is the weighted average of the various isotopes present. For example, chlorine contains 17 protons and 17 electrons, but it exists as two isotopes that contain 18 and 20 neutrons. The mass of 1 mole of each separate isotope is 35 and 37 g. Naturally occurring chlorine is a mixture of the two isotopes in proportions such that its experimental atomic mass is approximately 35.45 g/mole. The isotopes of chlorine are symbolized as ${}^{35}_{17}Cl$ and ${}^{37}_{17}Cl$, where the subscript represents the atomic number and the superscript the mass number.

5.3 MODELS OF THE ATOM

THE NUCLEUS

Once it was realized that atoms are not indivisible and are not the fundamental building blocks of matter, but rather that electrons, protons, and neutrons are more basic particles, it became of great interest to ascertain the structure of atoms. The electron and proton possess charges of equal magnitude but opposite in sign, whereas the neutron is electrically neutral. The proton and neutron are of essentially identical mass and are 1836 times heavier than the electron.

In 1898 Thomson suggested that atoms consist of spheres of positively charged matter (the protons) surrounded by the negatively charged and relatively small electrons embedded on the surface of the sphere. This suggestion does not seem unreasonable when it is recalled that (1) protons are more massive than electrons and (2) electrons can be removed from atoms to form positively charged ions.

Lord Rutherford proposed an experimental test of the Thomson model of the atom. The experiments, carried out in 1911 by two students of Rutherford, H. Geiger and E. Marsden, produced results that could not be interpreted in terms of the Thomson model. High energy α particles (the helium atom minus its two electrons) were used to bombard a thin sheet of gold (Figure 5.12). The gold atoms in the sheet were closely packed, and it was anticipated on the basis of the Thomson model that the α particles would move right straight through the uniformly dense sea of charged matter in the thin sheet. As expected, approximately 99.9 percent of the particles did go through. However, the remaining 0.1 percent were deflected from their pathway by quite large angles. Indeed, a very small percentage was reflected right back along its original pathway.

The extremely large angles of deflection were incredible in terms of the Thomson model of the atom. If the massive, positively charged portion of the atom were not considered to be uniformly distributed but rather concentrated in specific regions, then the experimental observations would make sense, because if the α particle encountered a dense, major obstruction in its path, it would be bounced back or deflected at a large angle. Rutherford deduced that there is a dense region of the atom called the *nucleus* that contains the protons and neutrons and is positively charged. The electrons occupy a large region of space around the nucleus and are

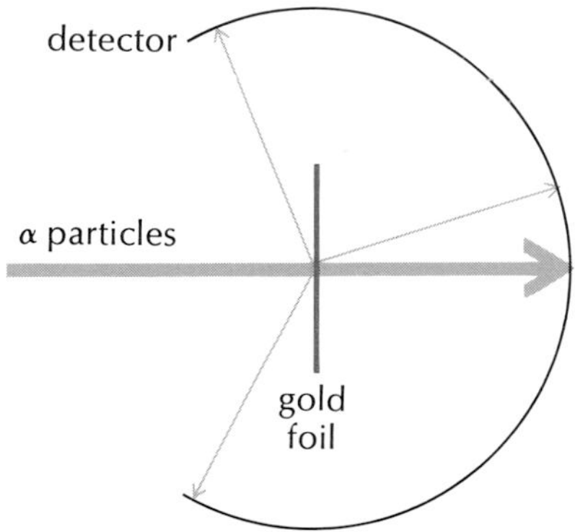

Figure 5.12
The Rutherford experiment.

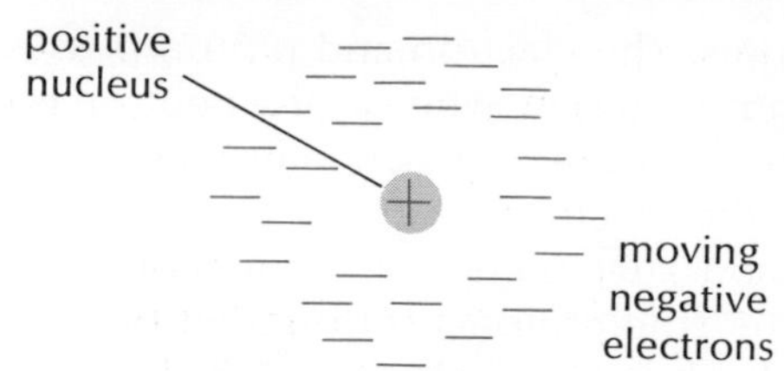

Figure 5.13
Rutherford model of the atom.

moving (Figure 5.13). It is an interesting historical coincidence in the progress of chemistry and physics that Dalton's original suggestion of an indivisible atom was made in Manchester, England, in 1806. Approximately one century later, Rutherford, who also was at Manchester, made the major contribution in the concept of the structure of the atom, considering it to be divisible.

The picture we have of the arrangement of electrons in the atom is reminiscent of the gaseous state. The electrons are not very massive, but because of their movement through space they occupy a large volume. The diameter of the nucleus of an atom is approximately 10^{-13} cm, as determined from quantitative scattering experiments such as Rutherford's. The diameter of an atom however, is approximately 10^{-8} cm, leaving the nucleus to occupy only a small fraction of the total volume of the atom.

At this point in the development of a picture of the atom it might be said that the atom resembles our solar system in some ways, with the nucleus corresponding to the sun and the electrons to the planets. However, the model can be refined further by seeking answers to questions that may be posed about atomic structure. How are the electrons located in the atom? Are they moving randomly or in fixed paths? Why are the electrons not drawn into the nucleus by the force of electrostatic attraction? Can electrons be distinguished from one another in a multielectron atom? How are the protons and neutrons arranged in the nucleus? Why do the protons in the nucleus not repel each other? The questions on the electrons are examined in this chapter, while those concerning the nucleus are discussed in Chapter 15.

THE BOHR ATOM

When atoms or molecules are made to absorb high energy, they often emit light of particular frequencies. For example, in a cathode-ray tube that can contain a variety of gases, or more specifically in the commercial neon signs, the gas emits a particular color of light when subjected to high electrical potentials. Similar observations can be made with elements when they are heated to sufficiently high temperatures. When the light from such a source is passed through a prism, a series of lines of light are observed that are characteristic of the element (Figure 5.14). Each line of light corresponds to a definite energy (E), which can be calculated from the equation

$$E = h\nu$$

where h is Planck's constant (1.59×10^{-34} cal sec) and ν is the frequency of the light. Blue light, which is of high frequency, corresponds to the high-energy side of the visible spectrum, while red light is of low energy.

Since each narrow line of light observed in the resolved emission spectrum of an element corresponds to a specific frequency and a specific energy, it must be concluded that the energy emitted by an atom can be of only definite discrete values.

In order to account for the line spectrum observed for hydrogen, Niels Bohr in 1913 proposed a theory of electronic energy levels in which he postulated that only definite discrete amounts of energy could be associated with the electron in the atom. These specific energy portions are called *quanta* (singular, *quantum*), and the energy of the electron is said to be quantized. In other words, the electron cannot have just any energy but is restricted to certain values. In order for the electron to change in energy, it can assume or release only that amount of energy necessary to reach some other allowed energy content. In this way the observations of line spectra can be rationalized. When energy is added to an atom by means of electrical potential or heat energy, the electrons are raised in energy from their normal values. In order to reachieve their normal values, the energy difference between high and low energy states must be emitted. Bohr's postulate also explains why the electrons do not collapse into the nucleus. Once the electrons are in their lowest energy state they can proceed no further.

The Bohr model for the hydrogen atom resembles a planetary system in which electrons can exist only in certain stable orbits. Using his model for the hydrogen atom and mathematically combining several laws of

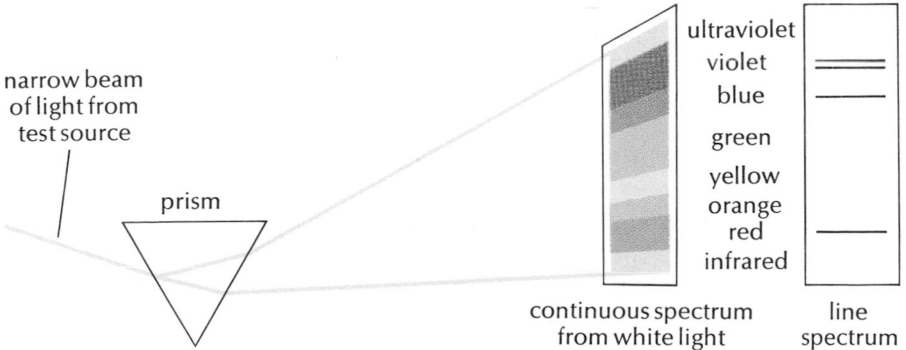

Figure 5.14
Continuous spectrum and line spectrum.

physics, Bohr was able to calculate a line spectrum for hydrogen that agrees with the actual spectrum with great accuracy.

5.4 ELECTRONIC CONFIGURATION OF ELEMENTS

Bohr's postulate has been elaborated and expanded into a highly developed field of study called quantum mechanics. The picture of the atom that has evolved is complex and resembles only slightly the original Bohr model, but the present-day model of the atom can be understood in sufficient detail to enable high school and college students to gain insight into the relationship between observed physical properties and the elementary aspects of the quantum theory. Rather than attempt to justify the basic working rules of electronic assignments of atoms as they were evolved historically, we shall present the rules and then show how they agree with physical properties of atoms. Again, it should be recalled that agreement between reality and a proposed model does not mean that the model is correct. However, no more plausible model has been evolved with the introduction of new experimental data. The data available today are so overwhelmingly in agreement with our model that the theory has become well accepted.

SHELLS AND SUBSHELLS

The basic principle of quantum mechanics is that there are specific *energy levels* possible for electrons in atoms. These energy levels also are called *shells* and are numbered $n = 1, 2, 3, \ldots$, starting with one as the lowest energy level. The letters *K, L, M,* etc., were used earlier and may still be found in some texts. The second principle that has evolved is that the total number of electrons allowed in an energy level n is $2n^2$, In the lowest energy level only two electrons are allowed, whereas in the second, third, and fourth energy levels, 8, 18, and 32 electrons, respectively, are allowed. The third principle is that electrons in elements occupy the lowest energy state available to them. For example, if two electrons are present, as is the case for helium, the electrons are located in the first energy level. Thus the elements lithium through neon, which are of atomic number 3 through 10, have their electrons in excess of two in the second energy level. For sodium, which has 11 electrons, one electron must be located in the third energy level. Now if the periodic table (Figure 5.3) is examined and compared to the concept of electronic energy levels, it can be seen that the periodic properties of elements are related to the number of electrons in the highest energy level. The elements lithium and sodium each have one electron in the highest energy level. Similarly, beryllium and magnesium each have two electrons in the highest energy level. The horizontal rows (*periods*) in the periodic table indicate the energy level being filled, and the position of an element in a given column (*group*) of the table indicates the number of electrons contained in the highest energy level. This correlation between the periodic table and electronic configuration is important, and it should be obvious that the electrons with the highest energy are the most likely to undergo changes. It must be this feature that contributes to the similarity of chemical and physical properties of the elements in a group. The third period contains only 8 elements rather than the 18 that would be expected from the expression $2n^2$. The fourth period contains only 18 elements instead of the expected 32. Apparently some further refinement in our rules is necessary.

It has been tacitly implied that all electrons in a given energy level are of the same energy. The apparent anomaly between the predicted value of 18 for the third period and the 8 observed can be explained on the basis that not all the electrons in a given shell are of equal energy. Indeed, if the line spectrum of an element is examined under high resolution, each apparent line is actually a group of closely spaced lines. This phenomenon provides part of the basis for the concept of *subshells*. Each shell consists of subshells or energy sublevels, the number of which is equal to the assigned number of the shell. Thus there is only one subshell in the first energy level, whereas the fourth energy level has four subshells. The subshells differ in ways other than energy. Because some of their characteristics are independent of the energy level in which the subshell is located, the subshells are designated according to their type by the letters *s, p, d,* and *f*. While there are other subshells theoretically possible in the fifth and higher energy levels, they are not needed to describe the elements known to date. The relative energies of the subshells within a shell increase in the order *s, p, d,* and *f*. The maximum number of electrons allowed in the *s, p, d,* and *f* subshells are 2, 6, 10, and 14, respectively. With this energy scheme in mind the energy levels and sublevels can be presented and the periodic table can be discussed. In Figure 5.15 the relative energies of the shells are pictured on the left, with the spacing between adjacent energy levels diminishing as higher energies are reached. Within each energy level the energy of the subshells is pictured. In the first energy level the *s* subshell is at the energy of the shell. In the second energy level there are an *s* and a *p* subshell, with the energy of the *s* being lower than the average energy of the shell and the *p* being higher than the average. The arrangement of the *s, p,* and *d* subshells according to their relative energies in the third energy level is uncomplicated. In the fourth energy level some complications do appear. The *s, p, d,* and *f* subshells are arranged in order, but the *s* subshell of the fourth energy level (4*s*) is of lower energy than the *d* subshell of the third energy level (3*d*). This overlap results from the close spacing between high energy levels, and it should be evident that if this picture is correct electrons are placed in the 4*s* subshell before the 3*d* subshell. While this appears to be an odd way to label energy levels and sublevels, there is a clear distinction between electrons in different subshells that becomes evident later, and these distinctions are independent of energy.

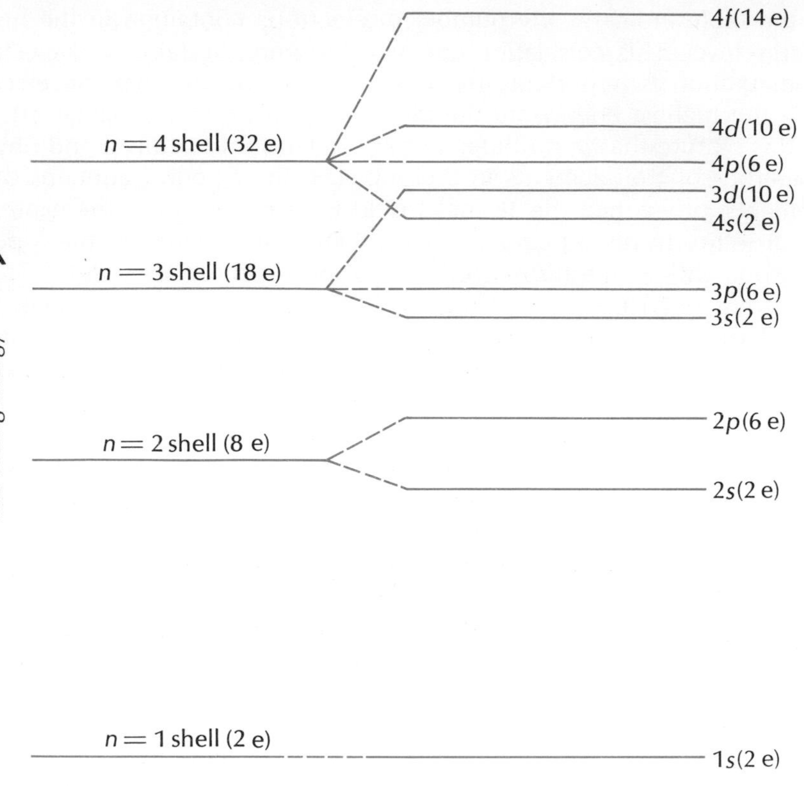

Figure 5.15
Energy of shells and subshells.

If the fifth, sixth, and seventh energy levels were added to Figure 5.15, a complicated overlap of subshells would result. The increasing order of energies is 1*s*, 2*s*, 2*p*, 3*s*, 3*p*, 4*s*, 3*d*, 4*p*, 5*s*, 4*d*, 5*p*, 6*s*, 4*f*, 5*d*, 6*p*, 7*s*, and 5*f*. This order does not need to be memorized since it always can be deduced from the periodic table, as shall be illustrated now. Although the actual ordering of the subshells is a function of the element chosen, most elements behave in the indicated fashion.

The order of filling the energy levels of atoms is shown by the periodic table in Figure 5.16. In the first period there are only two elements. These are hydrogen and helium, which have one and two electrons, respectively, in the 1*s* subshells. The elements lithium (at. no. 3) and beryllium (4) have electrons in the 2*s* subshell and are separated from boron and five other elements. Obviously, boron (5) through neon (10) involve the addition of six successive electrons to the 2*p* subshell. The 3*s* subshell is filled by the addition of electrons 11 and 12 to produce sodium and magnesium. Again, the six elements on the right of the periodic table involve the filling of a *p* subshell (3*p*). Elements 19 and 20 are in the same region in which s subshells were filled in earlier periods, which in this case is the 4s subshell. Now for the first time a center region of the periodic table contains elements and the electronic configurations of the ten elements scandium (21) through zinc (30) are successively pictured by the addition of electrons to the 3*d* subshell. Then elements 31 through 36 appear in the same region where *p* subshells have been filled previously. In this case it is the 4*p*, as it is in the fourth period. The remainder of the periodic table follows in an orderly fashion in the case of *s* and *p* subshells. The *d* subshells are always one behind the period number. Outside the general body of the periodic table are located two groups of 14 elements that represent filling of the 4*f* and 5*f* subshells. There are exceptions to the above presentation, which is revealed by a close inspection of Table 5.1—a list of actual electronic configurations of the elements. However, these need not concern us here. Some of these exceptions are discussed in Chapter 14.

The electronic configuration of silicon, atomic number 14, can be written down by examining the periodic table and going through all regions until silicon is reached. The configuration is $1s^22s^22p^63s^23p^2$, with the superscripts indicating the number of electrons in each subshell. In a similar manner the electronic configuration of palladium (at. no. 46) is $1s^22s^22p^63s^23p^64s^23d^{10}4p^65s^24d^8$. Not only is the entire electronic configuration of an element available from its position in the periodic table, but the number and type of electrons in the outermost energy level are available at a glance.

Technetium (at. no. 43) has five electrons in the 4*d* orbital. This fact is derived from the periodic table by noting that technetium is the fifth element in the region of the periodic table where the 4*d* orbital is filled.

ELECTRON SPIN AND ORBITALS

At this point one might be intrigued with the orderliness of the very complex atom: The number of subshells increases with the energy level; each subshell, *s, p, d,* and *f,* can accommodate a specific number of electrons; and the number increases in the order 2, 6, 10, and 14.

One final subdivision of the atom involves the number of *orbitals* in a subshell. An orbital can accommodate two electrons of opposite spins. There is only one orbital in an *s* subshell, three orbitals in a *p* subshell, five orbitals in a *d* subshell, and seven orbitals in an *f* subshell. In a free atom every orbital in a subshell is of equal energy, and the distinction between orbitals of a given subshell lies in the region of space about the nucleus in which the pair of electrons are located. (The geometries of orbitals are described in the next section.)

The experimental basis for the concept of electron spin and orbitals is provided by the behavior of silver atoms in a magnetic field, as observed by O. Stern and W. Gerlach in 1924. A vapor beam of electrically neutral silver atoms, passed through a magnetic field (Figure 5.17), was split into two separate beams of equal intensity. If the silver atom has an electron

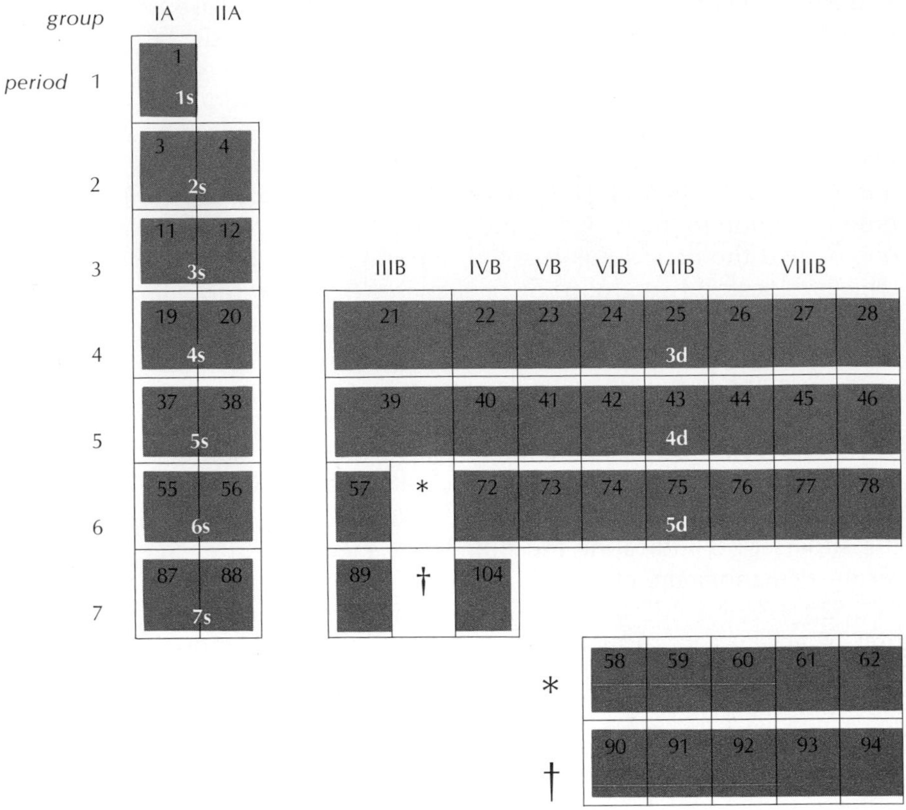

period	group IA	IIA	IIIB		IVB	VB	VIB	VIIB		VIIIB		subshell
1	1											1s
2	3	4										2s
3	11	12										3s
4	19	20	21		22	23	24	25	26	27	28	4s, 3d
5	37	38	39		40	41	42	43	44	45	46	5s, 4d
6	55	56	57	*	72	73	74	75	76	77	78	6s, 5d
7	87	88	89	†	104							7s

*	58	59	60	61	62
†	90	91	92	93	94

Figure 5.16
The relation between the periodic table and electronic configuration.

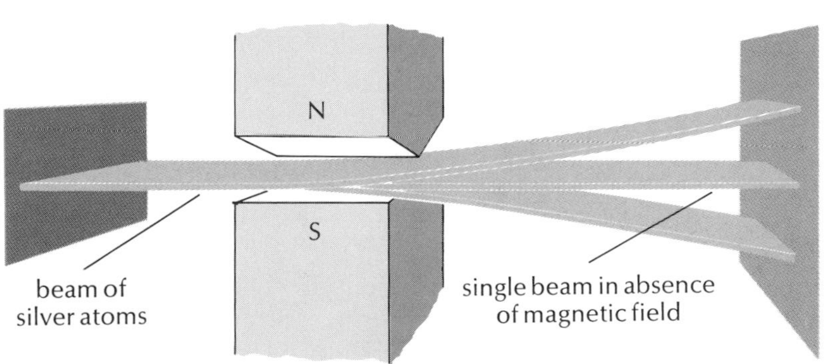

Figure 5.17
The Stern–Gerlach experiment.

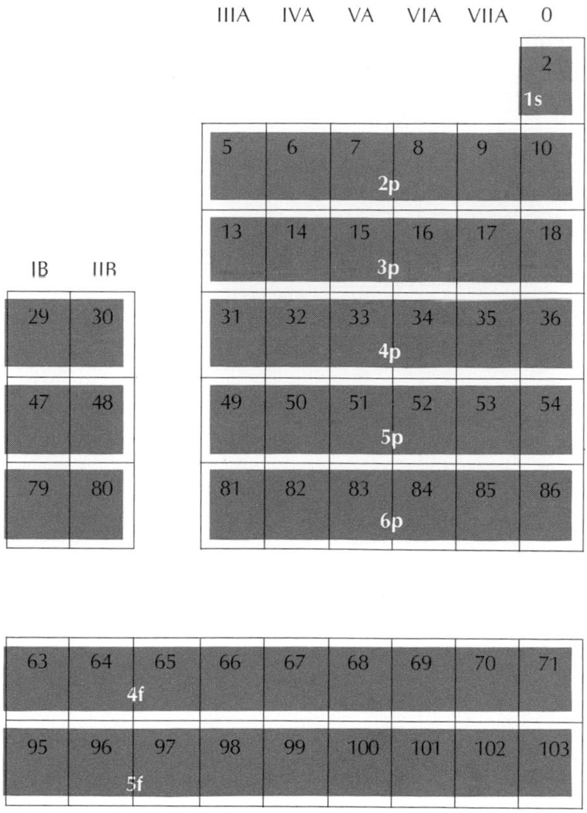

IB	IIB	IIIA	IVA	VA	VIA	VIIA	0
							2 1s
		5	6	7	8 2p	9	10
		13	14	15	16 3p	17	18
29	30	31	32	33	34 4p	35	36
47	48	49	50	51	52 5p	53	54
79	80	81	82	83	84 6p	85	86

63	64 4f	65	66	67	68	69	70	71
95	96 5f	97	98	99	100	101	102	103

that is spinning, a magnetic field would be generated as would be the case for any spinning charged entity. If two directions of spin, clockwise and counterclockwise, are possible, then the tiny atomic magnets will be oriented in opposite directions. Thus silver atoms with an electron spinning in one direction will be attracted to one pole and those with opposite spin to the other pole. Silver is said to be *paramagnetic* because it is affected by magnetic fields.

Many atoms are unaffected by the conditions of the Stern-Gerlach experiment and are called *diamagnetic*. Since all atoms contain electrons, this must mean that certain combinations of electrons are not affected by magnetic fields. The postulate of orbitals allows a convenient explanation for the absence of magnetic field effects. Two electrons of opposite spin in the same orbital would magnetically cancel each other. It automatically

Table 5.1
Electronic configuration of the elements

		1s	*2s*	*2p*	*3s*	*3p*	*3d*	*4s*	*4p*	*4d*	*4f*	*5s*	*5p*	*5d*	*5f*
H	1	1													
He	2	2													
Li	3	2	1												
Be	4	2	2												
B	5	2	2	1											
C	6	2	2	2											
N	7	2	2	3											
O	8	2	2	4											
F	9	2	2	5											
Ne	10	2	2	6											
Na	11	2	2	6	1										
Mg	12	2	2	6	2										
Al	13	2	2	6	2	1									
Si	14	2	2	6	2	2									
P	15	2	2	6	2	3									
S	16	2	2	6	2	4									
Cl	17	2	2	6	2	5									
Ar	18	2	2	6	2	6									
K	19	2	2	6	2	6		1							
Ca	20	2	2	6	2	6		2							
Sc	21	2	2	6	2	6	1	2							
Ti	22	2	2	6	2	6	2	2							
V	23	2	2	6	2	6	3	2							
Cr	24	2	2	6	2	6	5	1							
Mn	25	2	2	6	2	6	5	2							
Fe	26	2	2	6	2	6	6	2							
Co	27	2	2	6	2	6	7	2							
Ni	28	2	2	6	2	6	8	2							
Cu	29	2	2	6	2	6	10	1							
Zn	30	2	2	6	2	6	10	2							
Ga	31	2	2	6	2	6	10	2	1						
Ge	32	2	2	6	2	6	10	2	2						
As	33	2	2	6	2	6	10	2	3						
Se	34	2	2	6	2	6	10	2	4						
Br	35	2	2	6	2	6	10	2	5						
Kr	36	2	2	6	2	6	10	2	6						
Rb	37	2	2	6	2	6	10	2	6			1			
Sr	38	2	2	6	2	6	10	2	6			2			
Y	39	2	2	6	2	6	10	2	6	1		2			
Zr	40	2	2	6	2	6	10	2	6	2		2			
Nb	41	2	2	6	2	6	10	2	6	4		1			
Mo	42	2	2	6	2	6	10	2	6	5		1			
Tc	43	2	2	6	2	6	10	2	6	6		1			
Ru	44	2	2	6	2	6	10	2	6	7		1			
Rh	45	2	2	6	2	6	10	2	6	8		1			
Pd	46	2	2	6	2	6	10	2	6	10					
Ag	47	2	2	6	2	6	10	2	6	10		1			
Cd	48	2	2	6	2	6	10	2	6	10		2			
In	49	2	2	6	2	6	10	2	6	10		2	1		
Sn	50	2	2	6	2	6	10	2	6	10		2	2		
Sb	51	2	2	6	2	6	10	2	6	10		2	3		
Te	52	2	2	6	2	6	10	2	6	10		2	4		
I	53	2	2	6	2	6	10	2	6	10		2	5		
Xe	54	2	2	6	2	6	10	2	6	10		2	6		
		2	8		18			18				8			

		1	2	3	4s	4p	4d	4f	5s	5p	5d	5f	6s	6p	6d	7s
Cs	55	2	8	18	2	6	10		2	6			1			
Ba	56	2	8	18	2	6	10		2	6			2			
La	57	2	8	18	2	6	10		2	6	1		2			
Ce	58	2	8	18	2	6	10	2	2	6			2			
Pr	59	2	8	18	2	6	10	3	2	6			2			
Nd	60	2	8	18	2	6	10	4	2	6			2			
Pm	61	2	8	18	2	6	10	5	2	6			2			
Sm	62	2	8	18	2	6	10	6	2	6			2			
Eu	63	2	8	18	2	6	10	7	2	6			2			
Gd	64	2	8	18	2	6	10	7	2	6	1		2			
Tb	65	2	8	18	2	6	10	9	2	6			2			
Dy	66	2	8	18	2	6	10	10	2	6			2			
Ho	67	2	8	18	2	6	10	11	2	6			2			
Er	68	2	8	18	2	6	10	12	2	6			2			
Tm	69	2	8	18	2	6	10	13	2	6			2			
Yb	70	2	8	18	2	6	10	14	2	6			2			
Lu	71	2	8	18	2	6	10	14	2	6	1		2			
Hf	72	2	8	18	2	6	10	14	2	6	2		2			
Ta	73	2	8	18	2	6	10	14	2	6	3		2			
W	74	2	8	18	2	6	10	14	2	6	4		2			
Re	75	2	8	18	2	6	10	14	2	6	5		2			
Os	76	2	8	18	2	6	10	14	2	6	6		2			
Ir	77	2	8	18	2	6	10	14	2	6	7		2			
Pt	78	2	8	18	2	6	10	14	2	6	9		1			
Au	79	2	8	18	2	6	10	14	2	6	10		1			
Hg	80	2	8	18	2	6	10	14	2	6	10		2			
Tl	81	2	8	18	2	6	10	14	2	6	10		2	1		
Pb	82	2	8	18	2	6	10	14	2	6	10		2	2		
Bi	83	2	8	18	2	6	10	14	2	6	10		2	3		
Po	84	2	8	18	2	6	10	14	2	6	10		2	4		
At	85	2	8	18	2	6	10	14	2	6	10		2	5		
Rn	86	2	8	18	2	6	10	14	2	6	10		2	6		
Fr	87	2	8	18	2	6	10	14	2	6	10		2	6		1
Ra	88	2	8	18	2	6	10	14	2	6	10		2	6		2
Ac	89	2	8	18	2	6	10	14	2	6	10		2	6	1	2
Th	90	2	8	18	2	6	10	14	2	6	10		2	6	2	2
Pa	91	2	8	18	2	6	10	14	2	6	10	2	2	6	1	2
U	92	2	8	18	2	6	10	14	2	6	10	3	2	6	1	2
Np	93	2	8	18	2	6	10	14	2	6	10	4	2	6	1	2
Pu	94	2	8	18	2	6	10	14	2	6	10	6	2	6		2
Am	95	2	8	18	2	6	10	14	2	6	10	7	2	6		2
Cm	96	2	8	18	2	6	10	14	2	6	10	7	2	6	1	2
Bk	97	2	8	18	2	6	10	14	2	6	10	8	2	6	1	2
Cf	98	2	8	18	2	6	10	14	2	6	10	10	2	6		2
Es	99	2	8	18	2	6	10	14	2	6	10	11	2	6		2
Fm	100	2	8	18	2	6	10	14	2	6	10	12	2	6		2
Md	101	2	8	18	2	6	10	14	2	6	10	13	2	6		2
No	102	2	8	18	2	6	10	14	2	6	10	14	2	6		2
Lw	103	2	8	18	2	6	10	14	2	6	10	14	2	6	1	2
		2	8	18	32				32				9			2

follows that an element with an odd atomic number must be paramagnetic. However, it is not valid to conclude that atoms with an even number of electrons are diamagnetic. Some even-numbered atoms have two, four, six, or eight unpaired electrons.

If two electrons have opposite spins, they are magnetically attracted and might be expected to become paired in an orbital, but the like charges of the electrons should electrically repel. The electrical repulsion exceeds the magnetic attraction and, therefore, electrons are of lower energy if they occupy different orbitals in the same subshell and are unpaired. This generalization is known as *Hund's rule.* In the case of the three orbitals of a *p* subshell, the six possible electrons are arranged in the orbitals in a definite manner. All three orbitals are of equal energy and, therefore, the first electron may be in any of the three orbitals. A second electron, even with the opposite spin, will not be in the same orbital as the first because this arrangement is not energetically as favorable as the one that results when the electron is located in one of the other two vacant *p* orbitals. Similarly, a third electron will be in the remaining *p* orbital. Thus, the Group V elements, nitrogen, phosphorus, arsenic, antimony, and bismuth, all have three unpaired electrons. The addition of the other three electrons must involve pairing, because there is no other energetically favorable choice. Therefore, the elements of Groups VI and VII have two and one unpaired electrons, respectively.

A schematic way of illustrating the operation of Hund's rule involves the use of boxes to represent orbitals and of arrows to represent spin. A *p* subshell can be represented by three boxes. The order of filling of these equal energy boxes is illustrated in Figure 5.18.

The five *d* orbitals and the seven *f* orbitals also are filled according to Hund's rule. Therefore, iron is predicted to have four unpaired electrons. The number of unpaired electrons in a given element should be obvious from inspection of the periodic table because it is only necessary to know what type of subshell is being filled and then to determine how many electrons are in the subshell. Proper use of Hund's rule then will give the number of unpaired electrons.

ORBITAL GEOMETRY

Many commercial advertisements contain representations of the atom in terms of a solar system, in which electrons occupy orbits around a nucleus. Unfortunately, this picture (which corresponds to the Bohr model of atoms) is not even approximately correct. The space occupied by an electron about the nucleus can be expressed only in terms of probabilities that are calculated from complex mathematical expressions. The electron is moving through space very rapidly, and its specific position at a given time is less important than the probability that it will be in a specific region at some time.

The probability of locating an electron in a volume is more difficult to represent pictorially than the solar system model. If the probability of

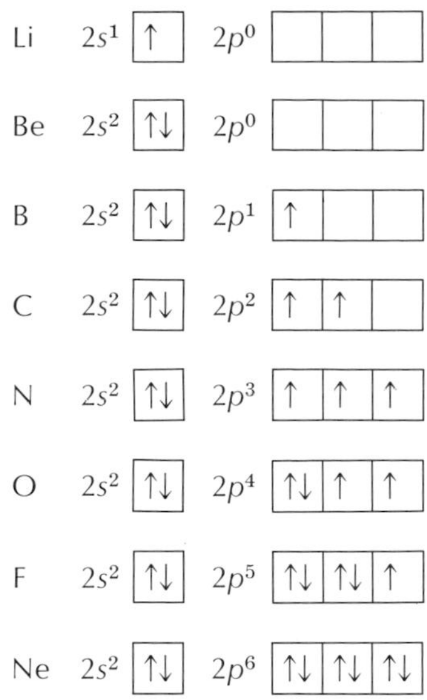

Figure 5.18
The application of Hund's rule to the second period.

finding an electron in a 1*s* orbital in hydrogen is calculated within equal small volumes at a given radius from the nucleus in various directions in space, the probabilities are found to be equal. Thus the probability of finding an electron in a thin spherical shell at a given radius from the nucleus is independent of direction. However, this probability for spherical shells at various distances from the nucleus is a function of radius, as shown in Figure 5.19. Since the 1*s* orbital is spherically symmetrical, it can be pictured

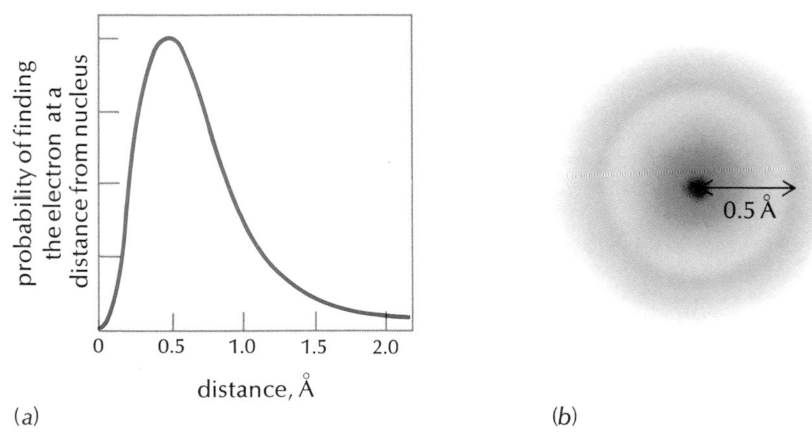

Figure 5.19
The s orbital. (a) The plot of probability vs. distance. (b) The 95 percent probability sphere.

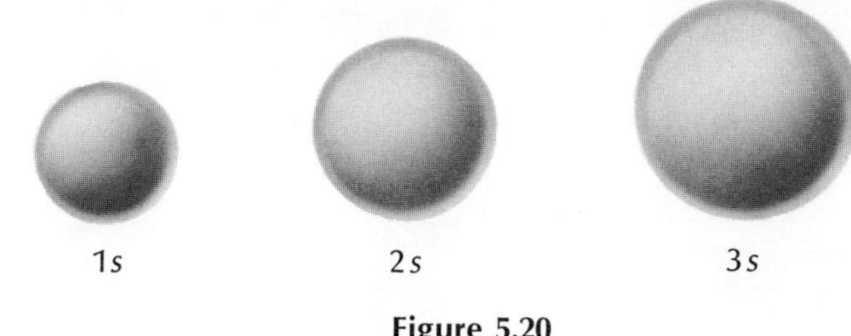

Figure 5.20
The s orbitals.

best as a spherical region of space in which a defined probability of finding an electron can be stated. If the total probability is 95 percent, then the 1*s* orbital is a sphere of a definite radius in which there is a 95 percent chance that the 1*s* electron may be found.

The other *s* orbitals also are spherically symmetrical, with the nucleus at the center of the sphere. There are two differences between a 1*s* electron and an *s* electron in some other energy level. First, the energies of the 2*s*, 3*s*, and 4*s* electrons are progressively higher than the energy of the 1*s* electrons. Second, the probability function varies so that the 95 percent probability sphere encompasses a larger volume of space (Figure 5.20).

The distribution of electrons in the *p* subshell is not spherically symmetrical. There are three orbitals in a *p* subshell; each corresponds to a different region in space. If a coordinate axis system with the nucleus at the center is considered, then the three *p* orbitals, p_x p_y and p_z, are located on the three perpendicular axes and are hourglass shaped (Figure 5.21). The three *p* orbitals are mutually perpendicular, and each can hold two electrons. As the energy level is increased, the size of the *p* orbitals increases in a manner similar to the increase in the size of the *s* orbitals.

The five *d* and seven *f* orbitals also are characterized in terms of geometric shapes. The geometry of the *d* orbitals is illustrated in Chapter

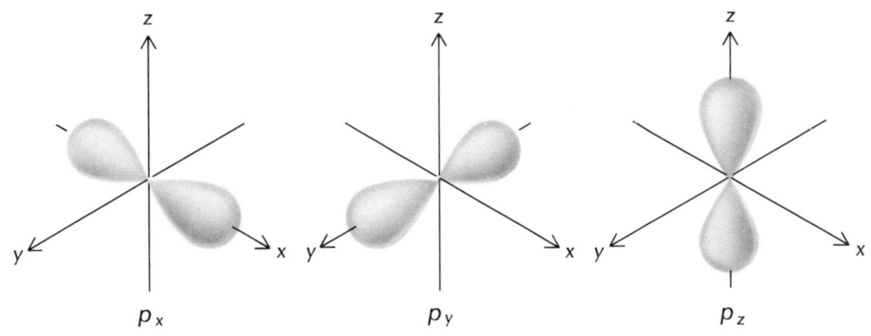

Figure 5.21
The p orbitals.

14. The *f* orbitals of even more complex shapes will not concern us in this text because the chemistry in which these orbitals play a part is very specialized.

5.5
PROPERTIES OF THE ELEMENTS

ATOMIC RADII

The definition of the size of an atom is difficult because the electron probability distribution is never zero at any distance from the nucleus. Furthermore, the distribution is known to be affected by neighboring atoms and it changes from compound to compound. Nevertheless, an arbitrary distance for the boundary of atoms is chosen in a consistent manner, and the resultant radii can then be compared qualitatively. Table 5.2 lists the sizes of atoms in angstrom units (1 angstrom, Å = 10^{-8} cm).

The atomic radii decrease from left to right in a period of the periodic table. Within a period the nuclear charge increases in the same direction, and this increase should tend to draw the electrons toward the nucleus; the probability of finding an electron then becomes higher in a smaller volume.

Going from top to bottom in a family of the periodic table, the atomic radii increase. Each successive member has one additional energy level containing electrons. Because the size of an orbital increases with the number of its energy level, the size of the atom tends to increase. The increase in orbital size is partially balanced by the increase in the nuclear charge.

Consideration of changes in atomic radii within periods and families is complicated by the fact that as successive elements are compared both the number of electrons and the nuclear charge increase. To simplify, we can compare series of ions in which the nuclear charge varies, but the electronic configuration is constant. Consider three common ions that have the neon configuration: (1) the O^{2-} ion, which results from the addition of two electrons to oxygen; (2) the F^- ion, which results from the addition of one electron to fluorine; and (3) the Na^+ ion, which results from the loss of one electron from sodium. The electronic configurations of O^{2-}, F^-, Ne, and Na^+ are identical, and their radii are as follows:

$\mathbf{O^{2-}}$	1.40 A
$\mathbf{F^-}$	1.36 Å
Ne	1.12 Å
$\mathbf{Na^+}$	0.95 Å

The decrease is clearly the result of increasing nuclear charge.

IONIZATION POTENTIALS

One of the best indications that the assigned electronic configurations of atoms are correct is derived from the ionization potential, an energy

Table 5.2
Atomic radii of the elements (A)

1 H **0.37**								
3 Li **1.23**	4 Be **0.89**							
11 Na **1.57**	12 Mg **1.36**							
19 K **2.03**	20 Ca **1.74**	21 Sc **1.44**	22 Ti **1.32**	23 V **1.22**	24 Cr **1.17**	25 Mn **1.17**	26 Fe **1.17**	27 Co **1.16**
37 Rb **2.16**	38 Sr **1.91**	39 Y **1.62**	40 Zr **1.45**	41 Nb **1.34**	42 Mo **1.29**	43 Tc	44 Ru **1.24**	45 Rh **1.25**
55 Cs **2.35**	56 Ba **1.98**	*	72 Hf **1.44**	73 Ta **1.34**	74 W **1.30**	75 Re **1.28**	76 Os **1.26**	77 Ir **1.26**
87 Fr	88 Ra	†						

*	57 La **1.69**	58 Ce **1.65**	59 Pr **1.65**	60 Nd **1.64**	61 Pm	62 Sm **1.66**	63 Eu **1.85**
†	89 Ac	90 Th **1.65**	91 Pa	92 U **1.42**	93 Np	94 Pu	95 Am

term. If sufficient energy is added to a gaseous atom, an electron may be removed to form a positive ion or cation. The process of electron removal is called *ionization,* and the amount of energy required to accomplish the removal is called the *ionization potential.* Ionization potentials are expressed in electron volts (Table 1.6) per atom. The ionization potential of gaseous lithium, Li(g), is 5.4 electron volts (eV) per atom:

$$\mathbf{Li}(g) \longrightarrow \mathbf{Li}^{+}(g) + 1e^{-}$$

The lithium cation has a superscript plus sign to indicate its unit positive charge. Since the ionization potential is a measure of the energy required to remove an electron from an element, it indicates the energy of the electron in the element. Higher energy electrons require less additional energy (lower ionization potential) to dissociate them from the element; conversely, low-energy electrons are closely bound to the atom and require more energy for ionization.

								2 He
			5 B **0.80**	6 C **0.77**	7 N **0.74**	8 O **0.74**	9 F **0.72**	10 Ne
			13 Al **1.25**	14 Si **1.17**	15 P **1.10**	16 S **1.04**	17 Cl **0.99**	18 Ar
28 Ni **1.15**	29 Cu **1.17**	30 Zn **1.25**	31 Ga **1.25**	32 Ge **1.22**	33 As **1.21**	34 Se **1.17**	35 Br **1.14**	36 Kr
46 Pd **1.28**	47 Ag **1.34**	48 Cd **1.41**	49 In **1.50**	50 Sn **1.41**	51 Sb **1.41**	52 Te **1.37**	53 I **1.33**	54 Xe
78 Pt **1.29**	79 Au **1.34**	80 Hg **1.44**	81 Tl **1.55**	82 Pb **1.54**	83 Bi **1.52**	84 Po **1.53**	85 At	86 Rn

64 Gd **1.61**	65 Tb **1.59**	66 Dy **1.59**	67 Ho **1.58**	68 Er **1.57**	69 Tm **1.56**	70 Yb **1.70**	71 Lu **1.56**
96 Cm	97 Bk	98 Cf	99 Es	100 Fm	101 Md	102 No	103 Lw

The ionization potentials can be correlated with the model of atomic structure. Among the elements of the second period, a general increase in ionization potential is noted in proceeding from lithium to neon, as indicated in Table 5.3. The electron removed always will be one from the second energy level, because the 1*s* electrons are of lower energy and more difficult to ionize. Why, then, does it become harder to pull off an electron within the period? The nuclear charge increases within the period from +3 to +10, and the electrostatic attraction of the nucleus increases proportionally. This factor would increase the energy required to ionize an electron. There also is a decrease in atomic radius within the period, which leads to the same effect.

There are two irregularities in the ionization potentials of the second period. The first occurs between beryllium and boron. Beryllium has a filled 2*s* subshell, whereas boron has one electron in the 2*p* subshell. The lower ionization potential for boron results from the fact that the electron to

Table 5.3
Ionization potential of the elements (eV)

1 H **13.6**								
3 Li **5.4**	4 Be **9.3**							
11 Na **5.1**	12 Mg **7.6**							
19 K **4.3**	20 Ca **6.1**	21 Sc **6.5**	22 Ti **6.8**	23 V **6.7**	24 Cr **6.8**	25 Mn **7.4**	26 Fe **7.9**	27 Co **7.9**
37 Rb **4.2**	38 Sr **5.7**	39 Y **6.4**	40 Zr **6.8**	41 Nb **6.9**	42 Mo **7.1**	43 Tc **7.3**	44 Ru **7.4**	45 Rh **7.5**
55 Cs **3.9**	56 Ba **5.2**	*	72 Hf **7**	73 Ta **7.9**	74 W **8.0**	75 Re **7.9**	76 Os **8.7**	77 Ir **9**
87 Fr	88 Ra **5.3**	†						

*	57 La **5.6**	58 Ce **6.9**	59 Pr **5.8**	60 Nd **6.3**	61 Pm	62 Sm **5.6**	63 Eu **5.7**
†	89 Ac **6.9**	90 Th	91 Pa	92 U **4**	93 Np	94 Pu	95 Am

be removed comes from a $2p$ orbital, which is of higher energy than the $2s$ orbital. The second irregularity occurs with nitrogen, whose ionization potential is higher than might have been expected owing to the fact that nitrogen with electron configuration $1s^22s^22p^3$ has a half-filled $2p$ subshell. This half-filled subshell has extra stability, apparently because of the fact that a half-filled subshell has one electron in each orbital. In addition to the two irregularities, the large jump in the ionization potential of neon should be noted. Again this reflects the great stability of a filled subshell (and energy level).

Within a group there is a trend of decreasing ionization potentials that is in agreement with the increasing size of the atoms from top to bottom within a group. However, the opposing nuclear charge effect also must be considered. If a nuclear charge change from +3 in lithium to +10 in neon causes an increase of approximately 16 electron volts (eV), then the +55 nuclear charge of cesium should lead to a dramatic increase

									2 He **24.6**
			5 B **8.3**	6 C **11.3**	7 N **14.5**	8 O **13.6**	9 F **17.4**	10 Ne **21.6**
			13 Al **6.0**	14 Si **8.2**	15 P **11.0**	16 S **10.4**	17 Cl **13.0**	18 Ar **15.8**
28 Ni **7.6**	29 Cu **7.6**	30 Zn **9.4**	31 Ga **6.0**	32 Ge **8.1**	33 As **9.8**	34 Se **9.8**	35 Br **11.8**	36 Kr **14.0**
46 Pd **8.3**	47 Ag **7.6**	48 Cd **9.0**	49 In **5.8**	50 Sn **7.3**	51 Sb **8.6**	52 Te **9.0**	53 I **10.5**	54 Xe **12.1**
78 Pt **9.0**	79 Au **9.2**	80 Hg **10.4**	81 Tl **6.1**	82 Pb **7.4**	83 Bi **7.3**	84 Po **8.4**	85 At	86 Rn **10.7**

64 Gd **6.2**	65 Tb **6.7**	66 Dy **6.8**	67 Ho	68 Er	69 Tm	70 Yb **6.2**	71 Lu **5.0**
96 Cm	97 Bk	98 Cf	99 Es	100 Fm	101 Md	102 No	103 Lw

in its ionization potential. That this increase is not observed is the result of electronic screening. The 6*s* electron of cesium that is to be removed does not effectively feel the +55 nuclear charge because 54 lower energy electrons are providing a balancing, intervening screen.

One last consideration in the relation of ionization potentials to electronic configuration is the effect of maintaining nuclear charge and altering the number of electrons. This effect can be observed by successive ionization of electrons. The ionization potential for the removal of a second electron is always higher than for the first. That this must be so is evident if we consider the following two equations:

$$\mathbf{Na}(g) \longrightarrow \mathbf{Na}^+(g) + 1e^- \qquad \text{ionization potential} = 5.1 \text{ eV}$$
$$\mathbf{Na}^+(g) \longrightarrow \mathbf{Na}^{2+}(g) + 1e^- \qquad \text{ionization potential} = 47.3 \text{ eV}$$

In the second ionization it is necessary to remove an electron from a positively charged ion. Thus the second ionization potential must be higher than the first even if the second electron is in the same energy level as the first. In the case for sodium the second ionization potential also reflects the difficulty of removing a 2*p* electron in Na^+ as compared to a 3*s* electron in Na. The ionization potentials for sodium and other elements of the third period are given in Table 5.4. The dramatic increase in ionization potential indicated by the "stair steps" is noted at the points at which it is necessary to remove electrons from the second energy level instead of the third.

Table 5.4
Ionization potentials of third period elements (eV)

	1st	*2nd*	*3rd*	*4th*	*5th*	*6th*	*7th*	*8th*
Na	5.1	47.3	71.7	98.9	138.6	172.4	208.4	264.2
Mg	7.6	15.0	80.1	109.3	141.2	186.9	225.3	266.0
Al	6.0	18.8	28.4	120.0	153.8	190.4	241.9	285.6
Si	8.1	16.3	33.5	45.1	166.7	205.1	246.4	303.9
P	11.6	20.0	30.2	51.4	65.0	220.4	263.3	309.3
S	10.4	23.4	35.0	47.3	72.6	88.0	281.0	328.8
Cl	13.0	23.8	39.9	53.5	67.8	96.7	114.3	348.3
Ar	15.8	27.6	40.9	59.8	75.0	91.3	124.0	143.5

ELECTRON AFFINITY

The energy released when an electron is added to an atom is called its *electron affinity*. When the fluorine atom accepts one electron, 3.6 eV are released and a negative ion (anion), the fluoride ion, is formed.

$$F + 1e^- \longrightarrow F^-$$

The electron affinity is a measure of the tightness with which the additional electron is bound to the atom. The greater the energy released, the greater the "willingness" of the atom to accept electrons. Elements in Group VII would be expected to have high electron affinities because the addition of one electron completes a subshell. In addition, the atomic radii of the elements decrease in going from left to right in a given period and, therefore, the addition of an electron to an element in Group VII involves a closer approach to the positive nucleus.

Unfortunately, electron affinities are difficult to measure. However, an empirical approach termed *electronegativity* has been devised. Electronegativity is the measure of the ability of an element to attract electrons in the presence of a second element. Its significance and the method of its determination are discussed in the next chapter.

Aston, F. W., "Isotopes and Atomic Weights," in W. C. Dampier and M. Dampier (eds.), *Readings in the Literature of Science*. New York: Harper & Row, 1959.

Berry, R. S., "Atomic Orbitals," *J. Chem. Educ.*, **43,** 283 (1966).

Birks, J. B., *Rutherford at Manchester*. New York: W. A. Benjamin, 1963.

Cohen, I., "The Shape of the 2*p* and Related Orbitals," *J. Chem. Educ.*, **38,** 20 (1961).

Devault, D. C., "Electronic Structure of the Atom," *J. Chem. Educ.*, **21,** 526 (1944), and **21,** 575 (1944).

Eve, A. S., *Rutherford*. New York: Macmillan, 1939.

Garrett, A. B., "The Flash of Genius: The Neutron Identified: Sir James Chadwick," *J. Chem. Educ.*, **39,** 638 (1962).

Garrett, A. B., "The Flash of Genius: The Bohr Atomic Model: Niels Bohr," *J. Chem. Educ.*, **39,** 534 (1962).

Garrett, A. B., "The Nuclear Atom: Sir Ernest Rutherford," *J. Chem. Educ.*, **39,** 287 (1962).

Hammond, C. R., "Collecting the Chemical Elements," *J. Chem. Educ.*, **41,** 401 (1964).

Redfern, R. N., and Salmon, J. E., "Periodic Classification of the Elements," *J. Chem. Educ.*, **39,** 41 (1962).

van Spronsen, J. W., "The Prehistory of the Periodic System of the Elements," *J. Chem. Educ.*, **36,** 565 (1959).

Thomson, J. J., *Recollections and Reflections*. New York: Macmillan, 1937.

Weeks, M. E., *Discovery of the Elements*. Easton, Pennsylvania: Chemical Education, 1956.

Terms and concepts

anion
Aston
atomic number
atomic radius
Bohr
cation
diamagnetic
Dobereiner's triads
electron
electron affinity
electronegativity
electronic configuration
energy level
Goldstein
group
Hund's rule
ion
ionization potential
isotope
mass spectrometer
Mendeleev
metal
Millikan
neutron
Newlands' octaves
nonmetal
nucleus
orbital
orbital geometry
paramagnetic
period
periodic law
periodic table
proton
quantum
Rutherford
shell
Stern and Gerlach
subshell
Thomson

1. What relationships exist between the members of the triads Li, Na, K; S, Se, Te; and Ca, Sr, Ba?

2. What is the mass of an atom of a lithium isotope that has three protons and four neutrons? If one electron is removed from the lithium atom, what will be the charge-to-mass ratio of the resultant lithium cation?

3. What is the q/m ratio for the fluoride ion F^- formed by the addition of one electron to the fluorine atom?

4. Deuterium, which is an isotope of hydrogen, has one neutron in its nucleus. Will the q/m ratio of the deuterium cation be equal to, greater than, or less than the q/m ratio of the hydrogen cation?

5. Write out the electronic configurations for aluminum, titanium, germanium, and palladium.

6. How many electrons are present in the indicated subshell of each of the following elements:

a. $2p$ of O
b. $4s$ of Ca
c. $3d$ of Ni
d. $5p$ of Sn
e. $4p$ of Te

7. How many unpaired electrons are there in the indicated subshells in Problem 6.

8. The electronic configuration of chromium is such that there is one electron in the $4s$ subshell and five electrons in the $3d$ subshell. How does this differ from the simple picture expected from discussions in this chapter? How can the observed configuration be rationalized as energetically favorable?

9. How can it be shown that naturally occurring chlorine consists of two isotopes?

10. What are the experimental observations on which the concept of quantized electronic energy levels is based?

11. In each of the following pairs of elements, which has the higher ionization potential?

a. O or S
b. Ge or Se
c. Mg or Rb
d. Xe or Cs
e. Ne or Kr
f. P or Si

12. Arrange the elements oxygen, chlorine, lithium, selenium, bromine, and rubidium according to increasing ionization potential on the basis of their positions in the periodic table.

13. Why does bromine have a higher electron affinity than iodine?

14. The existence of the neutron was suspected for several years prior to its discovery. What two facts indicated that the neutron must exist?

15. The first, second, third, and fourth ionization potentials for beryllium are 9.3, 18.2, 153.9, and 217.7 eV, respectively. What do these values indicate about the electronic configuration of beryllium?

16. The atomic radii of H^-, He, and Li^+ are 2.0, 0.9, and 0.6 Å, respectively. How can these values be rationalized in terms of atomic structure?

17. Arrange the atoms helium, chlorine, selenium, argon, and krypton according to decreasing atomic radii on the basis of their positions in the periodic table.

18. The gram-atomic volume of an element is defined as the volume occupied by 1 mole of the element in the solid state. The gram-atomic volumes parallel the atomic radii. Why is there a correlation between atomic radii and gram-atomic volume?

19. Naturally occurring silicon consists of a mixture of three isotopes. The masses of the three isotopes, which occur in 92.2, 4.7, and 3.1 percent relative abundances, are 27.98, 28.98, and 29.97 amu, respectively. Show why the accepted atomic mass of silicon is 28.08 amu.

20. Explain the reason why Hund's rule is followed in filling orbitals.

21. In terms of electronic structure, why do the elements in a given group resemble each other chemically?

22. Why are there not 18 elements in the third period of the periodic table?

23. It is believed that hydrogen is present in the sun. How could this be verified by instruments on earth?

24. What is meant by an atomic orbital? Describe the shapes of 2*s* and 2*p* orbitals and compare them with the 3*s* and 3*p* orbitals.

25. What observation makes it necessary to postulate orbitals and electron spins?

26. It is theoretically predictable that the fifth energy level has five subshells. Why would you intuitively feel this should be the case? How many orbitals should be in the subshell of highest energy in the fifth energy level? How many electrons can this new subshell contain? How many electrons can the fifth energy level contain?

6

CHEMICAL BONDING

Chapter 5 summarized the basic features of atomic structure and related them to the physical and chemical properties of the elements. With this background, the nature of compounds, which is the main concern of chemistry, can be presented. In chemical reactions elements combine to form compounds, a process that involves the formation of bonds. The term *bond* is used to describe the manner in which atoms are held together in polyatomic aggregates (molecules). Studying the vast number of known compounds, which exceeds several million and comprises the great majority of known substances, is a far more imposing problem than studying the 103 known elements. For this reason we shall look for general principles that describe the formation and properties of classes of compounds.

How do atoms combine to form compounds? Why do they combine rather than remain as unassociated elements? The objective of this chapter is to answer in part these two questions and to develop some principles to provide a foundation for predicting the formation of unknown compounds.

With the presentation of the accepted framework for molecular structure, the physical and chemical properties of compounds can be examined in terms of the models for microscopic complex matter. The physical properties are discussed in Chapter 7, while the chemical properties are treated throughout the remainder of this text.

6.1
EARLY CONCEPTS OF BONDING

In the seventeenth and eighteenth centuries elements were pictured as having a combining capacity related to their gross shapes—a concept akin to the relationship of a lock and key. Each atom was seen as shaped so that it formed a stable arrangement with some number of other atoms. However, no experimental verification for such a macroscopic analogy on a microscopic level has ever been found.

The beginning of our modern theory of bonding can be traced to the concept of *valence* introduced in 1850. Each element was said to have a valence equal to its *combining capacity*. The number of hydrogen or chlorine atoms with which another atom combines is called its combining capacity. The valence of these two reference elements was set as 1. Therefore, oxygen, which reacts with hydrogen to produce H_2O, was said to have a valence of 2. By using such definitions it was possible to predict the formulas of some compounds. For example, it would be expected that oxygen, with a valence of 2, and chlorine, with a valence of 1, would combine to produce OCl_2, which actually is a known molecule.

As a starting point the concept of valence was extremely useful, although some elements exhibit multiple valences. Oxygen combines with hydrogen to form a compound, H_2O_2, called hydrogen peroxide, in addition to H_2O. Nitrogen combines with hydrogen to form either ammonia, NH_3, or hydrazine, N_2H_4. Under such circumstances the combination of several elements having multiple valences can lead to many compounds, and the predictive aspect of the valence concept is lost. The series of compounds containing nitrogen and oxygen alone illustrates the point, because N_2O, NO, NO_2, N_2O_3, N_2O_4, and N_2O_5 are all nitrogen oxides.

The term *valence* is no longer used to describe the attachment of atoms to each other in a molecule because it explains nothing; it simply provided a means of restating what was already known, that is, the molecular formula of a compound.

6.2
WHY ATOMS COMBINE

To ask why atoms combine is much like asking why any physical phenomenon occurs. Invariably it is necessary to return to fundamental truths that must be accepted as a natural fact of life. One of the most basic of these is that all natural systems tend to lose potential energy and become more stable. Other things being equal, a system that has stored potential energy is less stable than a system that has none. The term *stable* refers to systems that do not easily undergo spontaneous change. Unstable systems either change spontaneously to stable systems or do so under conducive conditions. For example, water on the middle of a slope has a higher potential energy than water at the bottom of a hill. If there are no obstructions, the water will flow downhill to the more stable position, releasing energy in the process. In a similar manner elements have stored potential energy and may, under the proper conditions, release this energy.

Most reactions that form stable compounds liberate energy. The energy liberated or absorbed in a physical or chemical process is called the *enthalpy change* and is symbolized by ΔH. The symbol Δ, the Greek capital letter delta, is generally used to denote a difference. Thus ΔH is the difference between the enthalpy of the final state, H_f, and the enthalpy of the original state, H_i.

$$\Delta H = H_f - H_i$$

By definition, if energy is liberated, the sign of ΔH is negative. The enthalpy of a reaction represents the potential energy lost by the system in going from an unstable to a stable condition. Therefore, most chemical reactions that proceed to yield stable compounds have negative enthalpy changes.

There is another natural law that applies to all physical and chemical processes: Particles tend to occupy as much space as is available to them and to assume a state of maximum disorder. An example of this phenomenon is the expansion of a gas into an evacuated space. For a spontaneous change the disorder of a system increases and the *entropy S* is said to have increased in a positive sense. The change in entropy for a system is symbolized as ΔS. There is a tendency, all other things being equal, for chemical reactions to occur to produce a positive change in entropy. In chemical reactions as well as physical changes, the two effects of enthalpy (energy) loss and entropy gain contribute to produce a stable state. A piece of wood that is burned releases energy and loses its high degree of order. In some cases the two effects of enthalpy and entropy may oppose each other and still lead to a net change. Compounds are often more ordered than a collection of their elements. However, the energy released when the elements combine outweighs the drive to maintain disordered elements. The quantitative study of the balance of enthalpy and entropy terms is called *thermodynamics*. This subject entails complexities requiring entire texts to explain, which in turn require an extensive mathematical background. There is no need to proceed further with this subject in order to comprehend why chemical reactions occur. Chemical reactions are controlled by enthalpy and entropy changes as are all other observable processes.

The stability of gaseous compounds relative to the constituent gaseous atoms is indicated by the *bond dissociation energy,* the energy required to dissociate the bond. Usually the bond dissociation energy, which is a change in enthalpy term, is expressed in kilocalories per mole of bonds

Table 6.1
Bond dissociation energies (kcal/mole)

F_2	36	HF	135
Cl_2	59.2	HCl	103
Br_2	46.1	HBr	87.4
I_2	36.1	HI	71.4
H_2	103.2	CO	256
O_2	119	NO	150
N_2	226		

broken. In Table 6.1 the bond dissociation energies of some diatomic molecules are listed. The bond dissociation energy is indicative of the type of bonding present in the compound. Therefore, trends among related compounds such as those of hydrogen and the Group VII elements must be accounted for by the models proposed for bonding.

6.3
BONDING TYPES

NATURE OF CHEMICAL BONDS

As this chapter shows, the chemical bond results from a change in the electronic structures of atoms that are associated with each other in a molecule. Since there are 103 known elements, the number of combinations for possible bonds in compounds is enormous. It is convenient to describe the large variety of possible bonds in terms of two basic types or models of bonds. These models are called *covalent* and *ionic bonds* and represent limiting cases. Most real bonds fall in between these two models and are described in terms of one of the models depending on the degree to which the characteristics of that particular bond resemble the model. This approach is fraught with difficulties, as it is not unlike having to sort a warehouse full of potatoes into two separate piles of large potatoes and small potatoes. Although there will always be boundary decisions to make, the problem cannot be alleviated by creating a third classification. In a similar manner the classification of bond types is a difficult problem that cannot be made easier by describing additional models.

Because chemical bonding is an electronic phenomenon, it is instructive to consider the electronic structure of atoms with respect to their chemical reactivity. The gases of Group 0 are noted for their lack of chemical reactivity. There are no known compounds of helium, neon, and argon. The number of compounds of krypton and xenon is very small. Why are these elements so unreactive toward other elements? All these gases have electronic structures that consist of filled subshells. Except for helium, whose configuration is $1s^2$, the *s* and *p* subshells of the highest energy level contain a total of eight electrons. These elements require a large amount of energy to separate an electron from an atom; that is, they have a high ionization potential. The gases also have very low affinities for electrons, which would have to be located in a higher energy level. The particular stability of the filled level suggests that under the proper circumstances atoms of other elements might gain or lose electrons to achieve an inert gas electronic configuration. The tendency of atoms to achieve the inert gas electronic configuration, ns^2np^6, is referred to as the *Lewis octet rule*. Atoms of the second period, while they may form compounds with fewer than eight electrons associated in bonding, do not form compounds in which more than eight electrons are involved. Elements of other periods also tend to behave in such a manner as to achieve an octet of electrons; however, they often violate the octet rule and have 10 or 12 electrons involved in bonding.

Oxygen, fluorine, sodium, and magnesium can achieve the inert gas configuration of neon by, respectively, gaining two electrons, gaining one electron, losing one electron, and losing two electrons:

$$2e^- + \mathbf{O}\ (1s^22s^22p^4) \longrightarrow \mathbf{O}^{2-}\ (1s^22s^22p^6)$$
$$1e^- + \mathbf{F}\ (1s^22s^22p^5) \longrightarrow \mathbf{F}^-\ (1s^22s^22p^6)$$
$$\mathbf{Na}\ (1s^22s^22p^63s^1) \longrightarrow \mathbf{Na}^+\ (1s^22s^22p^6) + 1e^-$$
$$\mathbf{Mg}\ (1s^22s^22p^63s^2) \longrightarrow \mathbf{Mg}^{2+}\ (1s^22s^22p^6) + 2e^-$$

In a similar manner, sulfur, chlorine, potassium, and calcium can achieve the inert gas configuration of argon. By an extension of these examples to other elements, it is reasonable to postulate that any element of Groups I, II, VI, and VII could achieve an inert gas configuration if the necessary energy requirements were met.

The concept of the stability of filled subshells can be extended to other elements. The metals with electrons in the *d* subshells can achieve inert gas configurations by losing electrons. Zinc and the other members of the Group IIb can lose two electrons from the highest energy level, leaving the filled lower level unperturbed. Similarly, indium can lose three electrons:

$$\mathbf{Zn}\ (1s^22s^22p^63s^23p^64s^23d^{10}) \longrightarrow \mathbf{Zn}^{2+}\ (1s^22s^22p^63s^23p^63d^{10}) + 2e^-$$
$$\mathbf{In}\ (1s^22s^22p^63s^23p^64s^23d^{10}4p^65s^24d^{10}5p^1) \longrightarrow$$
$$\mathbf{In}^{3+}\ (1s^22s^22p^63s^23p^64s^23d^{10}4p^64d^{10}) + 3e^-$$

The above examples resemble inert gas electronic configurations except for the presence of an additional filled *d* subshell. The chemistry of the elements with unfilled *d* subshells is discussed in Chapter 14.

IONIC BONDING

Two atoms may react via an electron transfer process to create ions with inert gas configurations that are attracted to each other. In such a process an atom with a low attraction for its highest energy electron loses it to an atom with a high electron affinity. Typically this occurs most easily with elements of Groups I, II, VI, and VII, which are within two electrons of achieving an inert gas configuration. With electron transfer the atoms be-

Na· Mg: Al: ·Si: ·P: ·S: :Cl: :Ar:

Figure 6.1
Electron dot representation of third row elements.

come ions and the resulting force of attraction between the positive and negative ions is called an *ionic bond.*

Sodium reacts with chlorine to form the compound sodium chloride NaCl. The sodium atom may be represented by an electron dot notation Na·, in which the chemical symbol Na represents the nucleus and all of the lower energy level electrons. The dot represents the single 3*s* electron (Figure 6.1). Chlorine is written as :Cl·, where the seven dots represent the two electrons in the 3*s* and the five electrons in the 3*p* subshells. Sodium can achieve a filled-shell arrangement (Lewis octet) by the transfer of one electron to chlorine, which simultaneously achieves a filled-shell arrangement:

$$\text{Na}\cdot + :\ddot{\underset{..}{\text{Cl}}}\cdot \longrightarrow \text{Na}^+ :\ddot{\underset{..}{\text{Cl}}}:^-$$

Sodium chloride is more stable than the isolated sodium and chlorine atoms. This might suggest that the energy released by chlorine in gaining one electron is greater than the energy required to remove one electron from sodium. Actually this is not the case. The ionization potential for sodium is 5.1 eV, whereas chlorine liberates only 3.8 eV in gaining an electron. Thus 1.3 eV of energy should be required for the process depicted by the above equation. Sodium and chlorine react vigorously with each other to release energy, and this experimental observation suggests that some factor other than ionization potential or electron affinity must be responsible for the observed release of energy. The bringing together of positively charged and negatively charged species releases energy. This energy, when added to the ionization potential and electron affinity terms, provides the net total observed release of energy:

$\text{Na}\cdot$	$\longrightarrow \text{Na}^+ + 1e^-$	5.1 eV required
$1e^- + \cdot\ddot{\underset{..}{\text{Cl}}}\cdot$	$\longrightarrow :\ddot{\underset{..}{\text{Cl}}}:^-$	3.8 eV liberated
$\text{Na}^+ + :\ddot{\underset{..}{\text{Cl}}}:^-$	$\longrightarrow \text{Na}^+:\ddot{\underset{..}{\text{Cl}}}:^-$	7.0 eV liberated
$\text{Na}\cdot + :\ddot{\underset{..}{\text{Cl}}}\cdot$	$\longrightarrow \text{Na}^+:\ddot{\underset{..}{\text{Cl}}}:^-$	5.7 eV liberated

Magnesium and oxygen react to form magnesium oxide, MgO. The ratio in which these two elements combine is consistent with the electron dot notation represented below:

$$\text{Mg}: + \cdot\dot{\underset{..}{\text{O}}}: \longrightarrow \text{Mg}^{2+} :\ddot{\underset{..}{\text{O}}}:^{2-}$$

Note that the two valence electrons of magnesium are written together to emphasize the fact that they are paired in the 3*s* subshell. Similarly, oxygen is written with two separate electrons and two pairs to agree with the $2s^2 2p^4$ configuration in which there are two unpaired electrons.

In compounds containing ionic bonds, the proportions of the atoms of each element that will react can be predicted by the use of the periodic table. The position in the periodic table indicates how many electrons must be gained or released to achieve the inert gas structure. Elements of Groups I and II lose one electron and two electrons, respectively. Elements of Groups VI and VII gain two electrons and one electron, respectively:

$$2\text{Na}\cdot + \cdot\dot{\underset{\cdot\cdot}{\text{O}}}: \longrightarrow 2\text{Na}^+ + :\ddot{\underset{\cdot\cdot}{\text{O}}}:^{2-}$$

$$\text{Mg}: + 2:\ddot{\underset{\cdot\cdot}{\text{Cl}}}\cdot \longrightarrow \text{Mg}^{2+} + 2:\ddot{\underset{\cdot\cdot}{\text{Cl}}}:^-$$

COVALENT BONDING

The second major class of bonds cannot be explained by a complete electron transfer process. In the case of simple molecules such as H_2, F_2, and O_2, the bonds are holding together identical atoms. Obviously, since there is no difference in the ionization potentials or electron affinities, there is no energetic reason for electron transfer to occur.

The hydrogen molecule is composed of two hydrogen atoms, each with one electron in a 1*s* orbital. In order to achieve the helium electron configuration each hydrogen would have to accept one electron. The electron that one hydrogen needs cannot be obtained from another atom identical to itself. In such a case it is more feasible energetically for the two atoms to join in such a way that their two electrons are shared. Each of the two atoms can be said to possess both electrons as long as they are together. When atoms are associated by means of one or more pairs of electrons, the atoms are covalently bound. *Covalent bonds* can be regarded as shared electron pairs. The bonding of two hydrogen atoms to form a hydrogen molecule is pictured below, using the electron dot notation:

$$\text{H}\cdot + \text{H}\cdot \longrightarrow \text{H}:\text{H}$$

The fluorine molecule is pictured as two fluorine atoms joined by a mutually shared pair of electrons. Each fluorine atom requires one electron to fill the second energy level. Since there is no difference between the electron donating or accepting properties of the two atoms, each must contribute one electron:

$$:\ddot{\underset{\cdot\cdot}{\text{F}}}\cdot + :\ddot{\underset{\cdot\cdot}{\text{F}}}\cdot \longrightarrow :\ddot{\underset{\cdot\cdot}{\text{F}}}:\ddot{\underset{\cdot\cdot}{\text{F}}}:$$

In representing covalently bound molecules it is convenient to denote a pair of electrons with a dash. The hydrogen and fluorine molecules are more easily written using this convention:

$$\text{H—H} \qquad |\overline{\underline{\text{F}}}\text{—}\overline{\underline{\text{F}}}|$$

More than one pair of electrons can be shared between pairs of atoms if the sharing process leads to a stable molecule. If four or six electrons are shared the bonds are called *double* and *triple* bonds, respectively. The nitrogen molecule consists of two nitrogen atoms bound together by six shared electrons, or a triple bond:

$$\cdot\ddot{\underset{\cdot}{N}}\cdot + \cdot\ddot{\underset{\cdot}{N}}\cdot \longrightarrow :N:::N: \quad \text{or} \quad |N{\equiv}N|$$

The shorthand conventions for depicting bonding in molecules does not lend much insight into how the electrons are shared. Although an orbital picture will be developed shortly, at this point it is possible to represent many molecules in terms of a model to which greater detail can be added later.

In addition to molecules containing identical atoms, there are many molecules that contain nonidentical atoms bound together by a covalent bond. The molecule hydrogen fluoride can be pictured as consisting of two atoms, each of which needs one electron for a complete octet. Fluorine has a higher electron affinity than does hydrogen, but the difference in their relative electron donating and attracting powers is not sufficient to cause complete electron transfer to occur. Consequently, the necessary electrons are shared. The sharing is unequal, however, since only in the case of identical atoms can electrons be shared equally. The hydrogen fluoride molecule is represented by the conventional electron dash notation, even though the shared electron pair must be associated to a somewhat larger extent with fluorine:

$$^{\delta+}H{-}\overline{\underline{F}}|^{\delta-}$$

The result of this unequal sharing is that the fluorine end of the molecule has acquired a negative charge while the hydrogen end is positively charged. The symbol δ, the Greek lower case letter delta, is used to denote the fractional charge located at a site within a molecule. The HF molecule is said to possess a *dipole* (that is, two poles), and the bond is regarded as a *polar covalent bond.*

POLAR COVALENT BONDS

The bonds in H_2 and F_2 are nonpolar, whereas the bond in HF is polar. In the case of the molecules of the elements the electron distribution with respect to the center of the molecule is symmetrical. Not only are such molecules electrically neutral, but the equal sharing of electrons dictates that the centers of positive charge and negative charge must coincide in the individual atoms.

Although the HF molecule is electrically neutral, there are separate centers of net excess positive and negative charge. The electron pair that constitutes the bond is not symmetrically located on a time average with respect to the two atoms. Rather, the fluorine portion of the molecules has the electron pair more strongly associated with it than does hydrogen. There is a dissymmetry of electron distribution and, therefore, a positive end and a negative end of the molecule results.

A molecule in which there are centers of charge possesses a *dipole moment.* The dipole of a diatomic molecule consists of two charges separated by a distance. Its dipole moment is equal to the charge difference times the distance separating the charges. The dipole moment is a measure of the tendency of the dipole to become oriented with respect to an electric field. The behavior of dipoles in the absence and presence of an electric field is depicted in Figure 6.2. The molecule is represented by the charged objects. By quantitatively evaluating the behavior of molecules in the apparatus depicted, it is possible to calculate the dipole moment of the molecules. Dipole moments are expressed in Debye (D) units after the American chemist Peter Debye. The hydrogen and fluorine molecules exhibit no dipole moment consistent with our expectation of nonpolar molecules. Hydrogen fluoride has a dipole moment of 1.98 D.

The absolute magnitude of the dipole moments of diatomic molecules is discussed later in this chapter. For now it is sufficient to say that there is a distinction between polar and nonpolar bonds represented by the dipole moment and that the size of the dipole moment increases as the atoms are more dissimilar in their electron-attracting abilities.

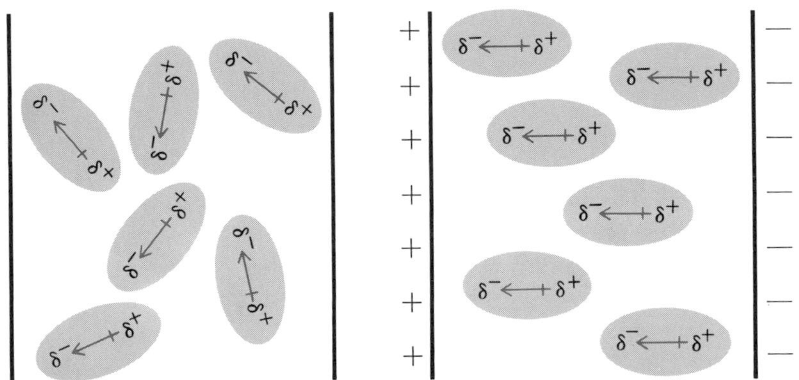

Figure 6.2
Idealized dipoles in the presence and absence of an electric field.

6.4
POLARITY AND MOLECULAR GEOMETRY

In a molecule of more than two atoms the individual bonds may be polar or nonpolar. However, even the presence of two or more polar bonds does not necessarily mean that the resultant molecule is polar. The polarity of a polyatomic molecule is dependent on the geometry of the molecule as well as on the electron distribution in the bonds.

In the triatomic molecule water the two hydrogen atoms are bonded to the same oxygen atom. Since oxygen has a higher affinity for electrons than does hydrogen, each of the two bonds must be polar, with the three atoms in either an angular or linear arrangement. The exact value of the angle describing the position of the two hydrogens and the oxygen is not important, as the conclusions to be derived from this argument are independent of the value. The $\longrightarrow$ represents the direction of the electron dissymmetry of a bond, with the arrow head indicating the negative charge and the cross on the arrow denoting the positive charge. The direction of the $\longrightarrow$ is always toward the more electronegative element:

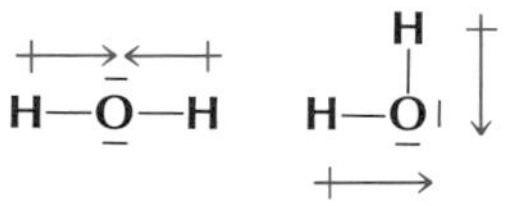

H—O—H   H—O (with H bonded above O)

For the structure on the left there can be no dipole moment, because the two individual polar bonds are colinear, canceling the individual bond moments. The electron distribution in the molecule as a whole is symmetrical. Since water has a dipole moment of 1.8 D, it must be an angular molecule.

In the carbon dioxide molecule CO_2 a double bond connects the central carbon atom to each of the two oxygen atoms. Oxygen attracts the four shared electrons more strongly than does carbon, causing the bonds to be polar. However, the molecule as a whole is nonpolar, as indicated by the fact that it does not line up in an electric field. Therefore, carbon dioxide is a linear molecule, as shown below on the right:

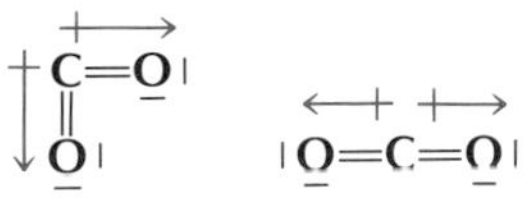

C=O (with O double-bonded below C)   O=C=O

The two examples of water and carbon dioxide serve to illustrate how it is possible to determine experimentally the geometry of a molecule. With a background in geometric principles it is possible to deduce the structure of very complicated molecules by the measurement of their dipole moments.

The bond angles in the hydrogen compounds of the Groups IV, V, and VI elements are given in Table 6.2. These values reflect the type of bonds present in the molecules and provide data to which models for bonding must conform (See Section 6.6).

Table 6.2
Bond angles (deg)

CH_4	109.5	NH_3	107.3	H_2O	104.5
SiH_4	109.5	PH_3	93.3	H_2S	92.2
GeH_4	109.5	AsH_3	91.8	H_2Se	91.0
SnH_4	109.5	SbH_3	91.3	H_2Te	89.5

6.5 ELECTRONEGATIVITY OF ATOMS

Dipole moments are just one experimental indication that the relative electron-attracting power of atoms in molecules vary. As indicated in Chapter 5, electron affinities provide a quantitative measure of the electron-attracting ability of the atoms. However, electron affinities are difficult to determine experimentally.

An empirical scale called *electronegativity* provides an indication of the relative electron-attracting ability of atoms. The dissymmetry of the bonding electron pair in a molecule is a function of (1) the energy required to separate an electron from one atom and (2) the energy released when the electron is added to another atom. Thus dipole moments can provide a scale of the electron-attracting powers of elements if they are compared to a standard reference element whose electron-donating ability is a constant. There are several means of establishing electronegativity scales, each slightly different in origin but yielding a consistent set of values. The electronegativity assigned to each atom can be interpreted readily in terms of atomic structure.

As an example of how measurements such as dipole moments could be used to set up an electronegativity scale, the hydrogen halides are instructive. The dipole moments of HF, HCl, HBr, and HI are 1.98, 1.03, 0.79, and 0.38 D, respectively. The electronegativities of fluorine, chlorine, bromine, and iodine are 4.0, 3.0, 2.8, and 2.5, respectively. The elements with the highest electronegativity with respect to the weakly electronegative hydrogen atom exhibit the largest dipole moments.

The electronegativity scale has been derived by quantitatively measuring the energy required to break a bond in a molecule to produce neutral atoms. In the nonpolar molecule H_2, the energy required to break the bond and separate the two atoms is 103.2 kcal/mole:

$$H_2 \longrightarrow 2H \qquad \Delta H = 103.2 \text{ kcal/mole}$$

Therefore it is reasonable to assume that each mole of hydrogen atoms that now contains 51.6 kcal/mole will release this energy when it combines with another mole of atoms of similar electronegativity. The bond energy of the fluorine molecule is 36 kcal/mole, and each mole of fluorine atoms

Table 6.3
Effect of electronegativity on bond energy

	bond dissociation energy (kcal/mole)			
bond	*X = F*	*X = Cl*	*X = Br*	*X = I*
H—H	103.2	103.2	103.2	103.2
X—X	36.0	57.2	46.1	36.1
H—X (nonpolar)	69.6	80.2	74.2	69.6
H—X (observed)	135	103	87.4	71.4
Difference	65	23	13	2

must release 18 kcal/mole of energy when it combines to form a nonpolar covalent bond:

$$F_2 \longrightarrow 2F \qquad \Delta H = 36 \text{ kcal/mole}$$

When 1 mole of hydrogen atoms combines with 1 mole of fluorine atoms it would be expected that 51.6 + 18 = 69.6 kcal of energy would be released if the electron pair were shared equally:

$$H + F \longrightarrow HF \text{ (nonpolar)} \qquad \Delta H = -69.6 \text{ kcal/mole}$$

Actually 135 kcal is released, indicating that a more stable arrangement than equal sharing has been achieved. The enhanced stability of the HF molecule is attributed to the unequal sharing of the electron pair. This additional bonding energy reflects the attraction of the positive and negative ends of the HF molecule, which in turn reflects the relative electron-attracting abilities of the atoms. Data for the other compounds of hydrogen and the elements of Group VII are given in Table 6.3. From such comparisons the scale of electronegativities has been established (Table 6.4).

6.6 ORBITALS AND BONDING

GEOMETRY OF MOLECULES

The concept of a shared pair of electrons in a covalent bond requires that the electrons occupy a region of space common to the bonding atoms. With such a requirement it is reasonable to expect that polyatomic molecules exist in geometric arrays that satisfy the bonding requirements of all atoms.

A hydrogen molecule can be visualized as being composed of two hydrogen atoms with interpenetrating or overlapping 1*s* orbitals (Figure 6.3). The resulting region of high electron probability is located between the two atoms. The strength of the covalent bond is the result of the attraction of the positively charged nuclei for the electron pair. Since the 1*s* orbital is spherically symmetrical, the same molecule would result regardless of the direction of approach of the two atoms.

Table 6.4
Electronegativity values of the elements

H								
2.1								
Li	**Be**							
1.0	1.5							
Na	**Mg**							
0.9	1.2							
K	**Ca**	**Sc**	**Ti**	**V**	**Cr**	**Mn**	**Fe**	**Co**
0.8	1.0	1.3	1.5	1.6	1.6	1.5	1.8	1.8
Rb	**Sr**	**Y**	**Zr**	**Nb**	**Mo**	**Tc**	**Ru**	**Rh**
0.8	1.0	1.2	1.4	1.6	1.8	1.9	2.2	2.2
Cs	**Ba**	**La-Lu**	**Hf**	**Ta**	**W**	**Re**	**Os**	**Ir**
0.7	0.9	1.1-1.2	1.3	1.5	1.7	1.9	2.2	2.2
Fr	**Ra**	**Ac-Lw**						
0.7	0.9	1.1-						

In hydrogen fluoride the shared electron pair consists of an electron from the 1*s* orbital of hydrogen and one from a fluorine 2*p* orbital. Owing to the shape of the 2*p* orbital of fluorine, which contains the single unpaired electron, it would be anticipated that the hydrogen atom would overlap the fluorine atom in the direction corresponding to the highest probability of finding an electron. This direction corresponds to approach along the longitudinal axis of the 2*p* orbital (Figure 6.4). Approach in any other direction to reach a region of as high probability of locating an electron would require a shorter distance between the hydrogen and fluorine nuclei, which repel each other. Of course, this diatomic molecule must be linear regardless of our model.

In the triatomic molecule water, the two hydrogen atoms must be bound to the oxygen atom by two electron pairs. Oxygen contributes one

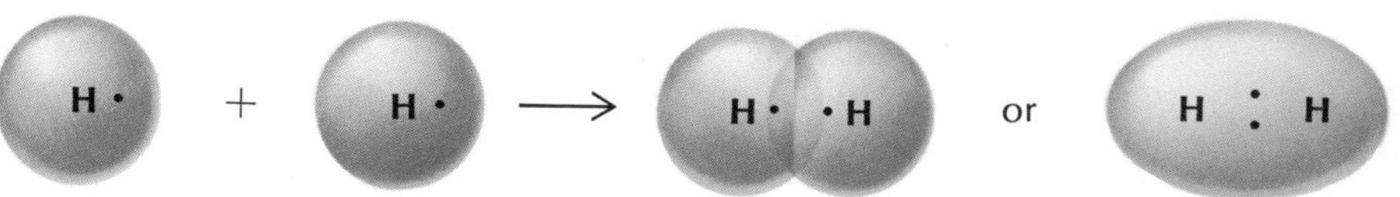

Figure 6.3
Interaction of 1s atomic orbitals of hydrogen atoms to form the hydrogen molecule

								He
			B	**C**	**N**	**O**	**F**	**Ne**
			2.0	2.5	3.0	3.5	4.0	
			Al	**Si**	**P**	**S**	**Cl**	**Ar**
			1.5	1.8	2.1	2.5	3.0	
Ni	**Cu**	**Zn**	**Ga**	**Ge**	**As**	**Se**	**Br**	**Kr**
1.8	1.9	1.6	1.6	1.8	2.0	2.4	2.8	
Pd	**Ag**	**Cd**	**In**	**Sn**	**Sb**	**Te**	**I**	**Xe**
2.2	1.9	1.7	1.7	1.8	1.9	2.1	2.5	
Pt	**Au**	**Hg**	**Tl**	**Pb**	**Bi**	**Po**	**At**	**Rn**
2.2	2.4	1.9	1.8	1.8	1.9	2.0	2.2	

electron to each of these pairs. Since the available electrons are in mutually perpendicular *p* orbitals, the shape of the water molecule would be expected to be angular with a bond angle of 90° (Figure 6.5). The actual bond angle is 104.5°, which represents a considerable deviation from the predicted value—a problem we will return to in a later section of this chapter.

Nitrogen and hydrogen combine to form ammonia, NH_3. The three nitrogen 2*p* orbitals, each of which contains one electron, must be involved in bonding. Since the three orbitals are mutually perpendicular, the shape

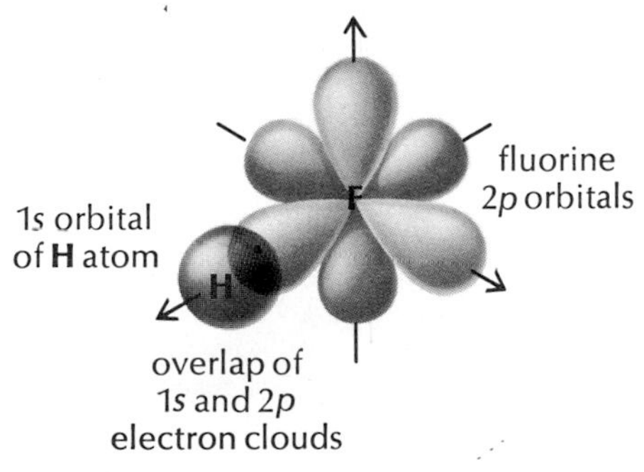

Figure 6.4
Representation of bonding in HF.

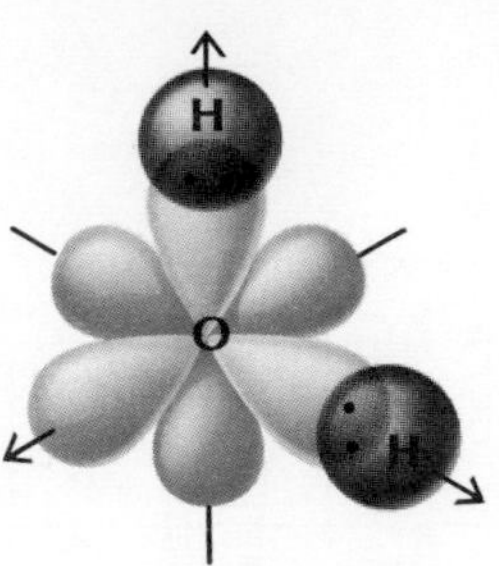

Figure 6.5
Model for bonding in water using p orbitals of oxygen.

of the ammonia molecule would be expected to be a trigonal pyramid with the three hydrogens forming the base (Figure 6.6). As in the water molecule, the H—N—H bond angle is larger than expected—107.3° instead of 90°. Clearly the orbital model for bond formation provides a first approximation of the shape of molecules such as NH_3 and H_2O, but other factors influence it as well.

HYBRIDIZATION

The electronic configuration of carbon, $1s^22s^22p^2$, suggests that the simplest compound of carbon and hydrogen is CH_2 and is angular. However, the simplest stable compound of carbon and hydrogen is CH_4, called methane. The explanation commonly advanced for the formation of CH_4 instead of CH_2 is that in this way carbon achieves an octet of electrons, rather than just the six available in CH_2. In terms of energy, CH_4 is a more reasonable structure, because four bonds are formed instead of two, thereby liberating more energy:

```
                H
                |
|C—H        H—C—H
 |              |
 H              H
```

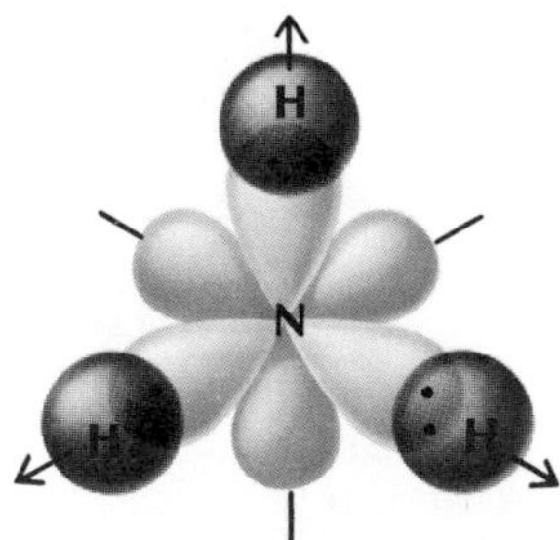

Figure 6.6
Model for bonding in ammonia using p orbitals of nitrogen.

The Lewis diagram of CH_4 does not indicate the orbitals involved in bonding.

In order to form four covalent bonds, carbon must contribute four electrons. In elemental carbon with configuration $1s^22s^22p^2$, two of the four electrons in the second energy level occupy an *s* orbital. They are paired and cannot be used in bonding unless they are somehow unpaired. The simplest way of achieving this would be to promote one *s* electron to the unoccupied *p* orbital (Figure 6.7). Even though energy is required to

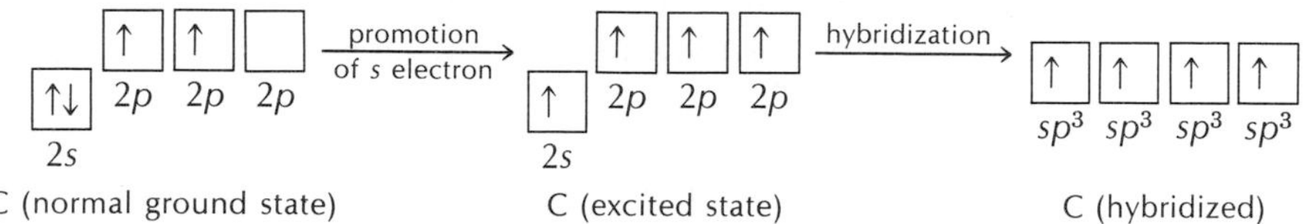

Figure 6.7
Changes in electronic configuration of carbon.

alter the electronic configuration of carbon from $1s^22s^22p^2$ to $1s^22s^12p^3$, the resultant state of carbon could form four bonds. Three of the bonds should be directed at right angles to each other. Since *s* orbitals are spherically symmetrical, the fourth bond should be undirected. Experimentally methane has been shown to contain four identical C—H bonds that are directed toward the corners of a regular tetrahedron with a bond angle

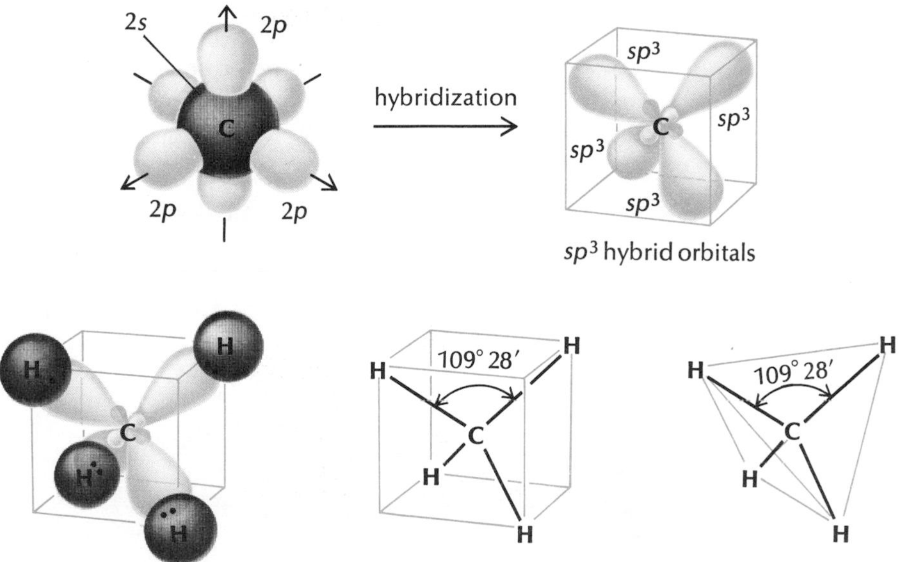

Figure 6.8
Hybridization of carbon and bonding in methane.

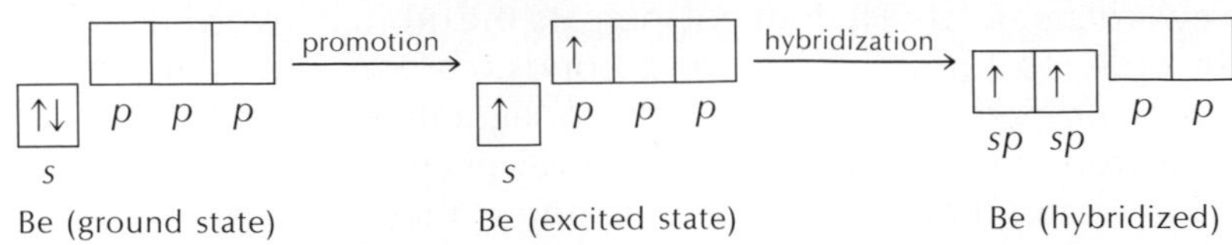

Figure 6.9
Changes in the electronic configuration of beryllium.

109°28′. The symmetry of the molecule in the tetrahedral arrangement and the identity of the bonds are indicated by the absence of any dipole moment.

The model of the bonding in methane is fashioned to agree with the observed experimental facts. Four orbitals of carbon are required, each containing one electron and each directed toward the corner of a tetrahedron. These four equivalent orbitals are termed *hybrid orbitals* and are designated as sp^3 since they must be derived from the three *p* and one *s* orbitals available (Figure 6.8). Four hybrid sp^3 orbitals are created from one *s* and three *p* orbitals by a process called *hybridization.*

Beryllium chloride is a linear, covalent molecule in the vapor phase. Beryllium itself, with a $1s^22s^2$ electronic configuration, would not be expected to share its electrons. If the two electrons in the 2*s* subshell were located in the 2*p* orbitals, then beryllium could share these electrons and form two covalent bonds at 90° to each other. In order to provide a model consistent with the experimental data, the two electrons are considered to be located in two *sp* hybrid orbitals (Figure 6.9). The geometry of the *sp* hybrid orbitals is shown in Figure 6.10.

Borane, BH_3, is a planar trigonal molecule with H—B—H angles of 120°. The bonding is pictured as consisting of three equivalent sp^2 hybrid orbitals of boron, each of which contains one electron and overlaps with the hydrogen 1*s* orbitals (Figures 6.11 and 6.12).

ELECTRON PAIR REPULSION AND MOLECULAR GEOMETRY

The geometry of a set of atoms about a central atom can be explained qualitatively on the basis of electrostatic repulsion between electron pairs. The electron pairs should be as widely separated as possible for minimum electrostatic repulsion. The bonding of $BeCl_2$ illustrates this separation when two electron pairs are present about a central atom.

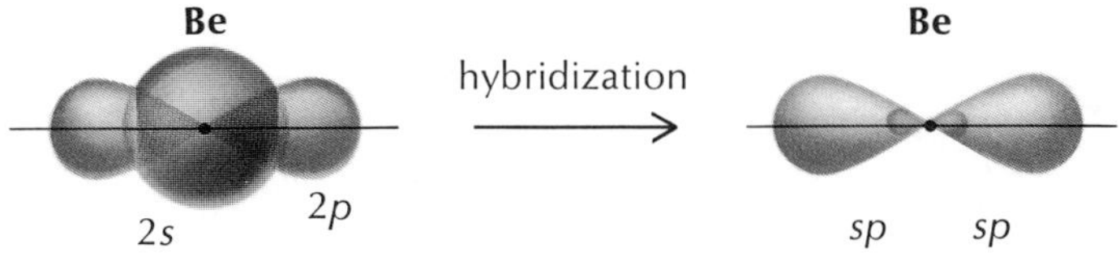

Figure 6.10
The sp hybrid orbitals of beryllium.

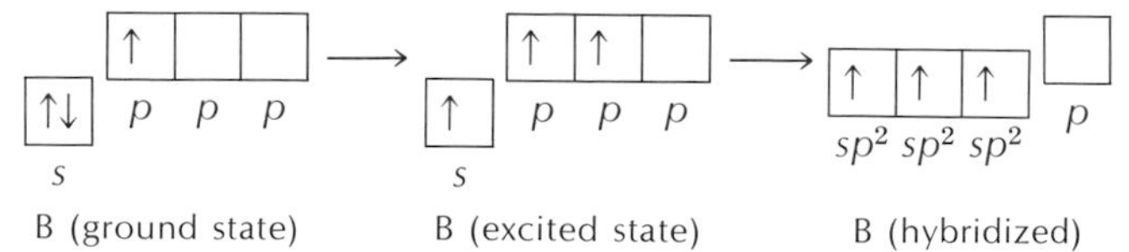

Figure 6.11
Changes in the electronic configuration of boron.

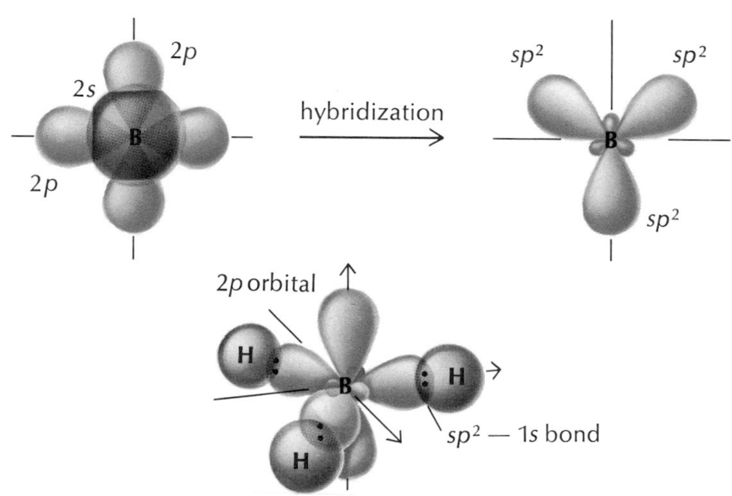

Figure 6.12
Hybridization in boron and bonding in borane.

Similarly, mercury, which has a $6s^2$ configuration, forms two linear covalent bonds to chlorine by means of *sp* hybrid orbitals:

Cl—Hg—Cl

Three electron pairs are farthest apart at the corners of an equilateral triangle. Thus BF_3 is triangular and planar. Each bond angle is 120°, and the orbitals that boron uses in bonding are sp^2.

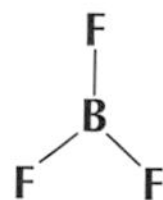

The four electron pairs about carbon in CH_4 are directed toward the corners of a tetrahedron, because this arrangement leads to minimum electrostatic repulsions. The H—C—H bond angles are all 109°28′:

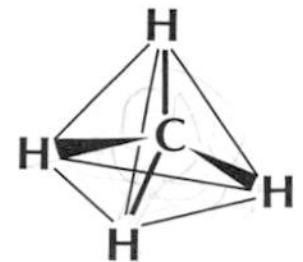

In each of the three examples presented there are enough atoms to bond to all available electron pairs. However, there are molecules in which some of the electron pairs are not involved in bonding. In such cases, will the electron pairs engaged in bonding be electrostatically equivalent to the *nonbonding* electron pairs? The NH_3 and H_2O molecules provide an answer to this question.

The three bonds in ammonia are more similar to the sp^3 hybridized bonds in methane than to the bonds expected from mutually perpendicular *p* orbitals. The 107.3° H—N—H bond angle is much closer to the tetrahedral angle of 109.5° than it is to 90°. Therefore, the ammonia molecule can be envisioned as tetrahedral with one of the positions occupied by an electron pair (Figure 6.13). If it is assumed that a nonbonding electron pair (*lone pair*) exerts a greater electrostatic repulsion on a bonded pair than does a bonded pair, then the three bonds in ammonia should be slightly closer than the ideal tetrahedral arrangement.

In the water molecule there are two bonded electron pairs and two lone electron pairs. If the four electron pairs were equal in terms of their mutual electrostatic repulsions, a tetrahedral molecule would result. The H—O—H bond angle in water is 104.5°, which is closer to 109.5° than to 90°. It is for this reason that sp^3 hybrid orbitals are thought to contribute to the bonding of oxygen in water (Figure 6.13). On the basis of the relative strengths of repulsions postulated to explain the geometry of ammonia, it should be anticipated that the lone pair-lone pair repulsion in water should be the largest of all repulsions. Thus the order of decreasing repulsions should be lone pair-lone pair $>$ lone pair-bonded pair $>$ bonded pair-bonded pair. In water the two lone pairs of oxygen lead to a decrease of the H—O—H bond angle of water as compared to the bond angles of the ammonia molecule, in which there is only one nonbonding electron pair.

RESONANCE

Unfortunately, there are many molecules for which an electron dot or dash representation consistent with all of the available experimental

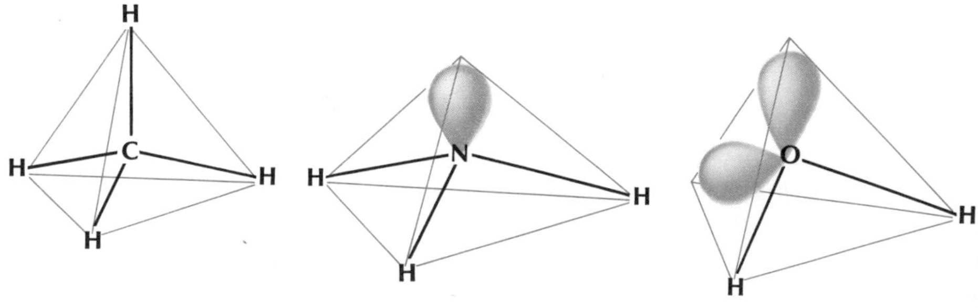

Figure 6.13
Similarities in the geometries of CH_4, NH_3, and H_2O.

evidence cannot be written, as in the sulfur dioxide molecule. Both the oxygen atoms and the sulfur atoms can be made to have an octet of electrons associated with them.

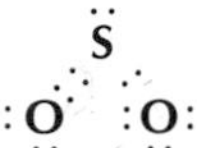

However, in such a representation the two S—O bonds are not equivalent, because one is a single bond and the other a double bond. The actual molecule has perfectly equivalent oxygen atoms and S—O bonds. Therefore, the representation is not an adequate model for the SO_2 molecule.

Two structures may be written for SO_2, neither of which corresponds to reality. These two structures differ only in that the locations of the single and double bonds have been interchanged:

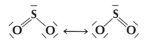

$$O{=}S{-}O \longleftrightarrow O{-}S{=}O$$

The actual molecule is said to be a *resonance hybrid* of the two indicated structures. The double-headed arrow ↔ does not indicate that the SO_2 molecule is a mixture of the two structures written. Rather, it indicates that there is no single structure that can be written by the use of conventional symbolisms. The bonds in SO_2 are identical and must be of an intermediate character, not single or double bonds.

It should be stressed that SO_2 is not a unique molecule. There are many molecules that cannot be represented by a single electron dot structure. The problem is not with any of the molecules but rather with our mode of representation. The terms *resonance* and *resonance hybrid* do not indicate anything associated with motion. They are merely terms used to breach the gap between real molecules and our attempts at representing them.

It is possible to describe molecules more accurately by using the principles of quantum mechanics. In such a treatment molecular orbitals that encompass more than two atoms are often employed. The molecular orbitals are a better representation of reality. However, in an introductory textbook it is easier to present molecules in terms of the interpenetration of atomic orbitals. No attempt will be made, therefore, to describe recent developments in the theories of molecular orbitals. Atomic orbitals can be used to represent many molecules in terms of a single conventional structure.

Suggested further readings

Barrow, G. M., *The Structure of Molecules.* New York: W. A. Benjamin, 1963.

Cohen, I., and Bustard, T., "Atomic Orbitals: Limitations and Variations," *J. Chem. Educ.* **43,** 187 (1966).

George, J. W., "Hybridization in the Description of Homonuclear Diatomic Molecules," *J. Chem. Educ.*, **42,** 152 (1965).
Gray, H. B., *Electrons and Chemical Bonding*. New York: W. A. Benjamin, 1965.
Greenwood, N. N., "Chemical Bonds," *Educ. Chem.*, **4,** 164 (1967).
Lagowski, J. J., *The Chemical Bond*. Boston: Houghton Mifflin, 1966.
Pauling, L., *The Nature of the Chemical Bond*. Ithaca, New York: Cornell University Press, 1960.
Ryschkewitsch, G. E., *Chemical Bonding and the Geometry of Molecules*. New York: Reinhold, 1963.
Sebera, H. H., *Electron Structure and Chemical Bonding*. Waltham, Massachusetts: Blaisdell, 1964.

Terms and concepts

bond angle
bond dissociation energy
covalent bond
Debye
dipole
dipole moment
electron pair repulsion
electronegativity
enthalpy
entropy
hybridization
ionic bond
Lewis octet rule
lone pair
molecular geometry
polar covalent bond
resonance
resonance hybrid
sp hybrid orbital
sp^2 hybrid orbital
sp^3 hybrid orbital
tetrahedron
thermodynamics
valence

Questions and problems

1. By using conventional chemical symbolism, diagram the octet structures for each of the following compounds: **a.** $BaCl_2$ **b.** Na_2S **c.** PH_3 **d.** NF_3 **e.** BrCl

2. What type of bonding is found in each of the compounds listed in Question 1?

3. The geometric arrangement of atoms in several molecules is given below. Using dashes to represent bonds and lone pair electrons, draw octet representations of the molecules with the number of allowed electrons.

a. H H
N O
H

b. H C C H

c. H C N

d. H Cl O

4. Which is the more polar, the BrCl or the BrF molecule?
5. What types of bond hybridization are present in each of the compounds listed below? What are their molecular geometries? **a.** BF_3 **b.** PF_3 **c.** $SiCl_4$ **d.** BeH_2 **e.** BrCl **f.** OF_2
6. Using the concept of electron pair repulsion, predict the shape of a molecule AX_n where $n = 5$, 6, and 8. (Contrary to the simple octet rule, such compounds are known and stable.)
7. Why is NF_3 pyramidal, whereas BF_3 is trigonal planar?
8. Ammonia can be made to gain or lose a proton, H^+. What bond angles would you expect in the ions NH_4^+ (ammonium ion) and NH_2^- (amide ion)?
9. The electronegativities of hydrogen, fluorine, chlorine, bromine, and iodine are 2.1, 4.0, 3.0, 2.8, and 2.5, respectively. Compare the electronegativity difference between hydrogen and each halogen with the dipole moments of the hydrogen halides. What correlation is observed?
10. Is carbon tetrachloride, CCl_4, polar? Explain. Is NF_3 polar? Explain.
11. The bond energies of Cl_2, Br_2, and I_2 are 57.2, 45.4, and 35.5 kcal/mole, respectively. Assuming equal sharing of electrons, predict the bond energies of ICl and IBr. Would the experimental values be greater or less than your values? For which molecule would you expect the closest agreement between experimental and predicted values?
12. Draw an electronic formula for SO_3. (All three oxygens are attached to sulfur and not directly to each other.)
13. Sulfur trioxide, SO_3, is a planar molecule in the gaseous state with the oxygens at the corners of an equilateral triangle. All bonds are identical. Reconsider your structure in Question 12 and state how the molecule can be better represented.
14. The ozone molecule, O_3, consists of three atoms of oxygen arranged in an angular manner. The bonds from the central oxygen to each terminal oxygen are equivalent. Draw an electron dot formulation for this molecule.
15. The molecule Cl_2O_7 consists of two chlorines bridged by a central oxygen atom; the remaining oxygens are bound three each to separate chlorine atoms. Draw an electron dot formulation for the molecule. What is the expected geometry of the molecule?
16. Sulfur forms the S^{2-} and S_2^{2-} ions. Describe these ions.
17. What types of orbitals are probably involved in the bonding of SbH_3 (see Table 6.2). Compare SbH_3 with NH_3 and suggest why the geometries of the two molecules are different.
18. Predict the shape of H_2O_2 and N_2H_4.

7

MOLECULAR STRUCTURE AND PHYSICAL PROPERTIES

In Chapters 2 and 3 the states of matter were discussed in a general manner with a few examples. The kinetic molecular theory was applied to all three states with appropriate modifications. Although the facts were presented that atoms have mass and also exert some attractive forces between themselves and neighboring molecules, there was no way to expand further on these molecular properties. Now that the electronic structure of atoms and molecules has been presented it is appropriate to reconsider the states of matter with relation to microscopic structure.

7.1 GASES

DEVIATIONS FROM THE IDEAL GAS LAW

The ideal gas law describes a model gas that according to the kinetic molecular theory consists of molecules of negligible volume that do not interact with their neighbors. The fact that real gases do have volume and do interact with neighboring molecules is established by deviations from predicted behavior at high pressure and low temperature.

At high pressure the dimensions of the molecules are comparable to the spaces between them. Therefore, agreement between experimental values and values calculated from the ideal gas law is not possible if molecular volumes are neglected. At low temperature the attractive forces between molecules become of comparable magnitude to the kinetic energies of the molecules. The molecules, therefore, cannot be considered to be free and independent of their neighbors. These attractive forces account for the fact that all gases can be liquefied at sufficiently low temperatures. The two effects of molecular volume and attractive forces cause deviations from idealized volumes in opposite directions. The first tends to increase the volume over that predicted, and the second tends to decrease the volume. The relative importance of the two terms is a function of molecular structure.

For an ideal gas the product PV is a constant at all pressures if the temperature is constant. The term PV/RT should be equal to the number of moles of gas present and be independent of pressure. A convenient way to visualize the deviations from ideality is to plot PV/RT vs. P at a specified temperature. An ideal gas yields a straight horizontal line at $PV/RT = n$. A real gas deviates from this line. At high pressure, as the gas is compressed, the intermolecular forces should cause the molecules to come closer together and to exert an attraction for each other. This would lead to a smaller volume than predicted from the ideal gas equation. At higher pressures the molecules become crowded and the observed volume is larger than predicted from the ideal gas equation. In Figure 7.1 the actual PV/RT products for 1 mole of several gases are plotted vs. the pressure in atmospheres at zero degrees Celsius. At higher temperatures the initial dip resulting from intermolecular attraction is not as pronounced. The compressibility PV/RT for nitrogen (N_2) is shown as a function of temperature in Figure 7.2.

To better describe the behavior of real gases, Johannes van der Waals suggested a modified gas equation in 1873. The van der Waals equation

$$\left(P + \frac{an^2}{V^2}\right)(V - nb) = nRT$$

is in a form similar to the ideal gas law, with the addition of terms to account for the effect of attractive forces an^2/V^2 and molecular volumes nb. The expressions in parentheses may be viewed as effective pressure felt by the molecules $(P + an^2/V^2)$ and effective free volume $(V - nb)$. Neglecting the actual derivation of the equation, we can relate deviations from ideality to molecular structure qualitatively. The proportionality constant a is related to the magnitude of intermolecular attractive forces that have the same effect as increasing the measured pressure. The propor-

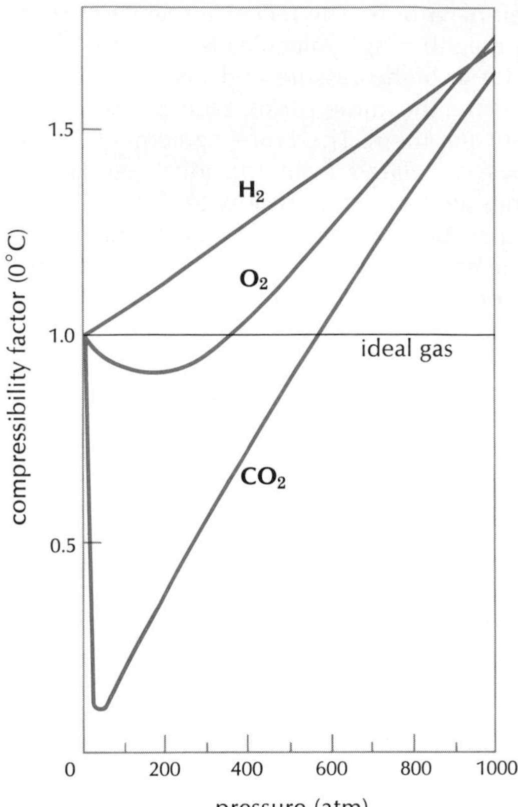

Figure 7.1
The compressibility factor PV/RT for gases at 0°C.

tionality constant b is related to the molecular volume of the molecules so that the second term $(V - nb)$ is the corrected volume.

A list of the numerical values for a and b as determined by experiment is given in Table 7.1.

It is easy to see why the numerical value of a is large for the polar molecules NH_3 and H_2O. These gases should deviate strongly from ideality as a result of intermolecular attractive forces. The positive end of one molecule will attract the negative end of a neighboring molecule.

The values for a in the case of helium and hydrogen are small. The values for N_2 and O_2, however, are quite large. Why should there be any net attractive forces between nonpolar molecules? Why do these forces vary as a function of the compounds? The forces in nonpolar molecules are imagined to result from instantaneous electronic asymmetry. If for a fraction of time the electronic distribution of an atom or molecule became unsymmetrical, the species would have a slightly negative and slightly positive end. Thus the atom or molecule has a temporary dipole.

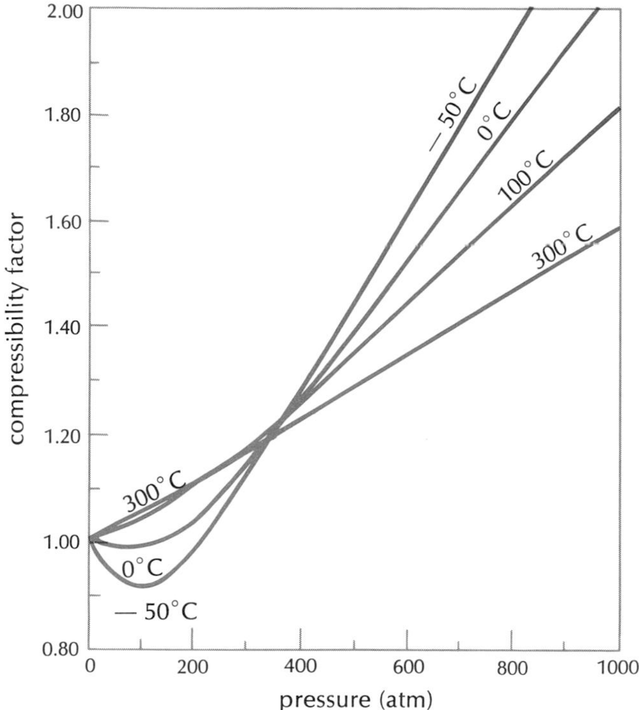

Figure 7.2
Compressibility factor of nitrogen as a function of temperature.

As a result of the asymmetry of its neighbor, a second molecule should have its electronic distribution polarized. The model of this phenomenon is illustrated in Figure 7.3.

A net attractive force results in a gas sample because there continually will be distorted electronic distributions with temporary instantaneous

Table 7.1
van der Waal's constants

substance	*a (liter²-atm/mole²)*	*b (liter/mole)*
H_2	0.244	0.0266
He	0.034	0.0237
Ar	1.34	0.0322
Kr	2.32	0.0398
N_2	1.39	0.0391
O_2	1.36	0.0318
CO_2	3.59	0.0427
NH_3	4.17	0.0371
H_2O	5.464	0.0305

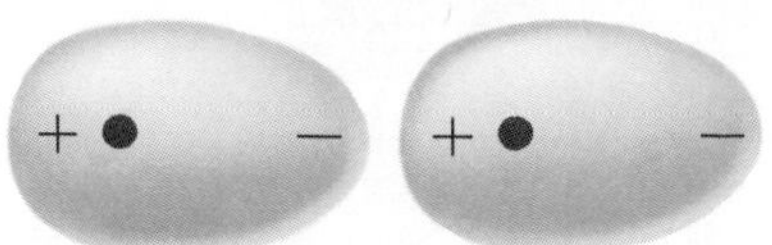

Figure 7.3
A model for the van der Waals attractive force between atoms.

dipoles. These dipoles are not long lasting because the atoms or molecules are moving rapidly and depart from the site of temporary dipoles. Nevertheless, if a small fraction of the gaseous sample is experiencing net attractive forces, there will be a net deviation from ideality. These forces are called *van der Waals attractive forces.* They become more important as the electrons become less tightly bound to the atom or molecule. It is for this reason that the atoms or molecules of larger volume exhibit large van der Waals attractive forces. For example, the electrons in hydrogen are more tightly held about the nucleus than those of chlorine. Note that the values for the constant *a* given in Table 7.1 for the gases helium, argon, and krypton are consistent with the expectation that the electrons of the higher atomic mass elements are under less restriction than those of the lower atomic mass elements.

CRITICAL TEMPERATURE

The liquefaction of a gas requires conditions that allow intermolecular forces to bind the molecules together in the liquid state. The molecules can be brought closer together by increasing the pressure. Decreasing the temperature will make the attractive forces significant compared to the kinetic energy. Thus a combination of high pressure and low temperature should serve to liquefy a gas. In general, the higher the temperature of the gas, the higher the pressure that must be used to liquefy it. Therefore, there are many combinations of pressures and temperatures under which the liquid state of a given substance can be achieved. However, for each gas there is some temperature called the *critical temperature* above which the gas cannot be liquefied no matter what pressure is applied. The minimum pressure required to liquefy the gas at its critical temperature is called the *critical pressure.* A list of the critical temperatures and pressures of some gases is given in Table 7.2.

The critical temperature provides some indication of the strength of the intermolecular attractive forces of a gas. If a substance is polar and has high intermolecular attractive forces, these forces will aid in overcoming the energetic molecular motion and the gas may be liquefied at relatively high temperatures. If a gas has weak attractive forces, it will have a low critical temperature since above this temperature the molecular motion is too energetic to be overcome by the intermolecular attractions.

Table 7.2
Critical temperature and critical pressure

substance	*critical temperature* (°*K*)	*critical pressure* (*atm*)
H_2	5.3	2.26
He	33.3	12.8
N_2	126.1	33.5
O_2	153.4	49.7
CO_2	304.2	73.0
NH_3	405.6	111.5
H_2O	647.1	217.7

Water with its strong attractive forces can be liquefied up to the temperature of 647.1°K. However, helium with its weak van der Waals attractive forces cannot be liquefied above 5.3°K.

7.2
LIQUIDS

VAPOR PRESSURE OF LIQUIDS

In Table 3.2 the vapor pressures of some common liquids were listed. The vapor pressure, which is a measure of the tendency of the substance to leave the liquid phase, is a function of both molecular mass and intermolecular attractive forces. All other things being equal, the lowest molecular mass materials should have the highest vapor pressure at a given temperature. For molecules of the same mass, the substance with the largest intermolecular forces should have the lowest vapor pressure.

The molecular mass and vapor pressure at 20°C of ether, chloroform, alcohol, and water are as follows:

ether ($C_4H_{10}O$)	74.0 amu	44.2 cm of **Hg**
chloroform ($CHCl_3$)	119.5 amu	14.5 cm of **Hg**
alcohol (C_2H_6O)	46.0 amu	4.3 cm of **Hg**
water (H_2O)	18.0 amu	1.8 cm of **Hg**

While there is the expected correlation between the molecular masses and vapor pressures of ether and chloroform, both alcohol and water are relatively nonvolatile, considering their low molecular masses. An indication of the reason for the low vapor pressures of these two substances is the presence of the O—H bond, which is quite polar:

```
    H  +↑          H   H   H  +↑
    |              |   |   |
 H—O| ↓         H—C—C—O| ↓
   ‾               |   |   ‾
 +——→              H   H+——→
```

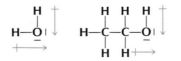

The substances with the highest vapor pressure at a given temperature have the lowest normal boiling point. Therefore, the normal boiling point of a liquid provides some indication of the strength of intermolecular attractive forces, providing account is taken of the molecular masses of the substances. The normal boiling points of ether, chloroform, alcohol, and water are 34.6, 61.3, 78.5, and 100.0°C, respectively. As was the case in the discussion of vapor pressures, the boiling points of alcohol and water indicate that strong intermolecular forces exist between molecules of these substances.

In general the *heat of vaporization* (Section 3.3) of a substance also indicates the strength of intermolecular attractive forces. The stronger the attractive forces, the higher the heat of vaporization. Dividing the heat of vaporization of a substance at its boiling point by its boiling point in degrees Kelvin yields Trouton's constant. Most nonpolar molecules without large intermolecular attractive forces have a value of 21 cal/mole-deg. However, the constants for water and alcohol and 26.0 and 26.2 cal/mole-deg, respectively. These high values attest to the presence of large intermolecular attractive forces in these polar liquids.

HYDROGEN BONDING

Two types of intermolecular attractive forces have been described to account for the observed properties of liquids and gases. These are the very weak van der Waals forces and the strong dipole-dipole type of interactions that exist in polar substances. The molecules of water and alcohol have been shown to exhibit strong intermolecular interactions, as reflected by the physical properties of these substances. However, there is a specific type of molecular interaction in these molecules that is stronger than many dipole-dipole interactions. This interaction is called the *hydrogen bond.*

The boiling points of the compounds of hydrogen and the elements of Groups IV, V, VI, and VII illustrate the basis for postulating the existence of a hydrogen bond. The compounds of Group IV, with hydrogen, CH_4, SiH_4, GeH_4 and SnH_4, are tetrahedral nonpolar substances. They exhibit a regular increase in the boiling point as the molecular mass increases. This typical trend that would be expected in the absence of any major differences in intermolecular interactions is illustrated in Figure 7.4.

A direct relationship between molecular mass and boiling point is observed for the compounds PH_3, AsH_3, and SbH_3 of Group V. However, NH_3 has a considerably higher boiling point than would be predicted on the basis of its molecular mass. Similarly, there are large deviations from a direct relationship for H_2O (from the Group VI compounds H_2S, H_2Se, and H_2Te), and for HF from the Group VII compounds (HCl, HBr, and HI). These deviations indicate that an intermolecular interaction stronger than a simple dipole-dipole interaction is present in NH_3, H_2O, and HF.

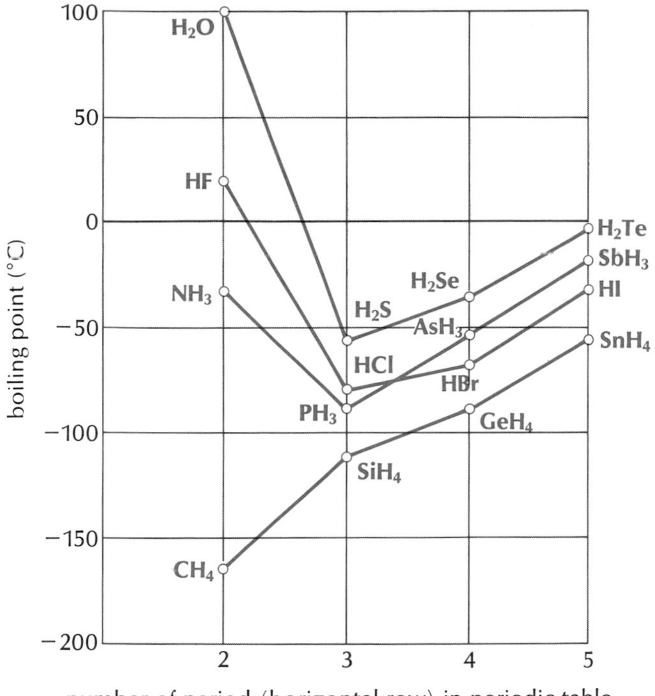

Figure 7.4
Illustration of the abnormally high boiling points of H_2O, HF, and NH_3.

The three molecules NH_3, H_2O, and HF have two structural features in common: Each has at least one hydrogen atom covalently bound to an electronegative element, and each electronegative atom has at least one unshared pair of electrons. It has been postulated, therefore, that the intermolecular attraction responsible for the observed physical properties involves a bridging of two or more molecules into aggregates by means of a hydrogen atom.

$$^{\delta-}\mathbf{F}\text{—}\overset{\delta+}{\mathbf{H}}\cdots\mathbf{F}^{\delta-}\text{—}\mathbf{H}^{\delta+}$$

Thus the very small positively charged hydrogen atom is attracted toward the electrons of a neighboring molecule. The hydrogen, in effect, is shared between two electronegative atoms. This type of interaction is called the hydrogen bond, and its energy is in the order of 3–10 kcal/mole. Typical covalent bond energies are at least ten times larger than hydrogen bond energies.

The order of decreasing hydrogen bond strength with the electronegative atoms might be expected to be F > O > N. However, in Figure 7.4

it appears that the anticipated order is incorrect. The molecule HF boils at a lower temperature than H_2O. Actually, the strengths of the hydrogen bonds in HF are larger than those in H_2O. The reason for the apparent inconsistency lies in the fact that although the hydrogen bond to fluorine is strong, the molecule HF can only form one hydrogen bond per molecule. The fluorine atom has three unshared electron pairs, but the hydrogen atoms available are limited. In the case of water there are two protons and two electron pairs per molecule. Each oxygen atom in water can be surrounded by four hydrogen atoms, and the extent of aggregation can be greater than in HF. Ammonia (NH_3) has three protons but only one electron pair to share. Water has the proper balance of protons and electron pairs to form the maximum number of hydrogen bonds per mole. For the bimolecular aggregates shown in Figure 7.5, the sites free on electronega-

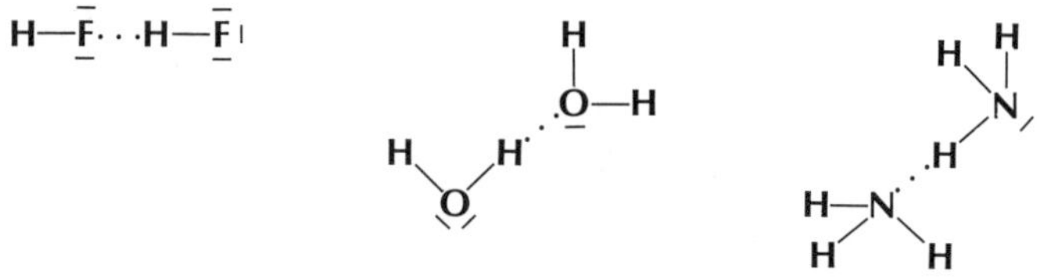

H—F···H—F

H—O···H—O—H (with H)

H$_2$N—H···NH$_3$

Figure 7.5
Hydrogen bonding in hydrogen fluoride, water, and ammonia.

tive atoms that could hydrogen bond are 5, 3, and 1 for $(HF)_2$, $(H_2O)_2$, and $(NH_3)_2$, respectively. The hydrogen atoms available for formation of additional hydrogen bonds are 1, 3, and 5 for $(HF)_2$, $(H_2O)_2$, and $(NH_3)_2$, respectively.

7.3 SOLIDS

CLASSIFICATION OF SOLIDS

The most noticeable characteristic macroscopic property of solids is the geometry of the crystals. The regular geometric shape of these solids suggests a regularity in the microscopic structure of the substance. Both the shape of the microscopic species making up the solid and the intermolecular forces between them might be expected to be important in determining the type of crystal produced.

It is useful to classify solids according to the material present in the crystal. The classification provides a means of systematizing and comparing the similarities and differences of the solid state. The classes of solids to be considered are *molecular, metallic, covalent network,* and *ionic* solids.

MOLECULAR SOLIDS

Molecular crystals consist of assemblies of uncharged atoms or molecules. The units making up crystals are bound together either by weak van der Waals attractive forces (in the case of atoms or nonpolar molecules) or by the somewhat stronger dipole–dipole attractive forces of polar molecules. In either case the energy holding the molecular crystals together is quite small. This small binding energy is manifested by the low melting point of molecular crystals.

Another physical consequence of the weak binding energy is the soft, compressible nature of molecular crystals. Application of an external force on molecular crystals merely shoves the molecules past each other without any major difference in the interaction between the adjacent molecules. The energies required to separate 1 mole of various molecular crystals are listed in Table 7.3. In general, molecular crystals do not conduct electricity, because the individual molecules consist of nuclei with localized electrons that cannot move under an applied electric potential. A representation of a

Table 7.3
Binding energy of crystals

molecular crystals	*energy required to separate atoms or molecules (kcal/mole)*
Ar	1.6
CH_4	2.0
Cl_2	4.9
CO_2	6.0
metallic crystals	*energy required to separate atoms (kcal/mole)*
Li	38
Ca	42
Al	77
Fe	99
covalent network crystals	*energy required to separate atoms (kcal/mole)*
C (diamond)	170
SiO_2	433
ionic crystals	*energy required to separate ions (kcal/mole)*
NaCl	186
LiF	247
ZnO	964

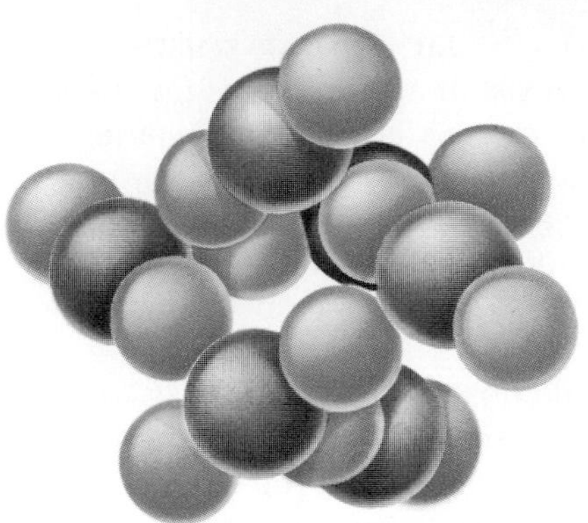

Figure 7.6
Carbon dioxide as an example of a molecular compound in a crystal.

crystal of carbon dioxide is shown in Figure 7.6. The individual molecules are only weakly associated.

METALLIC SOLIDS

The reflectivity, silvery luster, and high electrical and heat conductivity of metal crystals serve to distinguish them from other crystals. To account for these properties the atoms are envisioned as existing in a regular array without some of their highest energy electrons. The model consists of positive ions immersed in a sea of electrons that do not belong to any specific ion (Figure 7.7). These electrons are highly mobile and move in an applied electric field. Similarly, when heated, the electrons gain in kinetic energy and move rapidly through the crystal. Metals often can be drawn into wires and hammered into thin sheets. The ease with which metals can be worked suggests that the planes of metal ions can be moved without much energy. However, the energy required to separate the atoms from the crystal is somewhat greater than in the case of molecular crystals.

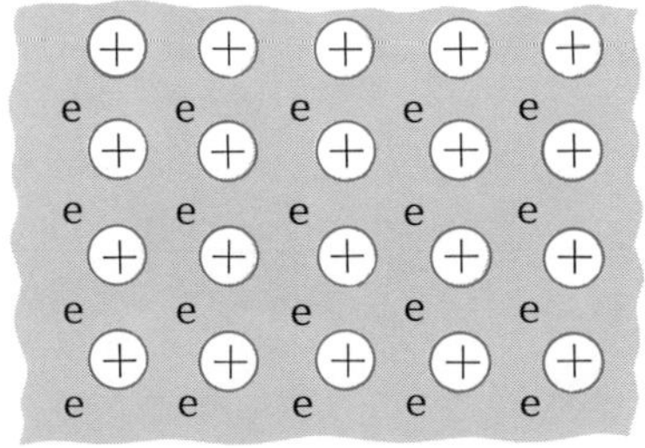

Figure 7.7
Model of a metallic crystal.

Both the high energy electrons and the ions must be removed from the sea of electrons and array of positive ions. The binding energy of some metals is listed in Table 7.3.

COVALENT NETWORK SOLIDS

There are some substances that consist of atoms held in a fixed three-dimensional network by covalent bonds to their immediate neighbors. This continuous system of atoms connected by covalent bonds makes the entire crystal a single molecule. The most common examples of such molecules are diamond and sand.

A diamond crystal consists entirely of carbon atoms covalently bonded to four other carbon atoms. The bond angle C—C—C is 109.5° and represents a typical tetrahedrally substituted carbon atom with sp^3 hybrid bonds. The structure of diamond is shown in Figure 7.8.

Sand consists of a network of silicon and oxygen atoms in the ratio of 1 to 2. While the formula for silica (sand) is written as SiO_2, there are no discrete molecules corresponding to that formula. Each silicon in the net-

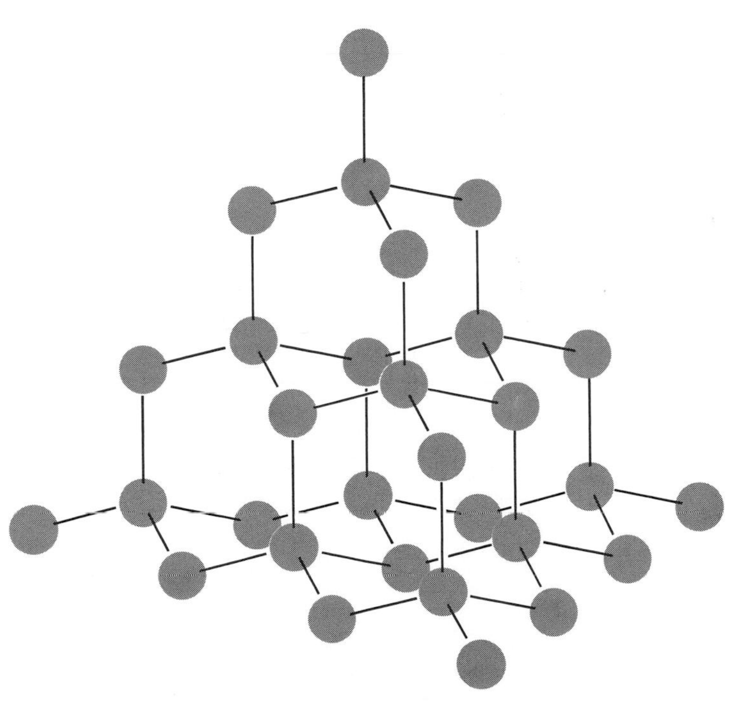

Figure 7.8
The diamond crystal network solid.

work is attached directly to four oxygen atoms, and each oxygen is attached directly to two silicon atoms, as shown in Figure 7.9.

The energy required to break apart network solids is very high, as shown in Table 7.3. Such solids have high melting points and are extremely nonvolatile. In order to work with a network solid it is necessary to disrupt covalent bonds and separate neighboring atoms from their rigidly defined positions in the network. Consequently, network solids are among the hardest substances known.

IONIC SOLIDS

Ionic solids consist of ions of a given charge in well-defined positions surrounded by ions of the opposite charge. The latter ions in turn are surrounded by ions of opposite charge. Thus any given ion is bonded by electrostatic attraction to all the ions of opposite charge immediately surrounding it. In a crystal of an ionic solid such as NaCl, there are no sodium chloride molecules as such. There is no single partner for a given sodium ion. Rather, it has six immediate neighbors equidistant from it. The chloride ion in turn has six immediate sodium neighbors. The only thing the formula

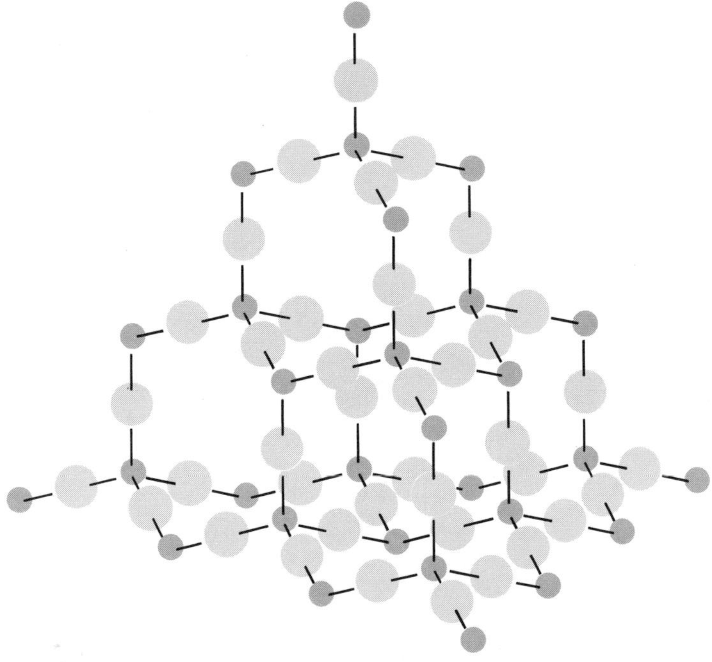

Figure 7.9
The silicon dioxide crystal network solid. Each silicon atom (small sphere) is attached to four oxygen atoms; each oxygen is attached to two silicon atoms.

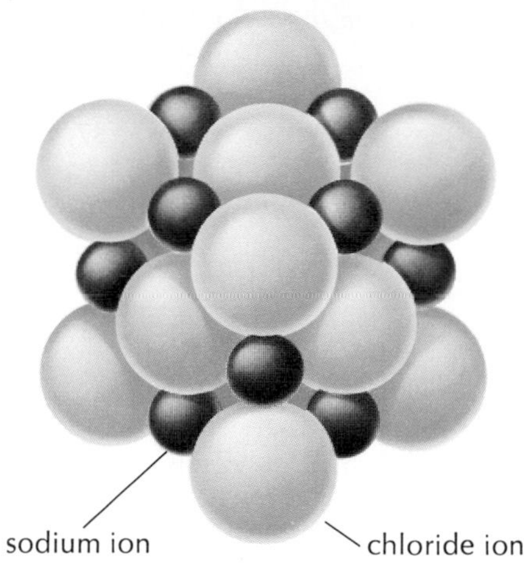

Figure 7.10
The ionic solid crystal of sodium chloride.

NaCl is meant to represent is the fact that a collection of sodium and chloride ions in a 1:1 ratio are present in a sodium chloride crystal. The crystal of NaCl is shown in Figure 7.10.

Ionic crystals have negligible vapor pressures at room temperature and melt and boil at high temperatures. They are very hard and brittle. All of these properties are consistent with the model of an ionic solid. A large amount of energy is required to alter or separate the ions from their positions in the crystal. When a crystal is subjected to an applied force, a plane of ions may be displaced relative to an adjacent plane. If such a displacement brings ions of the same charge close to each other, a strong repulsion between planes results and the crystal fractures.

All ionic crystals are electrical insulators because there is no way in which the charged ions can migrate in order to transport the current. However, if the solid is melted, the resultant liquid is a good electrical conductor, since in the liquid state the ions may move under the influence of an electric field.

Suggested further readings

Booth, N., "Chemical Bonds—Ionic Lattices," *Educ. Chem.* **1,** 151 (1964).
Bragg, L., "X-ray Crystallography," *Sci. Amer.,* 58 (January 1968).
Etzel, H. W., "Ionic Crystals," *J. Chem. Educ.* **38,** 225 (1961).
Gehman, W. G., "Standard Ionic Crystal Structures," *J. Chem. Educ.,* **40,** 61 (1963).
Revelle, R., "Water," *Sci. Amer.,* 92 (September 1963).

Runnels, L. K., "Ice," *Sci. Amer.*, 118 (December 1966).
Sproull, R. L., "The Conduction of Heat in Solids," *Sci. Amer.*, 92 (December 1962).
Verhoek, F. H., "What is a Metal?" *Chemistry,* **37,** 6 (1964).

Terms and concepts

covalent network solid
critical pressure
critical temperature
hydrogen bonding
ionic solids
metallic solids
molecular solids
van der Waals forces

Questions and problems

1. Which gas in each of the following pairs of gases should have the larger *a* term in the van der Waals equation? Explain the reason for your choice.

a. He and Kr
b. HF and HCl
c. SO_2 and CO_2

2. Are the *b* values for O_2 and N_2 reasonable in terms of the atomic volumes of the constituent atoms? Why?
3. Would you expect the van der Waals attractive forces in F_2 to be larger or smaller than those in Cl_2? Why?
4. Why might it be anticipated that the critical temperature of HF would be higher than that for HCl?
5. Can the following substances exist as liquids under the conditions specified?

a. He at $-260°C$
b. CO_2 at $0°C$
c. H_2O at $450°K$

6. Acetone (C_3H_6O) has a lower vapor pressure than ether ($C_4H_{10}O$) at 20°C. What statement can be made regarding the intermolecular forces in these two liquids?
7. Which one of the following liquids would be expected to boil at the higher temperature under the same atmospheric conditions: argon or krypton?
8. Why does H_2Se boil at a higher temperature than AsH_3?
9. The heat of vaporization of H_2S is 131.9 cal/g, and its boiling point is $-61.4°C$. Calculate Trouton's constant for H_2S and determine whether it is consistent with the lack of any significant degree of hydrogen bonding in this molecule.
10. The boiling point of octane is 125.6°C. Calculate its heat of vaporization.

11. Which element would be expected to have the higher molar heat of fusion, Li or Fe?

12. The heats of fusion of carbon tetrachloride (CCl_4) and alcohol (C_2H_6O) are 4.2 and 24.9 cal/g, respectively. Comment on the relative magnitudes of the molar heats of fusion in terms of the intermolecular forces present in the solid.

8

SOLUTIONS AND COLLOIDS

In presenting the properties of liquids and solids in Chapter 3 it was implied that the discussion pertained solely to pure substances. Only in Chapter 2 was the property of a mixture of substances discussed in connection with Dalton's law of partial pressures. While there is much to be learned about pure substances in all states of matter, there is broader knowledge to be explored when mixtures of substances are considered. In this chapter the properties of two specific types of mixtures called *solutions* and *colloids* are outlined.

Solutions are commonly encountered in nature. Water is a solution containing dissolved minerals, which to a large extent determine its palatability and which plants absorb from the water in the soil. The life processes of animals such as digestion and circulation involve chemical changes in solution. The atmosphere is a solution of gases, which must be in proper proportions to maintain life, just as dissolved gases in water support fish and marine life.

A colloid is an intermediate state of aggregation between pure substances and true solutions. This state of matter can be regarded as a suspension whose lifetime is finite. Eventually most colloids separate into their component substances. Among the commonly encountered colloids are whipped cream, mayonnaise, milk, aerosol sprays, glue, and starch.

8.1 SOLUTIONS

An exact definition of a solution is not easily stated in spite of the frequency with which solutions occur. Indeed it is often the case that the most common things are the most difficult to define in a precise and all-inclusive statement. *Solutions* are homogeneous mixtures of substances whose composition is continuously variable within limits. The homogeneity of a system can be ascertained by successive examination of smaller and smaller microscopic portions within the system. At some point it becomes necessary to cease subdividing the system and declare it a homogeneous mixture or a solution. Thus, the dividing line between heterogeneous mixtures and solutions may be arbitrary in certain cases. However, a true solution is a molecular dispersion whose properties are uniform down to the molecular level, whereas colloids, as shown later, consist of dispersions of aggregates of molecules.

The substance present in the largest quantity in a solution is referred to as the *solvent;* the substance dispersed in the solvent is called the *solute.* Here again the definitions often are more trouble than they are worth. There are substances that form mixtures with widely varying compositions. A drop of water will dissolve in a glass of alcohol, as will a drop of alcohol in a glass of water. Both systems are solutions of water and alcohol. In the former case the solvent is alcohol, while in the latter case the solvent is water. However, in using the terms *solute* and *solvent* in these solutions of water and alcohol, we lose sight of the fundamental fact that a system of two components exists for which the composition is variable and continuous over the entire possible range of quantities.

There are many types of solute–solvent systems possible, because matter exists in three states (Figure 8.1). Solutions of two gases were considered in Chapter 2. Dalton's law of partial pressures is the only property of such a solution that is significant to discuss in an introductory chemistry text. There are no true solutions of a liquid in a gas or a solid in a gas. Such systems are colloids and are discussed in the latter portion of this chapter. The most frequently encountered solutions are those of any of the three states of matter in a liquid solvent. Air dissolves in water the same as alcohol and salt. While liquid solutions are more common, solid solutions involving all three states of matter as solutes are known: liquid mercury dissolves in many metals, such as zinc, sodium, and gold, to form amalgams; hydrogen gas dissolves in certain metals, such as nickel, palladium, and platinum; and, finally, copper dissolves in gold to form an *alloy.*

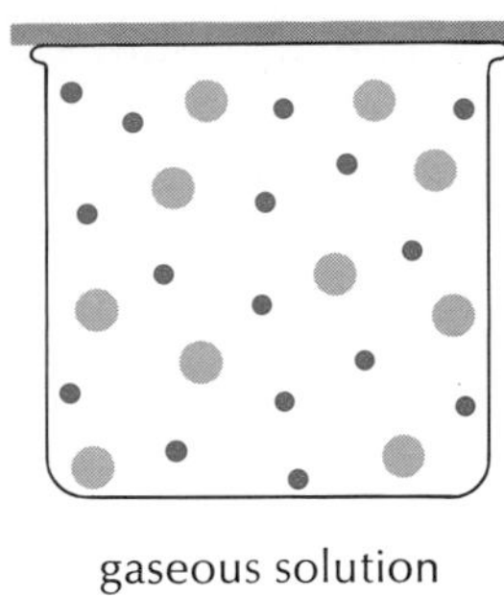

gaseous solution

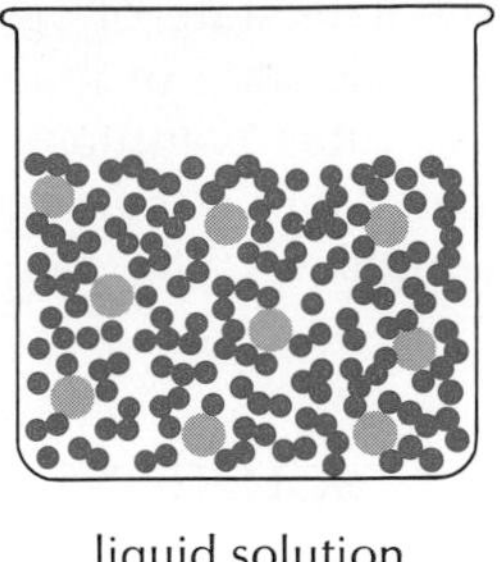

liquid solution

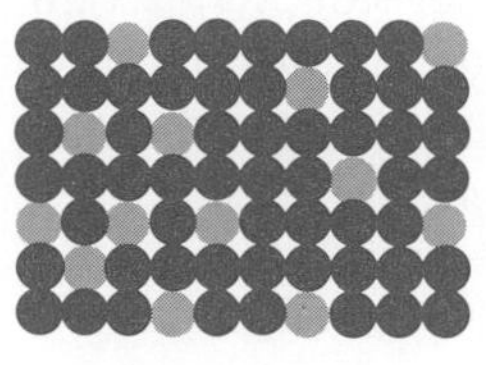

solid solution

Figure 8.1
Models of solutions.

There are many cases of solids dissolving in solids to form alloys. For example, a 16-carat gold piece consists of 16 parts of gold and 8 parts of copper by weight. Brass is a solution of copper and zinc, while bronze is a solution of copper and tin. Coinage metal always has been a solution of a precious metal and varying amounts of other metals to improve the durability of the coins.

8.2 CONCENTRATION UNITS

The terms *solute* and *solvent* do not indicate the specific amount of each component present in a solution. Because the properties of a solution are a function of its composition, the exact composition must be expressed. Concentration units indicate the closeness of the solute particles in the solution. When there is a large amount of solute dissolved in a quantity of solvent, the solution is said to be concentrated. If the quantity of solute is small with respect to the solvent, the solution is dilute. Concentration can be expressed in a variety of units.

In industry, hospitals, and other applied areas, concentration is measured most often in *mass percent.* The number of grams of solute contained in 100 g of solution is equal to the mass percent concentration of the solute:

$$\text{mass percent concentration} = \frac{\text{g of solute}}{\text{g of solution}} \times 100$$

If 5 g of dextrose, a sugar, is dissolved in 95 g of water, 100 g of a 5 percent dextrose solution suitable for intravenous therapy is obtained.

Since the basis of this method of expressing concentration involves a percentage, any quantity of solute and solvent can be dealt with conveniently. A solution of 6 g of sodium chloride in 594 g of water is a 1 percent salt solution. Although this method of expressing concentrations is a direct one involving measured quantities, it does not provide any immediate indication of the relative number of moles of solute or solvent present in the solution. Since many properties of solutions can be quantitatively correlated with the relative number of molecules or moles of substances making up the solution, mass percent concentrations are rarely used in the chemical laboratory. The units of concentration most commonly used there express the amount of solute in terms of moles.

The *molarity* of a solution is equal to the number of moles of solute present in 1 liter of solution:

$$\text{molarity} = \frac{\text{moles of solute}}{\text{liters of solution}} = M$$

Thus if two moles of alcohol are dissolved in enough water to produce 1 liter of an alcohol–water solution, the molarity of the alcohol is $2M$ (molar). Similarly, 1 mole of alcohol dissolved in sufficient water to produce 200 ml of solution results in a $5M$ alcohol solution. Molarity is a very convenient concentration unit because volumes of solutions can be easily measured with calibrated vessels, and the number of moles of solute is available by simple arithmetic calculation. Since chemical reactions and properties are related to the number of moles of substances, the molarity units are obviously very useful to the chemist.

EXAMPLE 8.1

Calculate the molarity of a solution prepared by dissolving 6.6 g of alcohol in sufficient water to obtain 100 ml of an alcohol solution. The molecular formula for alcohol is C_2H_6O.

The mass of 1 mole of alcohol is 44 g. The number of moles of alcohol dissolved in the solution is 0.15:

$$\frac{6.6\text{ g}}{44\text{ g/mole}} = 0.15\text{ mole}$$

The molarity of the solution is equal to the number of moles dissolved per liter of solution. In this case the volume is 100 ml or 0.1 liter. The molarity is $1.5M$.

$$\frac{0.15\text{ mole}}{0.1\text{ l}} = 1.5M$$

EXAMPLE 8.2

How many moles of alcohol are contained in 200 ml of a 0.2M solution?

The molarity of the alcohol solution indicates that 0.2 moles of alcohol are contained in 1 liter of solution. The 200-ml sample must contain a fraction of that number of moles because the sample is less than 1 liter in volume:

$$0.2M = \frac{0.2 \text{ mole}}{1 \text{ liter}}$$

$$\frac{0.2 \text{ mole}}{1 \text{ liter}} \times 0.200 \text{ liter} = 0.04 \text{ moles}$$

The above answer is correct as indicated both by the dimensional analysis and the magnitude of the answer.

The *molality* of a solution is defined as the number of moles of solute contained per 1000 g of solvent. A 1*m* (molal) solution of sodium chloride in water is one in which 1 mole (58.44 g) of sodium chloride is dissolved in 1000 g of water. Since the density of water is 1 g/ml, the molarity and molality of dilute aqueous solutions are nearly identical. In solvents whose densities are different from 1 g/ml the molarity and molality of a solution are not close in magnitude. Molarity is in terms of volume of a resultant solution, whereas molality is in terms of 1000 g of solvent.

The *mole fraction* of a component of a solution is the number of moles of that component divided by the total number of moles of all the components of the solution.

$$\text{mole fraction of solute } A = X_A = \frac{\text{moles of solute } A}{\text{moles of solute } A + \text{moles of solvent}}$$

A concentrated solution consisting of 1 mole of alcohol and 4 moles of water is 0.2 mole fraction in alcohol. The mole fraction of water is 0.8.

$$\text{mole fraction of alcohol} = \frac{1}{1+4} = 0.2 = X_{\text{alcohol}}$$

$$\text{mole fraction of water} = \frac{4}{1+4} = 0.8 = X_{\text{water}}$$

8.3
A MODEL FOR DISSOLUTION

The physical process that occurs when a solute is dissolved in a solvent can be pictured in terms of the kinetic molecular theory of matter. If a solid is placed in the proper solvent, it dissolves. This indicates that the forces maintaining the molecules in a lattice are exceeded by a somewhat stronger interaction between them and the solvent molecules, causing the solute to be dispersed throughout the solvent. If the quantity of solute relative to solvent is sufficiently large, the solid eventually ceases to dissolve. At this point the solution is said to be *saturated.* A solution is saturated when the concentration of dissolved solute is such that it can exist in equilibrium with excess undissolved solute.

While no further change is apparent in a saturated solution, two dynamic processes are continuing. Some of the solid is still dissolving in the solvent while some of the dissolved solute is returning to the solid lattice. The system is in a state of dynamic equilibrium, as can be demonstrated by placing an irregularly shaped crystal in the saturated solution and watching it change in shape. It retains its mass despite its obvious change. Some of the solid must be dissolving while a compensating precipitation of the solute is occurring (Figure 8.2).

The *solubility* of a solute in a solvent is its concentration in a saturated solution. This solubility can be shown experimentally to be a function of the nature of the solute, the nature of the solvent, the temperature, and the pressure. All of these factors must be considered in establishing a model for the dissolution process. The separation of solute particles from each other requires energy. Likewise, the solvent molecules must be separated in order to provide for the introduction of the solute particles, a process that also requires energy. These two facts suggest that there must be a solute–solvent interaction in which energy is released. If there were no net attraction between solvent and solute particles, energy would be required for the dissolution and the substance would not be expected to dissolve spontaneously. However, since the tendency for a spontaneous change is controlled not only by energy release but by entropy changes as well, there are solution processes in which energy is not released. A solution is more disordered than either the pure solvent or solute and, therefore, the solution is a more probable state for the sub-

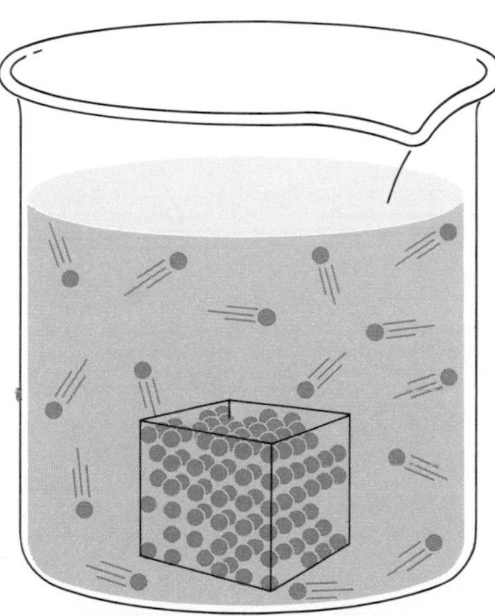

Figure 8.2
Model of a saturated solution.

stances. Dissolution frequently occurs because of an entropy increase in spite of the fact that solvent–solute attractions are not sufficiently large to counteract the energy required to separate solute particles from each other and solvent molecules from each other.

8.4 SOLUBILITY

EFFECT OF PRESSURE ON SOLUBILITY

As was pointed out in discussing the properties of the states of matter, the most dramatic physical changes that occur with pressure changes are encountered in the gaseous state. Similarly, the solubility of gaseous solutes in liquid and solid solvents exhibit the largest dependence on pressure. The solubility of a gas in a liquid can be expressed in terms of mass or volume of gas per unit volume of solvent or in mole fraction units. Whatever units are chosen, the solubility of a gas is directly proportional to the partial pressure of that gas above the surface of the solution. The change in solubility as a function of pressure is a property of the gas and the solvent. Therefore, k must be evaluated for each combination of solute and solvent:

$$X_{\text{gas}} = kP_{\text{gas}}$$

This relationship is known as *Henry's law* and is another phenomenon that can be interpreted in terms of Le Châtelier's principle. If the pressure above the surface of a liquid is increased, the strain imposed on the system can be relieved by diminishing the volume (Figure 8.3). Dissolution of the gas in the liquid is the physical option for such a system.

Carbonated beverages are practical examples of the operation of Henry's law. All carbonated beverages are bottled under pressure, and when the bottle is opened the pressure above the solution diminishes.

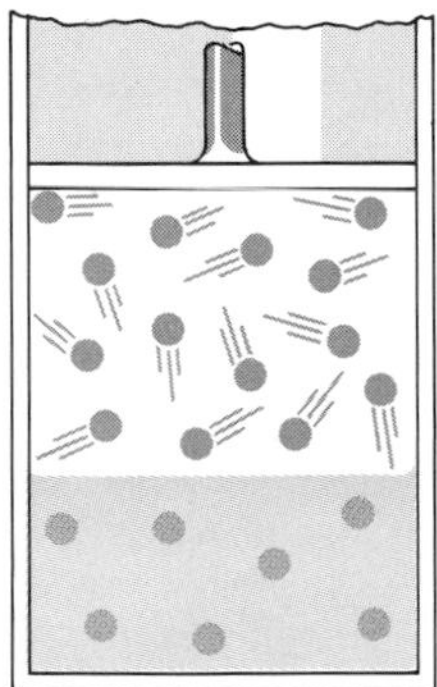

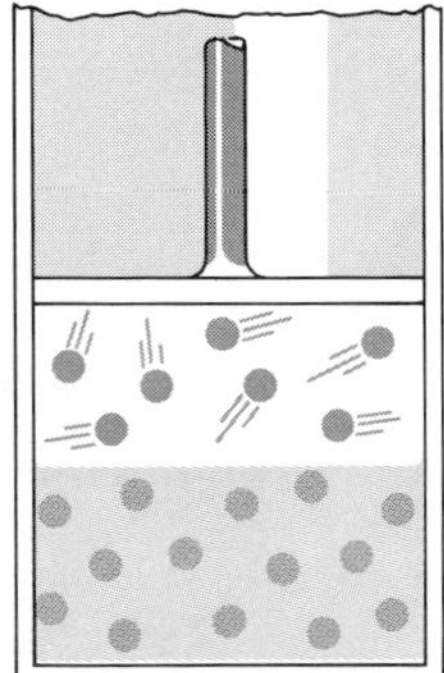

Figure 8.3
Model of the effect of pressure on the solubility of a gas.

Table 8.1
Solubility of gases in 1 liter of water at 20°C

gas	(g)
hydrogen	0.0019
nitrogen	0.029
oxygen	0.070
carbon dioxide	3.4
ammonia	1001

As a result the solubility of the gas decreases, the solution effervesces, and the dissolved carbon dioxide bubbles off. At 0°C a liter of water will dissolve 1.7 liters of carbon dioxide under standard conditions. If the pressure is increased to 5 atm the solubility of carbon dioxide increases fivefold. In Table 8.1 the solubilities of some common gases in water are listed.

There is virtually no change in solubility observed in solutions of liquids and solids when the pressure is slightly increased. However, if pressures in the order of thousands of atmospheres are applied on solutions of liquids and solids, solubility changes should result. The exact nature of the change can be predicted by Le Châtelier's principle. It is only necessary to know the relative volumes of the solutions and the pure substances. If the pure substances occupy a greater volume than the solution, a pressure increase will increase the solubility, and vice versa.

EFFECT OF TEMPERATURE ON SOLUBILITY

The solubility of many gases in water decreases with increasing temperature. As water is warmed, the dissolved air can be seen to form bubbles and escape from the surface of the liquid. While the solubility in other solvents often exhibits the same behavior, there are cases known where the opposite effect is observed.

There is no general rule for the solubility changes of liquids and solids with temperature. Quite often solubility increases with increasing temperature. For example, the solubility of potassium chloride (KCl) increases from 28 g/100 g of water at 0°C to 57 g/100 g at 100°C. Similarly, the solubility of sodium chloride (NaCl) increases from 35 to 40 g/100 g of water over the same temperature range. The solubilities of salts such as cerium sulfate [$Ce_2(SO_4)_3$] decrease with increasing temperature (Figure 8.4).

The relationship between solubility and temperature can be predicted from the heat of solution of the substance in the solvent. The heat evolved or absorbed when the solute dissolves to produce a saturated solution is called the *heat of solution*. If heat is evolved when solution occurs, the heat of solution is a negative quantity:

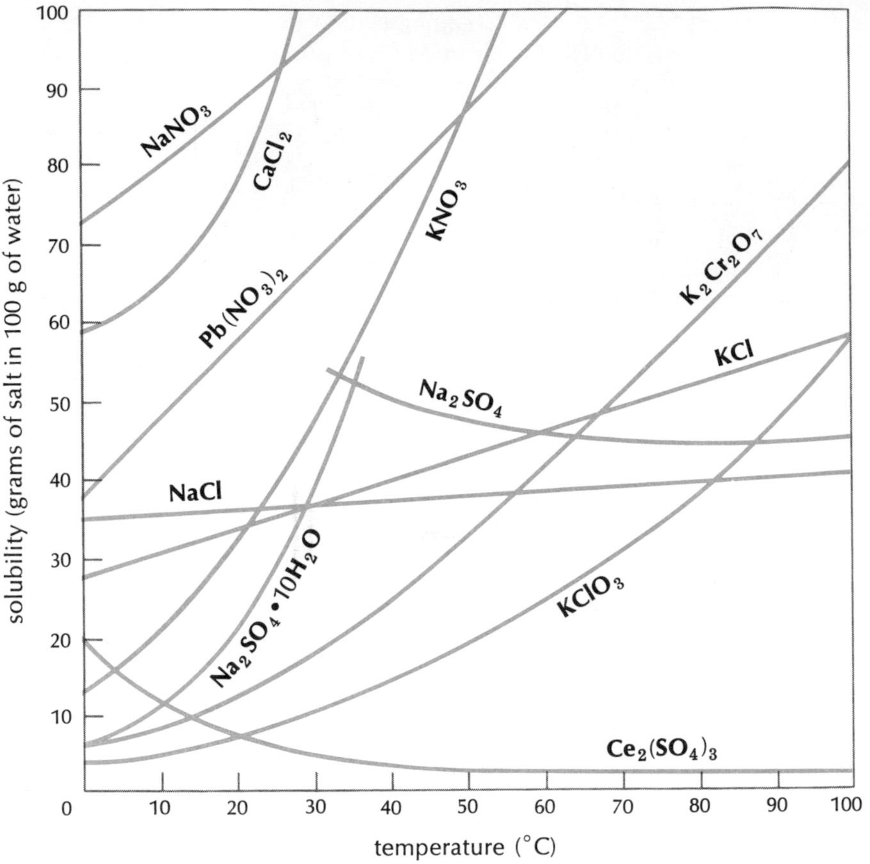

Figure 8.4
Solubility of solids in water as a function of temperature.

$$\text{solute} + \text{solvent} = \text{solution} + \text{heat} \qquad \Delta H < 0 \text{ negative heat of solution}$$
$$\text{heat} + \text{solute} + \text{solvent} = \text{solution} \qquad \Delta H > 0 \text{ positive heat of solution}$$

When potassium chloride dissolves in water the temperature of the solution decreases, indicating that heat is absorbed and that an endothermic process has occurred. The heat of solution of potassium chloride is positive. Thus, if heat is supplied to increase the temperature of the solution, Le Châtelier's principle predicts that the solubility of the potassium chloride should increase. The increased dissolution absorbs heat energy, thus decreasing the change imposed on the system. The heat of solution of cerium sulfate is negative, and its solubility decreases with increasing temperature, as would be expected from the proper application of Le Châtelier's principle.

Figure 8.5
Attraction between dipoles and ions.

EFFECT OF SOLVENT POLARITY ON SOLUBILITY

A maxim of the chemistry laboratory is that "likes dissolve likes." This generalization is reasonable since molecules of solute that are similar to molecules of solvent are expected to be better able to coexist in the same phase. Water is classified as a polar solvent since it consists of a collection of polar molecules. It is a good solvent for polar solutes, ionic compounds, and substances that can produce ions in water. Carbon tetrachloride (CCl_4), which can be used to remove grease spots from clothes, is a poor solvent for sodium chloride. However, fats and waxes readily dissolve in this nonpolar solvent because they are relatively nonpolar substances.

When water dissolves an ionic compound such as sodium chloride, the sodium and chloride ions are dislocated from their rigid lattice structure in which strong electrostatic attractions exist between neighboring ions. In order to overcome these forces water is postulated to be oriented about the ions in solution, as illustrated in Figure 8.5. In the dissolution process, water weakens the interionic attractions and separates the ions from the lattice by the development of attractions between the ions and the water dipole.

8.5
PROPERTIES OF SOLUTIONS

VAPOR PRESSURE OF SOLUTIONS

The addition of salt or sugar to water decreases the vapor pressure of water. This fact can be easily demonstrated by using the same apparatus for vapor pressure determination depicted in Figure 3.3 and substituting a solution of sugar in water for the capsule of solvent. However, the net observed difference in vapor pressure may be quite small if a dilute solution is used. A somewhat more dramatic experiment involves the direct comparison of the vapor pressures of water and a sugar solution, as shown in Figure 8.6. The volume of the sugar solution increases while that of the pure water decreases. Eventually complete transfer of the liquid to the beaker containing the sugar occurs. The only possible interpretation for observed results is that the escaping tendency of water, its vapor pressure, has been reduced by the fraction of solute contained in the beaker. Therefore, the solution, which is in contact with the water molecules in the gaseous phase, eventually captures the water molecules escaping from the beaker of pure water. Of course water molecules are escaping and being captured in the liquid phase of each beaker. The net transfer results from the slower rate of escape from the solution as compared to the pure liquid.

For solutions containing nonvolatile solutes such as sugar, the lowering of the vapor pressure of the solvent is directly proportional to the concentration of the solute in the solution. This relationship is known as Raoult's law.

vapor pressure of solution = vapor pressure of pure solvent × mole fraction of solvent

$$P_{\text{solution}} = P^{\circ}_{\text{solvent}} \, X_{\text{solvent}}$$

When a solution contains volatile solutes both the solvent and solute exist in equilibrium with their vapors. Therefore, the vapor pressure of such a solution cannot be predicted by Raoult's law as stated above. The

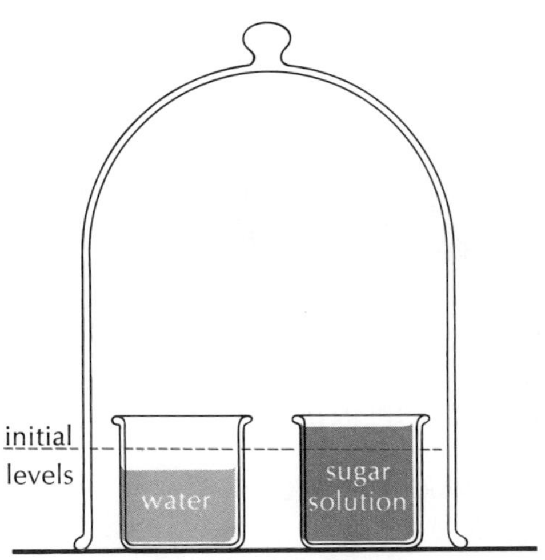

Figure 8.6
The difference in vapor pressure of a solution and a solvent.

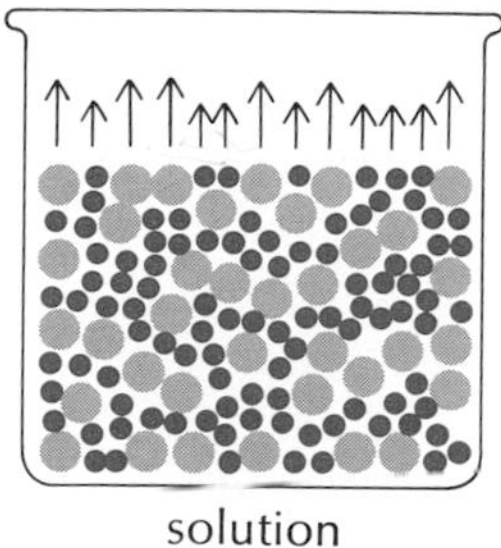

Figure 8.7
Model of the vapor pressure of a solution of a volatile solute.

individual vapor pressures or escaping tendencies of each component, however, should be proportional to the relative amount of that component in the system (Figure 8.7). It is observed often that the total vapor pressure of a solution is equal to the sums of the individual partial pressures of the component substances as calculated by Raoult's law. The partial pressure of each substance is the product of the mole fraction of that component and the vapor pressure of the pure substance:

$$P_x = P_x^\circ X_x$$
$$P_y = P_y^\circ X_y$$
$$P_{\text{total}} = P_x + P_y$$

If a solution contains equimolar amounts of substances *x* and *y* (that is, mole fraction of $x = X_x = 0.5$ and mole fraction of $y = X_y = 0.5$), the vapor pressure can be calculated if the vapor pressures of *x* and *y* at the same temperature are known. If the vapor pressures of *x* and *y* are 80 and 20 mm Hg, respectively, the partial pressures of *x* will be 40 mm Hg and that of *y* will be 10 mm Hg. The total vapor pressure of the solution can be predicted as 50 mm Hg.

The partial pressures of the components of a solution are related to the vapor pressures of the pure substances and their relative amounts in solution. Since the concentrations of substances in the gas phase are in the ratio of their partial pressures, the concentrations can be substantially different from those in the liquid solution. In the above example the mole fraction of *x* in the vapor phase is 0.8. There is an enrichment in the gaseous phase of the more volatile component. If the vapor were separated from the liquid and condensed, a partial separation of *x* from *y* would be effected. Repetition of the evaporation–condensation process would lead to a further increase in the mole fraction of *x*. Such a process is the basis for the distillation and separation of a volatile mixture into its components. The most volatile substance distills from the solution with the resultant residue having an increased concentration of the least volatile component.

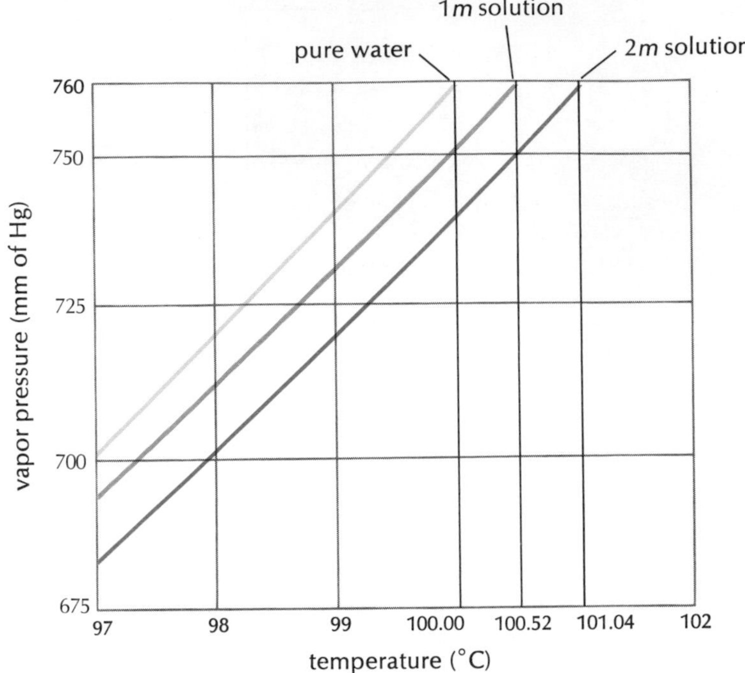

Figure 8.8
Effect of dissolved solute on the vapor pressure and boiling point of water.

BOILING POINT OF SOLUTIONS

The decreased vapor pressure of a solution of a nonvolatile solute means that the boiling point of the solvent must be elevated (Figure 8.8). It will require a higher temperature to raise the vapor pressure to atmospheric pressure. There is a direct relationship between the escaping tendency of the solvent molecules (vapor pressure) and the number of solute particles. Therefore, the increase in the boiling point of the solvent also is directly proportional to the concentration of solute.

By convention the concentration of solute is expressed in molal units. The elevation of the boiling point of a solution caused by the addition of sufficient solute to produce a 1m solution is called *the boiling point elevation constant*. In the case of water a 1m solution of a substance such as sugar boils at 100.52°C at atmospheric pressure. The boiling point elevation constant is 0.52°C. A solution containing 1 mole of sugar in 400 g of water will boil at 101.30°C because the solution is 2.5m.

$$\text{change in boiling point} = \Delta T_b = K_b m$$

$$K_b = 0.52°\text{C}/m \qquad \text{for water}$$

EXAMPLE 8.3

What is the molality of a sugar solution whose boiling point at 1 atm is 100.39° C?

The boiling point elevation of 0.39° is less than that of a 1*m* solution. Therefore, the molality must be equal to 1*m* times the factor 0.39/0.52, or 0.75*m*. The same answer can be obtained by direct substitution into the boiling point elevation equation:

$$T_b = K_b m$$
$$0.39° = (0.52°/m)m$$
$$0.75 = m$$

A 1*m* solution of sodium chloride in water does not boil at 100.52°C but at approximately 101°C. The boiling point elevation is twice that expected on the basis of the sugar example used above. The reason for the higher boiling point elevation in the case of sodium chloride is that a 1*m* solution contains 2 moles of particles: 1 mole of sodium ions and 1 mole of chloride ions. Therefore, the boiling point elevation constant actually is based on the number of moles of particles present in solution. A boiling point elevation of a solution of $CaCl_2$ of a given molarity is three times that of a solution of sugar of the same molality since 3 moles of ions are present in solution per mole of compound. Such observations aided the development of the model of an ionic compound. Alternatively, if the existence of ions is assumed, it can be said that the observed boiling point elevations of ionic compounds are consistent with the model.

FREEZING POINT OF SOLUTIONS

The escaping tendency of a solvent decreases upon addition of a solute both for the liquid-vapor and liquid-solid transformations. However, the observed physical change in the case of freezing points is a decrease rather than the increase noted for boiling points. This decrease in the freezing point can be seen qualitatively in Figure 8.9. The escaping tendency of solvent molecules in the solid state is unaffected by solute contained in the liquid phase. The tendency of the liquid to enter the solid is decreased because of the presence of the dissolved solute. If solute is added to an equilibrium system of solid and liquid at constant temperature, the solid will melt. In order to equalize the relative escaping tendencies of solvent between the two phases the temperature must be lowered. At some lower temperature the liquid and solid phases can coexist, and this temperature is the freezing point of the solution.

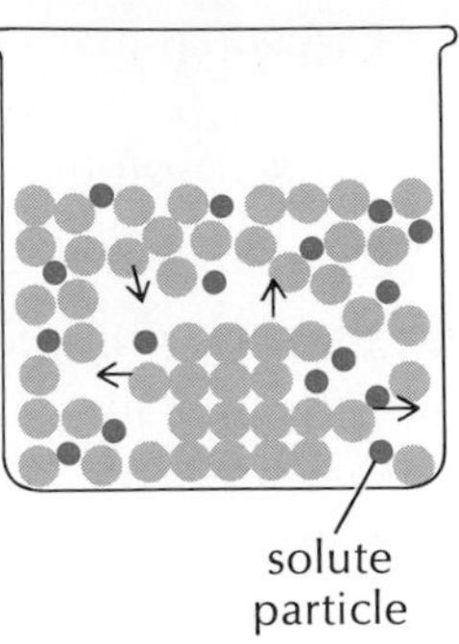

Figure 8.9
Model for the decreased freezing point of solvents containing solute.

Just as the boiling point elevation of a solution depends on the number of moles of solute per 1000 g of solvent, the freezing point depression of a solution also is a function of the number of particles present in solution:

$$\text{freezing point depression} = \Delta T_f = K_f m$$

The freezing point depression constant for water is 1.86. A solution of 1 mole of a covalent substance in 1000 g of water depresses the freezing point of water to −1.86°C. For an ionic substance such as sodium chloride the freezing point of a 1*m* solution reflects the presence of 2 moles of ions that are present in each mole of compound.

EXAMPLE 8.4

A 10-g sample of a covalent compound is dissolved in 250 g of water, and the freezing point of the resultant solution is −0.93°C. What is the molecular mass of the substance?

The freezing point of the solution indicates that the molality of the solution is less than 1*m*. The calculated molality is 0.5:

$$\begin{aligned} \Delta T_f &= (-1.86°/m)m \\ -0.93° &= (-1.86°/m)m \\ 0.5 &= m \end{aligned}$$

The amount of substance required to form the solution corresponds to 0.5 mole/1000 g of water. For the 250 g of water used in the problem a lesser amount (0.125 mole) is required:

$$0.5 \text{ mole}\left(\frac{250 \text{ g}}{1000 \text{ g}}\right) = 0.125 \text{ mole}$$

Therefore, the 10 g of compound is 0.125 mole. The mass of one mole is 80 g:

$$\frac{10 \text{ g}}{0.125 \text{ mole}} = 80 \text{ g/mole}$$

There are many practical applications of the freezing point depression of liquids. Salt is commonly spread on snow and ice in order to melt them. In the presence of salt, snow cannot exist at 0°C. Of course this method of melting ice is ineffective if the temperature of the ice is below that at which the freezing point of water can be depressed by the addition of salt. Another common example of freezing point depressions is the use of antifreeze in car radiators. The antifreeze consists of ethylene glycol ($C_2H_6O_2$), which is very soluble in water. The addition of antifreeze prevents the water from freezing at temperatures above that determined by the concentration of antifreeze.

OSMOTIC PRESSURE OF SOLUTIONS

If a solution such as sugar in water is separated from pure water by a *semipermeable membrane,* a membrane through which water can pass but not other large molecules such as sugar, (Figure 8.10), the height of the

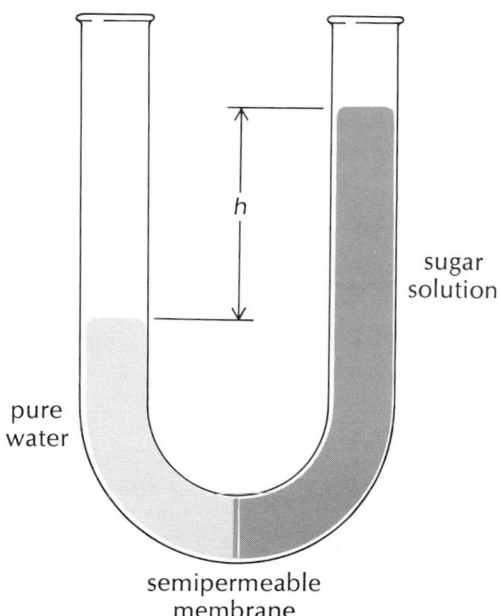

Figure 8.10
Demonstration of the osmotic pressure of a sugar solution.

solution increases at the expense of the pure water. The water remains pure because sugar molecules cannot pass through the membrane. The water molecules apparently can cross and do so up to a point by a process called *osmosis*. Eventually no further transfer occurs, and the level of the solution remains at a specific height above that of the water. The difference in the levels is a measure of the tendency of water to go through the membrane to dilute the solution. The height difference of the liquids exerts a pressure difference at the membrane that counterbalances the tendency of water molecules to continue to dilute the solution.

If pressure is applied on the side of the tube containing the solution, the flow of water molecules can be reversed. The pressure required to maintain equal levels of the water and the sugar solution is called the *osmotic pressure*. At constant temperature this pressure is directly proportional to the number of moles of particles present in solution. Therefore, the osmotic pressure of a sodium chloride solution of given molality is twice that of a solution of sugar of the same molality.

Osmosis of solvent or the passage through a semipermeable membrane is a process of great importance in maintaining life processes. Both animals and plants contain membranes through which water passes. If the concentrations of the solutes in the water solution are not properly balanced, water transport will occur and may lead to cell rupture. In blood the red cells may burst (hemolyze) if the concentrations of dissolved substances are not within certain narrow limits.

8.6 COLLOIDS

When water and sand are mixed a suspension of the sand results that separates upon standing. The time required for separation is a function of the coarseness of the sand. Most samples of sand settle in seconds or minutes. However, in the case of very fine sand resulting from the grinding action of glaciers, a long-lived suspension called *glacial milk* is pro-

Table 8.2
Colloids

dispersion medium	*dispersed phase*	*example*
gas	liquid	fog
gas	solid	smoke
liquid	gas	whipped cream
liquid	liquid	milk
liquid	solid	glue
solid	gas	foam rubber
solid	liquid	cheese
solid	solid	colored glass

duced. (The name was suggested from the appearance of the suspensions in glacial streams that are often like dilute, blue-colored milk.) The reason the particles remain suspended is that their sizes are comparable to atomic and molecular dimensions. The motion of the solvent molecules continually jostles the minute particles, counteracting the tendency of gravity to make them settle.

The size of atomic or molecular species is in the order of 10^{-7} cm in diameter. When aggregates of matter of 10^{-7} to approximately 10^{-5} cm in diameter are suspended in liquids, an intermediate state between heterogeneous mixtures and true solutions results. Substances in this state are referred to as *colloids, colloidal solutions,* or *colloidal suspensions.* Some examples of colloids are given in Table 8.2.

THE TYNDALL EFFECT

When a beam of light is passed through a true solution there is no evidence of it to an observer at right angles to the path of the light. This fact can be easily observed for a liquid solution, as illustrated in Figure 8.11. Similarly, a beam of light passing through dust- and moisture-free air cannot be observed at right angles to the beam. A colloid contains particles that are larger than the molecular species present in true solutions and that are large enough to scatter light. This scattering phenomenon is illustrated in Figure 8.11. The same observation may be made in air containing dust, moisture (fog), or smoke, which are gaseous colloids.

The *Tyndall effect* is controlled by the size and shape of the particles and the wavelength of the light, making it possible to measure the average size of colloidal particles by making quantitative measurements of the scattering pattern. A dramatic qualitative observation of the scattering of light as a function of its wavelength can be made by noting the color of the sky at noon and at dusk. The blue component (short wavelength) of light is very effectively scattered by dust particles in the atmosphere, so

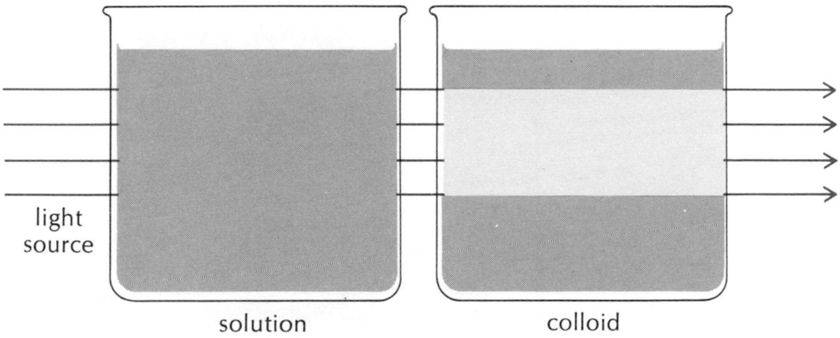

Figure 8.11
The Tyndall effect.

that an observer at noon sees a blue sky. However, since red light is not scattered to the same extent as blue light, an observer to the east at the same instant will see a red sky. This scattering phenomenon produces the vivid sunsets that can be seen in the desert or on the ocean. In each location the long distance to the horizon enables the observer to see light transmitted through miles of air containing colloidal matter.

ADSORPTION OF CHARGED MATTER

Colloids do not conduct an electric current, but when a current is passed through colloidal matter in a liquid, the particles usually move toward one of the charged electrodes in a manner characteristic of the individual colloid. At the electrode the particles are precipitated and no longer migrate. Therefore, the particles must have contained a common charge that was discharged, thereby allowing the previously identically charged species to

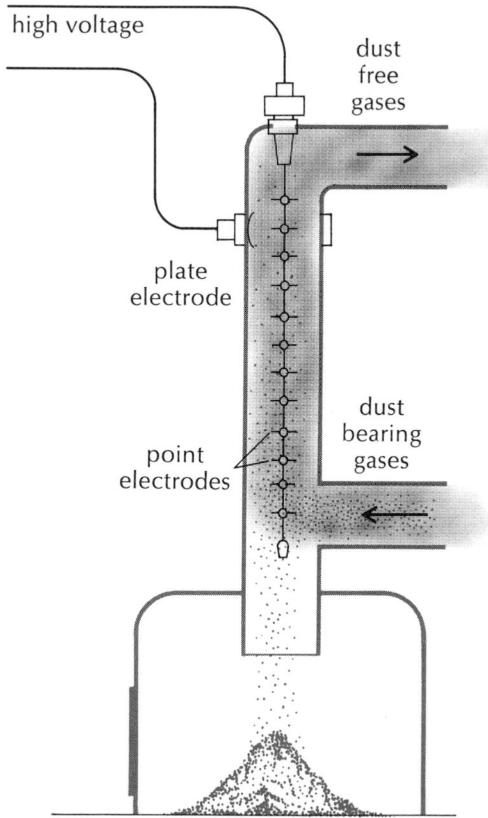

Figure 8.12
The Cottrell precipitator. Adapted from Figure 17-7 of General Chemistry *by W. A. Neville (New York, McGraw-Hill, 1968).*

aggregate. The charge of the colloidal particles is thought to result from adsorbed ions from the surrounding medium, and this charge is responsible for the stability of colloids; the force of repulsion prevents the particles from forming larger particles.

THE COTTRELL PRECIPITATOR

Smoke is a colloidal suspension of solid particles in air. In areas of high industrial concentrations the effluent from the factories and manufacturing concerns can be a considerable nuisance and in most cases a health hazard. A Cottrell precipitator (Figure 8.12) that discharges the colloidal material by passing the smoke through charged electrodes can be installed in the chimneys of commercial concerns. It is not uncommon for large industrial plants to recover tons of solids each day through the use of these precipitators. In such cases the value of the substances recovered may exceed the cost of maintaining the precipitation equipment.

DIALYSIS OF COLLOIDS

Ionic molecular matter of certain limited dimensions can be separated from a colloid by a process called *dialysis*. An animal membrane or, more recently, synthetic membranes with holes of a specific size are used. The dialysis process is illustrated in Figure 8.13. The colloid contained in the membrane bag cannot pass through the holes; ionic substances and molecules of certain dimensions, can pass through. Such substances diffuse through the membrane into the surrounding water and are washed away by the continuous flow of water. By dialysis it is possible to purify colloids.

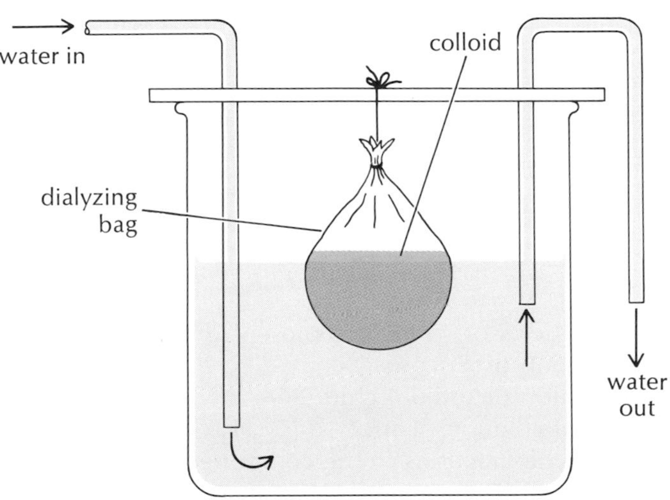

Figure 8.13
Dialysis of a colloid.

The artificial kidney machine is an important practical application of the dialysis process. Complete or partial kidney failure causes an increase in the level of poisonous materials in the blood. If the blood is circulated through tubes containing holes of the proper dimensions, these harmful substances can pass through to the surrounding aqueous solution. By this means the life of an individual can be extended until normal kidney functions are reestablished. In the case of complete kidney failure, life can be maintained by treatment approximately once a week with the artificial kidney.

Suggested further reading

Babbitt, J. D., "Osmotic Pressure," *Science,* **122,** 285 (1955).

Dreisbach, D., *Liquids and Solutions.* Boston: Houghton Mifflin, 1966.

Hauser, E. A., "The History of Colloid Science," *J. Chem. Educ.,* **35,** 271 (1958).

Neidig, H. A., and Yingling, R. T., "Relation of Enthalpy of Solution, Solvation Energy and Crystal Energy," *J. Chem. Educ.,* **42,** 473 (1965).

Sharpe, A. G., "Solubility Explained," *Educ. Chem.,* **1,** 75 (1964).

Snyder, A. E., "Desalting Water by Freezing," *Sci. Amer.,* **207,** 41 (1962).

Wilson, J. N., "Colloid and Surface Chemistry in Industrial Research," *J. Chem. Educ.,* **29,** 187 (1962).

Zingaro, R. A., *Nonaqueous Solvents.* Boston: D. C. Heath, 1968.

Terms and concepts

amalgam
boiling point elevation
colloid
concentrated
concentration
Cottrell precipitator
dialysis
dilute
heat of solution
Henry's law
mass percent
molality
molarity
mole fraction
osmotic pressure
Raoult's law
saturated solution
solubility
solute
solution
solvent
Tyndall effect

Questions and problems

1. How many particles of solute are present in a liter of each of these?
 a. a 1*M* solution of glucose
 b. a 1*M* solution of sodium chloride
 c. a 1 percent solution of sucrose ($C_{12}H_{22}O_{11}$)

2. Calculate the mole fractions of the components of a solution prepared from 92 g of alcohol (C_2H_6O) and 126 g of water.

3. What is the solute and what is the solvent in 12-carat gold?
4. The dissolution of potassium chloride in water is endothermic. How would you prepare a saturated solution at 25°C?
5. A solution of lithium fluoride (LiF) occupies a smaller volume than the LiF and water separately. What effect will an increase in pressure have on the solubility of the LiF?
6. Salt often is added to water used in cooking food. Calculate the boiling point of a salt solution prepared by dissolving 0.565 g of sodium chloride in 1 liter of water. Would such a solution allow food to cook significantly faster? For what purpose is the salt added?
7. The solubility of sodium sulfate (Na_2SO_4) is 488 g/1000 g of water at 40°C and 437 g/1000 g of water of 80°C. Is the dissolution of sodium sulfate exothermic or endothermic? Comment on the relative energy of hydration of the ions and the energy maintaining the ionic lattice.
8. Water and milk freeze at about the same temperature. However, a sugar solution of moderate concentration freezes well below 0°C. Explain why.
9. Calculate the freezing point of a solution containing 31 g of ethylene glycol ($C_2H_6O_2$) in 200 g of water.
10. What is the normal boiling point of an aqueous solution of a nonvolatile solute whose freezing point is −0.93°C? Does your answer depend on the nature of the solute?
11. Will the osmotic pressure of a 0.1*M* solution of sodium chloride be less than, the same as, or greater than (a) a 0.1*M* solution of potassium chloride (KCl), (b) a 0.1*M* solution of calcium chloride ($CaCl_2$), (c) a 0.2*M* solution of glucose?
12. The freezing point depression constants of acetic acid ($C_2H_4O_2$) and camphor ($C_{10}H_{16}O$) are 3.9 and 40.0, respectively. What advantages does the chemist have when he uses camphor instead of acetic acid in determining the nature of a solute?
13. The freezing point of an aqueous solution of a covalent substance containing 10 g of the solute and 500 g of water is −0.31°C. What is the molecular mass of the solute?
14. What effect might be observed in fish moved from a sea level environment to an aquarium in Mexico City? How could this effect be alleviated?
15. Divers suffer from the "bends" upon rising from great ocean depths at an excessively fast rate. What is responsible for this dangerous condition? Why does a slow ascent circumvent the condition?
16. A 1*m* aqueous sugar solution exerts an osmotic pressure of 26.5 atm. Red blood cells neither swell nor shrink in a 0.7 percent sodium chloride solution. Calculate the osmotic pressure of a red blood cell immersed in pure water.
17. Why is the effectiveness of automobile headlights diminished in fog. Why are fog lights usually yellow?
18. How can the charge on colloidal suspensions be determined?

PART TWO

CHEMICAL CHANGE

Common observation and experience testify to the constant change or interconversion of matter by chemical reactions. The rusting of iron, souring of milk, burning of wood, growth of plants and animals, and death of living matter all involve chemical changes. In order to begin to understand chemical reactions we will first examine in this part the principles that govern chemical changes in terms of energy, direction, and rate of change, without regard to the classification of reactions. Then two classes of reactions will be discussed in order to provide models for many of the chemical reactions discussed later in the book.

9

KINETICS AND EQUILIBRIA

In the previous chapters we dealt with the physical properties of substances, the composition of matter, and the relationship between composition and physical properties. Having studied the foundations of molecular structures, we now can examine the interaction of substances with one another from a new perspective. This chapter deals with the chemical transformations of substances—chemical reactions.

The number of chemical reactions that could occur is incalculable. There are known today more than 3 million different compounds; each of these substances could react with the others in many combinations. Although it is not practical to study all of these separately, there are certain basic features common to all chemical reactions; from these we can draw some general principles that provide an operational basis for the study of chemical reactions.

Kinetics is a study of the velocity of chemical reactions and the elucidation of the intimate processes that are involved in chemical change. The *reaction velocity* is a measure of the rate of conversion of reactants into products. The net reaction rate is not necessarily a reflection of a single step involving a molecular scrambling in which the interconversion of matter occurs. Many consecutive or parallel steps may be involved, and substances that are termed *intermediates* can be produced. The intermediates that may or may not be detectable are consumed ultimately in the formation of the product of the reaction. A description of the sequence of the individual steps involved in a chemical reaction is called the *reaction mechanism.*

In this chapter the factors that influence the reaction velocity of either single step or complex reactions are examined. These factors are the structure of the reactant, the temperature, the concentration of reactants, and a class of substances called catalysts. A theory for the rate of the molecular transformation of matter is outlined.

The conversion of reactants into products in theory is never complete. All reactions allowed to continue for an indefinite length of time eventually cease in the macroscopic sense. Although beyond this point no net change between reactants and products is observed, molecular interconversions are still occurring. For every transformation that produces a molecule termed a product, another molecule of product is reconverted to the molecule termed the reactant. Therefore, the rate of one reaction is equal to the rate of the reverse of the reaction. When this state is reached, *chemical equilibrium* is said to be attained. There are close analogies that can be drawn between chemical equilibria and the physical equilibria described for vapor pressure, boiling point, and melting point phenomena.

All equilibria can be interpreted in terms of a *mass action law* and an *equilibrium constant.* These and the effects of concentration and temperature on the position of equilibrium are described in this chapter.

9.1 CHEMICAL EQUATIONS

A reaction can be conveniently represented in terms of a chemical equation involving the symbols of the elements or compounds undergoing the chemical change. The reactants are indicated to the left and the products to the right of an intermediately placed arrow that replaces the customary equal sign in algebraic equations:

$$\text{reactants} \longrightarrow \text{products}$$

Since 1789, when Lavoisier first clearly stated the law of conservation of mass, it has been necessary to represent chemical reactions in terms of equations that comply with this law. In a chemical equation the number of atoms of every element in an elemental state or combined in compounds that appear to the left of the equation must equal the number of atoms to the right of the arrow.

The specific changes involved in a chemical reaction must be determined by experimentation. With the accumulation of many such results it has been possible to discover general classes of reactions that involve similar fundamental changes. This knowledge makes possible an educated guess about the potential products derivable from a particular set of reactants. Some of these reactions are discussed in the next two chapters.

It has been established that the solid element sodium reacts with the gaseous diatomic element chlorine to form the solid sodium chloride. If the proper symbols are used, the following equation can be written:

$$\mathbf{Na + Cl_2 \longrightarrow NaCl}$$

The equation does not represent an equality because there are two chlorine atoms on the left and only one on the right of the equation. The subscript of Cl_2 cannot be changed because this would constitute a misrepresentation of the known facts; chlorine exists as diatomic molecules. Similarly, sodium chloride cannot be represented as $NaCl_2$ in order to achieve an equality, because the substance $NaCl_2$ is unknown and sodium chloride is known to consist of equal numbers of constituent sodium and chloride ions. The only change that is both mathematically and chemically acceptable is to place a coefficient 2 in front of NaCl. The right-hand side of the equation now represents two chlorine and two sodium atoms. Finally, in order to achieve equality a coefficient 2 must be placed in front of Na, and the resultant equation is said to be balanced:

$$\mathbf{2Na + Cl_2 \longrightarrow 2NaCl}$$

A balanced chemical equation can be used to describe molecular changes in many units. The formation of sodium chloride can be viewed as the reaction of two sodium atoms and one chlorine molecule or any multiple of this basic relationship. Therefore, two times Avogadro's number of sodium atoms will react with Avogadro's number of chlorine molecules. For the same equation 2 moles of sodium atoms react with 1 mole of chlorine molecules. If only 0.1 mole of sodium atoms is available, then 0.05 mole of chlorine molecules will be required for complete reaction. Finally, on a laboratory scale the equations can be used to calculate the masses of the substances required for reaction. Two moles of sodium have a mass of 45.98 g and react with the mass of 1 mole of chlorine molecules or 70.90 g. From this mass ratio of reactants any other desired scale of masses may be calculated.

two atoms of sodium + one molecule of chlorine ⟶
two "molecules" of sodium chloride

two moles of sodium atoms + one mole of chlorine molecules ⟶
two moles of sodium chloride

2(22.99) g of sodium + 70.90 g of chlorine ⟶
2(58.44) g of sodium chloride

9.2 REACTION RATES

REACTANTS AND REACTION RATES

The transformation of reactants into products of necessity involves the rupture of some bonds and the formation of others. Therefore it is evident that the identity of the chemical substances is the most important variable controlling a reaction.

Reaction velocities vary considerably as a function of molecular structure. Many reactions between ions in aqueous solutions occur in microseconds and can be monitored only by the most sophisticated electronic devices. In sharp contrast are reactions that require centuries and even perhaps millions of years to reach completion. The formation of oil and related geological reactions that occur deep within the earth are examples of reactions whose velocities are so small as to be beyond our ability to detect. It is reasonable to examine reactions that occur at rates comparable to our time scale. From such information we can extrapolate conceptually to very fast and very slow reactions.

A study of reaction velocities in terms of molecular structure should show characteristics that are a function of chemical bonding and of the chemical potential energy stored within the molecule. Since there are many types of reactions, few generalizations can be made, and the incorporation of the effect of molecular structure into a theory of reaction rates has to be discussed at several points in this text. The specific structural contributions to the reactions of organic compounds are discussed in later chapters.

CONCENTRATION AND REACTION RATES

Two of the common reactions known to everyone are the burning of wood and the rusting of iron. These two reactions are heterogeneous in that they both involve the reaction of the oxygen of the air with the solid state. Obviously the reactants must come in contact with each other to react, and it is commonly known that the reaction velocity increases with an increase in surface area. If wood is chopped into fine kindling or if iron is ground into powder, the reaction velocity increases.

For homogeneous reactions a similar necessity for physical contact of reactants is noted. As the concentration of reactants in either a gaseous or liquid system is increased the reaction velocity increases. In a gaseous system the reaction velocity can be increased either by increasing the number of molecules in a constant volume or by decreasing the volume (increasing the pressure) of a system containing a fixed mass of matter. Similarly, in the liquid phase reactants may be added to increase their concentrations or solvent may be removed. In each case the result is to bring reacting particles closer together.

The actual quantitative change in reaction velocity as a function of concentration change depends on the individual reaction. There is no way to examine a balanced chemical equation and derive this information from the stoichiometry of the reaction. Only by experiment can the relationship between these two variables be determined. However, the rate velocity change is usually a simple integer or rational fraction power of the concentration change. Thus for a doubling of concentration of a specific reactant, the reactant velocity might increase by 2, 2^2, $2^{1/2}$, or $2^{1/3}$.

The experimental determination of the dependence of reaction velocity on concentration allows the statement of a mathematical relationship called a *rate law.* In the following generalized rate law for a hypothetical reaction between *A* and *B*, the square brackets represent the molar concentrations of the respective substances. A proportionality constant *k* is called the *rate constant* and is a function of temperature only.

$$\text{rate} = k[A]^m[B]^n$$

The powers *m* and *n* represent the powers to which the appropriate concentrations must be raised in order for the expression to properly summarize and represent the experimental data. As implied earlier, these powers may be integers or rational fractions of a simple number. Negative exponents are observed to be necessary for some reactions. In this case an increase in concentration decreases the reaction velocity.

TEMPERATURE AND REACTION RATE

While the majority of chemical reactions are exothermic, there are a number of endothermic conversions. In cases where heat energy is consumed there is an accompanying randomization of matter that increases the entropy of the chemical system. This entropy change resembles the one that accompanies physical processes. As will be seen in the latter portion of this chapter, all reactions must in some appropriate combination either give off energy or increase in randomness.

Regardless of the net energy difference between reactants and products or their respective entropy levels, all reaction velocities increase with a rise in temperature. A decrease in temperature decreases the reaction rate. The rate constant *k* that describes the reaction velocity is a function of temperature. The exact relationship between *k* and *T* has been found to be a linear one in log *k* vs. $1/T$. The general form for this relationship is

$$\log k = -\frac{m}{T} + b$$

where *m* is the slope and *b* is the intercept of the line. The value for *m* varies considerably as a function of molecular structure. Some reaction velocities are very sensitive to temperature changes, whereas others are only slightly affected. However, a general rule that can be used with some caution is that a 10°C rise in temperature usually doubles or triples the reaction rate. The increment is larger at lower temperatures than at higher temperatures for the same reaction.

CATALYSTS AND REACTION RATES

Many slow reactions can be speeded up by the addition of a substance called a *catalyst* to the reacting system. Such a substance is said to catalyze the reaction, and its effect is known as *catalysis.* At the termination of the reaction the catalyst, which is usually required only in trace amounts, can be recovered unchanged. Therefore, in a macroscopic sense the catalyst does not appear to be involved with the reactants. However, it is intuitively obvious that the catalyst must have been involved in a microscopic sense or else no change in the velocity of the reaction would have resulted. In those cases that have been examined in detail, it has been shown that the catalyst does interact either physically or chemically with one or several of the reactants but that any consumption of the catalyst at a given step in the reaction is always balanced by a regeneration step at a later point.

When potassium chlorate ($KClO_3$) is heated it slowly decomposes into oxygen and potassium chloride KCl.

$$2KClO_3 \longrightarrow 2KCl + 3O_2$$

If a small amount of manganese dioxide (MnO_2) is added, the decomposition of potassium chlorate is accelerated. At the conclusion of the reaction the $KClO_3$ is completely consumed, but all the MnO_2 remains.

In the chemical industry catalysts are widely used to facilitate the economical conversion of reactants into a desired product. The catalysts chosen are usually very specific; that is, they accelerate one chemical reaction while not facilitating other possible competitive reactions. The minimal requirements of catalysts and the nonconsumptive aspects of their function make the use of catalysts economically desirable in industrial processes. Sulfur dioxide can be converted into sulfur trioxide in the presence of finely powdered platinum. The catalyst is usually written over the arrow in chemical equations:

$$2SO_2 + O_2 \xrightarrow{Pt} 2SO_3$$

Depending on the catalyst employed and the experimental conditions chosen, a set of reactants can be made to produce different products. For example, the reaction of carbon monoxide and hydrogen can produce either methane (CH_4) or methanol (CH_4O), depending on the catalyst chosen.

$$CO + 3H_2 \xrightarrow{Ni} CH_4 + H_2O$$

$$CO + 2H_2 \xrightarrow{ZnO+Cr_2O_3} CH_4O$$

It is the use of such catalysts that makes the petroleum industry with its multiple product requirements so versatile.

Catalytic acceleration of reaction velocities also makes it possible to achieve a reasonable reaction rate at a lower temperature than without a catalyst. There are many catalysts in the plant and animal kingdoms that enable living organisms to function at temperatures that are not destructive to their growth. These catalysts are called *enzymes* and are required for biochemical processes. Each catalyst is highly specific and very efficient. The enzyme ptyalin allows the conversion of starch into sugar at a rate required for the maintenance of life processes. At body temperature the chemical conversion without ptyalin requires weeks, a rate much too slow to support life.

Among the many medical advances of recent years is the discovery of genetic diseases in which individuals cannot produce certain enzyme structures; this condition gives rise to debilitating conditions (Chapter 28). In some cases the design of proper diets has alleviated the dependence of patients on these enzymes. If the diet excludes the chemicals that the individual cannot metabolize, then the chemical cannot accumulate and poison the biological system.

9.3 THEORY OF REACTION RATES

MOLECULAR COLLISIONS AND REACTION RATES

Consider the hypothetical gas phase reaction between the diatomic molecules A_2 and B_2 to produce the molecule AB:

$$A_2 + B_2 \longrightarrow 2AB$$

The reaction must involve one or more steps in which the molecules A_2 and B_2 or some related species collide in order to form the molecule AB. The equation does not provide any information about the exact mechanism by which the reaction proceeds. However, if it is assumed that A_2 and B_2 collide to exchange atomic partners and form AB, then the reaction rate and the number of collisions per unit time must be related. The observed rate is usually only a small fraction of the total number of collisions, and it can be concluded that not all collisions are effective in producing a reaction. If all collisions produced a reaction, then the reaction velocity of gas phase reactions would be extremely rapid and would be equal for many pairs of different reactants under the same conditions of pressure and temperature. This requirement is a natural corollary to the kinetic molecular theory of general gas behavior.

There are two possible hypotheses for the observed inefficiency of molecular collisions with respect to a potential reaction. First, it is conceivable that the energy of collision may be important in determining whether a reaction will occur. In a gaseous sample containing a collection of molecules of varying energies the least energetic molecules upon collision may simply rebound without net change. The high-energy molecules could collide with such force as to allow the necessary scrambling of electrons between exchanging partners. The second hypothesis involves a proposed geometrical alignment of molecules in order to allow proper molecular transformation. The necessary alignment for A_2 and B_2 may be a side-by-side arrangement, and an end-to-end collision may not lead to a reaction (Figure 9.1).

The observed response of reaction velocities to variations in temperature is consistent with the hypothesis of only high-energy molecules leading to reaction upon collision with other molecules. For a change from 20 to 30°C the average velocity of a gas is increased by approximately 3 percent. Since the 10° temperature increase gives rise to a 200 to 300 percent increase in reaction rate, it is obvious that the molecules of average energy are not responsible for the increased reaction velocity. An examination of the Maxwell–Boltzmann distribution of molecular energies reveals that the fraction of high-energy molecules increases substantially at a higher temperature. The fraction at any intermediate energy increases

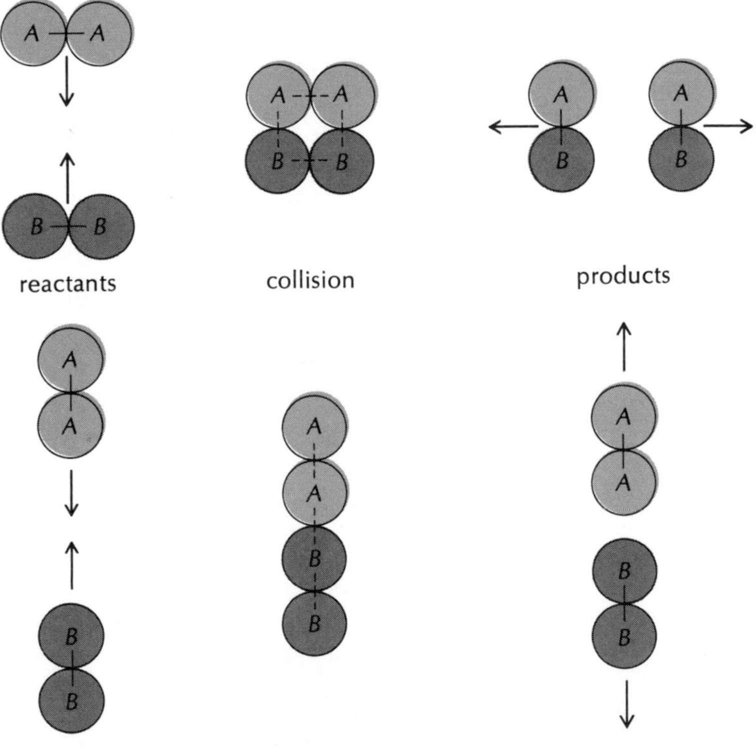

Figure 9.1
Progress of a hypothetical reaction.

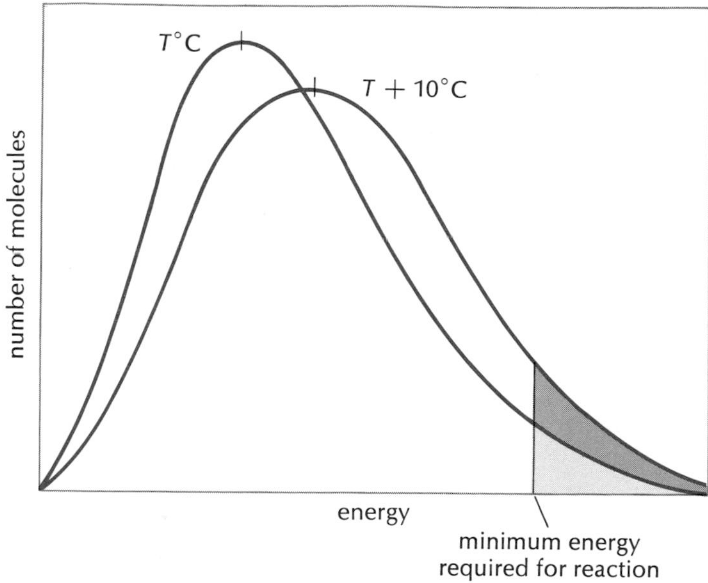

Figure 9.2
Fraction of molecules having energy for reaction as a function of temperature.

by a much smaller amount. This can be seen in Figure 9.2 for an arbitrarily chosen minimum energy required for reaction. The fraction of molecules having the requisite energy for reaction is equal to the shaded area under the distribution curve. For a higher temperature this fraction increases by a factor of two to three for a 10°-temperature increment. It is postulated that this fraction of molecules whose energy exceeds a certain minimum level is responsible for the collisions that give rise to a chemical reaction.

ACTIVATED COMPLEX AND REACTION RATES

The collision model of reaction rates has given rise to a theory of the activated complex. The *activated complex* is considered a transient species that is necessary for reaction. For a simple one-step reaction of molecules A and B to form products X and Y, the collision is assumed to produce a high energy activated complex that can split apart to form X and Y or, alternatively, the original components A and B. The activated complex may or may not resemble either the reactants or the products.

In Figure 9.3 a potential energy curve is plotted as a function of a reaction coordinate that represents the progress of the reaction. Initially A and B are far apart from each other and the sum of their potential energies is indicated by an arbitrary starting point. When A and B approach each other, repulsive forces come into play and the energy of the system increases. In order to force the molecule into proximity for reaction, energy must be added. At some distance of separation the potential energy reaches a maximum and the activated complex is said to be formed. This complex then can decrease the potential energy of the system by forming X and Y or by reverting to A and B. The energy difference between the starting material A and B and that of the activated complex E_a is called the *activation energy*. Only the more energetic molecules have enough kinetic energy to convert to the chemical potential energy of the activated complex. As the temperature is increased, the number of molecules that can undergo this energy transformation process increases and the number per unit time that can get over the potential energy hump increases, leading to a faster reaction rate.

Note that the total potential energy of the products X and Y is less than that of A and B by an amount ΔE. This is usually the case for chemical reactions that occur naturally or spontaneously. As a result of this energy difference, the reaction is exothermic. However, energy is required in order to achieve the activated complex. When the activated complex produces X and Y, this activation energy is recovered and some additional energy is liberated.

When a reaction is speeded by a chemical catalyst, the number of molecules converted from reactant to product per second increases sharply. Chemical catalysis does not occur by allowing molecules to proceed over the same potential energy hump with increased facility. Rather, the presence of a catalyst provides a new pathway for the reaction with lower energy requirements. Such a change implies that the inti-

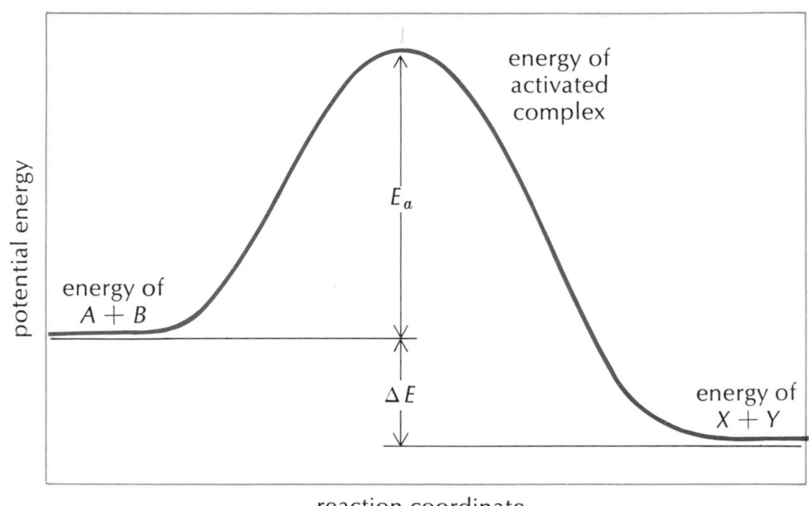

Figure 9.3
Energy and progress of a reaction.

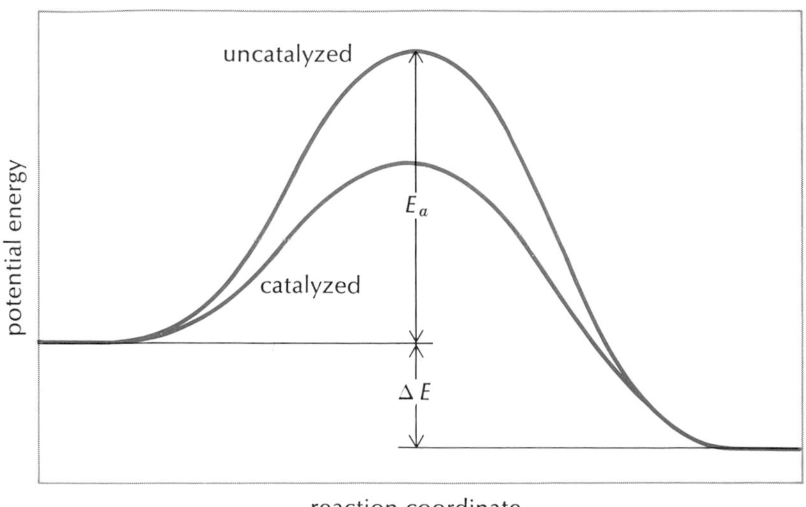

Figure 9.4
Effect of chemical catalyst on the energy requirement of a reaction.

mate details of the reaction, its mechanism, have been altered. Without specifying the details of this change, it is necessary to accept only that in the presence of a catalyst a new pathway becomes available to the molecules (Figure 9.4). The terminus of this pathway is the same, and the products are identical for both the catalyzed and uncatalyzed route. Naturally the number of molecules having sufficient energy to pass over the potential hump is greater for the catalyzed pathway, causing an increase in the reaction rate (Figure 9.5).

REACTION MECHANISMS

On the basis of observed rate equations and occasionally by detection of short-lived reaction intermediates, it is possible to propose a *reaction mechanism* consistent with the experimental evidence. Such mechanisms are only models for a real process and do not necessarily correspond to reality. A mechanism can never be proven but only shown to be consistent with all presently known facts.

It has been pointed out that the rate equation need bear no resemblance to the net reaction equation. The rate equations depend only on the steps up to and including the slowest step in the reaction. If a reaction involves several steps and several activated complexes, the final product can be formed no faster than the slowest step in the sequence. This slowest reaction therefore is called the *rate determining step*.

If reactants A and B form products X and Y, the observed rate law could assume many mathematical forms, depending on the reaction path-

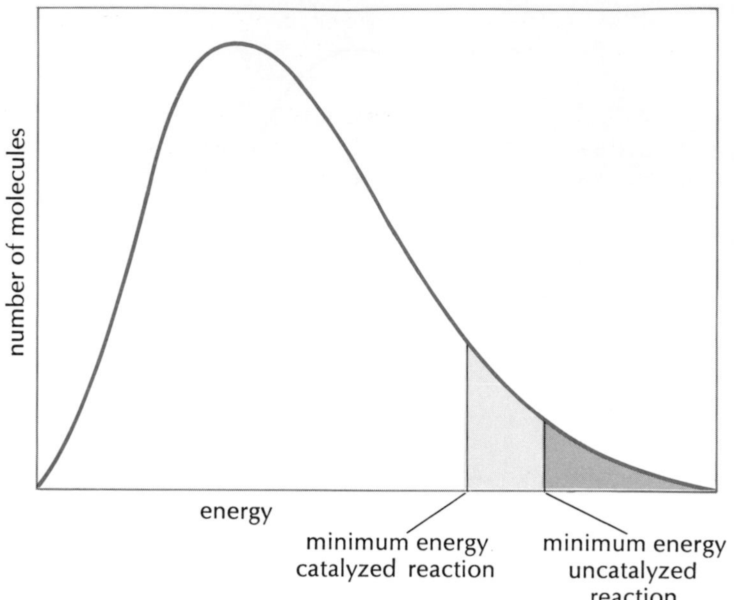

Figure 9.5
Comparison of the fraction of molecules that can react in catalyzed and uncatalyzed reactions.

way. A and B could collide and react via an activated complex to form X and Y:

$$A + B \longrightarrow X + Y$$

For such a single collision mechanism the rate law would be

$$\text{rate} = k[A][B]$$

and the rate would be directly proportional to the first power of the concentrations of both A and B.

If A undergoes a molecular rearrangement via an activated complex to produce M, which then reacts at a faster rate with B to produce X and Y, the rate would be dependent only on the concentration of A:

$$A \longrightarrow M \quad \text{(rate determining step)}$$
$$M + B \longrightarrow X + Y \quad \text{(fast step)}$$

Since the rate of conversion of A to M is slow relative to the subsequent reaction of M with B, the first step is the rate-determining step (rds). A disappears at the rate controlled by the first step. B cannot react until M is formed, and it disappears at a rate identical to the rate of disappearance of A. The products X and Y are formed only as fast as the rate-determining step will allow. The rate law for the proposed mechanism is

$$\text{rate} = k[A]$$

and it indicates possible experiments to support the pathway outlined. If the concentration of A were doubled, the rate of the reaction would double. However, if the concentration of B were doubled, the rate of the reaction would be unaffected because B does not appear in the rate law. Specific examples of reaction mechanisms are discussed in Chapters 16 through 21.

9.4 EQUILIBRIUM

REVERSIBLE REACTIONS AND CHEMICAL EQUILIBRIUM

So far we have considered unidirectional reactions in which reactants are converted to products. While some reactants do undergo essentially complete transformation into products, there are many in which the conversion is incomplete. No matter how much time is allowed for reaction the concentrations of both the reactants and products remain at some fixed concentration. When this condition is achieved the reaction is said to be in *chemical equilibrium*.

If reactants A and B are mixed in equimolar amounts and the products consist of X and Y, the chemical equilibrium eventually achieved is identical to that which can be reached by mixing equimolar amounts of X and Y and allowing sufficient time for equilibration. These observations indicate that there are two possible reactions that serve to interconvert these substances:

$$\begin{aligned} A + B &\longrightarrow X + Y \\ X + Y &\longrightarrow A + B \end{aligned}$$

The equations for the reaction can be combined into a single equation in which the arrows are written in opposing directions:

$$A + B \rightleftharpoons X + Y$$

The observed equilibrium results from the rate of the forward reaction being equal to the rate of the backward reaction at some combination of concentrations of A, B, X, and Y. When only A and B are present, the forward reaction occurs and begins to deplete the supply of A and B molecules, but as the concentrations of A and B decrease, their rate of conversion decreases. The rate of the backward reaction of X and Y is initially zero when only A and B are present. As X and Y accumulate, the rate of their reaction to yield A and B increases. When the rate of reaction of X and Y eventually equals that of the reaction of A and B, equilibrium is achieved.

Equilibrium in a chemical sense is a dynamic condition. Chemical activity has not lessened at equilibrium because microscopic processes are occurring, although the net apparent macroscopic appearance is that of a static condition.

For any reaction at equilibrium there is a definite mathematical equation that can be found experimentally and that serves to indicate the relationship between the concentrations of all substances at equilibrium. For the general balanced equation at equilibrium

$$mA + nB \rightleftharpoons pX + qY$$

the following expression is a constant at a specific temperature:

$$\frac{[X]^p[Y]^q}{[A]^m[B]^n}$$

and is called a *mass action expression.* It is important to realize that the mass action expression for a chemical reaction is a function that can be experimentally verified. There is no necessary dependence of the expression on any theoretical considerations. With the general acceptance of the mass action expression, theoretical models have been proposed.

Since it is found that the mass action expression as written is a constant, then it follows that the reciprocal of the expression also is a constant. However, by convention the expression is always written so that the materials on the right-hand side of the chemical equation appear in the numerator and those on the left-hand side appear in the denominator.

The mass action expression is valid independent of the source of the molecules *A, B, X,* and *Y.* An equilibrium state will be reached eventually if sufficient time is allowed, and the concentrations of the four molecules always will satisfy the mass action expression.

Mass action expressions for any reaction can be written by inspection of the equation

$$2N_2O \rightleftharpoons 2N_2 + O_2 \qquad \frac{[N_2]^2[O_2]}{[N_2O]^2}$$

$$2N_2O_5 \rightleftharpoons 4NO_2 + O_2 \qquad \frac{[NO_2]^4[O_2]}{[N_2O_5]^2}$$

$$N_2O_4 \rightleftharpoons 2NO_2 \qquad \frac{[NO_2]^2}{[N_2O_4]}$$

The numerical value of the mass action expression is called the *equilibrium constant* and is denoted by *K*:

$$K = \frac{[X]^p[Y]^q}{[A]^m[B]^n}$$

For a small equilibrium constant the numerator of the mass action expression must be smaller than the denominator. This condition means that the concentrations of the materials on the right-hand side of the chemical equation are smaller than those on the left-hand side. Conversely, if the equilibrium constant is large, the reaction system at equilibrium has proceeded toward completion in a left-to-right direction.

When the temperature of a reaction is changed, the value of the equilibrium constant changes in a manner that is a function of the energy of the chemical reaction. If the forward reaction is exothermic, an increase in temperature decreases the equilibrium constant because it favors the reverse reaction. For an endothermic forward reaction an increase in temperature gives rise to an increase in the equilibrium constant. These effects are discussed more thoroughly in a later section.

A catalyst does not change the position of a chemical equilibrium and, therefore, the equilibrium constant is independent of catalysts. The catalyst serves to facilitate the achievement of an equilibrium state and functions in allowing both the forward and reverse reactions to proceed at a faster rate. The increase in each rate is exactly equal, and no net positional difference in equilibrium can result.

EQUILIBRIUM AND RATE CONSTANTS

The law of mass action can be shown to be consistent with the kinetic concept of reactions. For the general reaction

$$A + B \rightleftharpoons X + Y$$

the equilibrium expression can be derived from the rate expressions. Of course the rate expressions must be experimentally obtained and are a function of the mechanism of the reaction. If it is assumed that A and B collide in a single step for the forward reaction and that X and Y collide in a single step for the backward reaction, then the rate expressions can be written

$$\text{forward rate} = k_f[A][B]$$
$$\text{backward rate} = k_b[X][Y]$$

At equilibrium the two rates are equal and the following identities can be obtained:

$$k_f[A][B] = k_b[X][Y]$$

$$\frac{k_f}{k_b} = \frac{[X][Y]}{[A][B]}$$

Therefore, the mass law expression is a constant equal to the ratio of two rate constants. Similar treatments of other equilibrium systems leads to the same relationship between K, k_f, and k_b. Although it will not be proven in this text, the equality can be derived regardless of the mechanism of the reaction.

9.5 EQUILIBRIUM CALCULATIONS

The equilibrium constant for the reaction of hydrogen and carbon dioxide at 750°C to yield water vapor and carbon monoxide is 0.77. This value is dimensionless because the concentration units cancel in the equilibrium expression:

$$H_2 + CO_2 \rightleftharpoons H_2O + CO$$

$$K = \frac{[H_2O][CO]}{[H_2][CO_2]} = 0.77$$

If 1 mole of H_2 and 1 mole of CO_2 are mixed in a 1-liter container, the concentrations of all chemicals can be calculated and expressed in moles per liter. Since the container is 1 liter, the concentrations will be numerically equal to the molar quantities.

At equilibrium the number of moles of H_2O may be set equal to a quantity x. Since one molecule of H_2O is produced simultaneously with every molecule of CO the number of moles of CO at equilibrium must be x. For every mole of H_2O produced a mole of H_2 must be consumed and the equilibrium concentration of H_2O must be $1 - x$; similarly, the equilibrium concentration of CO_2 must be $1 - x$. Substitution of these quantities into the mass action expression results in an equation that can be solved by use of the quadratic formula (Appendix 1):

$$K = \frac{[H_2O][CO]}{[H_2][CO_2]} = \frac{x \cdot x}{(1 - x)(1 - x)} = 0.77$$

$$\frac{x^2}{1 - 2x + x^2} = 0.77 \qquad 0.23x^2 + 1.54x - 0.77 = 0$$

$$x = 0.47 \quad x = -0.71$$

Of the two solutions only $x = 0.47$ corresponds to physical reality. The root $x = -0.71$ is a valid mathematical solution for the quadratic equation but is without chemical meaning.

The equilibrium constant at 100°C for the reaction

$$CO + Cl_2 \rightleftharpoons COCl_2$$

is 4.6×10^9 liters/mole. Note that the equilibrium constant has units associated with it because the concentration units in the denominator exceed those in the numerator of the mass action expression

$$\frac{[COCl_2]}{[CO][Cl_2]} = 4.6 \times 10^9 \text{ liters/mole}$$

When 1.0 mole of $COCl_2$ is placed in a 1-liter container at 100°C a small amount of CO and Cl_2 should be formed. The magnitude of the equilibrium constant indicates that the equilibrium concentrations of CO and Cl_2 must be much less than that of $COCl_2$. The number of moles of CO or Cl_2 may be set equal to x because they are produced in a 1:1 ratio. At equilibrium the concentration of $COCl_2$ is $1.0 - x$. Substitution of these quantities into the mass action expression again leads to a quadratic equation that can be solved as follows:

$$\frac{(1.0 - x)}{x \cdot x} = 4.6 \times 10^9$$

$$(4.6 \times 10^9)x^2 + x - 1.0 = 0$$

In this case the quadratic equation need not be solved. It is easier to use an approximate expression in which it is assumed that x is so much less than 1.0 that $1.0 - x$ may be written as 1.0 and the x is neglected. The magnitude of the equilibrium constant suggests that this approximation is valid within the desired accuracy of the answer. Upon solution of the resulting equation the approximation may be checked:

$$\frac{1.0}{x \cdot x} = 4.6 \times 10^9$$

$$x^2 = 2.2 \times 10^{-10}$$

$$x = 1.5 \times 10^{-5}$$

The concentration of CO and Cl_2 are both 1.5×10^{-5} mole/liter. Therefore, the concentration of $COCl_2$ is 0.999985 mole/liter, or essentially 1.0 mole/liter. The approximation used to facilitate solution of the mass action expression is shown to be valid.

9.6
LE CHÂTELIER'S PRINCIPLE

The principle of Le Châtelier as stated in Chapter 3 is that if an external force is applied on an equilibrium system, then the system, if possible, will readjust to reduce the stress imposed on it. While the principle has physical implications, it also can be applied to chemical systems. The application of the generalization of Le Châtelier is very simple and allows a qualitative evaluation of the alterations that occur in a chemical equilibrium when the conditions are changed.

Three changes in condition will be considered: concentration, pressure, and temperature. The first two changes do not affect the value of the equilibrium constant but rather only the individual concentrations of the substances in equilibrium with each other. Temperature changes alter the value of the equilibrium constant as well as the concentrations of the substances in equilibrium. Catalysts have no effect on either the value of the equilibrium constant or the concentrations because they affect the rates of the forward and backward reactions to an equal extent.

CONCENTRATION CHANGES

For the reaction of hydrogen and iodine to yield hydrogen iodide the equilibrium can be upset temporarily by the addition or removal of any component of the system. The equilibrium constant at 425° is 55:

$$H_2 + I_2 \rightleftharpoons 2HI$$

If additional HI is added to the system, the equilibrium will be upset and the system will seek to establish a new equilibrium condition. The reaction to the left tends to occur and increases the concentrations of H_2 and I_2 while using up some of the added HI. At equilibrium some of the HI will be used up, but the HI concentration still will be higher than it was prior to the addition. The increase in HI is such that it exactly counterbalances the increase in the H_2 and I_2 concentrations, leaving the equilibrium constant unaltered.

If I_2 were added to the system at equilibrium the H_2 would react to counteract the change imposed on the system. The concentration of H_2 would be diminished while that of HI would increase. The concentration of I_2 would decrease from its value immediately after addition to some new value higher than its original one. Again the reaction system adjusts to changes imposed on it, and the net effect is a shift of the reaction to the right. The equilibrium constant is unchanged because the concentration changes are mathematically balanced.

PRESSURE CHANGES

The reaction of H_2 and I_2 is not affected by changes in pressure or by changes in volume. The number of molecules on the left-hand side of the equation is two and is equal to the number on the right-hand side. The independence of the reaction from pressure is understandable because there is no way the molecules can counteract an increase in pressure or a decrease in volume. The atoms will be under the same strain as H_2 and I_2 as they would be when in the form of HI. The reaction of nitrogen and hydrogen to yield ammonia, on the other hand, is very dependent on pressure:

$$N_2 + 3H_2 \rightleftharpoons 2NH_3$$

The component atoms of the reagents on the left-hand side of the equation occupy four relative units of volume, whereas on the right-hand side they occupy only two units of volume. With an increase in pressure or a decrease in volume the reaction proceeds to the right to produce more ammonia.

The effect of pressure or volume changes on an equilibrium system is most dramatic for those reactions involving gaseous reactants. While pressure produces changes for liquid and solid components at equilibrium, they are small compared with gaseous components. If an equilibrium involves several phases, only the gaseous components need to be considered to predict the effect of pressure changes:

$$3Fe + 4H_2O \rightleftharpoons Fe_3O_4 + 4H_2$$

For the above reaction at temperatures exceeding 100°C there are four gas molecules on each side of the equation. The equilibrium is unaffected by pressure changes. The Fe and Fe_3O_4 need not be considered because they are not gaseous at ordinary temperatures.

If the energy of reaction is known, the effect of temperature on the equilibrium constant and the components can be predicted. The formation of ammonia from nitrogen and hydrogen is exothermic, and the backward reaction is endothermic. When the temperature is increased the system tends to counteract the change by using up the added energy. The reaction to the left is favored, causing a decrease in the equilibrium constant. As the system is cooled it tends to produce energy by the reaction that proceeds to the right. Therefore, the production of ammonia should be carried out at low temperatures in order to achieve the highest yield. However, it should be noted that at low temperatures the rate of the reaction will be slow.

9.7
ENTHALPY, ENTROPY, AND FREE ENERGY

The equilibrium state for both physical and chemical systems can be treated quantitatively by means of thermodynamics. While the concepts and implications of thermodynamics will not be covered rigorously in this text, there are certain basic features that can be qualitatively described. Indeed, some of these terms already have been implicitly or explicitly introduced.

There are many physical and chemical processes that occur naturally. The flow of water downhill and the expansion of a gas to fill a container are two examples of spontaneous physical processes. The reaction of hydrogen and oxygen to yield water and the burning of wood are examples of chemical processes. Analysis of all naturally occurring processes reveals that two fundamental features control them. Most systems tend to produce a state of lower energy and as a consequence release the net energy difference between the initial state and the final state. The reverse process is less common and can occur only if the second controlling feature, entropy change, is favorable. The flow of water downhill and liberation of heat in the formation of water from hydrogen and oxygen occur because a state of lower energy is achieved. In such cases the energy difference between initial and final states is called the change in *enthalpy* ΔH. By convention the liberation of energy is given a negative sign. For example, the reaction

$$\mathbf{CO_2 + H_2 \longrightarrow CO + H_2O}$$

liberates 9830 cal for each mole of CO_2 that reacts. The enthalpy change $\Delta H = -9380$ cal/mole reflects the fact that the products CO and H_2O are more stable than CO_2 and H_2 by 9830 cal/mole.

The second controlling feature of all processes is the tendency of a system to achieve the most random or disordered arrangement possible. This feature is important in the expansion of a gas or the vaporization of a liquid (Chapter 3). The degree of randomness or disorder is called the *entropy* of a system. By definition the entropy change (ΔS) is positive for increasing disorder. The entropy change counterbalances unfavorable enthalpy changes in some systems. If the increase in the degree of disorder is great, an endothermic process can occur.

The temperature is a significant factor in determining the relative importance of enthalpy and entropy contributions to a system that can undergo change. At absolute zero all substances are ordered and entropy differences between two or more substances are zero. Therefore, only their relative energies determine their relative stabilities. With an increase in temperature a variety of molecular motions become possible, and the tendency toward disorder changes from substance to substance. The differences in entropy can play a variable role as a function of temperature. At extremely high temperatures the entropy differences between various substances may play a dominant role in the course of a reaction.

The relationship between enthalpy changes and entropy changes is given by the following expression, in which ΔG symbolizes the change in *free energy* of a system at constant pressure: $\Delta H - T\,\Delta S = \Delta G$. The free energy change is a measure of the driving force of a reaction or the tendency of it to proceed spontaneously. When ΔG is negative a chemical or physical process occurs. The negative enthalpy change previously described as being important in determining the course of a reaction can be seen to contribute toward making ΔG more negative. Similarly, a positive entropy change contributes toward making ΔG negative. From the expression it can be seen that ΔH is more important at low temperatures and ΔS becomes more important at high temperatures.

The power of thermodynamics stems from the potential to be able to calculate whether or not a hypothetical reaction will proceed in a desired manner. Tables of enthalpies and entropies of substances are available, and by arithmetical addition and subtraction the free energy difference between two possible sets of substances can be calculated. If ΔG is negative the reaction can proceed spontaneously in the direction desired. However, the difference in free energy does not indicate the velocity of the reaction since the rate is dependent on the activation energy and not the difference in energy between reactants and products. Therefore, the spontaneous or naturally occurring processes may proceed at very slow rates.

The size of the free energy change between reactants and products determines the position of equilibrium. For a reaction with negative ΔG the equilibrium constant is larger than 1 and the reaction goes to the right. When ΔG is positive the equilibrium constant is less than 1. The magnitude of the equilibrium constant is related to ΔG by the expression:

$$\Delta G = -2.3RT \log K$$

where $R = 2$ cal/mole-deg.

Allen, C. R., and Wright, P. G., "Entropy and Equilibrium," *J. Chem. Educ.*, **41,** 251 (1964).
Ashmore, P. G., "Reaction Kinetics and the Law of Mass Action," *Educ. Chem.*, **2,** 160 (1965).
Benson, S. W., "Some Aspects of Chemical Kinetics for Elementary Chemistry," *J. Chem. Educ.*, **39,** 321 (1962).
Campbell, J. A., "Kinetics—Early and Often," *J. Chem. Educ.*, **40,** 578 (1963).
Campbell, J. A., *Why Do Chemical Reactions Occur?* Englewood Cliffs, New Jersey: Prentice-Hall, 1965.
de Heer, J., "Le Châtelier, Scientific Principle, or 'Sacred Cow,'" *J. Chem. Educ.*, **35,** 133 (1958).
King, E. L., *How Do Chemical Reactions Occur?* New York: W. A. Benjamin, 1963.
Lindauer, M. W., "The Evolution of the Concept of Chemical Equilibrium from 1775 to 1923," *J. Chem. Educ.*, **39,** 384 (1962).
Mahan, B. H., "Temperature Dependence of Equilibrium," *J. Chem. Educ.*, **40,** 293 (1963).
Miller, A. J., "Le Châtelier's Principle and the Equilibrium Constant," *J. Chem. Educ.*, **31,** 455 (1954).
Nash, L. M., "Elementary Thermodynamics," *J. Chem. Educ.*, **42,** 64 (1965).
Prettre, M., *Catalysis and Catalysts.* New York: Dover, 1963.
Sanderson, R. T., "Principles of Chemical Reaction," *J. Chem. Educ.*, **40,** 13 (1964).
Standen, A., "Le Châtelier, Common Sense and 'Metaphysics,'" *J. Chem. Educ.*, **35,** 132 (1958).

Terms and concepts

activated complex
activation energy
catalyst
equilibrium
equilibrium constant
enthalpy
entropy
enzyme
free energy
kinetic
law of mass action
molecular collision
product
rate constant
rate-determining step
rate law
reactant
reaction mechanism
reaction rate

Questions and problems

1. List the factors that influence reaction rates. Describe how these features are interpreted in terms of kinetic molecular theory.
2. Finely divided metals can burn in air, whereas large masses of the same metals do not burn. Why?

3. Many reactions that are exothermic nevertheless must be started by the addition of energy as heat or light. Explain.

4. Diamond is less stable than graphite, and the conversion reaction at atmospheric pressure is exothermic. Why does the diamond in an engagement ring not change to graphite?

5. Assume that the rate-determining step of a reaction is $A + 2B \longrightarrow C$. What effect on the initial rate will be observed if **a.** the concentration of A is doubled or **b.** the concentration of B is doubled?

6. The reaction rate of the reaction $A + B + C \longrightarrow X + Y$ is doubled when the concentration of either A or B is doubled. However, the reaction rate is unaffected by changes in the concentration of C. Write a possible mechanism and a rate expression consistent with these facts.

7. In the laboratory, the conversion of sugar into CO_2 and H_2O occurs only at high temperatures. However, in the body the reaction occurs at 37°C. Why?

8. The results of a kinetic investigation are given below for the reaction $A + B \longrightarrow X$:

experiment	$[A]$	$[B]$	*initial relative rate*
1	0.01	0.01	1
2	0.02	0.01	4
3	0.01	0.02	4
4	0.02	0.03	36

Write a rate law for the reaction. What would be the relative rate for $[A] = 0.04$ and $[B] = 0.05M$? Suggest a mechanism for the reaction and indicate the rate-determining step.

9. The rate law for the reaction $2O_3 \longrightarrow 3O_2$ is

$$\text{rate of decrease of } [O_3] = kK\frac{[O_3]^2}{[O_2]}$$

and the postulated mechanism involves a rapidly established equilibrium followed by a rate-determining step:

$$\mathbf{O_3} \underset{}{\overset{K}{\rightleftharpoons}} \mathbf{O_2 + O}$$
$$\mathbf{O + O_3} \xrightarrow{k} \mathbf{2O_2}$$

Show how the mechanism is consistent with the observed rate law.

10. List the factors that may shift an equilibrium.

11. What is the effect of increased temperature and pressure on each of the following equilibria in the gas phase?

$$\mathbf{N_2 + O_2 \rightleftharpoons 2NO} \qquad \Delta H = 32 \text{ kcal}$$
$$\mathbf{H_2 + Cl_2 \rightleftharpoons 2HCl} \qquad \Delta H = -44 \text{ kcal}$$
$$\mathbf{CO + Cl_2 \rightleftharpoons COCl_2} \qquad \Delta H = -49 \text{ kcal}$$

12. Write the mass action expression for the gas phase reaction

$$\mathbf{2SO_2 + O_2 \longrightarrow 2SO_3}$$

13. For the gas phase reaction below, the equilibrium constant is 0.022 at 225°C and 33 at 490°C:

$$PCl_5 \longrightarrow PCl_3 + Cl_2$$

Is PCl_5 more stable at high or low temperatures? Is this reaction exothermic or endothermic?

14. At very high pressures graphite can be converted into industrial diamonds. What does this fact indicate about the density of graphite and diamond? Explain why the reaction to form diamond is favored at high pressures.

15. What will be the effect of each of the following changes on the gaseous reaction

$$4NH_3 + 7O_2 \rightleftharpoons 6H_2O + 4NO_2 \text{ at } 200°C$$

- **a.** adding more NH_3 at constant pressure
- **b.** decreasing the volume of the system
- **c.** removing the water by the addition of an absorbing agent
- **d.** addition of a catalyst

16. The equilibrium constant for the gaseous reaction $A \rightleftharpoons 2B$ is 16. A container has 1 mole of A at equilibrium. How many moles of B are in the container?

17. Consider the reaction $X \rightleftharpoons 2Y + Z$. Initially 2 moles of X are placed in a 10-liter flask, and equilibrium is achieved when 0.5 mole of X remain. Calculate the equilibrium constant.

18. Initially 2 moles of HI were placed in a 10-liter container at 425°C. Calculate the concentration of HI after equilibrium is achieved. The equilibrium constant for the reaction $H_2 + I_2 \rightleftharpoons 2HI$ is 55.

19. Consider the following combination of enthalpy and entropy changes for hypothetical reactions. Which reactions will definitely occur? Which will not occur? Which might occur?

- **a.** positive enthalpy change and negative entropy change
- **b.** positive enthalpy change and positive entropy change
- **c.** negative enthalpy change and negative entropy change
- **d.** negative enthalpy change and positive entropy change

10

ACIDS AND BASES

In order to facilitate the learning process and derive some understanding of any group of facts it often is necessary to devise operational definitions to classify the information. From the classification of apparently related facts it usually is possible to derive some fundamental generalizations that provide a basis for understanding the phenomena.

Throughout chemical history a variety of concepts for classification of types of substances and the reactions they undergo have been proposed. In this chapter we discuss substances that have come to be known as *acids* and *bases*. The historical development of this classification scheme is traced to illustrate the growth of this concept. Several operational definitions for acids and bases have been proposed and used widely among chemists of different generations. All of the definitions serve useful purposes and provide basically the same insight into the behavior of acids and bases, and each definition has advantages under appropriate circumstances. The main difference between the definitions is one of scope, in that those developed more recently include more compounds in the classification.

10.1 EARLY CONCEPTS OF ACIDS AND BASES

Acids, as first characterized by Robert Boyle in 1680, were defined as substances that change certain naturally occurring dyes from blue to red and that lose this property when in contact with alkalies (bases). The term *acid* is derived from the Latin words *acidus,* which means sour, and *acetum,* which means vinegar. The term *alkali* is derived from the Arabic *alkali,* which means ashes of a plant. During the eighteenth century acids were considered to be substances that have a sour taste and liberate carbon dioxide when placed in contact with limestone. While many acids do have a sour taste, there are many sour-tasting substances that are not now considered acids.

In terms of elemental composition Antoine Laurent Lavoisier proposed in 1787 that acids are compounds of oxygen. The acids sulfuric acid (H_2SO_4), nitric acid (HNO_3), and perchloric acid ($HClO_4$) all contain oxygen. A few of the many acids that occur in living matter and contain oxygen are acetic acid ($C_2H_4O_2$) of vinegar, oxalic acid ($C_2H_4O_4$) of rhubarb, citric acid ($C_6H_8O_7$) of lemons, and tartaric acid ($C_4H_6O_6$) of grapes. In 1810 Sir Humphrey Davy showed that hydrochloric acid contained only hydrogen and chlorine and is HCl. There are many acids now known that do not contain oxygen, and the work of Davy gradually lead to the view that all acids contain hydrogen rather than oxygen. This operational definition is still useful, although there are a number of acids (see Section 10.5) that do not contain hydrogen. Likewise, there are many substances containing hydrogen that do not behave as acids.

10.2 THE ARRHENIUS CONCEPT OF ACIDS AND BASES

Solutions of some substances such as sodium chloride in water conduct electricity and are termed *electrolytes*. Substances whose solutions are nonconductors are called *nonelectrolytes*. Nonelectrolytes in water lower the freezing point or elevate the boiling point by an amount directly proportional to the molal concentration, whereas electrolytes produce a change in the freezing point or boiling point greater than that for nonelectrolytes at the same concentration (Chapter 8). This behavior indicates that there are more particles in solution per mole solute of electrolytes than for nonelectrolytes. In 1884 the Swedish chemist Svante Arrhenius proposed a theory that is still considered to explain correctly the behavior of electrolytes in water. He suggested that the molecules of electrolytes separate into particles of positive and negative charges called ions, which account for the ability of solutions of electrolytes to conduct electricity. The formation and/or separation of ions in water, *dissociation,* is fundamental to understanding acids and bases.

The Arrhenius definition of acids and bases was based on the ions that are formed upon dissolution of these substances in water. For the acid HCl both the chemical reactivity and electrolytic behavior can be explained by the formation of the hydrogen cation or proton H^+ and the chloride ion Cl^-. According to Arrhenius, an acid is a substance that yields hydrogen ions (protons) in water, for example,

$$\mathbf{HCl \longrightarrow H^+ + Cl^-}$$

Bases are substances that produce hydroxide ions OH^- in solution, as is the case for sodium hydroxide:

$$\mathbf{NaOH \longrightarrow Na^+ + OH^-}$$

When a base reacts with or neutralizes an acid in water the reaction involves the proton and the hydroxide ion:

$$\mathbf{H^+ + OH^- \longrightarrow H_2O}$$

The proton is responsible for acidic properties of acids and the hydroxide ion for the basic properties of bases.

When Arrhenius originally proposed this theory the distinction between ionic and covalent compounds was not clearly understood. Sodium chloride consists of ions in the solid state, and these ions are separated by the action of water when they dissolve in water. Some electrolytes are covalent substances that do not contain ions until they are placed in water. Hydrogen chloride, for example, is a covalent though highly polar gas. When hydrogen chloride is dissolved in a nonpolar solvent such as benzene, it remains undissociated and the solution will not conduct electricity. In water the very polar water molecules cause the hydrogen chloride to split into the hydrogen ion and the chloride ion. It might be imagined that in solution the positive portion of the water molecules clusters about the chloride ion while the negative portion of the water molecules clusters around the proton (Figure 10.1).

The attraction between a proton and water molecules is unusually strong. In fact, the proton is considered unique among ions. Since it has no electrons its radius is that of the nucleus, or 10^{-13} cm. For comparison, ions containing electrons are approximately 10^{-8} cm in radius. The concentration of charge in such a small volume makes the attraction between water and the proton quite different from that between water and other ions. The proton is very tightly bound to water, and "free" protons do not exist in aqueous solutions of acids.

The hydrated proton or *hydronium ion* H_3O^+ is a stable species that can be thought of as being derived from the reaction of "free" proton with water:

$$H^+ + \underset{H\quad H}{\ddot{O}} \longrightarrow \left[\underset{H\quad H}{\overset{H}{\dot{O}}}\right]^+$$

The hydronium ion has been found in the crystalline hydrated perchloric acid, which consists of H_3O^+ and ClO_4^-. In aqueous solution even H_3O^+ probably is further hydrated and may never exist separately. There is evidence that three additional water molecules are attached to the hydrated proton to yield $H_9O_4^+$:

$$\begin{array}{ccccc} H & & & & H \\ \ \ \ \diagdown & & & & \diagup \\ O\cdots & H-\overset{+}{O}-H & & \cdots & O \\ | & | & & & \diagdown \\ H & H & & & H \\ & \vdots & & & \\ & O-H & & & \\ & | & & & \\ & H & & & \end{array}$$

The notation $H_9O_4^+$ tends to clutter chemical equations with extra water molecules. Therefore, throughout the rest of this text H_3O^+ will be used to represent the proton in water.

When HCl dissolves in water the reaction may be represented by the following equation:

$$HCl + H_2O \longrightarrow H_3O^+ + Cl^-$$

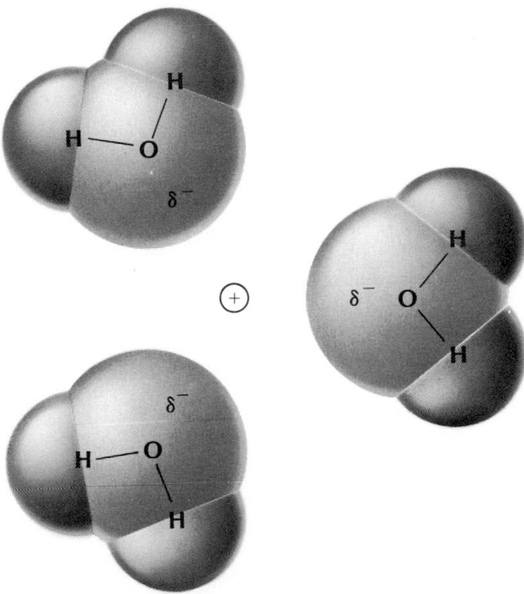

Figure 10.1
Hypothetical solution of proton in water.

The acid HCl functions by transferring or donating a proton to the water molecule. Water in turn is an acceptor of the proton of the acid HCl. This verbal description of the behavior of an acid in water is a very useful one and suggests an operational definition of an acid and a base.

10.3 THE SOLVENT SYSTEM CONCEPT OF ACIDS AND BASES

Although the acid concept of Arrhenius emphasized water as the solvent, his definition can be extended to apply to other solvents. With water as solvent an acid is a substance which produces a cation, the protonated solvent molecule. Similarly, a base is a substance that yields the anion related to the solvent by loss of a proton. Neutralization of the cation by the anion produces the solvent.

Ammonia is a liquid below −33.4°C at atmospheric pressure and resembles water in many ways. It is a polar molecule with hydrogen atoms attached to a central electronegative atom, is associated by hydrogen bonds, and is a good solvent for polar and ionic compounds.

If a compound produces the ammonium ion (NH_4^+) in liquid ammonia it could be classified as an acid. Hydrochloric acid behaves as an acid in ammonia in much the same way as it does in water:

$$HCl + NH_3 \longrightarrow NH_4^+ + Cl^-$$

The anion related to ammonia is NH_2^-, the amide ion. When NH_2^- and NH_4^+ react, neutralization occurs:

$$NH_4^+ + NH_2^- \longrightarrow 2NH_3$$

EXAMPLE 10.1

Sulfuric acid can serve as a solvent. What are the acid and base forms of the solvent?

If a compound were to donate a proton to H_2SO_4 the acid form would be $H_3SO_4^+$. Loss of a proton by sulfuric acid would yield HSO_4^-. The reaction of the acid and base forms yield the solvent:

$$\begin{aligned} H^+ + H_2SO_4 &\longrightarrow H_3SO_4^+ \\ H_2SO_4 &\longrightarrow HSO_4^- + H^+ \\ H_3SO_4^+ + HSO_4^- &\longrightarrow 2H_2SO_4 \end{aligned}$$

10.4
THE BRØNSTED-LOWRY CONCEPT OF ACIDS AND BASES

One of the problems with the Arrhenius concept of acids and bases is that it suggests that basic properties result from the presence of hydroxide ions that are necessary for the neutralization of acids. There are many other substances that react with and neutralize acids; ammonia, as indicated in the previous section, reacts with hydrogen chloride. The ammonia molecule acts as an acceptor of protons in the same way as the hydroxide ion. This common feature provides a broader definition of an acid and a base. An acid is a substance that can donate protons, and a base is a substance that can accept protons. This definition was proposed independently by Johannes Brønsted and Thomas Lowry in 1923.

Acids and bases can be either molecules or ions according to the Brønsted-Lowry concept. Indeed, the dependence on water that restricted the Arrhenius concept is completely removed. The base NH_3 accepts a proton from HCl to produce NH_4^+ and Cl^-:

$$\underset{\text{base}}{\mathbf{NH_3}} + \underset{\text{acid}}{\mathbf{HCl}} \rightleftharpoons \underset{\text{acid}}{\mathbf{NH_4^+}} + \underset{\text{base}}{\mathbf{Cl^-}}$$

In the reverse reaction the ammonium ion can function as an acid by donating a proton to the base chloride ion. NH_3 and NH_4^+ are related and are called a *conjugate pair;* NH_3 is the base and NH_4^+ is its *conjugate acid.* In like manner HCl is an acid and Cl^- is its *conjugate base.*

EXAMPLE 10.2

What are the conjugate acid and conjugate base of HS^-?

The ion HS^- can act as either an acid or a base. The conjugate acid of HS^- is obtained by protonation to yield H_2S. The conjugate base of HS^- is derived by removal of a proton to yield S^{2-}.

10.5
THE LEWIS CONCEPT OF ACIDS AND BASES

Another approach to defining acids and bases was presented by G. N. Lewis in 1923, although he did not fully develop his ideas until 1938. In the Brønsted-Lowry sense a molecule is a base if it can attract and hold a proton, and a molecule that supplies the proton is an acid. The base is capable of functioning as such because it possesses an electron pair that can be used to form a bond to the proton. Lewis suggested that the concept of acids and bases could be extended if the fundamental phenomenon was regarded from the viewpoint of electron pairs rather than protons. According to his definition a base is a substance that has an unshared pair of electrons that can form a covalent bond to some atom, molecule, or ion. The definition includes the hydroxide ion and all other substances previously defined as bases according to the Brønsted–Lowry concept as well as many substances not included in earlier schemes. An acid is a substance that can form a covalent bond by accepting an unshared electron pair from a base. This definition includes the proton of the Brønsted–Lowry concepts but allows many other substances to be defined as acids. The class of acids is expanded to the greatest extent by the Lewis definition.

One example of a Lewis acid that is not an acid under the Brønsted–Lowry definition is boron trifluoride (BF_3). This molecule does not have a complete octet of electrons about the boron atom and can function as a Lewis acid:

$$\underset{\text{Lewis acid}}{\mathrm{F{-}B(F)_2}} + \underset{\text{Lewis base}}{\mathrm{{:}NH_3}} \longrightarrow \mathrm{F_3B{-}NH_3}$$

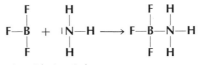

Fundamentally the reaction is the same as the reaction of hydrogen chloride with ammonia:

$$\mathrm{H^+} + \mathrm{H{-}\ddot{N}H_2} \longrightarrow \mathrm{H{-}N^+H_3}$$

Many substances with incomplete octets of electrons function in the same way as BF_3. Some compounds can function as Lewis acids by accepting electron pairs with the resultant formation of a substance in which an atom contains more than an octet of electrons (Section 6.3). An example of this type of behavior is silicon tetrafluoride (SiF_4), which can react with fluoride ions to yield SiF_6^{2-}:

$$\underset{\text{Lewis acid}}{\mathrm{SiF_4}} + \underset{\text{Lewis base}}{\mathrm{2F^-}} \longrightarrow \mathrm{SiF_6^{2-}}$$

Mercury bromide ($HgBr_2$) reacts with bromide ion to yield $HgBr_4^{2-}$ in a Lewis acid–base reaction. The bromide ion can function as a Lewis base by virtue of the electron pairs it contains in *p* orbitals. One of these pairs could be donated to the Lewis acid to form a bond. $HgBr_2$ contains two bonds utilizing electrons in *sp* hybrid orbitals (Chapter 6). There are two vacant *p* orbitals that could accept the electrons of the Br^- ions. (The hybridization of Hg in the ion $HgBr_4^{2-}$ is sp^3 because four equivalent species surround it. See Chapter 14.)

10.6
STRENGTHS OF ACIDS AND BASES

While the Lewis concept is more general than the Brønsted-Lowry concept, this section deals with the more limited approach to acid-base theory. The principles discussed in this section could be extended to include Lewis acids as well, but it is more convenient to restrict the subject of strengths of acids and bases to the Brønsted-Lowry types.

When hydrogen chloride is dissolved in water the resultant solution is called hydrochloric acid and contains virtually no covalent hydrogen chloride. The reaction

$$HCl + H_2O \rightleftharpoons H_3O^+ + Cl^-$$

proceeds essentially to completion to the right. Therefore, HCl must have a stronger tendency to lose protons than does H_3O^+ and is said to be a stronger acid than H_3O^+. In addition the direction of the equilibrium must be a reflection of the willingness of H_2O to accept protons as compared to Cl^-. The H_2O molecule is said to be a stronger base than Cl^-.

From consideration of equilibria between acids and bases and their conjugate bases and acids it can be concluded that a strong acid, with its great tendency to lose protons, must be conjugate to a weak base that has a low affinity for protons. The stronger the acid, the weaker its related conjugate base. Similarly, bases that attract protons strongly are conjugate to weak acids that do not readily lose protons. A list illustrating the relationship between common acid-base pairs is given in Table 10.1.

Hydrochloric, nitric, and perchloric acids are virtually completely dissociated in water. Therefore, their acid properties, which really reflect the presence of H_3O^+ in solution, are the same and their relative strengths cannot be determined. This phenomena is known as the *leveling effect* of the

Table 10.1
Strengths of acids and related conjugate bases

	acid	*conjugate base*	
strongest ↑	$HClO_4$	ClO_4^-	weakest ↓
	H_2SO_4	HSO_4^-	
	HCl	Cl^-	
	H_3O^+	H_2O	
	HSO_4^-	SO_4^{2-}	
	HF	F^-	
	$HC_2H_3O_2$	$C_2H_3O_2^-$	
	H_2S	HS^-	
	NH_4^+	NH_3	
	HCO_3^-	CO_3^{2-}	
	H_2O	HO^-	
	HS^-	S^{2-}	
weakest	HO^-	O^{2-}	strongest

solvent, which in this case is the base water. If a solvent that is a weaker base than water is used, then partial dissociation occurs and relative strengths can be compared. In methanol (CH_3OH) nitric acid is not completely ionized and is seen to be weaker than perchloric acid, which is completely ionized in the same solvent:

$$\mathbf{HNO_3 + CH_3OH \rightleftharpoons CH_3OH_2^+ + NO_3^-}$$

A leveling effect on bases can be caused by solvents that can function as acids. In water the strongest base that can exist is its conjugate base OH^-. If a base that is inherently stronger is placed in water it accepts protons from water and generates OH^-. The amide ion NH_2^- reacts in this manner:

$$\mathbf{H_2O + NH_2^- \rightleftharpoons NH_3 + OH^-}$$

Most acids are only partially ionized in water and are classified as weak acids. At 25° a 1*M* solution of acetic acid ($HC_2H_3O_2$) is approximately 0.4 percent ionized and the concentration of ions is very low:

$$\mathbf{HC_2H_3O_2 + H_2O \rightleftharpoons H_3O^+ + C_2H_3O_2^-}$$

Acetic acid is a weaker acid than H_3O^+, and $C_2H_3O_2^-$ is a stronger base than H_2O. In this reaction the equilibrium lies to the side containing the weaker acid and weaker base. This statement is general and quite logical because the proton must reside with the weaker acid or the substance that has the smallest tendency to lose it. Furthermore, the proton remains on the acid because the base on that side of the equation does not have much tendency to remove it.

In order to compare the strengths of acids it is necessary to measure their tendencies to transfer protons to a common reference base, usually water. The order of acid strengths can be established by measuring the acid dissociation constant. For the general acid *HA*, the equilibrium constant for dissociation, $HA + H_2O \rightleftharpoons H_3O^+ + A^-$, is

$$K = \frac{[\mathbf{H_3O^+}][A^-]}{[\mathbf{H}A][\mathbf{H_2O}]}$$

The concentration of water is so large, compared to the other components of the equilibrium, that its value changes very little on a percentage basis when the acid *HA* is added. Therefore, it is included in a constant called the *acid dissociation constant:*

$$K[\mathbf{H_2O}] = K_d = \frac{[\mathbf{H_3O^+}][A^-]}{[\mathbf{H}A]}$$

If an appropriate means of measuring the concentrations of H_3O^+, A^-, and *HA* is available, then the constant can be calculated. The dissociation constants of some acids are listed in Table 10.2.

Table 10.2
Dissociation constants for acids at 25°C

acid	reaction	K_d (*M*)
acetic	$HC_2H_3O_2 + H_2O \rightleftharpoons H_3O^+ + C_2H_3O_2^-$	1.8×10^{-5}
formic	$HCO_2H + H_2O \rightleftharpoons H_3O^+ + HCO_2^-$	1.8×10^{-4}
hydrocyanic	$HCN + H_2O \rightleftharpoons H_3O^+ + CN^-$	4.0×10^{-10}
hydrofluoric	$HF + H_2O \rightleftharpoons H_3O^+ + F^-$	6.7×10^{-4}
nitrous	$HNO_2 + H_2O \rightleftharpoons H_3O^+ + NO_2^-$	4.5×10^{-4}

EXAMPLE 10.3

Where should the equilibrium lie in the following reaction?

$$H_2O + S^{2-} \rightleftharpoons HS^- + OH^-$$

H_2O is a stronger acid than HS^- (Table 10.1). Therefore, H_2O should tend to donate protons to S^{2-} with greater facility that HS^- to OH^-. Furthermore, S^{2-} is a stronger base than OH^- and will accept protons with greater facility. The equilibrium should lie to the right.

10.7 WATER AND THE pH SCALE

Because concentrations of hydronium ions produced by the dissolution of acids in water often are very small, they are commonly expressed in negative powers of 10. As an example, the concentration of hydronium ions in a 1*M* solution of acetic acid is $4.3 \times 10^{-3}M$. It is more convenient to utilize a compact notation called *pH,* which avoids the use of the exponential notation:

$$\text{pH} = -\log\,[\mathbf{H_3O^+}]$$

The logarithmic relation was chosen arbitrarily for its ease of use. For example, a $1.0 \times 10^{-4}M$ concentration of hydronium ions has a pH = 4.0:

$$\begin{aligned}\text{pH} &= -\log\,[\mathbf{H_3O^+}] = -\log\,[1.0 \times 10^{-4}]\\ &= -(\log 1.0 + \log 10^{-4})\\ &= -(0.0 - 4)\\ &= +4\end{aligned}$$

A solution whose pH is 3.7 has a hydronium ion concentration of $2 \times 10^{-4} M$.

$$pH = 3.7 = -\log [\mathbf{H_3O^+}]$$
$$-3.7 = \log [\mathbf{H_3O^+}]$$
$$-4 + 0.3 = \log [\mathbf{H_3O^+}]$$
$$10^{-4} \times 10^{0.3} = [\mathbf{H_3O^+}]$$
$$2 \times 10^{-4} = [\mathbf{H_3O^+}]$$

In a similar way the hydroxide ion concentration of solutions can be conveniently expressed by pOH, which is the negative logarithm of the hydroxide ion concentration.

$$pOH = -\log [\mathbf{OH^-}]$$

The relationship between $[H_3O^+]$, $[OH^-]$, pH, and pOH is illustrated in Table 10.3.

Water can act as either an acid or as a base; it can donate protons to bases that are stronger than OH^-, and it can accept protons from acids that are stronger than H_3O^+. Pure water can assume both roles simultaneously to establish an equilibrium that proceeds as follows to a small but measurable degree:

$$\mathbf{H_2O + H_2O \rightleftharpoons H_3O^+ + OH^-}$$

The equilibrium expression that incorporates the concentrations of water in the ion product constant K_w is $1 \times 10^{-14}\ M^2$ at 25°C:

$$K = \frac{[\mathbf{H_3O^+}][\mathbf{OH^-}]}{[\mathbf{H_2O}]^2}$$

$$K_w = K[\mathbf{H_2O}]^2 = [\mathbf{H_3O^+}][\mathbf{OH^-}] = 1 \times 10^{-14} M^2$$

Table 10.3
pH and pOH of aqueous solutions

$[H_3O^+]$	pH	pOH	$[OH^-]$
10^{0}	0	14	10^{-14}
10^{-1}	1	13	10^{-13}
10^{-2}	2	12	10^{-12}
10^{-3}	3	11	10^{-11}
10^{-4}	4	10	10^{-10}
10^{-5}	5	9	10^{-9}
10^{-6}	6	8	10^{-8}
10^{-7}	7	7	10^{-7}
10^{-8}	8	6	10^{-6}
10^{-9}	9	5	10^{-5}
10^{-10}	10	4	10^{-4}
10^{-11}	11	3	10^{-3}
10^{-12}	12	2	10^{-2}
10^{-13}	13	1	10^{-1}
10^{-14}	14	0	10^{0}

Since the reaction involves the formation of equal quantities of H_3O^+ and OH^-, the concentration of each is $1 \times 10^{-7}M$. Therefore, the pH and the pOH of pure water at 25°C are both 7:

$$[\mathbf{H_3O^+}][\mathbf{OH^-}] = 10^{-14}M^2 \quad \text{and} \quad [\mathbf{H_3O^+}] = [\mathbf{OH^-}]$$
$$[\mathbf{H_3O^+}]^2 = 10^{-14}M^2$$
$$[\mathbf{H_3O^+}] = 10^{-7}M \qquad \text{pH} = 7$$
$$[\mathbf{OH^-}] = 10^{-7}M \qquad \text{pOH} = 7$$

The dissociation does not usually complicate our measurement of acid or base strength of substances dissolved in water. For example, when 0.1 mole of HCl is dissolved in sufficient water to make 1 liter of solution the HCl totally dissociates to produce a 0.1*M* solution of H_3O^+. The H_3O^+ produced from the dissociation of water is negligible compared to the amount produced from HCl. Furthermore, the concentration of H_3O^+ from the dissociation of water is actually less than $10^{-7}M$ because the equilibrium is affected by the change imposed on it by the addition of another source of H_3O^+. The equilibrium for the dissociation of water is shifted to the left, and some of the OH^- in solution is consumed. At equilibrium the K_w value is unchanged in the presence of the added H_3O^+, and the concentration of OH^- can be calculated:

$$K_w = [\mathbf{H_3O^+}][\mathbf{OH^-}] = 10^{-14}M^2$$
$$10^{-1}[\mathbf{OH^-}] = 10^{-14}M$$
$$[\mathbf{OH^-}] = 10^{-13}M$$

In the 0.1*M* solution formed by adding HCl to water the H_3O^+ concentration is much greater than that of pure water and the OH^- concentration is less. The pH is 1.

If 0.001 mole of sodium hydroxide is added to sufficient water to produce 1 liter of solution, the OH^- concentration derived from this completely ionized base is $10^{-3}M$. The OH^- produced from the dissociation of pure water is 10^{-7}, which is much less than that obtained by dissolving the sodium hydroxide. Furthermore, the dissociation of water is repressed by the addition of OH^- from another source. The OH^- and H_3O^+ concentrations derived from water are decreased:

$$[\mathbf{OH^-}] = 10^{-3}M$$
$$[\mathbf{H_3O^+}][\mathbf{OH^-}] = 10^{-14}M^2$$
$$[\mathbf{H_3O^+}]\ 10^{-3} = 10^{-14}M$$
$$[\mathbf{H_3O^+}] = 10^{-11}M$$

The solution has pH = 11 and pOH = 3.

From the above examples it can be concluded that the pH of acidic solutions is less than 7 and the pH of basic solutions is greater than 7. As a general rule the calculation of pH and pOH and the related hydronium and hydroxide ion concentrations in solutions of strong acids or bases can be done readily because they dissociate completely and are the prime contributors to the concentrations of H_3O^+ and OH^- in solution.

10.8 WEAK ACIDS

When a weak acid is dissolved in water the equilibrium concentration of H_3O^+ must be less than the calculated molarity based on the number of moles of the acid added to water. Acetic acid is such an example of a partially dissociated acid:

$$HC_2H_3O_2 + H_2O \rightleftharpoons H_3O^+ + C_2H_3O_2^-$$

$$K_d = \frac{[H_3O^+][C_2H_3O_2^-]}{[HC_2H_3O_2]} = 1.8 \times 10^{-5}M$$

If 1.0 mole of acetic acid is added to enough water to make 1 liter of solution, some of the acid will dissociate to produce H_3O^+ and $C_2H_3O_2^-$. The acid dissociation constant relates the concentrations $[H_3O^+]$, $[C_2H_3O_2^-]$, and $[HC_2H_3O_2]$. Since all of the quantities are unknown, some approximations and interrelationships must be used in order to convert the equation to a form that can be easily solved.

The first approximation is to assume that the hydronium ion concentration is essentially equal to that derived from the ionization of acetic acid. In other words, the dissociation of water is assumed to contribute an insignificant amount of H_3O^+. From the dissociation equation it can be seen that

$$[H_3O^+] = [C_2H_3O_2^-]$$

since the ions are produced in equal amounts. One unknown can be removed from the acid dissociation equilibrium expression by substituting $[H_3O^+]$ for $[C_2H_3O_2^-]$:

$$\frac{[H_3O^+][C_2H_3O_2^-]}{[HC_2H_3O_2]} = \frac{[H_3O^+]^2}{[HC_2H_3O_2]} = 1.8 \times 10^{-5}M$$

The second approximation is to assume that the concentration of acetic acid at equilibrium is equal to that which would be present if no acid dissociated. This approximation, which is suggested by the fact that the dissociation constant is small, can be validated after the problem has been solved:

$$[HC_2H_3O_2] \simeq 1.0M$$

$$\frac{[H_3O^+]^2}{[HC_2H_3O_2]} = \frac{[H_3O^+]^2}{1.0M} = 1.8 \times 10^{-5}M$$

$$[H_3O^+]^2 = 1.8 \times 10^{-5}M^2$$

$$[H_3O^+] = 4.3 \times 10^{-3}M$$

The first assumption is quite valid because the hydronium ion concentration is greater than the $10^{-7}M$ of pure water. In actuality the H_3O^+ that could be derived from the dissociation of water is less than 10^{-7} because the hydronium ion from acetic acid will repress the equilibrium. The second assumption also is valid since the amount of acetic acid dissociated is approximately 0.0043*M* and the concentration of undissociated acetic acid is approximately 0.996*M*.

The same procedure as described in this section can be applied to other acids and bases as well. The approximation can be made and checked by the final solution. If the approximations are shown not to be valid, the equation must be solved by using a quadratic expression (Appendix 1).

10.9 HYDROLYSIS

Table 10.1 shows that a definite relationship exists between the strengths of a conjugate acid–base pair. Since acetic acid is a moderately weak acid, the acetate ion $C_2H_3O_2^-$ must be a moderately strong base. If acetate ions are dissolved in water, from a source such as the ionic compound sodium acetate, they should abstract protons from water (Table 10.1):

$$\mathbf{C_2H_3O_2^- + H_2O \rightleftharpoons HC_2H_3O_2 + OH^-}$$

As a result some acetic acid and some hydroxide ions will be produced, and the pH of the solution will be greater than 7. Such a process is called a *hydrolysis reaction*. The equilibrium constant expression for the hydrolysis of acetate ion is

$$\frac{[\mathbf{HC_2H_3O_2}][\mathbf{OH^-}]}{[\mathbf{C_2H_3O_2^-}]} = K_h$$

where K_h is the hydrolysis constant that contains the constant term for the concentration of water. Hydrolysis constants are not tabulated because they can be calculated easily from the quotient K_w/K_d:

$$K_h = \frac{K_w}{K_d}$$

This identity can be shown to be valid in the following manner: If the equilibrium constant for hydrolysis is multiplied by the quotient $[H_3O^+]/[H_3O^+]$, the hydrolysis constant is unaltered.

$$\frac{[\mathbf{HC_2H_3O_2}][\mathbf{OH^-}][\mathbf{H_3O^+}]}{[\mathbf{C_2H_3O_2^-}][\mathbf{H_3O^+}]} = K_h$$

The product $[OH^-][H_3O^+]$ in the numerator is equal to K_w:

$$\frac{[\mathbf{HC_2H_3O_2}]K_w}{[\mathbf{C_2H_3O_2^-}][\mathbf{H_3O^+}]} = K_h$$

The remaining concentration terms are equal to the reciprocal of the acid dissociation constant:

$$\frac{[\mathbf{HC_2H_3O_2}]}{[\mathbf{C_2H_3O_2^-}][\mathbf{H_3O^+}]} = \frac{1}{K_d}$$

$$\left(\frac{1}{K_d}\right)K_w = K_h$$

For acetate ion the hydrolysis constant is $5.6 \times 10^{-10}M$:

$$K_h = \frac{K_w}{K_d} = \frac{10^{-14}}{1.8 \times 10^{-5}} = 5.6 \times 10^{-10}M$$

If 1 mole of sodium acetate is dissolved in enough water to produce 1 liter of solution, some of the acetate ions will hydrolyze and form a basic solution. Assuming that the concentration of OH^- from the hydrolysis reaction is greater than that from the dissociation of water, the value can be calculated. From the equation for hydrolysis it can be seen that $[OH^-] = [HC_2H_3O_2]$. Since the acetate ion is not a strong base, most of it will remain as the ion and very little will form acetic acid. The approximation is made that

$$[C_2H_3O_2^-] \simeq 1.0M$$

and the hydrolysis expression can be solved with only one unknown:

$$K_h = 5.6 \times 10^{-10}M = \frac{[OH^-][HC_2H_3O_2]}{[C_2H_3O_2^-]}$$

$$5.6 \times 10^{-10}M = \frac{[OH^-]^2}{1.0M}$$

$$2.4 \times 10^{-5}M = [OH^-]$$

The hydroxide ion concentration is somewhat larger than would be obtained from water under the conditions of slight repression by the addition of another source of OH^-. Therefore, the first assumption is valid. The second assumption also is very reasonable because the actual concentration of acetate ion will be 1.0–0.000024*M*, or approximately 1.0*M*.

10.10 BUFFER SOLUTIONS

In the two preceding sections the behavior of weak acids and the hydrolysis of the salts of weak acids were examined. Now solutions that contain both a weak acid and the salt of a weak acid can be treated by using the previously established procedures. These solutions are known as *buffers* because of their special properties.

If 1.00 mole of acetic acid and 0.55 mole of sodium acetate are added to sufficient water to produce 1 liter of solution, the concentration of H_3O^+ can be calculated from the following rearranged equilibrium expression for the ionization of acetic acid:

$$K_d = \frac{[C_2H_3O_2^-][H_3O^+]}{[HC_2H_3O_2]} = 1.8 \times 10^{-5}M$$

$$[H_3O^+] = \frac{[HC_2H_3O_2]}{[C_2H_3O_2^-]} \times 1.8 \times 10^{-5}M$$

It is only necessary to know the equilibrium concentrations of acetic acid and acetate ions in order to calculate the concentration of H_3O^+.

The acetic acid added to the solution can dissociate into ions. However, it has been shown that this dissociation is very small for the acid, and it will be even less in the presence of added acetate ions. Acetic acid can be produced from the hydrolysis of acetate ions. However, it was shown in Section 10.9 that this quantity of acetic acid is very small. Furthermore, the presence of added acetic acid represses the hydrolysis of acetate ions. It can be concluded that the acetic acid concentration in the solution is essentially 1.0*M*.

The acetate ion concentration is diminished somewhat by hydrolysis, but this change is small for a pure solution of sodium acetate and will be even less in the presence of the added acetic acid. A slight increase in the acetate ion concentration can result from the ionization of acetic acid. However, from previous calculations for a solution of the pure acid the degree of ionization was shown to be small, and the presence of added acetate ions diminishes this increase further. Therefore, the acetate ion concentration is essentially 0.55*M*.

The concentration of H_3O^+ now can be calculated from the approximate values of $[HC_2H_3O_2]$ and $[C_2H_3O_2^-]$:

$$[\mathbf{H_3O^+}] = \left(\frac{1.0}{0.55}\right)(1.8 \times 10^{-5})M = 3.3 \times 10^{-5}M \qquad (\text{pH} = 4.48)$$

This value is lower than that of the hydronium ion concentration ($4.3 \times 10^{-3}M$) of a solution of acetic acid alone (pH = 2.47). The decrease in the hydronium ion concentration, as a result of added acetate ions, is consistent with Le Châtelier's principle.

A buffer solution such as this one of acetic acid and sodium acetate is very useful since it is capable of maintaining a nearly constant pH despite the addition of acids or bases. The acetic acid will react with any strong base that might be added. In addition, the acetate ions will react with any strong acid that might be added. For example, if 0.01 mole of HCl were added to water to make a 0.01*M* solution, the hydronium ion concentration would be 0.01*M* (pH = 2). However, the acetate ion of the buffer solution will react with the H_3O^+ introduced to yield additional acetic acid. The concentrations of acetic acid and acetate ion after the addition of HCl are given by:

$$[\mathbf{HC_2H_3O_2}] \simeq 1.01M \qquad [\mathbf{C_2H_3O_2^-}] \simeq 0.54M$$

The hydronium ion concentration is not changed drastically:

$$[\mathbf{H_3O^+}] = \frac{1.01}{0.54}(1.8 \times 10^{-5}M) = 3.4 \times 10^{-5}M \qquad (\text{pH} = 4.47)$$

In a similar manner the addition of sodium hydroxide does not decrease the hydronium ion dramatically in a buffer solution. The acetate ion concentration is increased, and the acetic acid is decreased.

Buffers may be prepared from any ratio of concentrations of weak acids and the salt of the weak acid. A 1:1 ratio of acid to salt is the most efficient in handling the addition of either acid or base. If the ratio becomes too large the buffer will be less efficient in handling acid, and if the ratio is too small the buffer will not be efficient in handling base.

A variety of pH values may be achieved for various ratios of a weak acid and the salt of the weak acid. For a 1:1 ratio of acid to salt the pH of the acetic acid–acetate buffer is 4.7. For a 10:1 and a 1:10 ratio of acid to salt the pH values are 3.7 and 5.7, respectively.

Buffers are very important in industrial processes such as the plating of metals and the manufacture of dyes. Many processes proceed at a maximum rate at a specific pH, and the buffer facilitates the maintenance of a rapid and hence economical process. Human blood is buffered by a combination of substances which maintain the pH at 7.4. If these buffers fail and the pH either increases or decreases by 0.2 pH units, death results. The importance of buffers in human blood is reflected by the 0.02 pH unit difference between arterial and venous blood, which is phenomenal considering the many acid–base reactions that take place in the cells of the body.

10.11 INDICATORS

The degree of dissociation of an acid or the constitution of a buffer solution can be determined if the hydronium ion concentration or the pH of the solution is known. Although mechanical pH meters exist, these quantities can be readily determined by the use of chemical *indicators* that are themselves weak acids or bases. Each member of the conjugate pair of the indicator is a different color so that an indicator changes color with a change in pH. For example, phenolphthalein is colorless in acid solution and pink in basic solution.

The general formula for an indicator may be given by H*In*. In a basic solution the conjugate base In^- of the indicator will be present. The dissociation constant for the indicator can be written in the same form as for any other acid:

$$\mathbf{H}In + \mathbf{H_2O} \rightleftharpoons \mathbf{H_3O^+} + In^-$$

$$K_d = \frac{[\mathbf{H_3O^+}][In^-]}{[\mathbf{H}In]}$$

When only a small quantity of the indicator is added to a solution of unknown pH the concentration of H_3O^+ is essentially unaffected. Therefore, the ratio of the concentrations of H*In* and In^- is controlled by K_d for the indicator and the hydronium ion concentration of the solution:

$$\frac{[In^-]}{[\mathbf{H}In]} = \frac{K_d}{[\mathbf{H_3O^+}]}$$

If the hydronium ion concentration is high (an acidic solution) the concentration of In^- is small with respect to H*In*, and the color of the solution reflects the predominance of H*In*. When the hydronium ion concentration (a basic solution) is low the concentration of In^- is increased over that of H*In*, and the color of the solution corresponds to that of In^-. The use of an indicator to determine hydronium ion concentrations is limited only by the eye of the observer. When the ratio of the concentrations of the two colored species changes from 0.1 to 10 the color change of the indicator conjugate acid-base pair is at a maximum. Therefore, the usual range of sensitivity of an indicator is a hundredfold range of H_3O^+ concentrations or two pH units. The exact pH range over which the indicator is sensitive is determined by the K_d of the indicator. Many indicators are available, and their K_d values allow the determination of the pH of solutions over the entire experimental range.

10.12 MULTIPLE EQUILIBRIA

Only solutions of a single weak acid or base in water or a mixture of the related acid-base conjugate pairs have been considered in this chapter. There are many systems, such as human blood, in which several acids and bases are present. Mathematical equations for multiple equilibria can be solved by carefully analyzing the system and interrelating all concentrations by the appropriate individual equilibrium expressions. While the solution involves nothing more than algebra, there is little to be served by a complete treatment in this text. The principles necessary for solution have been described for the simple equilibrium situations, and the more complex equilibria involve only a facility for problem solving.

One example of an important multiple equilibria system is phosphoric acid (H_3PO_4) in water. Here 1 mole of acid potentially can furnish 3 moles of hydronium ions. The acid is said to be *triprotic*. Phosphoric acid is not a strong acid, and it dissociates in a stepwise manner:

	K_d
$\mathbf{H_3PO_4 + H_2O \rightleftharpoons H_2PO_4^- + H_3O^+}$	7.5×10^{-3}
$\mathbf{H_2PO_4^- + H_2O \rightleftharpoons HPO_4^{2-} + H_3O^+}$	6.2×10^{-8}
$\mathbf{HPO_4^{2-} + H_2O \rightleftharpoons PO_4^{3-} + H_3O^+}$	10^{-12}

Therefore, in a solution of phosphoric acid in water the species $H_2PO_4^-$, HPO_4^{2-}, and PO_4^{3-} are all present, but in progressively smaller amounts—which reflects the K_d values for each dissociation step.

Suggested further readings

Clever, H. L., "The Hydrated Hydronium Ion," *J. Chem. Educ.*, **40,** 637 (1963).

Hall, F. M., "The Theory of Acids and Bases," *Educ. Chem.*, **1,** 91 (1964).

Herron, F. Y., "Models Illustrating the Lewis Theory of Acids and Bases," *J. Chem. Educ.*, **30,** 199 (1953).

Luder, W. F., "Contemporary Acid-Base Theory," *J. Chem. Educ.*, **25,** 555 (1948).

Sisler, H. H., *Chemistry in Non-Aqueous Solvents.* New York: Reinhold, 1961.

Szabadvary, F., "Development of the pH Concept," Trans. R. E. Oesper, *J. Chem. Educ.*, **41,** 105 (1964).

Terms and concepts

acid	conjugate acid	ion
acid dissociation constant	conjugate base	Lavoisier
Arrhenius	Davy	leveling effect
base	hydrolysis	Lewis
Brønsted	hydronium ion	Lowry
buffer	indicator	pH

Questions and problems

1. What experimental evidence shows that water is only slightly dissociated?

2. Why do solutions having a high concentration of H_3O^+ always have a low OH^- concentration?

3. Why do some salts when dissolved in water yield solutions whose pH is not 7?

4. When sulfuric acid is dissolved in pure acetic acid the following reaction occurs:

$$\mathbf{H_2SO_4 + HC_2H_3O_2 \longrightarrow HSO_4^- + H_2C_2H_3O_2^+}$$

Which species are the Brønsted-Lowry acid, base, conjugate base, and conjugate acid?

5. Would the degree of dissociation of acetic acid be larger or smaller in methanol as compared to water as solvent?

6. Which of the following species can act as Lewis acids? as Lewis bases? H_2O, NH_3, BF_3, O^{2-}, NH_2^-, OH^-

7. The pH of lemons and grapes are 2.3 and 4.0, respectively. What gives rise to these observed pH values? Which substance is the more acidic?

8. Calculate the pH and pOH of solutions whose hydronium ion concentrations are 1×10^{-3}, 1×10^{-5}, 4×10^{-8}, and 2×10^{-9}.

9. Powdered limestone ($CaCO_3$) is added to soil to compensate for excess acidity. Why?

10. Compare the Arrhenius, Brønsted-Lowry, and Lewis definitions of acids and bases.

11. HF is a strong acid in liquid ammonia but is a weak acid in water. Explain.

12. The hydrogen ion concentration of a 0.02*M* solution of HNO_2 is $2.8 \times 10^{-3}M$. Calculate the K_d for HNO_2.
13. Calculate the concentration of OCl^- in a 0.1*M* solution of HOCl. The K_d for HOCl is $3.2 \times 10^{-8}M$.
14. The equilibrium constant for the reaction of ammonia with water is $1.8 \times 10^{-5}M$:

$$\mathbf{NH_3 + H_2O \longrightarrow NH_4^+ + OH^-}$$

Calculate the concentration of NH_4^+ in a solution prepared from 0.1 mole of ammonia and sufficient water to make 1 liter of solution.
15. Calculate the concentrations of $HC_2H_3O_2$, H_3O^+, $C_2H_3O_2^-$, and OH^- in a solution prepared from 0.2 mole of HCl and 0.1 mole of $HC_2H_3O_2$ and enough water to make 1 liter of solution.
16. Calculate the hydrolysis constant for the following reaction and determine the OH^- concentration of a 0.1*M* solution of HCO_2^- (see Table 10.2):

$$\mathbf{HCO_2^- + H_2O \longrightarrow HCO_2H + OH^-}$$

17. What would be the pH of a solution prepared by dissolving 0.1 mole of HCO_2H and 0.18 mole of $NaHCO_2$ in sufficient water to make 1 liter of solution?
18. The K_d for HPO_4^{2-} is $1 \times 10^{-12}M$. Calculate the hydrolysis constant for PO_4^{3-}.
19. Explain the principle of a buffer solution. When will a buffer solution become ineffective?
20. Suggest a reason for the fact that the second dissociation constant of H_3PO_4 is less than the first dissociation constant.

11

OXIDATION-REDUCTION

In the preceding chapter one of the major types of chemical reactions, acid–base reactions, was discussed. In this chapter a second important type of reaction, *oxidation-reduction,* will be examined. In acid-base reactions the common basic feature is proton transfer between pairs of acids and their conjugate bases. Oxidation-reduction reactions involve the transfer of one or more electrons between pairs of substances.

Oxidation-reduction or *redox* reactions that occur spontaneously are accompanied by the energy change characteristic of all spontaneous reactions (Section 9.7). This change involves the liberation of energy, $\Delta H < 0$, which exactly equals the difference in potential energy between the reactants and products. As is the case with many other reactions, the energy liberated can be observed as heat energy. However, oxidation-reduction reactions also can be used to produce electrical energy under the proper experimental conditions.

Because oxidation and reduction reactions involve electron transfer, they always must occur in pairs. The substance that loses electrons is said to be *oxidized* or to undergo an oxidation reaction. The substance that gains electrons is *reduced.* The electron transfer either may be simple and direct from the oxidized atom to the reduced one or may occur in a more complex way. If the transferred electrons can be made to flow along a suitable pathway, a conductor, the electric current generated can be utilized to perform useful work. The methods by which electrical energy can be obtained are discussed later in this chapter, but first it is necessary to lay a proper foundation by examining oxidation-reduction reactions.

11.1 OXIDATION AND REDUCTION

TERMINOLOGY

Oxidation involves the loss of electrons in either a simple or complex way from a substance. Reduction involves the gain of electrons by a substance. An oxidation reaction cannot occur without the simultaneous occurrence of a reduction reaction. The relationship of these two interdependent processes also is described in terms of agents that are responsible for the oxidation and reduction. The substance that causes a second substance to lose its electrons is termed an *oxidizing agent.* Oxidizing agents accept electrons and are reduced as a result of this acceptance. A substance that supplies electrons to a second substance and enables it to be reduced is called a *reducing agent.* Reducing agents, by supplying electrons, become oxidized. These relationships are summarized in Table 11.1.

OXIDATION NUMBERS

In discussing oxidation-reduction reactions, it is convenient to devise a term to indicate (on some arbitrary scale) the state of oxidation of substances. The *oxidation state* or *oxidation number* of an atom in a monatomic ion, complex ion, or neutral compound is indicated by the combi-

Table 11.1
Oxidation-reduction terminology

term	*electron change*	*oxidation number change*
oxidation	loss of electrons	increase
reduction	gain of electrons	decrease
substance oxidized	loses electrons	increase
substance reduced	gains electrons	decrease
oxidizing agent	accepts electrons	decrease
reducing agent	donates electrons	increase

nation of a sign and a number. The oxidation number reflects the number of electrons lost or gained relative to the elemental state. The assignment of oxidation numbers involves a few rules that can be easily learned. For a simple monatomic substance the oxidation number is the charge of the species. In comparing Na and Na^+ it is obvious that a process in which Na is converted to Na^+ must involve a loss of one electron by Na, which becomes oxidized. The oxidation number of the element Na is defined as 0, whereas that of Na^+ is +1.

In polyatomic species containing a variety of covalent bonds it is less obvious exactly what change in electron distribution has occurred in an oxidation-reduction reaction. What is the state of oxidation of nitrogen in NO_2 or NO relative to elemental nitrogen? Lacking knowledge of exact electron distribution in the covalent bonds of NO_2 and NO, we must define the oxidation state in an arbitrary manner. Even though the assignment of oxidation states of atoms in polyatomic molecules or ions is arbitrary, it is internally consistent and is an aid in organizing a great deal of information into easily grasped, fundamental principles.

OXIDATION NUMBER CONVENTIONS

1. *The oxidation of an element in its elemental form is zero.* Bonds between two atoms of the same element do not alter the oxidation number of the elements because the electrons in the bonds are shared equally and hence cannot be said to be lost or gained. Thus, O_2, P_4, and S_8, which represent the natural state of the elements, are all in the zero oxidation state.

2. *The oxidation state of a monatomic anion or cation is equal to its charge.* The charge on a cation, such as Na^+, indicates that the ion is in a higher oxidation state (+1) than the free element, resulting from the loss of one electron from the element. Similarly, Cl^- is in a lower oxidation state (−1) than the free element.

3. *The algebraic sum of the oxidation numbers of the atoms in a neutral molecule is zero.* The sum of the electrons gained or lost by each atom in a compound, relative to the free elements, must balance in order that the molecule be uncharged. In H_2O the oxidation numbers of hydrogen and oxygen are +1 and −2, respectively (see 7 and 9 below). The sum of the oxidation numbers $2(+1) + (-2) = 0$.

4. *In a complex ion, the algebraic sum of the oxidation numbers of the constituent atoms is equal to the charge on the ion.* The charge on the ion represents the net excess of electrons gained or lost by each atom relative to the free elements. For example, the oxidation numbers of sulfur and oxygen in the sulfate ion SO_4^{2-} are +6 and −2, respectively. The sum is given by $+6 + 4(-2) = -2$, which is the charge on the ion.

5. *The oxidation numbers of the alkali metals (Group I) and the alkaline earth metals (Group II) in compounds are +1 and +2, respectively.*

6. *The oxidation state of fluorine in combination with any other element is −1.*

7. *The oxidation state of oxygen in compounds is usually −2.* The only exceptions occur when oxygen is combined with the more electronegative fluorine and in substances containing oxygen-oxygen bonds. For example, in peroxides, which have the —O—O— grouping, the oxidation number is −1 (Chapter 13).

8. *The oxidation number of a halogen (Group VII) in compounds is −1,* unless the halogens are bonded to elements that are more electronegative [such as oxygen (Chapter 13)].

9. *In the majority of compounds hydrogen is in the +1 oxidation state.* Exceptions occur when hydrogen is combined with certain metals, such as the alkali and alkaline earth elements. In these cases the oxidation state of hydrogen is −1 because the metals are more electropositive.

EXAMPLE 11.1

What is the oxidation number of Mn in $KMnO_4$?

The sum of the oxidation numbers in the neutral substance must be equal to zero. K is an alkali metal (Group I), and its oxidation number is +1, according to the convention listed in 5 above. The oxidation number of O is −2, according to 7 above. In order for the oxidation numbers to equal 0, the oxidation number of Mn must be equal to +7:

$$(+1) + (x) + 4(-2) = 0$$
$$x = -1 + 8$$
$$x = +7$$

EXAMPLE 11.2

What is the oxidation number of Cl in the ion ClO_3^-?

The sum of the oxidation numbers must equal the charge on the ion, −1. Chlorine is combined with the more electronegative element oxygen. Therefore, the oxidation number of chlorine is not −1. The oxidation number of oxygen is −2 and that of chlorine must be +5:

$$(x) + 3(-2) = -1$$
$$x = -1 + 6$$
$$x = +5$$

11.2 OXIDATION-REDUCTION REACTIONS

BALANCING SIMPLE EQUATIONS

The simplest oxidation-reduction reaction involves direct union of two elements to form a compound, such as the reaction between sodium and chlorine to produce sodium chloride:

$$2Na + Cl_2 \longrightarrow 2NaCl$$

In this reaction sodium loses one electron per atom and is oxidized, and each chlorine of the chlorine molecule accepts one electron and is reduced. The reaction is described by the above equation, which can be balanced by inspection. Since each molecule of sodium chloride contains one sodium ion and one chloride ion, it is necessary to supply equal numbers of atoms of each element for the reaction. Chlorine exists as a diatomic molecule, and sodium is monatomic. Therefore, two atoms of sodium are required for every molecule of chlorine to provide the 1:1 ratio of ions in sodium chloride.

A second type of oxidation-reduction reaction involves the displacement of a metal ion from solution by a second metal. When a strip of zinc metal is immersed in a solution of the copper ion Cu^{2+}, the zinc becomes coated with copper metal and the blue color characteristic of the Cu^{2+} solution is diminished. Chemical analysis of the solution shows the presence of Zn^{2+}:

$$Zn + Cu^{2+} \longrightarrow Zn^{2+} + Cu$$

Each atom of zinc must lose two electrons to form Zn^{2+}. Each Cu^{2+} must gain two electrons in order to form metallic copper. Again the equation can be balanced by inspection, remembering that both the total number of atoms and the charges must be balanced. In the reaction described, only the positive ions are considered. The solution must contain negative ions, but they are not changing in oxidation number and are not considered in the equation describing the oxidation-reduction process.

Oxidation-reduction reactions also may occur between two ions in aqueous solution. One example is the oxidation of Sn^{2+} by Fe^{3+}. The oxidizing agent is Fe^{3+}, and the reducing agent is Sn^{2+}:

$$Sn^{2+} + 2Fe^{3+} \longrightarrow Sn^{4+} + 2Fe^{2+}$$

Each Fe^{3+} gains one electron when reduced to Fe^{2+}, and each Sn^{2+} loses two electrons in being oxidized to Sn^{4+}. Therefore, two Fe^{3+} are required for each Sn^{2+}. Again the simplicity of the reaction allows one to balance the equation by inspection. However, there are many complicated oxidation-reduction reactions which require a more organized approach. Many methods have been devised to balance complex equations, but the only one that will be described in this text is the *half-reaction method.*

In oxidation-reduction reactions there are often several atoms of a molecule that do not undergo changes in oxidation numbers. All species must be shown in the balanced equation, but the presence of species that do not undergo changes in oxidation number appears to complicate the problem of balancing the equation. The basic principles that govern the *half-reaction method* for balancing redox equations are (1) only the species that undergo changes in oxidation state are given primary consideration and (2) the total number of electrons lost by all the species being oxidized must be gained by the species reduced. In the half-reaction method cognizance is taken of the fact that oxidations and reductions must occur simultaneously in order that the reaction not acquire a net charge. However, each process is considered separately and is regarded as one-half of the overall reaction. The separate equations describing the oxidation and reduction processes are first balanced, and then the two equations are adjusted by numerical factors so that the number of electrons produced in one process is equal to the number required in the other process. Then the two *half-reactions* are summed to produce the final balanced equation.

A half-reaction must be balanced with regard to both the number of atoms and the charges. Balance of the number of atoms can be accomplished if compounds or ions containing the necessary atoms are present on both sides of the equation. It is often necessary to introduce protons (H^+), hydroxide ions (OH^-), or water to provide elemental balance of hydrogen and oxygen. This is permissible if the reaction occurs in the presence of water. Electrical balance of a half-reaction requires the use of electrons that will be eliminated in the final balanced equation.

An unbalanced equation describing the oxidation of hydrogen sulfide by permanganate ions in acidic solution is

$$H_2S + MnO_4^- + H^+ \longrightarrow Mn^{2+} + S + H_2O$$

and the two half-reactions that must be considered involve sulfur and manganese:

$$H_2S \longrightarrow S + H^+$$
$$MnO_4^- + H^+ \longrightarrow Mn^{2+} + H_2O$$

A proton is placed on the right-hand side of the oxidation reaction because the conversion of H_2S into S (oxidation) must lead to some form of hydrogen in the same oxidation state as in the original compound. A proton is placed on the left-hand side of the reaction, in which MnO_4^- is converted into Mn (reduction), because it must be available to react with the oxygen contained in MnO_4^-.

The number of electrons involved in the oxidation and reduction half-reactions are calculated. In the oxidation of H_2S to S, two electrons are released:

$$H_2S \longrightarrow S + H^+ + 2e^-$$

In the reduction of MnO_4^-, in which the oxidation state of manganese is +7, to Mn^{2+} five electrons must be added:

$$5e^- + MnO_4^- + H^+ \longrightarrow Mn^{2+} + H_2O$$

Next the half-reactions are balanced atomically, but the numbers of species specifically involved in the oxidation and reduction cannot be altered because they have been balanced relative to the number of electrons lost and gained. Electrical balance and atomic balance are achieved simultaneously as indicated:

$$H_2S \longrightarrow S + 2H^+ + 2e^-$$
$$5e^- + MnO_4^- + 8H^+ \longrightarrow Mn^{2+} + 4H_2O$$

In order to combine the two half-reactions into a final balanced equation, the number of electrons produced in one half-reaction must be made equivalent to the number used in the other half-reaction. This can be accomplished by multiplying the first equation by 5 and the second equation by 2. The two half-reactions then can be added, and the 10 electrons are eliminated:

$$5H_2S \longrightarrow 5S + 10H^+ + 10e^-$$
$$10e^- + 2MnO_4^- + 16H^+ \longrightarrow 2Mn^{2+} + 8H_2O$$
$$\overline{5H_2S + 2MnO_4^- + 16H^+ \longrightarrow 2Mn^{2+} + 8H_2O + 10H^+ + 5S}$$

The final correct equation is obtained by cancellation of the excess protons common to both sides of the equation. Inspection of the final equation indicates that it is balanced atomically and electrically:

$$5H_2S + 2MnO_4^- + 6H^+ \longrightarrow 2Mn^{2+} + 8H_2O + 5S$$

11.3 ELECTRODE POTENTIALS

While the balancing of oxidation-reduction equations provides the same type of challenge that some individuals find in solving puzzles, the inquisitive person might ask if it is possible to predict in advance whether a reaction will occur as written or in the reverse direction. A reaction will proceed if the potential energy of the products is less than that of the reactants.*

The oxidation and reduction half-reactions can be used to provide an answer to whether a reaction will proceed as written. Half-reactions can be described in general as follows:

$$e^- + \text{oxidant} \longrightarrow \text{reductant}$$

*As noted in Chapter 9, the rate of the reaction is independent of the difference in potential energy of reactants and products but rather is controlled by the activation energy. The remainder of this chapter is concerned only with the tendency of reaction to proceed and does not consider the rate of the reaction.

The term *oxidant* refers to an oxidized state of some species, and the term *reductant* refers to the reduced state of the related species. Two oxidants and two reductants are involved in any oxidation-reduction equation. The direction that a reaction takes must be controlled by the relative tendencies for the species involved to become oxidized or reduced. If it were possible to determine the relative tendencies for the individual half-reactions to occur, then it should be possible to predict the direction of a reaction.

In the previous chapter it was shown that acids and bases cannot be classified as strong or weak independent of other bases and acids; pairs of conjugate acid-base related substances must be considered. Similarly, the direction of an oxidation-reduction reaction must depend on pairs of half-reactions containing oxidants and reductants. How, then, can a single half-reaction be discussed? A reference reaction must be chosen, and the oxidizing or reducing tendency of all other reactions must be assigned relative to it. For convenience the reduction of hydrogen to hydrogen gas has been chosen as a reference and has been assigned a potential of 0.00V:

$$2e^- + \mathbf{2H^+}\text{ (solution)} \longrightarrow \mathbf{H_2}\text{ (gas)}$$

The potentials of all other half-reactions involving a reduction and written in the same direction are defined as *standard electrode potentials* ($\mathcal{E}^0$). By convention the potential of a given half-reaction is defined as positive if the oxidant has a greater tendency to gain electrons than does a proton in aqueous solution. A negative sign means that the oxidant has a smaller tendency to lose electrons than a proton. As the magnitude of the electrode potential increases algebraically, the strength of the oxidant increases. A partial list of standard electrode potentials is given in Table 11.2. At this time only the technique of predicting the direction of an oxidation-reduction reaction is considered. The method of measuring the standard electrode potentials is discussed in the next section. The significance of the electrode potentials is discussed in Chapters 12 and 13.

The oxidation of zinc by the copper(II) ion Cu^{2+} was described earlier in this chapter. Experimentally it has been shown that the reaction occurs in the direction shown in the following equation:

$$\mathbf{Zn + Cu^{2+} \longrightarrow Zn^{2+} + Cu}$$

The half-reactions for the equation and their standard electrode potentials can be found in Table 11.2. The copper ion is a better oxidizing agent than Zn^{2+} and has a stronger tendency to gain electrons, in the equation as written, than does Zn^{2+}:

$$\mathbf{Cu^{2+}} + 2e^- \longrightarrow \mathbf{Cu} \qquad \mathcal{E}^0 = +0.34\text{ V}$$
$$\mathbf{Zn^{2+}} + 2e^- \longrightarrow \mathbf{Zn} \qquad \mathcal{E}^0 = -0.76\text{ V}$$

Therefore, the direction of reaction is predictable from the standard oxidation potentials. Any oxidant with an algebraically larger electrode potential than a second oxidant will oxidize the reduced form (reductant) of the second oxidant. An oxidant with a high electrode potential will react in

Table 11.2
Electrode potentials

half reaction	$\mathcal{E}^0$
$K^+ + 1e^- \longrightarrow K$	−2.92
$Na^+ + 1e^- \longrightarrow Na$	−2.71
$Mg^{2+} + 2e^- \longrightarrow Mg$	−2.37
$Al^{3+} + 3e^- \longrightarrow Al$	−1.66
$Zn^{2+} + 2e^- \longrightarrow Zn$	−0.76
$Fe^{2+} + 2e^- \longrightarrow Fe$	−0.44
$Cd^{2+} + 2e^- \longrightarrow Cd$	−0.40
$Sn^{2+} + 2e^- \longrightarrow Sn$	−0.14
$Pb^{2+} + 2e^- \longrightarrow Pb$	−0.13
$2H^+ + 2e^- \longrightarrow H_2$	0.00
$Sn^{4+} + 2e^- \longrightarrow Sn^{2+}$	+0.15
$Cu^{2+} + 2e^- \longrightarrow Cu$	+0.34
$O_2 + 2H_2O + 4e^- \longrightarrow 4OH^-$	+0.40
$I_2 + 2e^- \longrightarrow 2I^-$	+0.53
$Fe^{3+} + 1e^- \longrightarrow Fe^{2+}$	+0.77
$Ag^+ + 1e^- \longrightarrow Ag$	+0.80
$Hg^{2+} + 2e^- \longrightarrow Hg$	+0.85
$Br_2 + 2e^- \longrightarrow 2Br^-$	+1.09
$Cl_2 + 2e^- \longrightarrow 2Cl^-$	+1.36
$4H^+ + PbO_2 + 2e^- \longrightarrow Pb^{2+} + 2H_2O$	+1.46
$H_2O_2 + 2H^+ + 2e^- \longrightarrow 2H_2O$	+1.77
$F_2 + 2e^- \longrightarrow 2F^-$	+2.87

the direction written in Table 11.2, and all half-reactions with lower electrode potentials will occur in the reverse direction in its presence.

EXAMPLE 11.3

Will Mg reduce Zn^{2+}?

The electrode potentials to consider are

$$2e^- + Mg^{2+} \longrightarrow Mg \qquad \mathcal{E}^0 = -2.37 \text{ V}$$
$$2e^- + Zn^{2+} \longrightarrow Zn \qquad \mathcal{E}^0 = -0.76 \text{ V}$$

Zn^{2+} has a larger tendency to accept electrons and become reduced than does Mg^{2+}. Therefore the Zn^{2+} should accept electrons from Mg.

Will Cu^{2+} be reduced by Hg?

The electrode potentials to consider are

$$2e^- + \mathbf{Cu^{2+}} \longrightarrow \mathbf{Cu} \qquad \mathcal{E}^0 = +0.34\text{ V}$$
$$2e^- + \mathbf{Hg^{2+}} \longrightarrow \mathbf{Hg} \qquad \mathcal{E}^0 = +0.85\text{ V}$$

Hg^{2+} has a larger tendency to accept electrons than Cu^{2+}. Therefore, Hg should retain its electrons in the presence of Cu^{2+} and a reaction should not occur.

11.4 GALVANIC CELLS

Most oxidation-reduction reactions can be carried out, with proper experimental modifications, to produce electrical energy. In order to accomplish this it is necessary to separate the oxidation and reduction half-reactions and provide a *conductor* through which electrons may flow. Transfer of electrons through the conductor allows the oxidation and reduction half-reactions to occur and provides a means of utilizing the electric current. The experimental device that produces electric current from half-reactions in solution is called a *galvanic cell.*

The oxidation of zinc metal by copper(II) ion Cu^{2+} can be carried out in a cell constructed so that the zinc metal and Cu^{2+} are separated (Figure 11.1). A container is partitioned into two compartments by a porous divider that allows solutions placed in the separate compartments to be in contact but retards the rate of mixing. In one compartment a bar of zinc is immersed in a solution of a soluble zinc salt, such as zinc nitrate, $Zn(NO_3)_2$. In the other compartment a bar of copper is immersed in a solution of $Cu(NO_3)_2$. If the two metal bars are then connected to a voltmeter, a potential difference of 1.10 V is observed if the solutions are both 1*M*. (The voltage observed is a function of concentration, but this detail need not be elaborated upon. All electrode potentials listed in Table 11.2 are for 1*M* solutions.)

The zinc bar slowly dissolves, the copper bar thickens, and the copper nitrate solution becomes a lighter blue. Chemical analysis of the zinc nitrate solution indicates that the concentration of zinc ions has increased. The two half-reactions are occurring in the two compartments.

The two metal bars in the galvanic cell are called *electrodes.* At the zinc electrode, zinc is oxidized and enters the solution as zinc ions; this electrode is called the *anode.* In the process of being oxidized, the zinc leaves two excess electrons per atom on the zinc bar. Electrons travel to the copper bar where copper ions accept them and are plated out on the

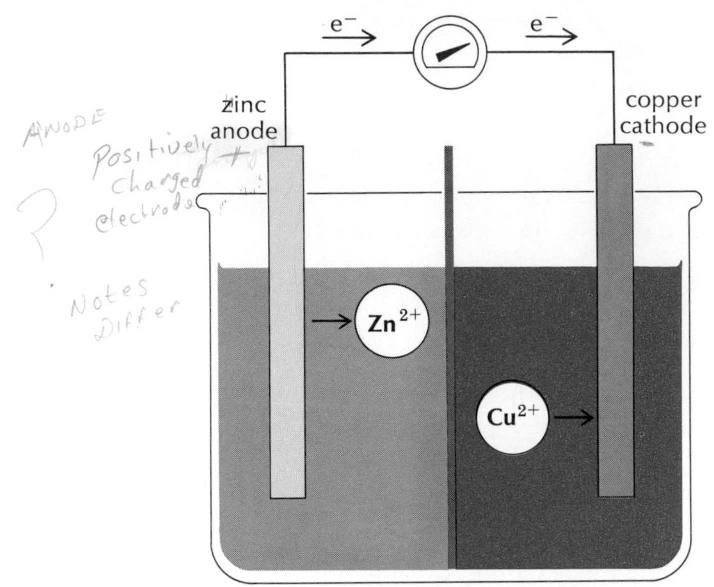

Figure 11.1
The zinc-copper galvanic cell.

copper bar. The copper bar is the electrode at which reduction occurs and is called the *cathode*.

The 1.10 V obtained from the galvanic cell is a measure of the tendency of Cu^{2+} ions to be reduced relative to Zn^{2+} ions. The 1.10 V is the algebraic sum of standard potentials for the two half-reactions:

$$\begin{array}{lll} \mathbf{Zn} \longrightarrow \mathbf{Zn^{2+}} + 2e^- & & \mathcal{E}^0 = +0.76 \\ \mathbf{Cu^{2+}} + 2e^- \longrightarrow \mathbf{Cu} & & \mathcal{E}^0 = +0.34 \\ \hline \mathbf{Zn} + \mathbf{Cu^{2+}} \longrightarrow \mathbf{Zn^{2+}} + \mathbf{Cu} & & \mathcal{E}^0 = +1.10 \end{array}$$

The oxidation potential for zinc is obtained from the electrode potential of zinc ion by sign reversal. The voltage for any given set of half-reactions can be obtained in a similar manner from the standard electrode potentials.

EXAMPLE 11.5

What is the voltage of a cell constructed such that Mg will reduce Pb^{2+}?

The half-reactions for the oxidation and reduction are

$$\begin{array}{ll} \mathbf{Mg} \longrightarrow \mathbf{Mg^{2+}} + 2e^- & \mathcal{E}^0 = +2.37\ \text{V} \\ 2e^- + \mathbf{Pb^{2+}} \longrightarrow \mathbf{Pb} & \mathcal{E}^0 = -0.13\ \text{V} \end{array}$$

The sum of the voltages of the half-reactions is +2.24 V.

11.5
THE HYDROGEN ELECTRODE

Earlier in this chapter the standard electrode potentials were defined relative to the reduction of protons to hydrogen gas. The standard electrode potentials recorded in Table 11.2 indicate the relative tendencies of substances to be reduced and are measured by a galvanic cell in which one of the electrodes is the hydrogen electrode. In theory any substance could be used as a reference, and if the Cu^{2+} ions of the electrode potential listed in Table 11.2 had been chosen as the standard reaction, the half-reaction of the reduction of Zn^{2+} to Zn would have an electrode potential of −1.10 V.

For the measurement of standard electrode potentials a hydrogen electrode consisting of a strip of platinum (which is resistant to oxidation) immersed in a solution of hydronium ions is used. Hydrogen gas is bubbled over the surface of the platinum where the gas and hydronium ions may come in contact. If hydronium ions gained electrons to produce hydrogen, the electrode then would be acting as a cathode. Alternatively, hydrogen could give up electrons and produce hydronium ions, in which case the hydrogen electrode would be acting as an anode. The actual behavior of the hydrogen electrode depends upon the electrode coupled to it. In the presence of zinc and zinc ions the hydrogen electrode would function as a cathode. The voltage of the hydrogen and zinc galvanic cell is 0.76 V or the standard electrode potential of zinc (Figure 11.2):

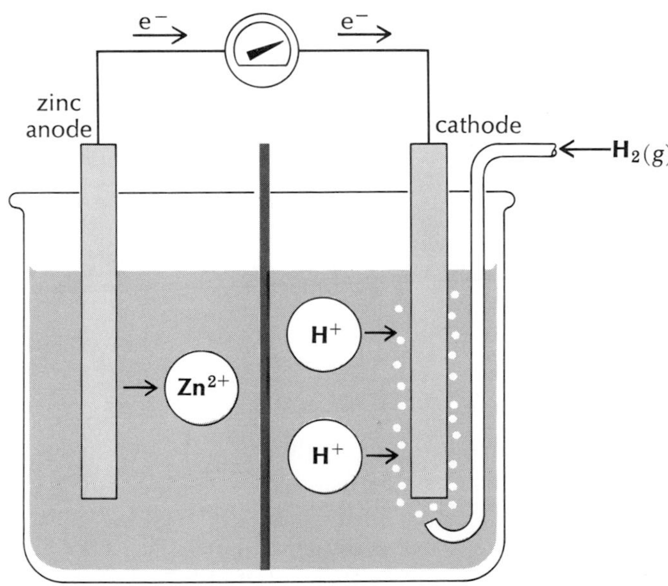

Figure 11.2
The zinc–hydrogen galvanic cell.

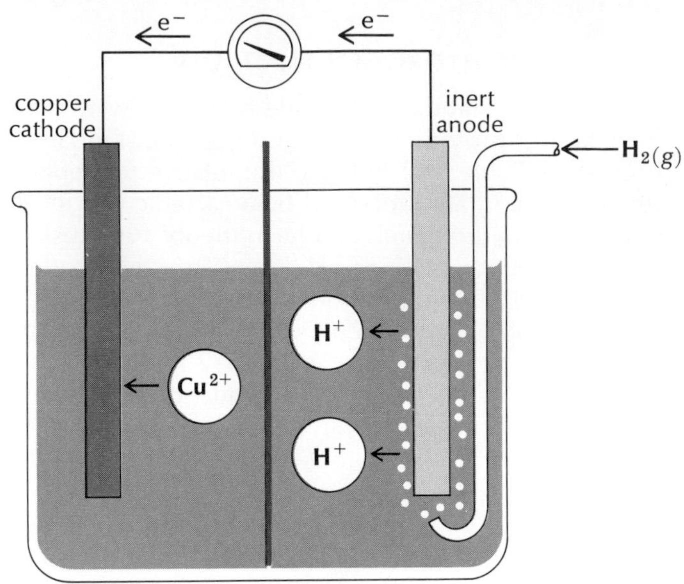

Figure 11.3
The copper-hydrogen galvanic cell.

$$
\begin{array}{rcll}
\mathbf{Zn} & \longrightarrow & \mathbf{Zn^{2+}} + 2e^- & \mathcal{E}^0 = 0.76 \\
\mathbf{2H_3O^+} + 2e^- & \longrightarrow & \mathbf{2H_2O + H_2} & \mathcal{E}^0 = 0.00 \\
\hline
\mathbf{Zn + 2H_3O^+} & \longrightarrow & \mathbf{2H_2O + H_2 + Zn^{2+}} & \mathcal{E}^0 = 0.76
\end{array}
$$

If copper and copper(II) Cu^{2+} ions are coupled to the hydrogen electrode, the copper ions will be reduced, because the electrode potential for the hydrogen electrode is smaller than for the Cu^{2+} ion to copper half-reaction. In this case the hydrogen electrode acts as an anode and the voltage for the cell is 0.34 V (Figure 11.3):

$$
\begin{array}{rcll}
2e^- + \mathbf{Cu^{2+}} & \longrightarrow & \mathbf{Cu} & \mathcal{E}^0 = 0.34 \\
\mathbf{2H_2O + H_2} & \longrightarrow & \mathbf{2H_3O^+} + 2e^- & \mathcal{E}^0 = 0.00 \\
\hline
\mathbf{Cu^{2+} + H_2 + 2H_2O} & \longrightarrow & \mathbf{Cu + 2H_3O^+} & \mathcal{E}^0 = 0.34
\end{array}
$$

11.6 ELECTROLYSIS

Spontaneous reactions occurring in a galvanic cell can be used as a source of voltage and electric power. *Electrolysis* involves the opposite process: An external source of voltage and electric power is used to produce chemical changes that are not spontaneous. Such processes are extremely important in the industrial preparation of many metals and nonmetals.

Many ions of metals are difficult to reduce because there are no readily available chemical reducing agents for them. As an example, magnesium ions could be reduced only by sodium or potassium metal, which have higher electrode potentials, but such a reduction process involves many difficulties in the handling of the very active metals (Chapter 12). The commercial preparation of magnesium involves electrolytic reduction of Mg^{2+} at a cathode and electrolytic oxidation of chloride ion at the anode in the molten salt $MgCl_2$ (Figure 11.4):

$$\begin{array}{rl} \textit{cathode} & \mathbf{Mg^{2+}} + 2e^- \longrightarrow \mathbf{Mg} \\ \textit{anode} & \mathbf{2Cl^-} \longrightarrow \mathbf{Cl_2} + 2e^- \end{array}$$

The net reaction decomposes the compound $MgCl_2$ into its constituent elements. It is the reverse of the process of compound formation, which in this case is spontaneous. The electrolytic reaction proceeds because a voltage source forces it to occur. In effect electrolysis is an application of Le Châtelier's principle to an electrochemical system.

Electrolytic processes are used to obtain pure metals from less pure samples. If two strips of copper are connected to the terminals of an appropriate voltage source and are placed in a solution of copper sulfate ($CuSO_4$), copper can be made to dissolve from one electrode and become deposited on the other. The copper that is removed from the anode is changed into Cu^{2+} ions, which then are reduced at the cathode to elemental copper:

$$\begin{array}{rl} \mathbf{Cu} & \longrightarrow \mathbf{Cu^{2+}} + 2e^- \\ \mathbf{Cu^{2+}} + 2e^- & \longrightarrow \mathbf{Cu} \end{array}$$

If the anode is 99.0 percent "pure" copper, the copper deposited at the cathode is approximately 99.98 percent "pure" copper since the voltage

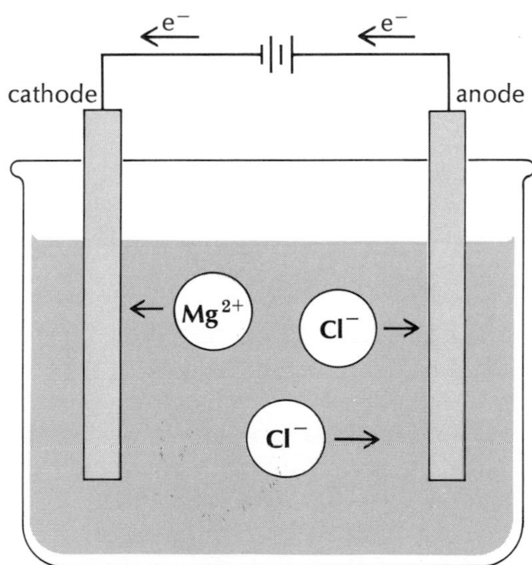

Figure 11.4
The electrolysis of molten magnesium chloride.

for the electrolytic cell can be chosen so that most of the impurities in the copper anode are not oxidized. Those that are oxidized are not reduced at the cathode.

11.7 SOURCES OF ELECTRICAL ENERGY

THE STORAGE BATTERY

The automobile storage battery is one example of a device whose operation depends on a chemical reaction that generates energy. The cathode is a grid containing lead oxide (PbO_2) in which the oxidation number of lead is +4; the anode is a lead metal grid. Both electrodes are immersed in a solution of sulfuric acid. Lead from the anode is oxidized and forms Pb^{2+}, which is deposited as $PbSO_4$, while at the cathode PbO_2 is reduced to Pb^{2+}, which also is deposited as $PbSO_4$. The half-reactions and the overall reaction are:

$$\mathbf{Pb + SO_4^{2-} \longrightarrow PbSO_4} + 2e^-$$
$$\mathbf{PbO_2 + 4H^+ + SO_4^{2-}} + 2e^- \longrightarrow \mathbf{PbSO_4 + 2H_2O}$$
$$\mathbf{Pb + PbO_2 + 4H^+ + 2SO_4^{2-} \longrightarrow 2PbSO_4 + 2H_2O}$$

Only a small concentration of Pb^{2+} is present in solution in the course of battery operation, and at both electrodes lead sulfate is formed. Eventually both electrodes are completely converted to lead sulfate if the battery is not recharged. Sulfuric acid is consumed and water is generated during the operation of the battery, resulting in a decrease in the density of the liquid. Since the densities of sulfuric acid solutions are known, the extent of the depletion of sulfuric acid can be determined by measuring the density of the battery liquid.

The lead storage battery can be recharged by the generator of a car or by application of an external electrical source. In either case, the reverse of the spontaneous process is made to occur.

FUEL CELL

The thermal energy derived from fuels such as coal and gasoline have been used to operate electrical generators, but such processes are inherently inefficient because of mechanical losses. Recent research has led to the development of fuel cells in which the energy of a combustion process (a redox reaction) is directly obtained as electrical energy instead of thermal energy. In principle a fuel cell can convert chemical potential energy into electrical energy with 100% efficiency.

The fuel cells produced to date have utilized gaseous reactants, and both hydrogen-oxygen and methane-oxygen fuel cells have been developed. In the hydrogen-oxygen fuel cell, hydrogen and oxygen are forced under pressure through separate porous carbon rods that serve as inert electrodes. These electrodes are immersed in a solution of potassium hydroxide. The two half-reactions that occur are:

$$H_2 + 2OH^- \longrightarrow 2H_2O + 2e^-$$
$$O_2 + 2H_2O + 4e^- \longrightarrow 4OH^-$$

Electrical energy thus is produced from a reaction that is in total simply the oxidation of hydrogen to form water.

Suggested further readings

Anson, F. C., "Electrode Sign Conventions," *J. Chem. Educ.*, **36,** 394 (1959).
Austin, L. G., "Fuel Cells," *Sci. Amer.*, 72 (October 1959).
Garrett, A. B., "Nuclear Batteries," *J. Chem. Educ.*, **33,** 446 (1956).
Jolly, W. L., "The Use of Oxidation Potentials in Inorganic Chemistry," *J. Chem. Educ.*, **43,** 198 (1966).
Raisbeck, G., "The Solar Battery," *Sci. Amer.*, 102 (December 1955).
Sanderson, R. T., "On the Significance of Electrode Potentials," *J. Chem. Educ.*, **43,** 505 (1966).
Weissbart, J., "Fuel Cells—Electrochemical Converters of Chemical to Electrical Energy," *J. Chem. Educ.*, **38,** 267 (1961).

Terms and concepts

anode
cathode
electrode
electrolysis
fuel cell
half-reaction
hydrogen electrode
galvanic cell
oxidation
oxidation number
oxidizing agent
redox reaction
reducing agent
reduction
standard electrode potential

Questions and problems

1. Define the terms *oxidation number, oxidation, reduction, oxidizing agent,* and *reducing agent.*
2. What are the oxidation numbers of all the elements in the listed species? **a.** N_2O_3 **b.** H_2SO_3 **c.** IO_3^- **d.** CrO_4^- **e.** PH_3 **f.** Fe **g.** $HClO_2$ **h.** SnO_2 **i.** $PbCl_2$ **j.** NO_3^-
3. What are the oxidation numbers of oxygen in OCl_2 and in OF_2? Why do they differ?

4. Identify the oxidizing agent and the reducing agent in each of the following equations:

a. $2Al + 3CoCl_2 \longrightarrow 2AlCl_3 + 3Co$
b. $2Na + H_2 \longrightarrow 2NaH$
c. $2H_2S + SO_2 \longrightarrow 3S + 2H_2O$
d. $Cu + Br_2 \longrightarrow Cu^{2+} + 2Br^-$
e. $3H_2SO_3 + 2NO_3^- \longrightarrow 2NO + 4H^+ + H_2O + 3SO_4^{2-}$
f. $2H^+ + 2Fe^{2+} + H_2O_2 \longrightarrow 2H_2O + 2Fe^{3+}$

5. What is meant by a standard electrode potential?

6. What limitation exists in predicting that a chemical reaction will occur from standard electrode potentials?

7. If the electrode potential for the conversion of Fe^{3+} to Fe^{2+} had been chosen as the standard reference, what would be the standard electrode potential for the following half-reactions?

a. $e^- + Na^+ \longrightarrow Na$
b. $2e^- + 2H^+ \longrightarrow H_2$
c. $e^- + Ag^+ \longrightarrow Ag$
d. $2e^- + Br_2 \longrightarrow 2Br^-$

8. Balance the following equations by using the half-reaction method:

a. $Fe^{3+} + Sn^{2+} \longrightarrow Fe^{2+} + Sn^{4+}$
b. $Br_2 + SO_2 + H_2O \longrightarrow H_2SO_4 + HBr$
c. $Cu + H^+ + NO_3^- \longrightarrow Cu^{2+} + NO_2 + H_2O$
d. $ClO_3^- + H^+ + Fe^{2+} \longrightarrow Fe^{3+} + Cl^- + H_2O$
e. $Cr_2O_7^{2-} + I_2 \longrightarrow Cr^{3+} + IO_3^-$

9. Which of the following reactions will proceed as written?

a. $Co^{2+} + Sn \longrightarrow Co + Sn^{2+}$
b. $Cu^{2+} + 2Br^- \longrightarrow Cu + Br_2$
c. $2I^- + 2Fe^{3+} \longrightarrow I_2 + 2Fe^{2+}$
d. $2Br^- + Cl_2 \longrightarrow 2Cl^- + Br_2$
e. $2Fe + 6H^+ \longrightarrow 3H_2 + 2Fe^{3+}$
f. $10F^- + 16H^+ + 2MnO_4^- \longrightarrow 2Mn^{2+} + 5F_2 + 8H_2O$

10. If the calculated potential for a reaction is negative, what does this mean?

11. What chemical process occurs at the anode in a galvanic cell?

12. Which of the following pairs of half-cell reactions when coupled with the $Zn \longrightarrow Zn^{2+} + 2e^-$ half-reaction would produce the highest voltage?

a. $Fe^{2+} + 2e^- \longrightarrow Fe$
b. $Cd^{2+} + 2e^- \longrightarrow Cd$
c. $Pb^{2+} + 2e^- \longrightarrow Pb$

13. Consider the species Mg, Cu^{2+}, Br_2, Br^-, I^-, Na^+, and F_2. Which of these seven is **a.** the best oxidizing agent, **b.** the poorest oxidizing agent, **c.** the best reducing agent, and **d.** the poorest reducing agent?

14. Distinguish between a voltaic cell and an electrolytic cell.

15. Electrolysis often is used to plate metal onto objects from aqueous solutions of a salt. Copper and silver can be plated out but magnesium and aluminum cannot. Indicate why these facts are observed.

16. Given four half-cell reactions, how many cells could be constructed?

17. Fuel cells are of interest in space exploration. What are the advantages of such cells over conventional storage cells?

PART THREE

THE ELEMENTS

Chemical principles and theories are meaningful only if the chemical alphabet (the elements) is known. There is no better way to understand chemistry than to interrelate the trends in both the chemical and physical properties of matter in terms of periodic relationships. Accordingly, in this part the elements are presented in chapters dealing with metals, nonmetals, transition metals, and the radioactive elements, which are the main subdivisions of the periodic table.

12

MAIN GROUP METALS

Historically the elements have been divided into two broad classes, *metals* and *nonmetals*. Approximately 80 percent of all elements are metals. Metals are electrical and thermal conductors and are characterized by their metallic luster and tendency to donate electrons and become oxidized. In this chapter elements that traditionally have been classified as metals will be examined. Since present knowledge about the physical and chemical properties of metals would fill many books, this chapter only can indicate some general properties of metals and their compounds. Any chemistry library contains volumes on the individual elements.

Only the metals in the main groups (IA, IIA, IIIA, and IVA) are discussed in this chapter. Elements with partially filled *d* and *f* subshells, which are called transition metals, are discussed in Chapter 14.

12.1
NATURAL OCCURRENCE OF METALS

Some of the unreactive metals exist in nature in their elementary form and can be obtained by physical separation from the other substances with which they are found. However, most metals are chemically combined with other elements such as oxygen and silicon. These naturally occurring materials are known as *ores,* and various industries exist solely to find ways of profitably extracting the pure metal from the available material.

Metallurgy is the science of obtaining metals and using them. Although there are as many ways to extract metals from ores as there are metals, there are three general operations necessary for obtaining pure metals, regardless of the specific method used. First the desired component of the ore is separated from other components in a physical treatment; then reduction of the metal compound to the free metal occurs; and finally the metal is refined to the desired state of purity. The problems involved in each step for various metals will not be outlined in this text.

Metal ores as mined are usually contaminated with a variety of materials such as granite and clay. This unwanted material called *gangue* often can be separated physically from the desired chemical compounds. Then the metal must be extracted by some chemical process that is as convenient and economical as possible. Since the majority of ores consist of compounds of oxygen (oxides), the process necessary for achieving the pure metal is reduction. Either carbon or hydrogen may be used to reduce a metal oxide. The products of the reaction in each case are the metal and a gaseous product at the temperature of the reaction. The loss of the gaseous product drives the reaction to the right:

$$\mathbf{ZnO + C \longrightarrow Zn + CO}\ (g)$$
$$\mathbf{WO_3 + 3H_2 \longrightarrow W + 3H_2O}\ (g)$$

Naturally, the higher the concentration of metal in the ore, the more desirable the ore and the less expensive should be the process. The exact concentration of metal below which it is deemed economically unfeasible to process is controlled both by the scarcity and value of the metal. For copper the limit may be as low as 1 percent, whereas for aluminum it is approximately 30 percent.

12.2
GROUP IA METALS

PROPERTIES

The elements of Group IA are called the *alkali metals.* The Arabic word *alkali* means "plant ashes," which are rich in salts of sodium and potassium. In addition, the term *alkali* refers to the hydroxides of this group of metals, which are all strong bases.

Table 12.1
Properties of Group IA metals

property	*Li*	*Na*	*K*	*Rb*	*Cs*
melting point (°C)	180	80	63	39	28
boiling point (°C)	1336	880	760	700	670
atomic radius (Å)	1.23	1.57	2.03	2.16	2.35
ionic radius, M^+ (Å)	0.67	0.98	1.33	1.48	1.68
ionization potential (eV/atom)					
first	5.4	5.1	4.3	4.2	3.9
second	75.6	47.3	31.8	27.4	23.4
electrode potential (volts)	−3.05	−2.71	−2.93	−2.93	−2.92

All of the metals of Group IA are excellent conductors of electricity and heat. They are relatively soft, *ductile* (can be drawn out into wires), and *malleable* (can be beaten into sheets). The melting points and boiling points are quite low for metals as a class (Table 12.1). In spite of the excellent metallic properties and ease with which the alkali metals can be physically manipulated, they are not widely used since they are expensive and are too chemically reactive to allow their use in unprotected places. Both oxygen and water react vigorously with the alkali metals:

$$4Li + O_2 \longrightarrow 2Li_2O$$
$$2Li + 2H_2O \longrightarrow 2LiOH + H_2$$

The excellent electrical and thermal-conducting properties of the alkali metals are the result of the mobility of the highest energy electron of the metals. The metal core, consisting of the nucleus and its lower energy electrons, remains stationary at a definite position in the crystal, while a fluidlike arrangement of mobile electrons surrounds all the nuclei (Chapter 7). These electrons have relatively little tendency to associate with a particular metal core and will move under the influence of even a small electric field. Liquid sodium has found a use as a conductor of heat from a nuclear reactor. The molten metal circulates through a closed system of pipes and is heated in the nuclear reactor, after which the heat energy is transported to another point and is converted to useful work.

OCCURRENCE AND PREPARATION

The alkali metals occur in nature exclusively as the +1 ions. Both sodium and potassium are abundant constituents of the earth's crust, whereas lithium, rubidium, and cesium are quite rare. The radioactive element francium is almost nonexistent in nature, although small quantities have been produced by nuclear reactions.

Almost all of the compounds of the alkali metal ions are soluble in water. The concentration of the salts of alkali metals in the ocean makes sea water a reasonable source for commercial extraction of the metals. The deposits of alkali metal salts on the earth's surface may have resulted from the evaporation of various ancient bodies of water.

The alkali metals can be prepared by electrolytic reduction of the +1 ions. This process usually involves the molten salts rather than a solution because the presence of water would lead to the generation of hydrogen rather than the desired metal. The commercial preparation of sodium from sodium chloride is carried out at 600°C and produces chlorine as well as metallic sodium:

$$2NaCl\ (l) \longrightarrow 2Na\ (l) + Cl_2\ (g)$$

REDUCING PROPERTIES

The electronic configurations of the alkali metals correspond to those of the inert gases plus a single *s* valence electron. These elements can lose the valence electron relatively easily to yield the +1 ions because their first ionization potentials are lower than those of any other element (Table 12.1). Since the second ionization potential is high, there is no tendency for the metals to exist in any higher oxidation state in the presence of other elements.

In the gaseous phase the reactivity of the metals with respect to each other can be predicted from their ionization potentials. For example, the redox reaction

$$Li^+ + Na \rightleftharpoons Li + Na^+$$

should proceed to the right since Li has less tendency to lose an electron than does sodium. The reaction can be divided into two gas phase half-reactions whose enthalpies are known ionization potential terms (one eV/atom is equal to 23 kcal/mole).

$$\begin{array}{ll} 1e^- + Li^+ \rightleftharpoons Li & \Delta H = -12.4 \text{ kcal/mole} \\ Na \rightleftharpoons Na^+ + 1e^- & \Delta H = +11.7 \text{ kcal/mole} \\ \hline Li^+ + Na \rightleftharpoons Li + Na^+ & \Delta H = -\ \ 0.7 \text{ kcal/mole} \end{array}$$

The first half-reaction, which is the reverse of the ionization equation, is exothermic and liberates an energy in calories corresponding to 5.4 eV/atom of Li^+. The second reaction requires 5.1 eV/atom of Na and is endothermic. In total, the reaction liberates 0.7 kcal/mole. There is no substantial difference between the entropies of the reactants or products because they are quite similar. Therefore, the free energy (Chapter 9) of the reaction is negative and products are favored over reactants. The above discussion can be applied to combinations of other alkali metals and ions in the gaseous phase. The conclusion from such considerations is that the reducing properties of the metals in the gaseous phase increases as the ionization potential decreases. However, since most chemical reactions are studied in solution, these conclusions may not apply.

The electrode potentials of the alkali metal ions that correspond to reactions in water are listed in Table 12.1. When these data are compared to the gaseous phase ionization potentials there seems to be an inconsistency. Lithium ion is reduced with the greatest difficulty in water, whereas it is the most easily reduced in the gaseous phase. The apparent discrepancy is a reflection of the difference in medium. In water it is necessary to take into account the interactions that occur between the positive ions and the polar solvent.

The reduction of the alkali metal ions (M^+) in water can be set equal to the sum of three consecutive steps whose individual energies must equal that of the single observed reduction reaction as a consequence of the law of conservation of energy. The parenthetical terms *g*, *s*, and *aq* refer to gas, solid, and aqueous, respectively:

$$\mathbf{M}\,(g) \longrightarrow \mathbf{M}\,(s) \qquad (1)\ \Delta H = -\text{sublimation energy}$$
$$1e^- + \mathbf{M}^+\,(g) \longrightarrow \mathbf{M}\,(g) \qquad (2)\ \Delta H = -\text{ionization potential}$$
$$\mathbf{M}^+\,(aq) \longrightarrow \mathbf{M}^+\,(g) + \mathbf{H_2O} \qquad (3)\ \Delta H = -\text{hydration energy}$$
$$\overline{1e^- + \mathbf{M}^+\,(aq) \longrightarrow \mathbf{M}\,(s) + \mathbf{H_2O}}$$

The three equations do not correspond to physical reality because the reduction of an alkali metal ion does not proceed stepwise. The only reason for choosing these steps is to provide some insight into the separate factors that determine the ultimate energy difference between reactants and products. In the first step the gaseous metal is condensed into a solid; the energy liberated corresponds in magnitude to the sublimation energy but with the sign changed. The energy for this step depends upon the specific metal, but the energy difference is relatively minor compared with those of steps 2 and 3. In step 2 the gaseous metal ion accepts an electron to yield the gaseous metal and liberates energy. This step is the reverse of the ionization potential. The third step involves the separation of associated water molecules from the metal ion, which then is restored to the gaseous phase. Since water is strongly attracted to positive ions step 3 requires the addition of energy to effect separation. The opposite reaction is called *hydration,* and the energy is the *hydration energy.*

In Table 12.2 the enthalpy change corresponding to each step for the individual alkali metals are listed in kilocalories per mole. The energy liberated in step 2 decreases with increasing atomic number, but the

Table 12.2
Enthalpy changes for Group IA metals (kcal/mole)

step	*Li*	*Na*	*K*	*Rb*	*Cs*
1 (−sublimation energy)	−37	−26	−22	−20	−19
2 (−ionization potential)	−124	−118	−100	−96	−89
3 (−hydration energy)	+121	+95	+76	+69	+62
total	−40	−49	−46	−47	−46

energy required in step 3 also decreases so that there is a compensatory change in both energy terms that eliminates to a large degree the differences between the reactions of the alkali metals in solution. From the values obtained by summing the individual energy terms, it can be seen that the reduction of the sodium ion liberates the most energy and should proceed with the greatest facility. The fact that the electrode potential of the sodium ion is the least negative is in agreement with the energy terms. Similarly, the lithium ion liberates the smallest quantity of energy and should be the least easily reduced, which is in agreement with its high negative electrode potential.

The *dehydration* (removal of water) of the alkali metal ions requires less energy for the ions of larger atomic dimensions. All of the ions have the same charge, and at first glance it might be expected that their attraction for water would be identical. However, it is the charge-to-size ratio that is important. The lithium ion is the smallest ion of the group, and the attraction for a negative species such as water is great for this ion of concentrated charge. As the charge-to-size ratio decreases for the ions of larger size the attraction for water decreases.

12.3 GROUP IIA METALS

PROPERTIES

The Group IIA metals are often called the *alkaline earth metals,* a name which derives its meaning from the era of alchemists. The term *earth* originally encompassed all nonmetallic substances that were unchanged by fire. The oxides of the Group IIA metals such as magnesia (MgO) and lime (CaO) fitted into the category of earths. The oxides, when placed in contact with water, gave rise to alkaline solutions, and the term *alkaline earth* was a natural consequence.

As a group the elements beryllium, magnesium, calcium, strontium, barium, and radium are harder than the elements of Group IA. In fact, beryllium is hard enough to scratch glass. The alkaline earth metals are gray-white metals with a silvery luster and are all conductors of electricity and heat. The boiling points and melting points of the Group IIA elements (Table 12.3) are higher than those of the Group IA elements. These latter physical properties suggest that the attraction between the M^{2+} ions of these metals immersed in a sea of mobile valence electrons is greater than for the M^+ ions of the alkali metals.

Of the Group IIA elements only magnesium has been widely used commercially. It is the lightest of the metals that are used for structural purposes. While the pure metal does not possess great structural strength, its low density is so desirable that it is alloyed with other metals such as aluminum, manganese, and zinc to improve its strength. The alloys of magnesium are used in construction so diverse that both cooking utensils

Table 12.3
Properties of Group IIA metals

property	*Be*	*Mg*	*Ca*	*Sr*	*Ba*
melting point (°C)	1280	650	850	800	850
boiling point (°C)	1500	1100	1440	1360	1540
atomic radius (Å)	0.89	1.36	1.74	1.91	1.98
ionic radius M^{2+} (Å)	0.32	0.65	0.99	1.12	1.35
ionization potential (eV)					
first	9.3	7.6	6.1	5.7	5.2
second	18.2	15.0	11.9	11.0	10.0
third	154	80	51	—	—
electrode potential (volts) from M^{2+}	−1.85	−2.36	−2.87	−2.89	−2.91

and aircraft make use of them. Magnesium burns rapidly if it is finely divided or is in the form of thin wires, as in photographic flash bulbs.

Beryllium is extremely toxic and is not widely used. The elements calcium, strontium, and barium are more reactive toward oxidizing agents than magnesium and have found few uses of wide applicability partly because the oxides of these metals are prone to chip away from the surface of the metal, which then becomes further oxidized. Magnesium oxide that does not chip as readily forms a protective coating on the metal that retards further oxidation.

OCCURRENCE AND PREPARATION

The alkaline earth metals occur in nature exclusively as the +2 ions. Calcium is the most abundant element of both Groups IA and IIA. It occurs as calcium carbonate ($CaCO_3$) in marble and limestone and as calcium phosphate $Ca_3(PO_4)_2$ in teeth, bones, and seashells. Magnesium also is abundant in the earth's crust and in seawater. The best-known mineral containing magnesium is asbestos ($CaMg_3Si_4O_{12}$). Strontium and barium are relatively rare. Radium is radioactive (Chapter 15) but is a product of another radioactive element, uranium. The concentration of radium in uranium ore is approximately 0.000003 percent of the weight.

The +2 ions of the alkaline earth metals must be reduced by electrolytic methods (Chapter 11) to produce the metals. The reduction is usually done on the molten salts of the chloride or bromide compounds. The major source of magnesium is seawater, which contains approximately 0.1 percent of this metal in ionic form.

REDUCING PROPERTIES

The electronic configurations of the alkaline earth metals correspond to those of the inert gases plus two *s* valence electrons. These metals can lose one of their valence electrons with only slightly greater difficulty than the alkali metals. Although their second ionization potentials are approximately twice the first ionization potentials, these energies are not excessively high, and the +2 ion should be easily produced in the gaseous phase even though the +1 ion is more easily attainable. The third ionization potential corresponds to the removal of an electron from a complete shell that has an inert gas electronic configuration. It therefore requires high energy.

In the gaseous phase the reduction of Mg^{2+} by Ca will proceed to the right:

$$\mathbf{Mg^{2+} + Ca \rightleftharpoons Mg + Ca^{2+}}$$

Magnesium has a higher ionization potential than Ca and loses electrons less easily. The reaction can be divided into two gas phase reactions whose enthalpy changes are known from ionization potentials:

$$\begin{array}{rcll} 2e^- + \mathbf{Mg^{2+}} & \rightleftharpoons & \mathbf{Mg} & \Delta H = -520 \text{ kcal/mole} \\ \mathbf{Ca} & \rightleftharpoons & \mathbf{Ca^{2+}} + 2e^- & \Delta H = 414 \text{ kcal/mole} \\ \hline \mathbf{Mg^{2+} + Ca} & \rightleftharpoons & \mathbf{Ca^{2+} + Mg} & \Delta H = -106 \text{ kcal/mole} \end{array}$$

The first half-reaction is exothermic and releases energy equal to the sum of the first and second ionization potentials of magnesium. The second reaction requires energy equal to the sum of the first and second ionization potentials of calcium. In total, the enthalpy change for both half-reactions is −106 kcal/mole.

The +1 ions of the alkaline earth metals are more stable than the corresponding +2 ions. The gas phase reduction of Mg^{2+} by Mg is exothermic and the reaction proceeds to the right:

$$\mathbf{Mg^{2+} + Mg \rightleftharpoons 2Mg^{+}}$$

In the usual manner the reaction can be divided into half-reactions:

$$\begin{array}{rcll} 1e^- + \mathbf{Mg^{2+}} & \rightleftharpoons & \mathbf{Mg^{+}} & \Delta H = -345 \text{ kcal/mole} \\ \mathbf{Mg} & \rightleftharpoons & \mathbf{Mg^{+}} + 1e^- & \Delta H = +175 \text{ kcal/mole} \end{array}$$

The first step involves the liberation of energy corresponding in magnitude to the second ionization potential of magnesium. The second step requires energy equal to the first ionization potential of magnesium. Since the second ionization potential is larger than the first ionization potential, energy is liberated in the total reaction. In this case the exothermic reaction produces 170 kcal/mole.

The fact that only the +2 ions are present in solution appears to be inconsistent with the gas phase data. Again, as was observed in the case of the alkali metal ions, the hydration of ions plays an important role in determining relative stabilities. For the formation of the +1 ion in solution the equations that can be considered involve an ionization step in the gas phase followed by a hydration step:

$$\mathbf{Mg}\ (g) \longrightarrow \mathbf{Mg}^{+}\ (g) + 1e^{-} \qquad \Delta H = 175\ \text{kcal/mole}$$
$$\mathbf{H_2O} + \mathbf{Mg}^{+}\ (g) \longrightarrow \mathbf{Mg}^{+}\ (aq) \qquad \Delta H = ?$$

The first step requires 175 kcal/mole. Energy should be released in the second step, but the actual quantity is unknown because Mg^+ does not exist in solution. However, as an approximation the hydration energy of the closely related Na^+ ion can be used. The Na^+ is of the same charge as Mg^+ but should be of slightly different radius. Therefore, the hydration energy of Mg^+ ought to be close to -118 kcal/mole. The total process for the formation of the Mg^+ in solution from the gaseous Mg should be endothermic with an enthalpy change of approximately $+57$ kcal/mole.

The formation of Mg^{2+} in solution from the gaseous Mg can be broken down into three steps:

$$\mathbf{Mg}\ (g) \rightleftharpoons \mathbf{Mg}^{+}\ (g) + 1e^{-} \qquad \Delta H = 175\ \text{kcal/mole}$$
$$\mathbf{Mg}^{+}\ (g) \rightleftharpoons \mathbf{Mg}^{2+}\ (g) + 1e^{-} \qquad \Delta H = 345\ \text{kcal/mole}$$
$$\mathbf{H_2O} + \mathbf{Mg}^{2+}\ (g) \rightleftharpoons \mathbf{Mg}^{2+}\ (aq) \qquad \Delta H = -590\ \text{kcal/mole}$$
$$\mathbf{H_2O} + \mathbf{Mg}\ (g) \rightleftharpoons \mathbf{Mg}^{2+}\ (aq) + 2e^{-} \qquad \Delta H = -70\ \text{kcal/mole}$$

The first two steps require energy equal to the sum of the first and second ionization potentials of magnesium and are in total endothermic by 520 kcal/mole. The energy released in the solvation of Mg^{2+} is approximately 590 kcal/mole. Therefore, the total process is exothermic by 70 kcal/mole and is more favorable than the reduction of Mg to Mg^+ in solution.

The reason for the stability of the $+2$ ions of the alkaline earth metals as compared to the $+1$ ions is the large hydration energy difference between the two ions. The hydration energy difference more than counterbalances the higher energy required to form the ion of higher charge.

In general the hydration energy of a $+2$ alkaline earth metal ion is about five times larger than that of the $+1$ alkali metal ion. This increase reflects both the higher charge of the alkaline earth metal ions and their smaller radius. The radii of Na^+ and Mg^{2+} are 0.95 and 0.65 Å, respectively. These radii correspond to a difference in their surface areas of 2.1 times.

HARD WATER

Because limestone is an abundant mineral in the earth's crust, most groundwater comes in contact with it and dissolves small amounts of calcium ion. The presence of this ion leads to many household and industrial problems. Groundwater containing dissolved Ca^{2+}, called *hard water,* forms precipitates when heated or when soap is added.

When soap, which is the sodium salt of a complex acid called *stearic acid* ($C_{18}H_{35}O_2H$; Chapter 21), is added to a solution containing Ca^{2+}, an insoluble salt called *calcium stearate* is formed. Its more familiar name is *scum* or *bathtub ring.*

$$\mathbf{Ca^{2+} + 2C_{18}H_{35}O_2^{-} \longrightarrow Ca(C_{18}H_{35}O_2)_2}$$

If enough soap is added, eventually all the Ca^{2+} is precipitated, and the residual soluble sodium salt will clean household goods. However, the precipitate is unpleasant, and hard water is uneconomical to use if the concentration of Ca^{2+} is too high.

Hardness in water creates problems in industry when heating the water leads to precipitation of $CaCO_3$ on the inside of equipment. The formation of $CaCO_3$ results from the displacement of the equilibrium by loss of CO_2, which is driven off when water is heated:

$$Ca^{2+} + 2HCO_3^- \rightleftharpoons CaCO_3 + CO_2 + H_2O$$

If the bicarbonate ion HCO_3^- is present, this reaction can be used to remove the calcium ion before the water is used. Hard water containing bicarbonate ion therefore is called *temporary* hard water. However, this process is impractical for large-scale industry. Both temporary and permanently hard water can be softened by the addition of sodium carbonate ($NaCO_3$), commonly called washing soda. The carbonate ion (CO_3^{2-}) precipitates out the calcium ions:

$$Ca^{2+} + CO_3^{2-} \rightleftharpoons CaCO_3$$

Temporary hardness can be eliminated by the addition of base that reacts with bicarbonate in an acid–base reaction to produce carbonate ions that then precipitate the Ca^{2+}:

$$HCO_3^- + OH^- \rightleftharpoons H_2O + CO_3^{2-}$$

The use of silicate minerals called *zeolites* to soften water by an ion exchange process has provided a remarkably convenient way of removing Ca^{2+} and replacing it by ions such as Na^+. The ion exchanger material consists of a large, covalently bonded, solid substance containing many negatively charged sites. The electrical neutrality of the substance is maintained by the presence of sodium ions. When water containing Ca^{2+} is passed through the ion exchanger the Na^+ is replaced by Ca^{2+}. The hard water is softened and now contains sodium ions.

12.4 GROUP IIIA METALS

PROPERTIES

Except for boron, the elements of Group IIIA are classified as metals. Boron possesses some of the properties of a metal but is a borderline case in the division between metals and nonmetals and is sometimes referred to as a *metalloid* or a *semimetal.* Boron is a hard, brittle substance with a rather dull metallic luster. While it is a poor conductor of electricity its conductivity increases as it is heated. This behavior is probably a reflection of increased motion or freeing of electrons at higher temperatures.

The physical properties, ionization potentials, and electrode poten-

Table 12.4
Properties of Group IIIA elements

property	*B*	*Al*	*Ga*	*In*	*Tl*
melting point (°C)	2300	660	30	155	304
boiling point (°C)	2600?	2300	2000	1450	1450
atomic radius (Å)	0.8	1.25	1.25	1.50	1.55
electrode potential (volts) from M^{3+}	—	−1.66	−0.53	−0.34	—

tials of the Group IIIA elements are listed in Table 12.4. The metals (excluding boron) are soft and have relatively low melting points. The melting point of gallium is unusually low: It will melt below body temperature.

Of the Group IIIA elements only aluminum is used widely in commercial products. Like magnesium it is very soft and is structurally weak when pure, but it is so light that it is advantageous to alloy it with other metals in order to increase its structural strength. Aluminum is reactive toward oxygen and water, but the oxide formed resists further oxidation of the underlying metal after surface oxidation has occurred.

OCCURRENCE AND PREPARATION

Boron is rare (0.0003 percent in the earth's crust) and occurs in the form of oxides such as borax ($Na_2B_4O_7 \cdot 10H_2O$). The element can be prepared by reduction with magnesium metal.

Aluminum is the most abundant metal in the earth's crust and occurs in the form of silicates such as felspar ($KAlSi_3O_8$). The oxide of aluminum known as bauxite ($Al_2O_3 \cdot nH_2O$) provides the most convenient source of uncontaminated aluminum ore, which is electrolytically reduced to produce aluminum. The process for reduction is quite costly in terms of the electric power consumption and, consequently, the plants for the production of aluminum are located near cheap electric current sources.

Gallium, indium, and thallium are very rare. Gallium occurs in aluminum ore; indium, in zinc and lead ores. The wide liquid range of gallium (30–2000°C) enables it to be used as a liquid in some thermometers. Both gallium and indium form protective oxide coatings that would allow them to be used structurally in the form of alloys were they more abundant. Thallium is oxidized readily by air but does not form a protective oxide coating. Its salts are poisonous and are used in some rodent poisons.

REDUCING PROPERTIES

All of the metals of Group IIIA have two electrons in an *s* subshell and one in a *p* subshell. They can achieve the electronic configuration of an inert gas by losing three electrons to form +3 ions. However, most of the metals in the +3 oxidation state are covalently bound to electronegative elements.

Aluminum is a good reducing agent in spite of the necessity of providing sufficient energy to remove three electrons. The successive ionization potentials of aluminum are 6.0, 18.8, and 28.4 eV/atom. The Al^{3+} ion has a high hydration energy that is over 1000 kcal/mole. This energy, which reflects the +3 charge and the small radius of 0.52 Å, is the main reason that the ion can be formed in solution.

12.5 GROUP IVA METALS

PROPERTIES

Of the elements of Group IVA only tin and lead can be classified as metals. Germanium is a borderline case and is classed as a semimetal. The elements carbon and silicon are nonmetals and are discussed in Chapter 13.

Both tin and lead are soft metals and have relatively low melting points (Table 12.5). While they are good electrical conductors, germanium is a semiconductor that is widely used in transistors. Although tin is relatively inert it does form a protective oxide coating whenever it reacts with oxygen or water. This inertness allows tin to be used in coating some steel products and in the production of "tin cans." The principal common use of lead is in the manufacture of lead storage batteries that serve as the electrical source in automobiles. Alloyed with other metals, lead is used as solder and in type for printing. Lead compounds are poisonous and tend to accumulate in the body, adversely affecting the central nervous system. Vessels made of alloys containing lead were used in ancient times for food, and the resultant prolonged contact and ingestion of lead salts undoubtedly shortened life. For this reason pipes in water systems now are made of copper instead of lead.

Table 12.5
Properties of Group IVA elements

property	*C*	*Si*	*Ge*	*Sn*	*Pb*
melting point (°C)	3500	1400	960	230	330
boiling point (°C)	4200	2400?	2700?	2250?	1600?
atomic radius (Å)	0.77	1.17	1.22	1.41	1.54
electrode potential (volts) from M^{2+}	—	—	—	−0.14	−0.13

OCCURRENCE AND PREPARATION

Germanium is the least abundant of the Group IVA metals, but it can be obtained as a by-product of zinc ores. Its use in transistors has spurred interest in this semimetal. The pure metal is obtained by reduction of the oxide GeO_2.

Tin occurs in nature as the oxide SnO_2, which can be reduced by carbon to yield the free metal:

$$SnO_2 + C \longrightarrow Sn + CO_2$$

Lead occurs as the mineral galena, PbS. This sulfide ore is oxidized in air to produce a lead oxide, which then can be reduced by carbon:

$$2PbS + 3O_2 \longrightarrow 2PbO + 2SO_2$$
$$2PbO + C \longrightarrow 2Pb + CO_2$$

OXIDATION STATES

The elements of Group IVA have two electrons in both the *s* subshell and the *p* subshell. The two oxidation states observed for germanium, tin, and lead are +2 and +4. While the +2 ions can exist in aqueous solutions, the +4 oxidation states occur only in combination of the metal with electronegative elements by means of covalent bonds. As an example of the difference between the +2 and +4 oxidation states, the compound $PbCl_2$ is an ionic solid substance whereas $PbCl_4$ is a volatile covalent compound.

Suggested further readings

Block, M. R., "The Social Influence of Salt," *Sci. Amer.,* 89 (July 1963).

Cooper, D. G., *The Periodic Table.* Washington, D.C.: Butterworth, 1964.

Cottrell, A. H., "The Nature of Metals," *Sci. Amer.,* 90 (September 1967).

Fleischer, M., "The Abundance and Distribution of the Chemical Elements in the Earth's Crust," *J. Chem. Educ.,* **31,** 446 (1954).

Heilbrunn, L. V., "Calcium and Life," *Sci. Amer.,* 60 (June 1951).

Holmes, H. N., "The Story of Hall and Aluminum," *J. Chem. Educ.,* **7,** 232 (1930).

Mero, J. L., "Minerals on the Ocean Floor," *Sci. Amer.,* 64 (December 1960).

Payne, Jr., D. A., and Fink, F. H., "Electronegativities and Group IVA Chemistry," *J. Chem. Educ.,* **43,** 654 (1966).

Rich, R. L., *Periodic Correlations.* New York: W. A. Benjamin, 1965.

Schubert, J., "Beryllium and Berylliosis," *Sci. Amer.,* 27 (August 1958).

Wagner, G. H., and Gitzen, W. H., "Gallium," *J. Chem. Educ.,* **29,** 162 (1952).

alkali metals
alkaline earth metals
ductile
gangue
hard water
hydration energy
malleable
metallurgy

Questions and problems

1. What steps are involved in preparing a metal from its ore?
2. What are the advantages of alloys as compared to their constituent metals?
3. What properties of aluminum and magnesium make these metals useful for structural purposes?
4. How would you expect the ionization potentials, atomic radius, and ionic radius of francium to compare with the other alkali metals?
5. Lithium ion has the lowest electrode potential of the alkali metal ions. How can this fact be reconciled with the low ionization potential of lithium?
6. Which of the following reactions will proceed in the gas phase in the indicated direction?

a. $Li^+ + Rb \longrightarrow Rb^+ + Li$
b. $K + Rb^+ \longrightarrow Rb + K^+$
c. $K + K^{2+} \longrightarrow 2K^+$

7. Which alkali metal is the best reducing agent in the gas phase? in the aqueous phase?
8. Account for the order of the electrode potentials for the alkaline earth metal ions.
9. What is the difference between permanently and temporarily hard water?
10. What is an ion exchanger? How could a combination of treatments with an anion exchanger and a cation exchanger be used to remove the ions from a salt solution?
11. Which of the following reactions will proceed in the gas phase?

a. $2Ca^+ \longrightarrow Ca + Ca^{2+}$
b. $Mg^{2+} + Ca \longrightarrow Mg + Ca^{2+}$
c. $2Na^+ + Mg \longrightarrow 2Na + Mg^{2+}$

12. Considering the following electrode potentials, will the reaction $3Tl^+ \longrightarrow 2Tl + Tl^{3+}$ proceed as written?

$$Tl^{3+} + 2e^- \longrightarrow Tl^+ \quad \mathcal{E}^\circ = +1.25 \text{ volts}$$
$$Tl^+ + e^- \longrightarrow Tl \quad \mathcal{E}^\circ = -0.34 \text{ volts}$$

13. Why is Sn^{2+} more strongly hydrated in solution than Pb^{2+}?
14. Why does iron corrode so much more rapidly than aluminum?
15. Why does aluminum form +3 ions in aqueous solution but not the +2 and +1 ions?

16. Contrast the structure and properties of CO_2 and SiO_2.
17. Plumbane (PbH_4) is thermally unstable, whereas germane (GeH_4) is stable. Suggest a reason for the comparative stabilities. What could be predicted about the stability of stannane (SnH_4)?
18. Suggest a structure for the compound Ge_2H_6.

13

THE NONMETALS

About 20 percent of the known elements have chemical and physical properties so different from the metals that they are classed as nonmetals. These elements are extremely important in spite of their small number. Carbon, hydrogen, and oxygen are perhaps the most important of all. Carbon forms the basis for the compounds contained in living organisms. Almost all of these organic compounds contain oxygen and hydrogen as well. The elements nitrogen and phosphorus of Group VA also are important in life processes.

13.1 HYDROGEN

PROPERTIES AND REACTIONS

Hydrogen is an odorless, colorless, tasteless gas that has the lowest density of any element. It occurs as the free element to a limited extent, but for the most part it is combined in many diverse compounds ranging from simple water molecules to complex molecules such as DNA, the genetic controller of life. Approximately two-thirds of the human body is water. Most of the other compounds in the human body contain hydrogen combined with carbon.

The hydrogen molecule is diatomic, with the two hydrogen atoms bound by a nonpolar covalent bond. The intermolecular forces of attraction are of the weak van der Waals type (Chapter 7), as indicated by the normal boiling point of −252.7°C.

In addition to the isotope $^{1}_{1}H$, which constitutes 99.985 percent of naturally occurring hydrogen, there is a heavier isotope $^{2}_{1}H$ known as *deuterium.* A third isotope $^{3}_{1}H$ (*tritium*) is radioactive and is present in trace amounts.

Most reactions of elemental hydrogen require high temperatures because the bond energy of the hydrogen–hydrogen bond is 103 kcal/mole. Once the bond is broken, however, the resulting hydrogen atoms are extremely reactive. If an electron is added to the hydrogen atom a *hydride ion* H^- results. If an electron is removed from the hydrogen atom a proton results.

The hydride ion is isoelectronic with helium, and the configuration is stable in the absence of oxidizing agents. In order to produce hydrides it is necessary to have a reactive metal whose low ionization potential allows it to transfer an electron to hydrogen. The hydrides of Group IA metals and the elements Ca, Sr, and Ba are salts with high ionic character, which can be produced by direct reaction with H_2:

$$\mathbf{2Li + H_2 \longrightarrow 2LiH}$$
$$\mathbf{Ca + H_2 \longrightarrow CaH_2}$$

Hydrogen combines with most of the nonmetals, yielding compounds in which hydrogen is in the +1 oxidation state. In all cases hydrogen is bound to the electronegative element by a polar covalent bond. The bare proton cannot exist except in the absence of other elements in the gaseous phase.

The reaction of hydrogen with oxygen to produce water vapor is exothermic, to the extent of 115.6 kcal/mole of oxygen:

$$\mathbf{2H_2}\,(g) + \mathbf{O_2}\,(g) \longrightarrow \mathbf{2H_2O}\,(g) \qquad \Delta H = -115.6 \text{ kcal}$$

However, this reaction is slow at ordinary temperatures. Only at high temperatures or in the presence of an electric spark does the reaction occur. Under either condition the reaction takes place with explosive violence.

The reaction of hydrogen with nitrogen is carried out at temperatures ranging from 400 to 600°C and at pressures from 500 to 1000 atm. In addition, a catalyst is necessary, and the ammonia formed must be removed by liquefaction to shift the equilibrium toward a more complete conversion of reactants to products:

$$3H_2 + N_2 \longrightarrow 2NH_3$$

HYDROGEN BONDING

The attraction between some compounds, each containing hydrogen bound to highly electronegative elements of small radius, is known as hydrogen bonding (Chapter 7). In such compounds the strong attraction of the electronegative element for the bonding electrons causes the hydrogen to acquire a partial positive charge. This hydrogen is strongly attracted to the lone pair electrons of the electronegative atoms of neighboring molecules. These bonds between hydrogen and the atoms N, O, and F have energies averaging 5 kcal/mole. This energy gives rise to the observed differences in boiling points noted in Section 7.2.

Hydrogen bonding accounts for the solubility of polar covalent compounds containing fluorine, oxygen, and nitrogen. For example, ammonia is very soluble in water, a fact that is consistent with the model of extensive hydrogen bonding.

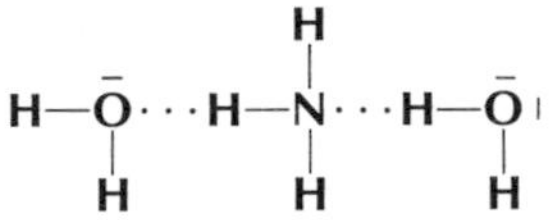

```
                H
                |
H—Ō···H—N···H—Ō|
   |        |       |
   H        H       H
```

Hydrogen bonding is one of the important structure-determining features in large biological molecules such as enzymes and DNA (Chapter 27).

13.2 THE INERT GASES

PROPERTIES

Before 1962 no chemical compounds of the elements of Group 0 were known to exist. Since that time approximately a dozen compounds of Kr, Xe, and Rn have been prepared. The lack of reactivity accounts for the name *inert* given to this group of elements, which are also called *noble* gases and *rare* gases.

Each element of Group 0, except for helium, has eight electrons in the highest filled energy level; two electrons are in the *s* subshell, and six are in the *p* subshell. Helium has a complete first energy level with two electrons in the 1*s* subshell. There is no tendency for these elements to combine with each other, and they all exist as monatomic gases.

In Table 13.1 the physical properties of the Group 0 elements are listed. The low melting points and boiling points are a reflection of the weak van der Waals forces of attraction between atoms. The strength of these forces increases with the increase in the number of electrons of the

Table 13.1
Properties of Group O elements

property	*He*	*Ne*	*Ar*	*Kr*	*Xe*	*Rn*
melting point (°C)	—	−249	−189	−157	−112	−71
boiling point (°C)	−269	−246	−186	−153	−107	−62
first ionization potential (eV)	24.6	21.6	15.8	14.0	12.1	10.7
radius (Å)	1.40	1.54	1.88	2.02	2.16	—

atoms. With an increase in the size of the atoms of Group 0 the electrons are less tightly held by the attraction of the nucleus. The electron cloud becomes more polarizable when the electrons are less under the influence of the nucleus, with the result that the intermolecular forces are larger.

COMPOUNDS

Both the high ionization potential and the complete *s* and *p* subshells of the inert gases make it quite unlikely that many compounds will be synthesized. Indeed, few serious efforts were made to react the inert gases until the 1960s. The impetus for investigations into the reactivities of inert gases came from the observation of N. Bartlett that O_2 and PtF_6 react to form a solid compound that consists of O_2^+ and PtF_6^-. The ionization potential of the oxygen molecule is 12.2 eV, which is close to the 12.1 eV ionization potential of xenon. This suggested that Xe and PtF_6 ought to be examined for potential reactivity. The yellow solid $XePtF_6$ formed is currently thought to consist of Xe^+ and PtF_6^- ions.

Once the first compound of an inert gas was produced there was a rapid acceleration in the study of ways to produce others. To date only compounds of Kr, Xe, and Rn have been prepared. These gases have the lowest ionization potential and largest radius, which makes them the most reactive. The very high ionization potentials of helium and neon suggest that they will remain truly inert elements. Argon probably will serve as the boundary between the reactive and nonreactive inert gases.

The only binary compounds of the inert gases that have been prepared involve oxygen and fluorine, the two most electronegative elements. The direct reaction of various mixtures of xenon and fluorine at 400° has led to the formation of XeF_2, XeF_4, and XeF_6, which are all crystalline solids. All these compounds react with water to produce compounds such as XeO_3, $XeOF_4$, and $Xe(OH)_4$. The tetrafluorides RnF_4 and KrF_4 also have been prepared.

All of the binary compounds of the noble gases are covalently bound substances. Therefore, the central Group 0 element has more than eight electrons about it. These electrons are accommodated by utilizing the empty 5*d* orbitals of xenon. In XeF_2 the linear molecule is thought to arise from

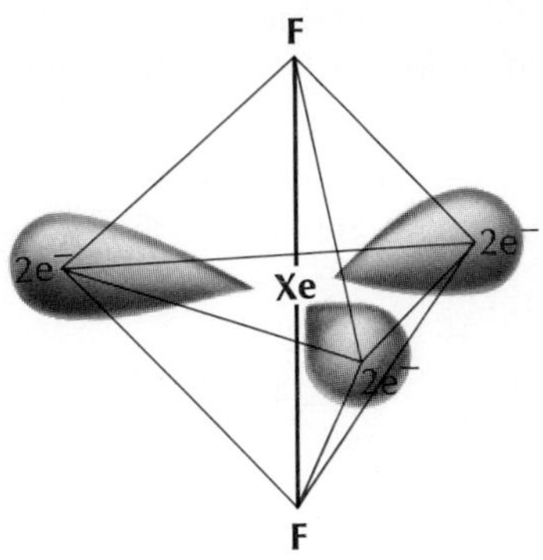

Figure 13.1
Structure of XeF_2.

five sp^3d hybrid orbitals that are derived from one 5*s*, three 5*p*, and one 5*d*. Xenon has eight electrons in the fifth energy level, which along with one electron from each of the fluorine atoms yields a total of 10 electrons for the complete filling of the five sp^3d orbitals. The hybrid orbitals are directed toward the corners of a trigonal bipyramid (Figure 13.1). In this structure the fluorine atoms are separated by the maximum possible distance while remaining bonded to xenon. The electron pairs in three of the sp^3d orbitals also are as far apart as possible.

Xenon tetrafluoride is a square planar molecule. The eight electrons of xenon and one each from the four fluorines require six orbitals. These orbitals may be derived from one 5*s*, three 5*p*, and two of the empty 5*d* orbitals leading to six sp^3d^2 hybrid orbitals that are directed toward the corners of an octahedron. Four of the fluorine atoms are in a plane bound to the central xenon by covalent bonds. There are two unshared electron pairs of xenon that are postulated to be directed toward the remaining two corners of the octahedron (Figure 13.2).

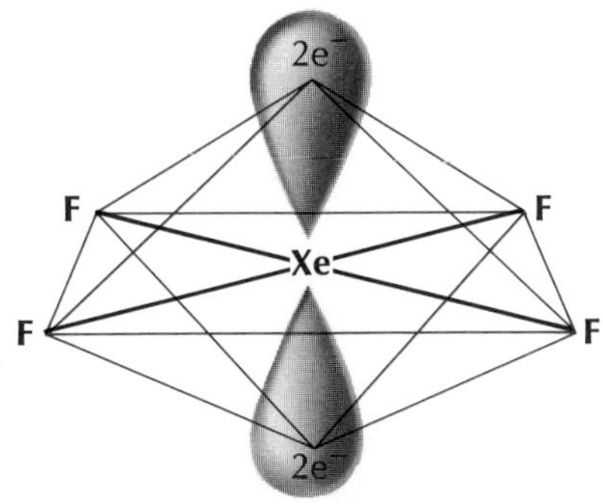

Figure 13.2
Structure of XeF_4.

13.3 THE GROUP VIIA ELEMENTS

PROPERTIES

The elements fluorine, chlorine, bromine, iodine, and astatine are collectively known as the *halogens*. The term *halogen* is derived from the Greek words *halos* (salt) and *genes* (born). The combination of words implies that the elements of Group VIIA are salt producers. With the exception of the rare, radioactive element astatine, the halogens are abundant in nature as the halide salts of metal cations.

The elements of Group VIIA are diatomic and consist of two atoms bound by a single covalent bond. The properties of these elements excluding the rare element astatine, are listed in Table 13.2. The melting points and boiling points of the halogens are a reflection of the van der Waals forces of attraction between the temporary dipoles of the molecules as a result of electron polarization. The iodine molecule has the largest molecular dimensions and the most electrons. Therefore, its electrons are the most polarizable of the common halogens (excluding astatine), and the intermolecular attractive forces are the strongest. The fact that iodine is solid, bromine is liquid, and chlorine and fluorine are gases at standard temperature and pressure is consistent with diminishing van der Waals forces with decreasing molecular size.

USES OF HALOGENS

Fluorine is so reactive that its commercial applications are limited. However, the fluoride anion is used widely in the fluoridation of water, which under controlled conditions inhibits tooth decay. Covalently bound fluorine and carbon compounds are known as fluorocarbons. These substances are very inert, resisting acids and strong oxidizing agents. They have found wide use in industry where other materials would rapidly deteriorate.

Table 13.2
Properties of Group VIIA elements

property	F_2	Cl_2	Br_2	I_2
color	pale yellow	yellow-green	red	violet-black
melting point (°C)	−220	−102	−7	113
boiling point (°C)	−187	−35	+59	+183
atomic radius (Å)	0.72	0.99	1.14	1.33
ionic radius (X^-) (Å)	1.36	1.81	1.95	2.16
electronegativity	4.0	3.0	2.8	2.5
first ionization potential (eV)	17.4	13.0	11.8	10.4
bond energy (kcal)	37	58	46	36
electrode potential to X^-	2.87	1.36	1.08	0.54

One of the more popular commercial products containing fluorine is the *polymer* known commercially as Teflon. (A polymer is a high molecular mass substance made from the reaction of a large number of simple molecules called *monomers.*) Teflon is similar in structure to the more common polymer known as polyethylene (Chapter 17), which is considerably less expensive than Teflon and is used to produce many housewares. However, Teflon's unique nonstick characteristics and greater thermal stability has made it useful as a coating in cooking ware. The thermal stability of Teflon is a reflection of the carbon fluorine bond energy and the high electronegativity of fluorine, which prevents the molecule from being oxidized.

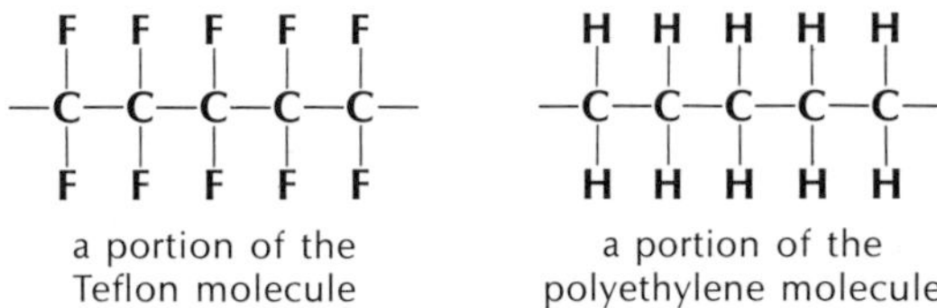

a portion of the Teflon molecule

a portion of the polyethylene molecule

Chlorine, which is produced in huge commercial quantities, finds use in the production of bleach, in paper processing, and in the disinfection of water supplies. Solutions of sodium hypochlorite ($NaOCl$) are used as household laundry bleaches. Potassium chlorate ($KClO_3$) is used as the oxidizing agent in matches and fireworks.

The most important commercial use of bromine is as silver bromide, one of the components of photographic emulsions. Exposure to light causes AgBr to break down into free silver that makes the dark areas of a negative. Bromine also is used in the production of ethylene dibromide ($C_2H_4Br_2$), which is added to gasolines containing the antiknock additive tetraethyl lead. In the process of burning gasoline, lead is produced from the additive that would deposit in the engine if ethylene dibromide were not present. In the presence of ethylene dibromide the lead is converted into $PbBr_2$, which is eliminated in the exhaust.

Iodine is not widely used. It was formerly in an alcohol solution as an antiseptic. However, the alcohol slowly evaporates, and the resultant concentrated solutions can cause skin burns that do not heal readily. Other preparations now have largely displaced the old tincture of iodine from medicine cabinets. The other halogens can cause severe burns and even death if inhaled in high concentration. For this reason the evacuation of wide areas is necessary when tank car quantities of chlorine are involved in transportation accidents on roads, rivers, or rail lines.

The halide ions Cl^-, Br^-, and I^- are vital to the human system. The proper balance of chloride and bromide ions in blood is required to help maintain the osmotic pressure balance that affects the transfer of nutrients from blood plasma into other cells. The chloride ion concentration in blood is approximately 20 times that of bromide ion. Only small quantities of iodide are needed in the human diet. The necessary amount can be obtained from a food source such as fish. Sodium iodide is customarily added to table salt (NaCl) to the extent of 0.001 percent in order to ensure that the iodide ion is available.

REDUCING PROPERTIES

The electrode potentials for the conversion of the halogens to the halide ions indicate that the halogens are oxidizing agents. Fluorine is the strongest oxidizing agent known. The oxidizing power of the halogens decrease in the order $F_2 > Cl_2 > Br_2 > I_2$, consistent with their electronegativities. The hydration energies of the halide ions also play an important role, but unlike the opposing contributions observed with the alkali metals, the contributions of electronegativity and hydration energy are in the same direction for the halogens. Fluorine has the highest affinity for electrons, and the resultant charged ion is very small and strongly hydrated. Both these factors contribute to the oxidizing ability of fluorine.

The halogens form an interesting series in terms of their ability to oxidize other halide ions whose elemental form has a lower electrode potential. For example, chlorine will oxidize bromide ions or iodide ions:

$$\mathbf{Cl_2 + 2Br^- \longrightarrow 2Cl^- + Br_2} \qquad \Delta\mathcal{E}^0 = 0.25$$
$$\mathbf{Cl_2 + 2I^- \longrightarrow 2Cl^- + I_2} \qquad \Delta\mathcal{E}^0 = 0.82$$

Fluorine can oxidize chloride, bromide, and iodide ions. Bromine can oxidize iodide ions.

OCCURRENCE AND PREPARATION

Fluorine exists only as the −1 ion in nature and occurs in the minerals fluorspar (CaF_2) and cryolite (Na_3AlF_6). Because no chemical oxidizing agent can remove electrons from fluoride ions, electrolytic methods must be used in order to produce fluorine. Once fluorine is produced it must be handled with care because fluorine reacts with so many other substances, often violently, as with both hydrogen and oxygen:

$$\mathbf{H_2 + F_2 \longrightarrow 2HF}$$
$$\mathbf{O_2 + 2F_2 \longrightarrow 2OF_2}$$

Chlorine in the form of the chloride ion is the most abundant of the halogens. In theory fluorine could be used to oxidize chloride ion but because fluorine presents handling difficulties, chlorine too is best obtained by electrolytic means.

The ratio of bromide ion to chloride ion in seawater is approximately 1:100. Bromide ion is oxidized to bromine by bubbling elemental chlorine gas through seawater. The bromine is removed from the resulting solution by bubbling air through it. The bromine is quite volatile and is driven out of solution and condensed out of the air.

Iodine is the only halogen that occurs naturally in a positive oxidation state as well as the −1 ion. Sodium iodate ($NaIO_3$) occurs mixed with Chile saltpeter ($NaNO_3$) and can be used as a source of elemental iodine. Iodine is also produced by chlorine oxidation of I^-.

Astatine, which is very rare in nature, is produced by nuclear reaction from bismuth. The known chemistry of astatine is limited, but the astatide ion At^- has been made.

HYDROGEN HALIDES

The hydrogen halides can be prepared directly from hydrogen and the halogens. The energy evolved in these reactions decreases from fluorine to iodine:

$$\mathbf{H_2 + X_2 \longrightarrow 2HX}$$

Hydrogen fluoride and hydrogen chloride are made in industry by the action of concentrated sulfuric acid on the halide salts:

$$\mathbf{NaCl + H_2SO_4 \longrightarrow NaHSO_4 + HCl}$$
$$\mathbf{CaF_2 + H_2SO_4 \longrightarrow CaSO_4 + 2HF}$$

However, hydrogen bromide and hydrogen iodide must be made from phosphoric acid (H_3PO_4) and the halide salts, because sulfuric acid, being a strong oxidizing agent, will produce Br_2 and I_2 from Br^- and I^-, respectively:

$$\mathbf{NaBr + H_3PO_4 \longrightarrow HBr + NaH_2PO_4}$$
$$\mathbf{NaI + H_3PO_4 \longrightarrow HI + NaH_2PO_4}$$

Solutions of the hydrogen halides in water are called *hydrohalic* acids. The covalent hydrogen chloride, when dissolved in water, yields a solution containing H_3O^+ and Cl^- and is called hydrochloric acid. Because of hydrofluoric acid's unique ability to react with glass, it is stored in wax-coated glass bottles or plastic containers.

OXYHALOGEN ACIDS

Chlorine, bromine, and iodine can exist in positive oxidation states when combined with oxygen, which is more electronegative than these halogens. The acids containing both oxygen and halogens are known as *oxyhalogen* acids. The known oxyhalogen acids are listed in Table 13.3. The general structures of the oxyhalogen acids and the oxidation numbers of the halogens are:

H—O—X	H—O—X—O	H—O—X(—O)—O	H—O—X(—O)(—O)—O
(+1)	(+3)	(+5)	(+7)

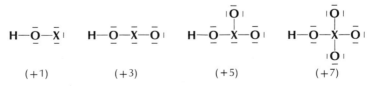

Table 13.3
Oxyhalogen acids

formula of acid	*name of acid*	*anion derived from acid*
$HClO$ $HBrO$ HIO	hypohalous acid	hypohalite ion
$HClO_2$ — —	halous acid	halite ion
$HClO_3$ $HBrO_3$ HIO_3	halic acid	halate ion
$HClO_4$ $HBrO_4$ HIO_4	perhalic acid	perhalate ion

The oxidizing power of the oxyhalogen acids in general decreases in the order $HXO_4 > HXO_3 > HXO_2 > HXO$. When compounds of the same oxidation number are compared, the oxychlorine acids are stronger oxidizing agents than the oxybromine acids, which in turn are better oxidizing agents than the oxyiodine acids.

The order of increasing acidity of the oxyhalogen acids is $HXO < HXO_2 < HXO_3 < HXO_4$. For acids of similar structure the chlorine acids are stronger than the bromine acids, which in turn are stronger than the iodine acids. The higher acidity with increasing oxidation number of the halogen can be attributed to the effect of the added oxygen atoms on the strength of the O—H bond with respect to dissociation. The addition of oxygen atoms tends to drain electrons away from the less electronegative halogen atom, which tends to pull electrons away from the O—H bond. As a result of decreased electron density near hydrogen in the O—H bond, a proton can leave with greater facility.

13.4 OXYGEN

PROPERTIES

Oxygen is the earth's most abundant element. On a weight basis approximately 50 percent of the earth's crust and 22 percent of the earth's atmosphere is oxygen. Water is 89 percent oxygen. Oxygen is an important constituent of animals and plants, and it is combined with carbon, hydrogen, sulfur, nitrogen, and phosphorus.

At standard temperature and pressure oxygen is an odorless, colorless, and tasteless gas. The normal boiling and melting points of oxygen are -183 and $-218°C$, respectively. Both liquid and solid oxygen are pale blue materials. The three naturally occurring isotopes of oxygen are $^{16}_{8}O$ (99.76 percent), $^{17}_{8}O$ (0.04 percent), and $^{18}_{8}O$ (0.20 percent).

Oxygen is paramagnetic (Section 5.4) and has two unpaired electrons. The oxygen molecule contains a single bond between the oxygen atoms and, therefore, violates the octet rule:

$$|\bar{\underset{\cdot}{O}}-\bar{\underset{\cdot}{O}}| \qquad |\bar{O}=\bar{O}|$$

I II

Although Structure II, a Lewis structure with a double bond between the oxygen atoms, can be written, it is incorrect since it predicts no unpaired electrons. While the paramagnetic structure for oxygen might appear superficially to be less stable than the Lewis structure, it is possible to account for its stability by molecular orbital concepts (which will not be outlined in this text).

Elemental oxygen can exist in more than one molecular form and therefore is said to be *allotropic*. When energy is added to the diatomic form O_2, a triatomic molecule ozone O_3 can be produced:

$$3O_2 \longrightarrow 2O_3 \qquad \Delta H = 68 \text{ kcal}$$

Ozone is formed by the energy released by lightning, sparks from electric motors, and ultraviolet light. Its sharp, rather penetrating smell can be noted after lightning storms and around some electrical motors. Unlike diatomic oxygen, ozone is diamagnetic, which means that its structure must have all of the electrons paired. The molecule is angular and consists of a central oxygen atom identically bound to the other two oxygen atoms. Two structures that satisfy the octet rule can be written. Ozone is not adequately represented by either of these structures in which the bonds are not equivalent and is considered a resonance hybrid for which the two Lewis structures are contributing resonance forms:

O O O ⟷ O O O

OXIDATION STATES

Except for the compounds O_2F_2 and OF_2, in which oxygen has a positive oxidation number, the oxidation states of oxygen are $-\frac{1}{2}$, -1, and -2, the -2 state being by far the most common. In ionic substances such as CaO (calcium oxide), oxygen is present in the form of the O^{2-} (oxide) ion. The nonmetals do not form ionic compounds with oxygen but rather share electrons. In most compounds of nonmetals except for fluorine the oxidation state of oxygen is -2. A few examples are water, sulfur dioxide, and carbon dioxide:

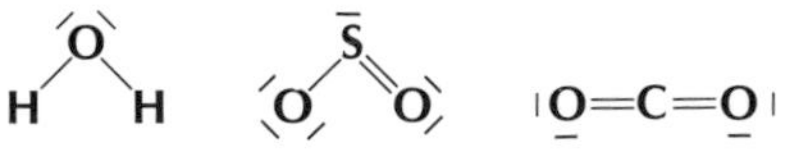

H–O–H O–S=O $|\underline{O}=C=\underline{O}|$

The elements potassium, rubidium, and cesium form compounds of the general type MO_2, called superoxides, which contain the M^+ ion and the superoxide ion O_2^{1-}, which is paramagnetic. Since the superoxide ion has one more electron than O_2, its formal oxidation number must be assigned as $-\frac{1}{2}$ because it contains two equivalent oxygen atoms. Two resonance contributors can be written for the superoxide ion:

$$\left[|\bar{\underline{O}}-\bar{\underline{O}}\cdot\right]^- \longleftrightarrow \left[\cdot\bar{\underline{O}}-\bar{\underline{O}}|\right]^-$$

The elements sodium, strontium, and barium form compounds that contain the peroxide ion O_2^{2-}. The formulas for these compounds are Na_2O_2, SrO_2, and BaO_2. Only one Lewis structure can be written for the peroxide ion:

$$|\bar{\underline{O}}-\bar{\underline{O}}|^{2-}$$

The compound hydrogen peroxide is a covalent substance and is unstable with respect to the many substances it can oxidize:

$$H-O-O-H$$

13.5 GROUP VIA ELEMENTS

PROPERTIES

The most abundant and important element of Group VIA is oxygen, which has been discussed. The remaining elements of Group VIA are sulfur, selenium, tellurium, and polonium. Polonium, a metal, is a product of the radioactive decay of radium and is itself radioactive. Very little of its chemistry is known. The properties of the Group VIA nonmetals are given in Table 13.4. As a group the elements are nonmetallic with in-

Table 13.4
Properties of Group VIA elements

	O	*S*	*Se*	*Te*
molecular formula	$\mathbf{O_2}$	$\mathbf{S_8}$	$\mathbf{Se_8}$ **(Se)**$_n$	**(Te)**$_n$
melting point (°C)	−218	119	218	452
boiling point (°C)	−183	445	685	1390
atomic radius (Å)	0.74	1.04	1.17	1.37
ionic radius (Å)	1.40	1.84	1.98	2.21
first ionization potential (eV)	13.6	10.4	9.8	9.0
electronegativity	3.5	2.5	2.4	2.1
electrode potential (volts) (to H_2X)	1.23	0.14	−0.40	−0.72

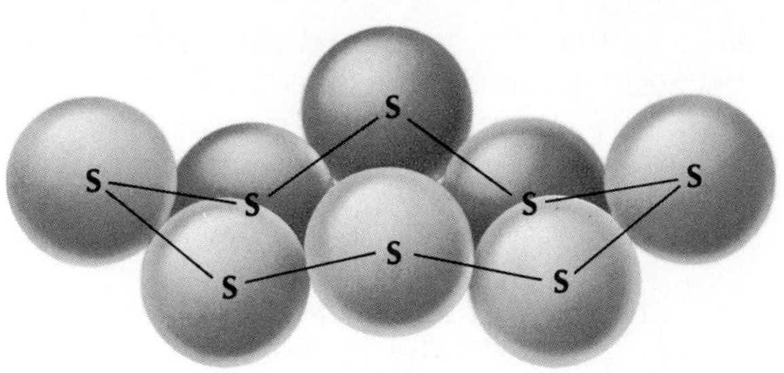

Figure 13.3
The sulfur molecule.

creasing metallic character being exibited by the elements of higher atomic mass.

The most dramatic difference between oxygen and the other elements of Group VIA is their molecular form. The significant increase in both the melting and boiling points between oxygen and sulfur is indicative of more than an increase in van der Waals attractive forces. This change in physical properties is much larger than shown by the halogens, whose molecular formulas are identical. Sulfur exists as S_8 molecules in the shape of a crown ring (Figure 13.3). Each of the atoms is bonded to its neighboring atoms by a covalent single bond. It is usual practice to use S for sulfur in equations for the sake of simplicity. Both selenium and tellurium exist as long chains of connected atoms. Each atom is covalently bonded to two neighboring atoms.

USES OF GROUP VIA ELEMENTS

Sulfur is used in its elemental form in rubber, insecticides, and fertilizers. However, its largest commercial use is as sulfuric acid (H_2SO_4), which is one of the most widely used materials in the production of an almost unlimited variety of products. By contrast, selenium and tellurium are little used. Selenium is a poor conductor of electricity, but it does conduct in proportion to the intensity of light on its surface. This property is utilized in exposure meters used in photography.

OXIDATION STATES

The Group VIA elements form compounds containing the -2 oxidation state expected for these nonmetals. The elements can exist as the ionic -2 species or as the covalent H_2X compounds. The stability of these compounds toward oxidizing agents is given by the electrode potentials listed in Table 13.4. Hydrogen sulfide, hydrogen selenide, and hydrogen telluride are all colorless, foul-odored, highly toxic gases and are in dramatic contrast to water, which is of the same general molecular formula, H_2X. If selenium or tellurium is ingested in nonfatal quantities into the human body, the organism very slowly gives it off as H_2X in the perspiration and breath. This feature alone has created a barrier to the investigation of the chemistry of these elements. No commercial mouthwash or deodorant can effectively counter the smell of the gas.

Positive oxidation states +4 and +6 are common when sulfur, selenium, or tellurium combine with a more electronegative element. Only the compounds of sulfur will be considered because those of selenium and tellurium are quite similar although in general less important.

Sulfur dioxide, which can be obtained by burning sulfur, is a colorless gas that has a sharp, irritating odor and is poisonous. The SO_2 molecule contains sulfur in the +4 oxidation state and is represented as a resonance hybrid of two contributing resonance forms:

$$\text{O}-\text{S}=\text{O} \longleftrightarrow \text{O}=\text{S}-\text{O}$$

When sulfur dioxide dissolves in water a solution of a weak acid called *sulfurous acid* (H_2SO_3) is produced. Two series of salts are derived from sulfurous acid and are known as *sulfites* (such as Na_2SO_3, sodium sulfite) and *bisulfites* (such as $NaHSO_3$, sodium bisulfite):

$$SO_2 + H_2O \rightleftharpoons H_2SO_3$$

Sulfur dioxide is called the *anhydride* of sulfurous acid because it produces the acid when it reacts with water.

Sulfur trioxide, which contains sulfur in the +6 oxidation state, can be produced by oxidizing sulfur dioxide at temperatures above 400°C in the presence of platinum as a catalyst:

$$2SO_2 + O_2 \xrightarrow{Pt} 2SO_3$$

The compound is very reactive and is a strong oxidizing agent. The reaction of sulfur trioxide with water is an exothermic process in which sulfuric acid is produced. Therefore, sulfur trioxide is the anhydride of sulfuric acid:

$$H_2O + SO_3 \longrightarrow H_2SO_4$$

Sulfuric acid dissociates in two steps, the first of which is virtually complete. The second step is not complete. The acid forms *sulfate* salts (such as Na_2SO_4, sodium sulfate) and *bisulfate* salts (such as $NaHSO_4$, sodium bisulfate):

First dissociation: $H_2SO_4 + H_2O \longrightarrow H_3O^+ + HSO_4^-$

Second dissociation: $HSO_4^- + H_2O \longrightarrow H_3O^+ + SO_4^{2-}$

Table 13.5
Properties of Group VA elements

	N	*P*	*As*	*Sb*	*Bi*
molecular formula	$\mathbf{N_2}$	$\mathbf{P_4}$	$\mathbf{As_4\ (As)_n}$	$\mathbf{Sb_4\ (Sb_n)}$	$\mathbf{Bi_n}$
melting point (°C)	−210	44		630 (Sb_n)	271
boiling point (°C)	−196	280		1350	1500
atomic radius (Å)	0.74	1.10	1.21	1.41	1.52
ionic radius (Å) (X^{3-})	1.4	1.85			
first ionization potential (eV)	14.5	11.0	10.0	8.6	8.0
electronegativity	3.0	2.1	2.0	1.9	1.9

13.6 GROUP VA NONMETALS

PROPERTIES

The elements nitrogen, phosphorus, arsenic, antimony, and bismuth as a group are less nonmetallic than the elements of Group VIA or VIIA. While nitrogen and phosphorus are nonmetals, arsenic and antimony are semimetals and bismuth is a metal, although its heat and electrical conductivity are low. The Group VA elements exhibit a wide range of properties in their transition from nonmetal to metal. The properties of these elements are listed in Table 13.5.

The first and second elements of Group VA differ greatly in properties. The change in physical properties reflects the differences in the molecular formulas for nitrogen and phosphorus. Nitrogen is a diatomic molecule containing a triple bond |N≡N|, whereas phosphorus exists as P_4, a tetrahedral molecule (Figure 13.4).

Phosphorus, arsenic, and antimony exist in allotropic forms. White phosphorus is a waxy solid that consists of crystals of P_4 molecules. Red phosphorus is prepared by heating white phosphorus to 250°C in the absence of air. The red form is a high molecular mass polymer in which the phosphorus atoms are linked in a network. Black phosphorus is a crystalline substance that resembles graphite and is a network of covalently bound phosphorus atoms. Arsenic and antimony both exist as tetrahedral

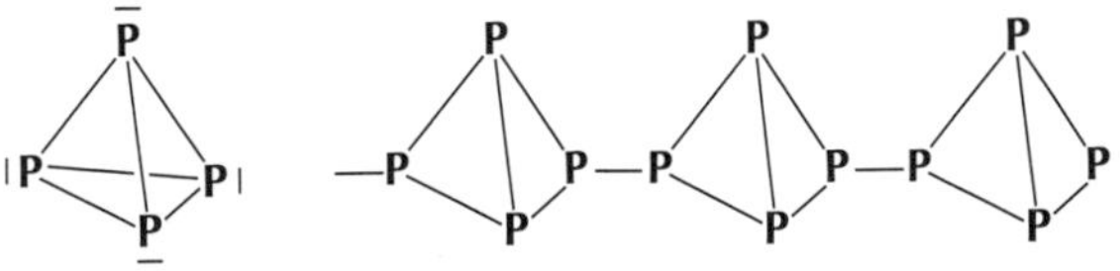

white phosphorus　　red phosphorus

Figure 13.4
Phosphorus.

molecules As_4 and Sb_4, which are soft, yellow solids, but they are unstable in this form and change readily into stable, gray, metal-like forms in which the atoms are joined in long chains.

OXIDATION STATES

The Group VA elements form compounds in which their oxidation states are any number between -3 and $+5$. The ions N^{3-} (nitride) and P^{3-} (phosphide) exist in ionic compounds. However, the most commonly occurring -3 oxidation states involve covalently bound material. All of the elements form XH_3 compounds called ammonia (NH_3), phosphine (PH_3), arsine (AsH_3), stibine (SbH_3), and bismuthine (BiH_3). As would be expected on the basis of the increasing metallic character of the larger elements, the stability of these compounds decreases in the series from NH_3 to BiH_3. Arsine and stibine decompose when warmed, and bismuthine is unstable at room temperature.

Hydrazine (N_2H_4) contains nitrogen in the -2 oxidation state. It is a liquid that is a strong reducing agent used in some rocket fuels. Hydroxyl amine (NH_2OH; Chapter 20) contains nitrogen in the -1 oxidation state:

$$\mathrm{H{-}\bar{N}(H){-}\bar{N}(H){-}H} \qquad \mathrm{H{-}\bar{N}(H){-}\bar{O}(H)|}$$

hydrazine   hydroxyl amine

The positive oxidation states of nitrogen are easily illustrated by the oxides of this element. Nitrous oxide (N_2O) is a colorless, relatively unreactive gas in which nitrogen is in the $+1$ oxidation state:

$$\mathrm{|\bar{N}{=}N{=}\bar{O}|}$$

Nitric oxide (NO) contains nitrogen in the $+2$ oxidation state and is unstable with respect to nitrogen and oxygen. However, the decomposition occurs at a very slow rate at room temperature. It is paramagnetic because it contains an odd number of electrons. Two contributing resonance forms for NO can be written:

$$\mathrm{\cdot\bar{N}{=}\bar{O}|} \longleftrightarrow \mathrm{|\bar{N}{=}\bar{O}\cdot}$$

Examples of the $+3$, $+4$, and $+5$ oxidation states are dinitrogen trioxide (N_2O_3), nitrogen dioxide (NO_2), and dinitrogen pentoxide (N_2O_5), respectively. Dinitrogen trioxide is the anhydride of nitrous acid. Dinitrogen pentoxide is the anhydride of nitric acid. The salts of nitrous and nitric acid are *nitrites* and *nitrates*, respectively:

$$N_2O_3 + H_2O \rightleftharpoons 2HNO_2$$
$$N_2O_5 + H_2O \rightleftharpoons 2HNO_3$$

Phosphorus acids H_3PO_4 (phosphoric acid), H_3PO_3 (phosphorous acid), and H_3PO_2 (hypophosphorous acid) are triprotic, diprotic, and monoprotic, respectively. Only the hydrogen atoms bound to oxygen dissociate, and those bound to phosphorus are not acidic:

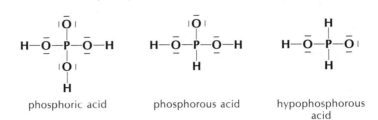

```
        |Ō|                    |Ō|                       H
         |                      |                        |
   H—Ō—P—Ō—H              H—Ō—P—Ō—H               H—Ō—P—Ō|
         |                      |                        |
        |Ō|                     H                        H
         |
         H
```

phosphoric acid — phosphorous acid — hypophosphorous acid

The salts of PO_4^{3-}, HPO_3^{2-}, and $H_2PO_2^{-}$ are *phosphates, phosphites,* and *hypophosphites,* respectively.

13.7 GROUP IVA NONMETALS

PROPERTIES

The only nonmetals of Group IVA are carbon and silicon. (The semimetal germanium and the metals tin and lead were discussed in Chapter 12.) The variation in the properties of the Group IVA elements reflects their middle position in the periodic table. The properties of the Group IVA elements are listed in Table 12.5 in the previous chapter.

Carbon and silicon exist in giant three-dimensional networks in which every atom of the crystal is bound by covalent bonds. The energy required to break these bonds is large, and the melting points and boiling points of these elements reflect the high stability of the solid structures.

Carbon exists in two allotropic forms, diamond and graphite. In diamond the carbon atoms are bonded through sp^3 hybrid orbitals and are arranged in a network of tetrahedrons. Diamond is an extremely hard, highly stable, nonconducting substance. It is a colorless, transparent material that appears to be colored if small traces of impurities are present in the crystal.

Graphite, a soft, black solid that feels slippery, is often used as a dry lubricant. The graphite crystal is composed of layers of atoms that are attracted to each other by van der Waals attractive forces. Each layer consists of a network of hexagonal rings in which each carbon is bound to three other carbon atoms (Figure 13.5). This bonding requires three electrons from each carbon atom, which leaves one electron in a p orbital perpendicular to the plane of the attached carbon atoms. Pairs of adjacent carbon atoms are attached by means of a bond that results from a side-by-side overlap of the p orbitals. Such a bond, called a π (pi) bond, is discussed further in Chapter 17. There are many possibilities for forma-

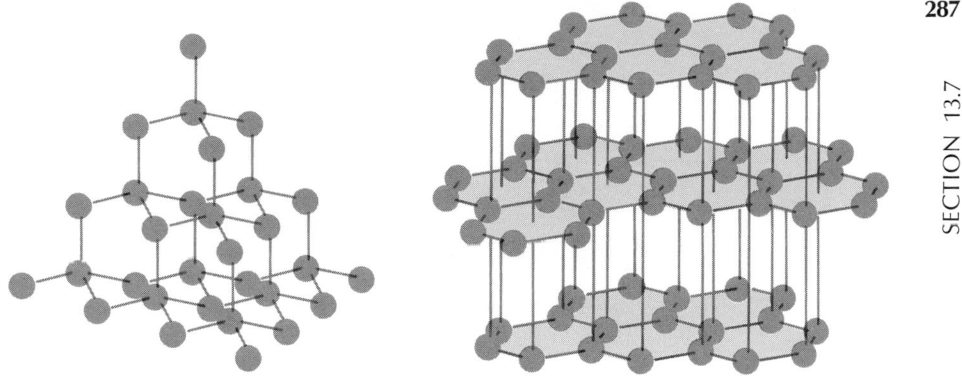

Figure 13.5
Diamond and graphite.

tion of π bonds that lead to a variety of differently positioned double bonds in conventional Lewis structures. The layer of carbon atoms as a whole must be described as a resonance hybrid structure:

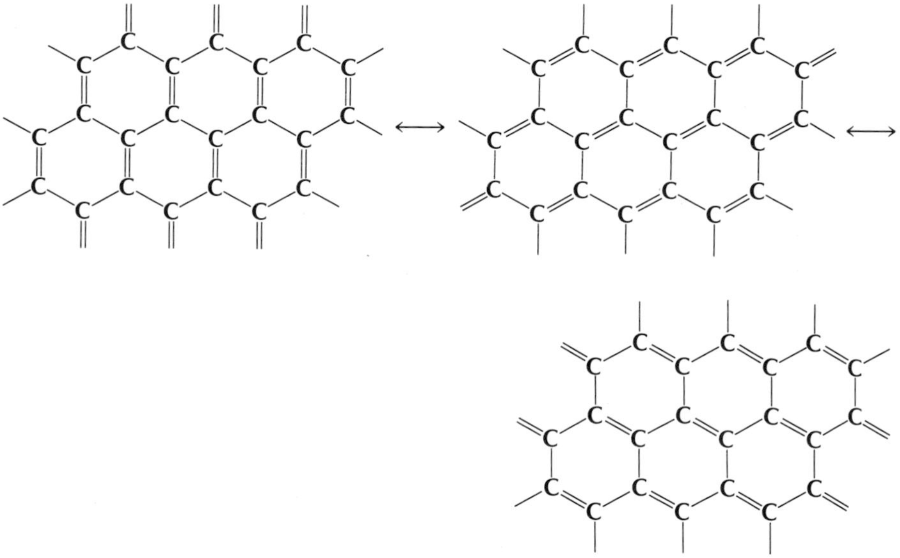

Silicon, below carbon in the periodic table, is not found free in nature. In compounds it constitutes 28 percent of the earth's crust and is the second most abundant element, preceeded only by oxygen. Silicon most often is found in combination with oxygen in nature. One of the more common silicon–oxygen substances is sand, or silicon dioxide (SiO_2), which is a stable, nonvolatile substance and a three-dimensional network crystal (Figure 13.6). The SiO_2 crystal resembles the diamond structure

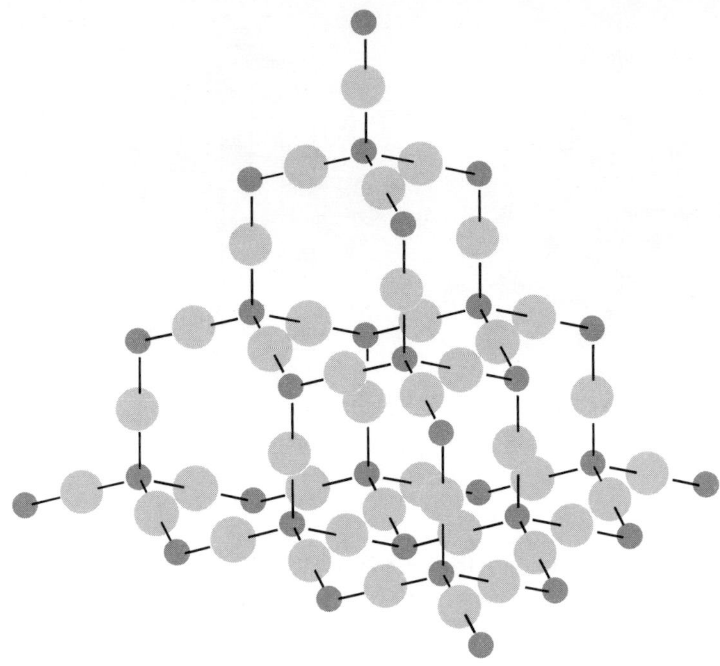

Figure 13.6
Silicon dioxide.

and can be derived by replacing the carbon atoms with silicon and oxygen atoms so that each oxygen atom is attached to two silicon atoms. Each silicon atom is attached to four oxygen atoms, which in turn are bonded to other silicon atoms.

The silicon oxygen bond energy is very high. Therefore, any structure containing Si—O bonds will have high thermal stability. Silicones, which contain long chains of a Si—O—Si arrangement, are used as lubricants at high temperatures. One typical silicone is illustrated below, in which the covalent bonds from silicon not containing oxygen are to a CH_3 (methyl) group:

```
     CH3    CH3    CH3    CH3    CH3
      |      |      |      |      |
O —  Si — O — Si — O — Si — O — Si — O — Si —
      |      |      |      |      |
     CH3    CH3    CH3    CH3    CH3
```

CARBON AND ORGANIC CHEMISTRY

Carbon differs from the other elements of Group IVA and all other elements in its ability to form stable compounds in which carbon atoms are bonded to each other in chains and rings. This property, which is called *catenation,* is found in other elements as well. However, the bond energies of other elements are lower and they do not form structures as stable as carbon. The carbon–carbon bond energy is 82 kcal/mole, whereas the silicon–silicon bond energy is 43 kcal/mole.

Carbon forms bonds to many other elements that are very stable. This feature allows it to incorporate many atoms into structures of an almost infinite variety. The carbon–hydrogen bond energy is 99 kcal/mole, and this combination of elements gives rise to a series of compounds called *hydrocarbons,* of which methane is the simplest example. The hydrocarbons are but one class of substances that are derived from carbon. The study of hydrocarbons and other classes of compounds in which carbon plays the foundation role is known as *organic chemistry.* Chapters 16 through 24 are devoted to this subject.

Suggested further readings

Addison, W. E., "The Physical Basis of Allotropy," *Educ. Chem.,* **1,** 144 (1964).

Bevington, J. C., "Heavy Hydrogen and Light Helium," *Educ. Chem.,* **3,** 196 (1966).

Chernick, C. L., "Fluorine Compounds of Xenon and Radon," *Science,* **138,** 136 (1962).

Frey, J. E., "Discovery of the Noble Gases and Foundations of the Theory of Atomic Structure," *J. Chem. Educ.,* **43,** 371 (1966).

Hall, H. T., "The Synthesis of Diamond," *J. Chem. Educ.,* **38,** 484 (1961).

Kaufmann, J. J., "Bonding in Xenon Hexafluoride," *J. Chem. Educ.,* **41,** 183 (1964).

Postgate, J. R., "The Sulfur Cycle," *Educ. Chem.,* **2,** 58 (1965).

Selig, J., Malm, J. G., and Claassen, H. H., "The Chemistry of the Noble Gases," *Sci. Amer.,* 66 (*May 1964*).

Terms and concepts

allotropy
anhydride
catenation
deuterium
diamond
graphite
halogens
hydride ion
organic chemistry
ozone
polyethylene
polymer
Teflon
tritium

1. What reaction would occur if NaH were placed in contact with water?
2. Are oxidation states other than +1, 0, and −1 possible for hydrogen? Why or why not?
3. How many different molecular species are present in a sample of hydrogen gas?
4. What kind of physical evidence is there for hydrogen bonding?
5. Suggest a geometric shape for the molecule XeO_3.
6. Why is it unlikely that $XeBr_2$ will be a stable molecule?
7. Teflon is very stable toward oxidizing agents. Why?
8. Which of the following reactions will proceed as written in aqueous solution?

  a. $Br_2 + 2Cl^- \longrightarrow 2Br^- + Cl_2$
  b. $Cl_2 + 2I^- \longrightarrow 2Cl^- + I_2$
  c. $I_2 + 2At^- \longrightarrow 2I^- + At_2$

9. Which member of each of the following pairs of acids is the more acidic? Why?

  a. $HClO_3$ and $HClO_4$
  b. $HClO_3$ and $HBrO_3$
  c. HIO_4 and $HClO$

10. Why does liquid oxygen have a higher boiling point than liquid nitrogen?
11. Why is the boiling point of H_2O_2 higher than that of H_2S_2?
12. What are the oxidation states of oxygen in OCl_2 and OF_2?
13. Why is the boiling point of H_2S lower than that of H_2Te?
14. Draw a Lewis structure of SO_3.
15. Which is the stronger oxidizing agent, Na_2SO_4 or Na_2SeO_4?
16. Which is the stronger acid, H_2SeO_3 or H_2SeO_4?
17. Will the following reaction proceed as written? Why or why not?

$$PH_3 + Bi \longrightarrow P + BiH_3$$

18. The atomic arrangement in N_2O_5 is as given below. Draw the electronic structure of the molecule.

19. Which is the stronger acid, HNO_2 or HNO_3?
20. What is the geometry of SiO_4^{2-}?

14
TRANSITION ELEMENTS

The groups of elements between Groups IIA and IIIA in the periodic table are known as the *transition elements*. In addition, the lanthanides and actinides, which are placed outside the general body of the periodic table, are also classified as transition elements. Separate classification of the transition elements is a reflection of their unique chemical properties. This chapter considers some of the structural and chemical characteristics of the transition elements and their compounds.

14.1 TRANSITION ELEMENTS

ELECTRONIC CONFIGURATION

The first row of transition elements, which occurs in the fourth period, consists of the elements scandium (at. no. 21) through zinc (at. no. 30).

Table 14.1
Electronic configurations

element	*symbol*	*at. no.*	*1s*	*2s*	*2p*	*3s*	*3p*	*3d*	*4s*	*4p*	*4d*	*4f*	*5s*
first row transition elements													
scandium	**Sc**	21	2	2	6	2	6	1	2				
titanium	**Ti**	22	2	2	6	2	6	2	2				
vanadium	**V**	23	2	2	6	2	6	3	2				
chromium	**Cr**	24	2	2	6	2	6	5	1				
manganese	**Mn**	25	2	2	6	2	6	5	2				
iron	**Fe**	26	2	2	6	2	6	6	2				
cobalt	**Co**	27	2	2	6	2	6	7	2				
nickel	**Ni**	28	2	2	6	2	6	8	2				
copper	**Cu**	29	2	2	6	2	6	10	1				
zinc	**Zn**	30	2	2	6	2	6	10	2				
second row transition elements													
yttrium	**Y**	39	2	2	6	2	6	10	2	6	1		2
zirconium	**Zr**	40	2	2	6	2	6	10	2	6	2		2
niobium	**Nb**	41	2	2	6	2	6	10	2	6	4		1
molybdenum	**Mo**	42	2	2	6	2	6	10	2	6	5		1
technetium	**Tc**	43	2	2	6	2	6	10	2	6	6		1
ruthenium	**Ru**	44	2	2	6	2	6	10	2	6	7		1
rhodium	**Rh**	45	2	2	6	2	6	10	2	6	8		1
palladium	**Pd**	46	2	2	6	2	6	10	2	6	10		
silver	**Ag**	47	2	2	6	2	6	10	2	6	10		1
cadmium	**Cd**	48	2	2	6	2	6	10	2	6	10		2

In these elements the third principal energy level is gradually filled by the progressive addition of 10 electrons to the third subshell (Section 5.4). Exceptions to the progressive addition of electrons occur in the case of chromium (at. no. 24) and copper (at. no. 29), which are thought to indicate the extra stability of half-filled and filled subshells (Table 14.1). In chromium the electronic configuration is $4s^13d^5$, which would not be expected from the addition of one electron to the vanadium configuration. Energetically the 4*s* and 3*d* subshells are very close, and increased stability must be gained in achieving a half-filled 3*d* subshell. In copper a $4s^13d^{10}$ configuration is preferred over $4s^23d^9$, which indicates that copper finds it energetically advantageous to complete its third energy level. While these slight deviations in the orderly buildup of the electronic configuration of the transition metals are intriguing and lend insight into the nature of the atom, it should be pointed out that the observed electronic configurations have been determined for the elements in the gas phase. In other states or in the presence of other substances these anomalous configurations can be altered.

In the second and third row transition elements, progressive addition to the 4*d* and 5*d* subshells occurs in a fashion similar to the filling of the 3*d* subshell of the first row transition elements. However, the elements of atomic numbers 41 through 45, 78, and 79 all have s^1 configurations instead of the expected s^2 configurations. The lanthanides and actinides involve 4*f* and 5*f* subshell buildup. There are several elements whose electronic configurations are not predicted from simple successive addition of electrons to the *f* subshells. However, these exceptions are not of concern in an introductory course in chemistry.

OXIDATION STATES

The most striking characteristic of transition elements is the multitude of oxidation states that they exhibit in their compounds. Table 14.2 lists the common oxidation states of the transition elements. Some of the values are in parenthesis and are for compounds that are not stable in aqueous solutions and are difficult to obtain. However, such compounds have been prepared and isolated from nonaqueous solvents. There has been great interest in the chemistry of transition elements in recent years, and the list of observed oxidation states undoubtedly will grow in the future.

Table 14.2
Oxidation states of transition metals

Sc	Ti	V	Cr	Mn	Fe	Co	Ni	Cu	Zn
		(1)[a]	(1)	(1)	(1)	(1)	(1)	1	
	2	2	2	2	2	2	2	2	2
3	3	3	3	(3)	3	3	(3)		
	4	4	(4)	4	(4)	(4)			
		5	(5)		(5)				
			6	6	(6)				
				7					
Y	**Zn**	**Nb**	**Mo**	**Tc**	**Ru**	**Rh**	**Pd**	**Ag**	**Cu**
						(1)		1	
	(2)	(2)	(2)	(2)	2	(2)	2	(2)	2
3	(3)	(3)	3	(3)	3	3	(3)	(3)	
	4	(4)	4	4	4	4	4		
		5	5	(5)	(5)	(5)			
			6	(6)	(6)	(6)			
				7	(7)				
					(8)				
La	**Hf**	**Ta**	**W**	**Re**	**Os**	**Ir**	**Pt**	**Au**	**Hg**
						(1)		1	1
		(2)	(2)		(2)	(2)	2		
3	3	(3)	(3)	(3)	(3)	3	(3)	3	
	4	(4)	4	4	4	4	4		
		5	5	(5)	(5)	(5)	(5)		
			6	(6)	6	(6)	(6)		
				7					
					8				

[a] Numbers in parentheses represent unstable oxidation states in water.

The first noteworthy feature of Table 14.2 is the common oxidation state of +2 in the first row transition elements, except for scandium, and the instability of the +1 oxidation state, with the exception of copper. The +2 oxidation state results from the loss of the $4s^2$ electrons—a process that does not seem reasonable on the basis of the energy levels described in Chapter 5. Since the $3d$ electrons are added after the $4s$, in the orderly model for electron arrangement in atoms, they would be expected to ionize more easily than the $4s$ electrons. However, the experimental fact is that in every transition element the s electrons are removed before the d electrons are ionized. It is incorrect to compare the buildup of the electrons in atoms with the ionization process. In the mental process of constructing atoms both the number of electrons and protons are increased simultaneously. In the ionization process one or more electrons are removed while the nuclear charge remains unchanged. Therefore, the ionization process is not the reverse of the buildup process.

The second feature to be noted in the list of oxidation states is the general increase in the maximum oxidation state for stable species toward the middle of each row of transition metals. From the middle to the end of each row the maximum oxidation state for stable species decreases until the +2 state is the highest. The +2 oxidation states result from removal of the s electrons and are found in ionic compounds. The higher oxidation states result from subsequent removal or sharing of additional electrons in covalent bonds. In the case of the +7 oxidation state of manganese in MnO_4^-, the two $4s$ and the five $3d$ electrons are utilized in covalent bonds to oxygen. The falloff in the oxidation states beyond this point suggests that the $3d$ subshell becomes more and more stable with the increasing number of electrons after the half-filled configuration is reached. The electrons therefore are less available for bonding. In the case of the iron ion Fe^{3+}, two $4s$ and one $3d$ electrons are ionized, leaving a $3d^5$ electronic arrangement in compounds stable in water.

The third observation to be made about the oxidation states of transition elements is the increase in the maximum oxidation state within a column of the periodic table. While the normal oxidation states for iron are the +2 and +3, ruthenium (Ru) and osmium (Os) in the same group occur in +4, +6, and +8 oxidation states. For example, one of the common stable oxides of iron is Fe_2O_3, whereas in the case of osmium, OsO_4 is a stable oxide.

COLOR OF TRANSITION METAL IONS

The chemistry of transition metal ions is visually more spectacular than the chemistry of any other group of elements since their compounds are often brightly colored in both the solid state and in solution. In order to understand the origin of this color, the concept of complementary colors should be reviewed. A substance is colored if the interaction of white light leads to absorption of a portion of the visible spectrum. Then the

Table 14.3
Color of transition metal ions in water

ion	configuration	color
Sc^{3+}	d^0	colorless
V^{3+}	d^2	green
Cr^{2+}	d^4	blue
Fe^{2+}	d^6	green
Ni^{2+}	d^8	green
Zn^{2+}	d^{10}	colorless

light transmitted by the substance is no longer white but instead is stronger in the color complementary to the absorbed color. If a portion of the red end of the spectrum is absorbed, the substance looks blue. Conversely, absorption of the blue part of the spectrum by the substance will make it appear to be red. The eye sees the complement of the color absorbed.

The fact that most transition metal ions are colored indicates the existence of excited electronic states that are above the ground state by an amount corresponding to the energy of visible light. Transitions to these excited states, due to changes in the *d* electrons and *d* orbital configurations, are accompanied by the absorption of light of a particular wavelength. Since several electronic configurations are possible in the excited state, ions of a single transition metal may occur in several different colors. Oxidation or reduction of transition metal ions is commonly accompanied by a vivid color change. In certain cases the ions are colorless, as is the case for Sc^{3+} (Table 14.3), which does not have any 4*s* or 3*d* electrons. The promotion of the 3*p* electrons of a complete subshell requires more energy than is available in the visible spectrum. The Zn^{2+} ion also is colorless in spite of the fact that it has 3*d* electrons. Its electronic configuration is $3d^{10}$, and the entire third energy level is completed. The promotion of an electron from this stable species also requires more energy than is available from visible light.

If elements are covalently bound to the transition metal ions, the energy required for electronic transitions also changes. This is the result of the effect of the attached elements on the *d* orbital energies. This phenomenon is discussed in Section 14.3.

It is difficult to predict the color of many transition metal ions as some promote more than one electron to higher energy states. As a result of multiple promotions there are more excited states possible. Absorption of several portions of the visible spectrum occurs, and the resultant complementary color is not easily predictable.

PARAMAGNETISM

In transition metal ions paramagnetism commonly is encountered because the partially filled *d* subshell contains unpaired electrons. The un-

Table 14.4
Magnetic moments of hydrated transition metal ions

ion	*configuration*	*magnetic moment* calculated	*magnetic moment* observed
Sc^{3+}	d^0	0	0
Ti^{3+}	d^1	1.73	1.75
Ti^{2+}	d^2	2.84	2.76
V^{2+}	d^3	3.87	3.86
Cr^{2+}	d^4	4.90	4.80
Mn^{2+}	d^5	5.92	5.96

paired electrons, as a result of their spin, possess a magnetic moment and are attracted in a magnetic field (Section 5.4). The magnitude of the magnetic strength of the transition metal ion is related to the number of unpaired electrons and is reflected by its behavior in a magnetic field. The magnetic moment in Bohr magnetons can be calculated from the formula

$$\text{spin magnetic moment} = \sqrt{n(n+2)}$$

where n is the number of unpaired electrons. The formula expressing the relation between the spin magnetic moment and the number of unpaired electrons is valuable in determining the electronic configurations in transition metal ions and compounds. In Table 14.4 the calculated spin magnetic moments of ions of the first row transition metals are compared with the experimental values for the hydrated ions. In agreement with the number of unpaired electrons predicted by Hund's rule, the magnetic moment increases until a d^5 configuration is reached. After this point the presence of additional electrons results in pairing with an accompanying decrease in the magnetic moment.

As illustrated in the next section, the knowledge of the number of unpaired electrons is useful in determining the structures of complex transition metal ions. The presence of elements bonded to the ion alters the energy of the *d* orbitals. The result of this alteration is a splitting of the five *d* orbitals into two or more classes of energetically nonequivalent orbitals. Application of Hund's rules leads to electronic configurations that are different from the arrangement predicted by considering all five *d* orbitals to be of equal energy. The measurement of the magnetic moments allows assignment of the electronic arrangements in these altered *d* orbitals.

14.2 COMPLEX IONS

A transition metal complex ion consists of a central metal cation to which anions or molecules are attached in a defined geometry with respect to the center. The anions or molecules about the metal ions are called *ligands* and have one or more electron pairs that can be used in bonding to the metal ion. The number of groups of atoms bonded to the metal ion defines the *coordination number* of the ion. Complexes with coordination numbers from two to nine are known, but the majority are four- and sixfold coordinate as in $PdCl_4^{2-}$ and CrF_6^{3-}.

The sum of the oxidation numbers of the constituent parts of a complex ion equals its total charge. Thus, in Pd $(NH_3)_4^{2+}$ the palladium ion is in the +2 oxidation state because the ammonia molecules are electrically neutral and the individual oxidation numbers of nitrogen and hydrogen must total zero. Palladium is also in the +2 oxidation state in $PdCl_4^{2-}$, but since each chlorine is present as Cl^- the total charge of the complex is $+2 + 4(-1) = -2$.

Six-coordinate complexes are octahedral, that is, the ligands are located at the corners of a regular octahedron. The most convenient way of depicting the ligand position is by the representation in Figure 14.1, in which *L* denotes a ligand. There is no difference between the solid direction lines of the vertical axis and the other lines that appear in a plane perpendicular to the vertical axis. All of the lines to the "corners" of the octahedron are identical. The colored lines indicate the geometry of an octahedron.

Four-coordinate complexes may be either square planar or tetrahedral. The arrangement of the ligands about the central metal ion for each of these complexes is given in Figure 14.2.

Ligands are often strongly bonded with the transition metal. For example, the $PdCl_4^{2-}$ complex ion behaves as a chemical unit, as do other complex ions involving nonmetals such as SO_4^{2-}. Another example is the ferrocyanide ion, $Fe(CN)_6^{4-}$, in which the cyanide units do not behave as free cyanide ions (CN^-) but rather remain associated as the complex ion when dissolved in water. The questions of interest are (1) what deter-

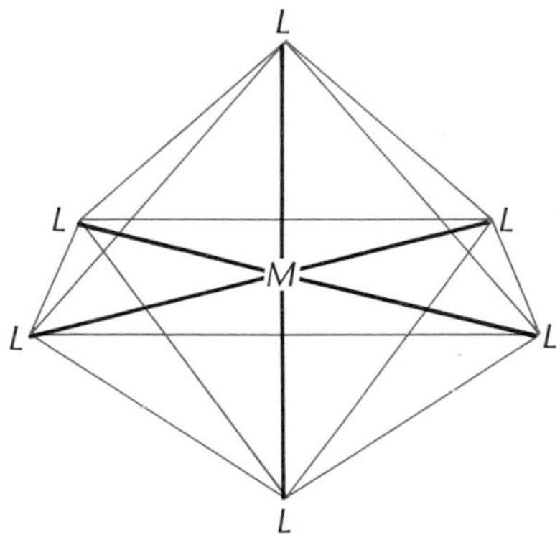

Figure 14.1
Representation of an octahedral complex.

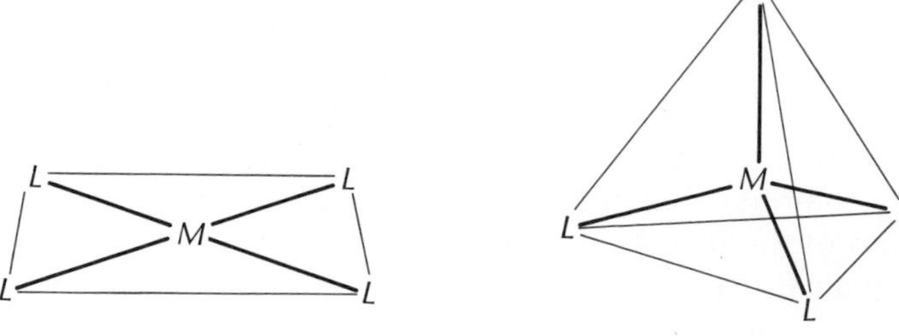

Figure 14.2
Square planar and tetrahedral complex ions.

mines the number of ligands which are associated with the transition metal, (2) why are the ligands arranged in a variety of geometric shapes, and (3) how are the ligands bonded to the metal?

14.3 GEOMETRY OF *d* ORBITALS

To explain the directional character of bonding in transition metals, geometric representations of the *d* orbitals must be examined. There are five *d* orbitals which, in the absence of the influence of other elements, are of equal energy. The placement of electrons in these orbitals is governed by Hund's rule. A single electron in the 3*d* subshell may be located in any of the five energetically equal *d* orbitals. The orientation of the five *d* orbitals with respect to a set of coordinate axes is illustrated in Figure 14.3. Three of the orbitals are identical in shape but are located in mutually perpendicular planes. The regions of high electron density in these orbitals are described by four lobes along 45° lines between the axes. The three orbitals are named d_{xy}, d_{xz}, and d_{yz}; the subscripts indicate the plane that contains the reference axes. Of the two remaining orbitals the $d_{x^2-y^2}$ orbital resembles the d_{xy}, d_{xz}, and d_{yz} orbitals except that the regions of high electron density are located along the *x* and *y* axes. The d_{z^2} orbital, which is quite different in shape than the other four *d* orbitals, has its electron density principally along the *z* axis.

From the geometry of the *d* orbitals it can be seen that the energy of electrons in these orbitals should be affected in different manners by the approach of a ligand containing electrons. A repulsion between the electron pair of the ligand and the negative charge of the electrons in the *d* orbitals may arise. A ligand approaching along the *z* axis is repelled by electrons in the d_{z^2} orbital, whereas the electrons in the other orbitals are not as adversely affected. Similarly, approach of ligands along the *x* and *y* axes is energetically unfavorable if an electron is contained in the $d_{x^2-y^2}$ orbital. It should be expected that the set of energetically equivalent *d* orbitals in the absence of ligands could separate into subsets of different

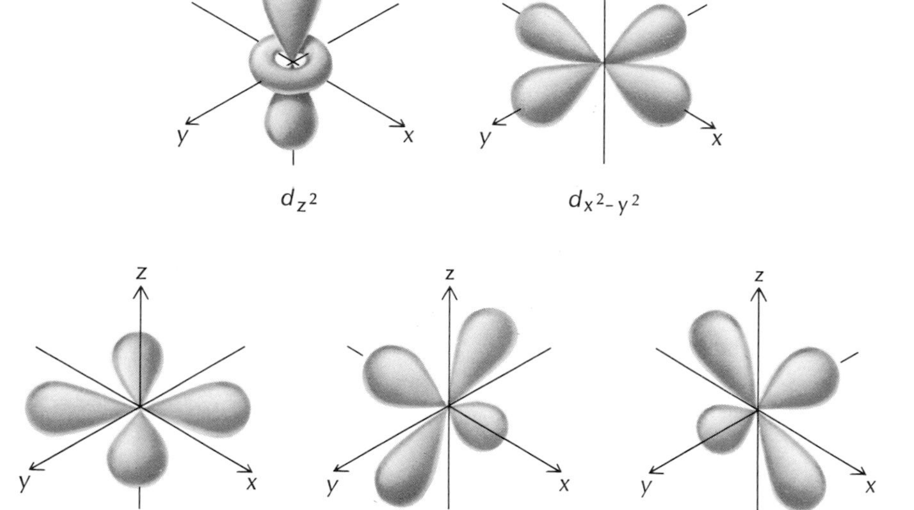

Figure 14.3
Geometry of d orbitals.

energies whenever approached by a ligand. The electrons in the *d* orbitals should distribute themselves in a different manner in the presence of ligands than in the absence of ligands. They should be distributed according to the relative energies of the *d* orbitals; and in those cases where the orbitals are of equal energy, Hund's rule should apply.

14.4
BONDING IN COMPLEX IONS

LIGAND FIELD THEORY

A *ligand field theory* has been proposed to describe the bonding of the ligands to the central metal atom by means of hybridized orbitals. The metal ion has a number of unoccupied orbitals or orbitals containing one electron that can be shifted into other orbitals, thus providing a means of bond formation with a ligand that has two available electrons. A vacant orbital of a metal ion overlaps with the filled orbital of the ligand to form a covalent bond with some degree of polarity. This model also accounts for observed magnetic moments of complex ions by detailing the manner in which the *d* orbitals become energetically nonequivalent in the presence of ligands. The theory is developed in considering the octahedral, square planar, and tetrahedral complexes.

The most common complex ion of transition metals is octahedral. Six ligands, such as H_2O, NH_3, or halide ions, are located symmetrically about the transition metal to form an octahedron with the ligands at the corner positions. Two examples of this type of complex ion involving chromium in the +3 oxidation state are $Cr(H_2O)_6{}^{3+}$ and $CrF_6{}^{3-}$ (Figure 14.4). In the absence of ligands the Cr^{3+} ion has three electrons in the 3*d* subshell. Since the five *d* orbitals are of equal energy, the electrons may occupy any of the orbitals providing they conform to Hund's rule (Section 5.4) and remain unpaired.

In the presence of six ligands approaching along the *x*, *y*, and *z* axes, the orbital energies are altered. Orbitals whose electron densities are directed along the axes are destabilized by the presence of an electron in those regions of space defined by the orbitals. Therefore, the d_{z^2} and $d_{x^2-y^2}$ orbitals are of higher energy than the d_{xy}, d_{xz}, and d_{yz} orbitals, whose regions of electron density are not directed along an axis. The three electrons in a Cr^{3+} complex ion are located in the orbitals of lowest energy and, taking into account Hund's rule, each of the d_{xy}, d_{xz}, and d_{yz} contain one electron (Figure 14.4).

Now that the most energetically favorable location of the electrons has been considered, the problem of how the ligands are attached to the transition metal will be examined. Note that the octahedral arrangement is a favorable one from simple spatial considerations. The approach of six ligands about the transition metal can be accomplished with a minimum of repulsion between the neighboring ligands if they are octahedrally

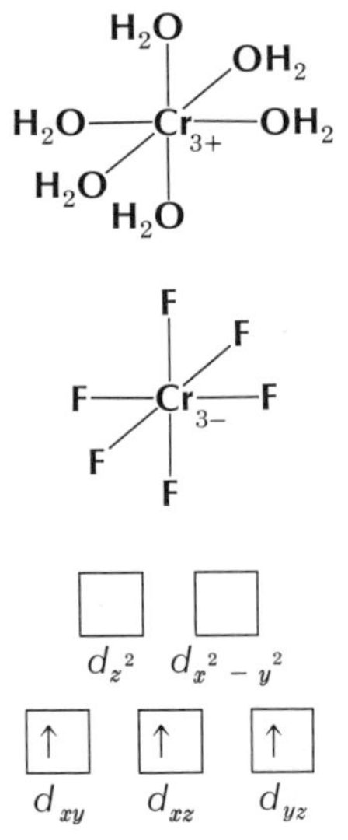

Figure 14.4
Geometry and electronic configuration of Cr^{3+} complex ions.

situated. Any other geometric arrangement would involve greater repulsion between neighboring ligands. This argument is similar to those described previously in accounting for the geometry of *sp*, sp^2, and sp^3 hybridized molecules (Section 6.6).

Ligands such as H_2O or F^- have filled *p* orbitals available for bond formation; the metal must provide empty orbitals. Therefore, six unoccupied orbitals are required to form the six bonds to the ligands. There are only two unfilled *d* orbitals, $d_{x^2-y^2}$ and d_{z^2} in ions such as Cr^{3+}. Additional orbitals are required for the proper number of bonds. The empty 4*s* and three 4*p* orbitals can be employed to make up the deficit. The six ligands bound to the transition metal ion are identical chemically and, therefore, the six orbitals used by the metal in bonding must be identical. Six hybrid orbitals mentally created from the two 3*d*, one 4*s*, and three 4*p* orbitals are postulated in order to account for the observed experimental facts. The hybrid orbitals d^2sp^3 are considered to be directed toward the corners of an octahedron in suitable positions for bonding to six ligands.

The bonding in ferrocyanide and ferricyanide ions is predicted correctly by the ligand field theory. The ferrocyanide ion $Fe(CN)_6^{4-}$ contains iron in its +2 oxidation state because the cyanide ion has a charge of −1. From Hund's rule Fe^{2+} in the absence of ligands should contain four unpaired electrons in its $3d^6$ configuration—a fact that can be experimentally verified by measuring the magnetic moment of the ion. However, in the ferrocyanide ion there are no unpaired electrons. This is consistent with the expected electronic configuration in the presence of six ligands arranged in an octahedral structure. Three *d* orbitals, d_{xy}, d_{xz}, and d_{yz}, are available for the six electrons of Fe^{2+}. Placing the six electrons in these three orbitals pairs all the electrons (Figure 14.5). The ferricyanide ion $Fe(CN)_6^{3-}$ is also an octahedral complex. It contains one less electron in the 3*d* subshell than ferrocyanide, and the oxidation state of iron is now +3. Distribution of five electrons in the d_{xy}, d_{xz}, and d_{yz} orbitals leads to an electronic configuration that has one unpaired electron (Figure 14.5). This configuration can be experimentally determined and is easily distinguishable from that of Fe^{3+}, which contains five unpaired electrons.

SQUARE PLANAR COMPLEXES

In the $PdCl_4^{2-}$ complex ion, which is square planar, the four chlorine atoms are located symmetrically in a square about the central palladium ion, which is in the +2 oxidation state. The complex ion is diamagnetic, whereas palladium in the +2 oxidation state in the absence of ligands has two unpaired electrons in the $4d^8$ electronic configuration and is paramagnetic. Therefore, the presence of the four chlorine ligands in $PdCl_4^{2-}$ must lead to an alteration in the electron configuration about the palladium ion.

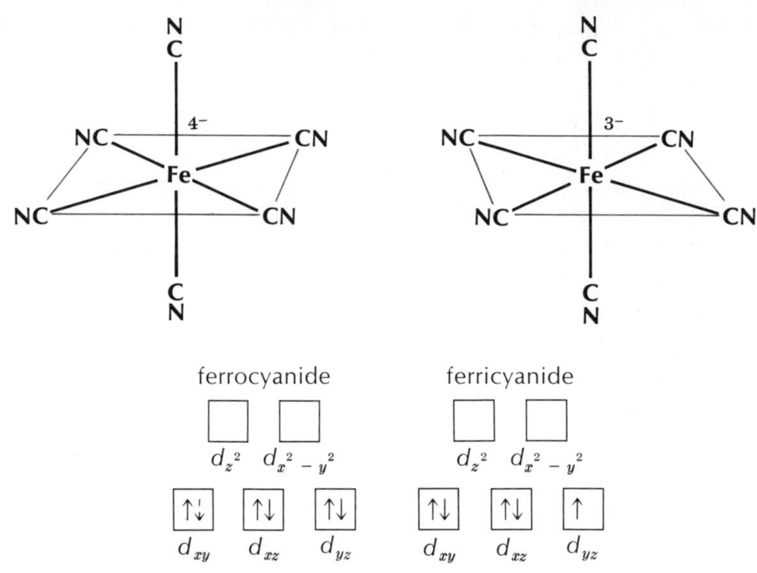

Figure 14.5
Geometry and electronic configuration of ferrocyanide and ferricyanide ions.

When four ligands are near a transition metal ion in the *xy* plane, the $d_{x^2-y^2}$ orbital is of higher energy than the d_{xy}, d_{xz}, d_{yz}, and d_{z^2} orbitals. (The d_{xy}, d_{xz}, d_{yz}, and d_{z^2} orbitals are not of equal energy because they are nonequivalently situated with respect to the ligands. However, since they all contain electrons, the small differences in energy will not be discussed.) Placement of eight electrons in such an altered *d* orbital arrangement leads to four sets of paired electrons (Figure 14.6). Thus the diamagnetism of $PdCl_4^{2-}$ can be rationalized.

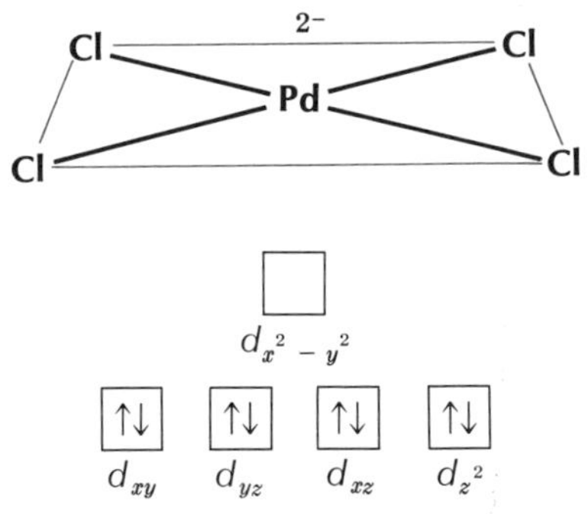

Figure 14.6
Geometry and electronic configuration of $PdCl_4^{2-}$. The four d orbitals shown as being of equal energy actually have slightly different energies.

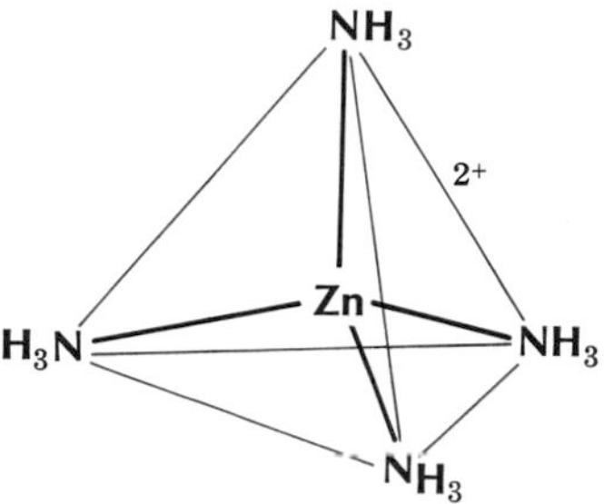

Figure 14.7
Tetrahedral $Zn(NH_3)_4^{2+}$ complex ion.

Four orbitals are required to bond to the four ligands, each of which has filled *p* orbitals available for bonding. Only the $d_{x^2-y^2}$ orbital is unoccupied in the Pd^{2+} ion in the presence of four ligands in the *xy* plane and, therefore, three additional orbitals are required for bonding with the ligands. The available orbitals of lowest energy are the empty 4*s* orbital and two of the three empty 4*p* orbitals. All four of the chlorines in $PdCl_4^{2-}$ ion are identically bonded, so the orbitals used by the transition metal must be identical. These orbitals are postulated to be a hybrid dsp^2. They are directed toward the corners of a square to provide the proper arrangement for the formation of bonds to the ligands.

TETRAHEDRAL COMPLEXES

The last type of transition metal complex to be discussed is the tetrahedral arrangement in which the four ligands are located at the corners of a regular tetrahedron. An example of such a complex is $Zn(NH_3)_4^{2+}$ (Figure 14.7). In Zn^{2+} all of the *d* orbitals are filled, and only the 4*s* and three 4*p* orbitals of zinc are empty and available for bond formation. The four identical orbitals required are the sp^3 hybrid type, which were discussed in Chapter 6. Other elements such as cadmium and mercury in the +2 oxidation state have filled *d* orbitals and form tetrahedral complexes.

14.5
ISOMERISM IN COMPLEXES

Each of the three types of complex ions discussed in the previous section involved sets of identical ligands bonded to a central ion. It is not necessary for all ligands surrounding a transition metal to be identical. If one of the H_2O molecules in $Cr(H_2O)_6^{3+}$ is replaced by F^-, the $Cr(H_2O)_5F^{2+}$ ion results. Since the octahedral complex contains six identical positions in $Cr(H_2O)_6^{3+}$, only one possible $Cr(H_2O)_5F^{2+}$ ion can be produced (Figure 14.8). However, when a second F^- replaces a water molecule, two possible compounds can be formed. Four of the water positions in

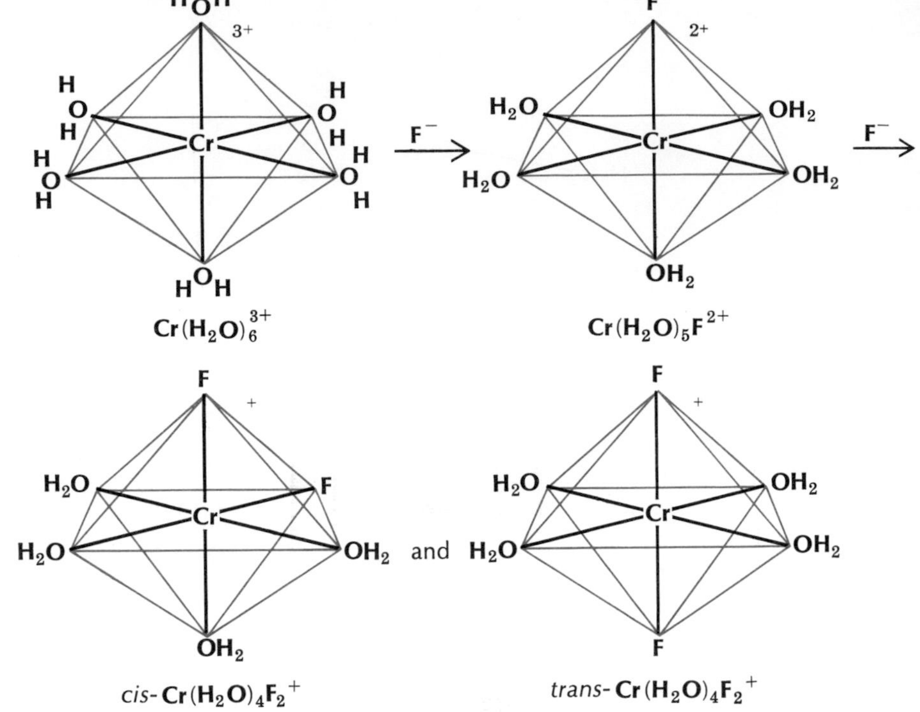

Figure 14.8
Octahedral geometrical isomerism.

$Cr(H_2O)_5F^{2+}$ are located symmetrically with respect to the first fluorine. Replacement of any of these H_2O molecules by F^- leads to the same $Cr(H_2O)_4F_2^+$ compound. The water molecule at the position directly opposite the fluorine is different from the other four H_2O molecules, and replacement of this H_2O leads to a compound that is geometrically different from the first compound obtained. The two compounds have the same molecular formulas but exhibit different chemical reactivities. Compounds that have the same molecular formula but differ in structure are called *isomers*. The isomer in which two identical groups occupy the same edge of a face of the octahedral complex is referred to as *cis*. The isomer with the ligands on the opposite sides of the structure is called *trans*.

The square planar complex also can lead to isomers upon successive replacement of two of the equivalent ligands (Figure 14.9). If one of the chlorines in $PdCl_4^{2-}$ is replaced by one NH_3, only one $Pd(NH_3)Cl_3^-$ ion can result since all four positions in $PdCl_4^{2-}$ are identical. However, replacement of a chlorine in $Pd(NH_3)Cl_3^-$ leads to two isomeric $Pd(NH_3)_2Cl_2$ compounds. Two of the chlorines in $Pd(NH_3)Cl_3^-$ are adjacent to the ammonia, and replacement of either one by NH_3 produces the same *cis*-isomer. The single chlorine opposite the ammonia is different from the

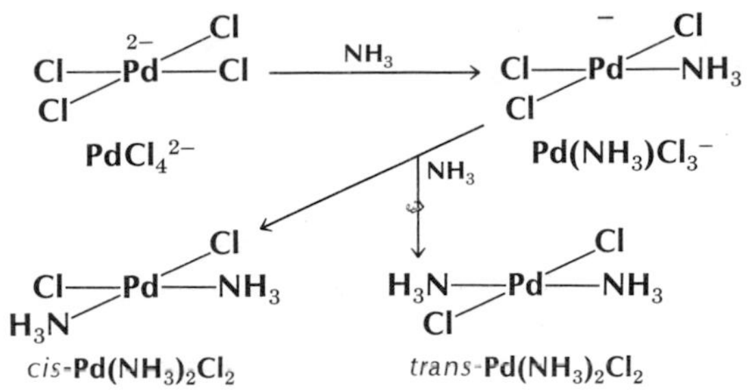

Figure 14.9
Square planar isomerism.

other two chlorines in $Pd(NH_3)Cl_3^-$, and replacement of it leads to the *trans*-isomer.

In the tetrahedral complex containing two different types of ligands, isomerism is not possible. Consider the $Zn(NH_3)_3Cl^+$ complex ion (Figure 14.10). The three NH_3 molecules are identically situated with respect to the chlorine. Replacement of a second NH_3 by Cl^- can lead to only one $Zn(NH_3)_2Cl_2$ molecule.

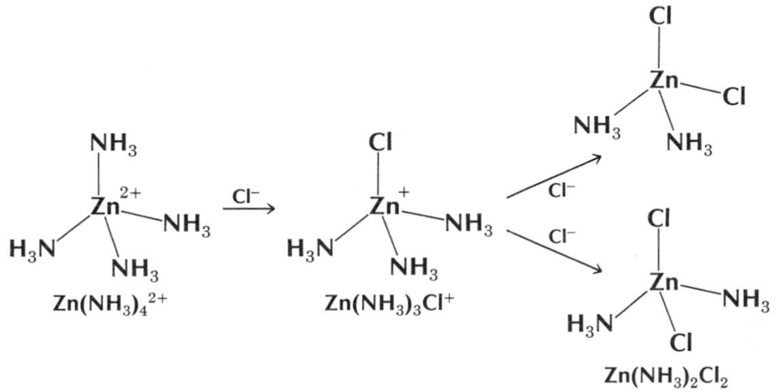

Figure 14.10
The lack of geometrical isomers in tetrahedral complex ions.

14.6
POLYDENTATE LIGANDS

Some molecules or ions have two or more electron pairs that can bond to transition metals. If a group can bond to two positions in a transition metal complex, it is called *bidentate*. Examples of bidentate ligands are ethylene diamine and the dianion of oxalic acid (Figure 14.11). Both substances contain carbon-carbon bonds and are called *organic* compounds, which are discussed in Part Four. Ethylene diamine can be regarded as two molecules of ammonia bound together by a chain of carbon atoms. The

$H_2\bar{N}-CH_2-CH_2-\bar{N}H_2$

ethylene diamine

$^{-}O(O=)C-C(=O)O^{-}$ $^{2-}$

oxalate ion

Figure 14.11
Bidentate ligands.

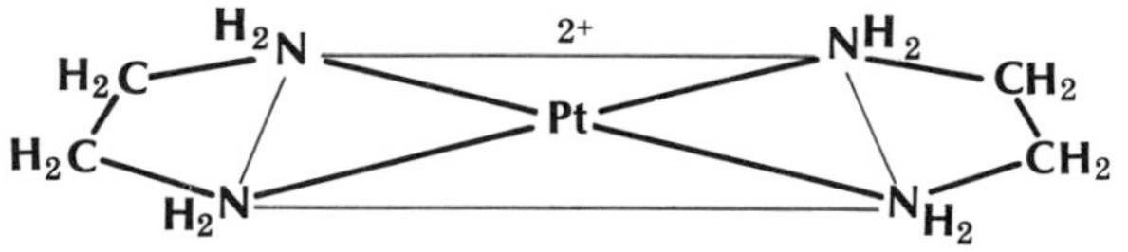

$[Pt(H_2NCH_2CH_2NH_2)_2]^{2+}$

Figure 14.12
Square planar coordination with bidentate ligands.

chain is sufficiently long to allow the two nitrogen sites to reach adjacent positions in a complex ion, as is illustrated with Pd^{2+} in Figure 14.12.

The dianion of oxalic acid, oxalate ion, also can bridge adjacent positions in complex ions, for example, three oxalate ions can serve as ligands to Fe^{3+} (Figure 14.13). Use of this bidentate ligand in forming an octahedral complex is valuable in removing rust.

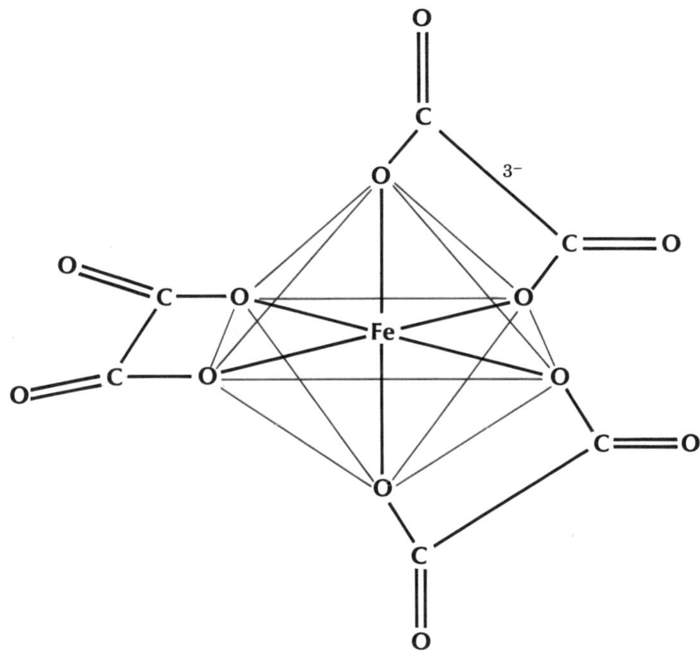

Figure 14.13
Octahedral complex between Fe^{3+} and oxalate ions.

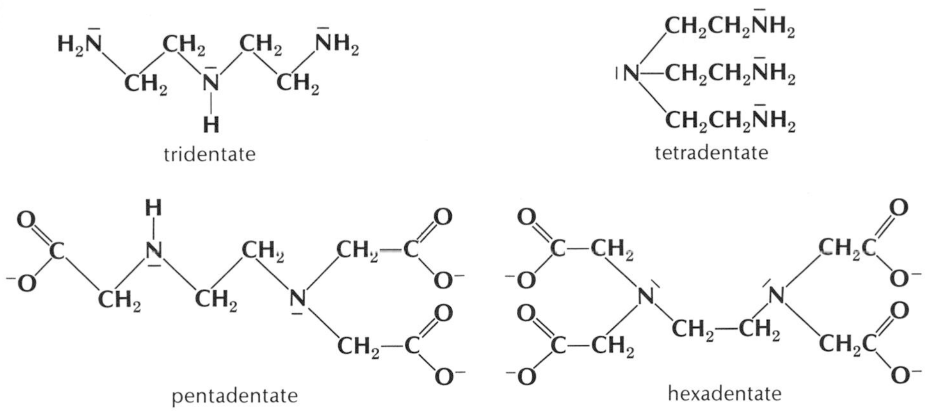

Figure 14.14
Polydentate ligands.

Tridentate, tetradentate, pentadentate, and hexadentate ligands have been made, and these ligands can be used to bond to several positions in transition ions if their geometry is suitable. Examples of polydentate ligands are shown in Figure 14.14.

14.7
OPTICAL ISOMERISM

In some complex ions a form of isomerism quite different from the geometrical isomerism discussed earlier in this chapter is possible. Geometrical isomers, while having the same molecular formula, have different physical and chemical properties. In *optical isomerism* the isomers are two compounds that differ only in that they are mirror images of each other. Their chemical properties are identical, and all of their physical properties except one are identical, the difference being their action on a beam of polarized light (Chapter 23). An example of such isomerism is the complex of Co^{3+} with three ethylene diamine ligands (Figure 14.15). The ligands can be arranged in two different ways to form two isomers that bear the same relationship to each other as do our right and left hands. They are mirror images of each other. The reflection of a right hand in a mirror is a left hand. In a similar manner, reflection of one isomer in the plane of a mirror produces the equivalent of the other isomer. Because no amount of turning from right to left or up to down will allow the superimposition of the two isomers, they are not identical. The inability to superimpose mirror images is a requirement for optical activity and optical isomers. The physical consequences of this phenomenon are discussed in Chapter 23. All substances have a mirror image, but many times the mirror image is superimposable on the original. Such compounds do not exhibit optical isomerism.

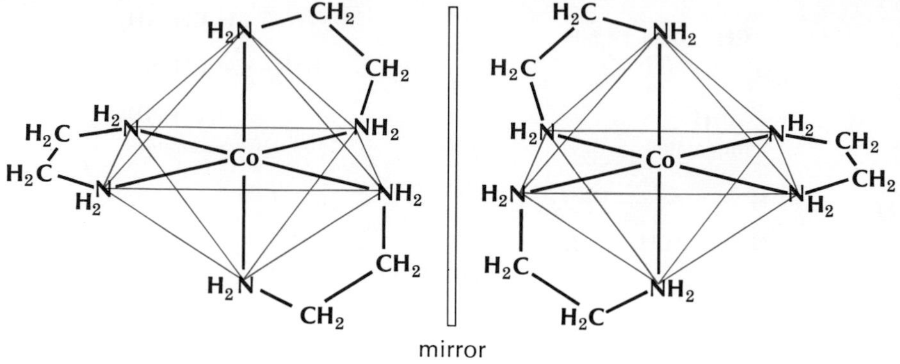

Figure 14.15
Optical isomerism.

14.8 COMPLEX IONS IN BIOCHEMISTRY

Several key substances in biological systems are complex ions. Many of these consist of transition metals surrounded by large organic ligands. Chlorophyll, which is responsible for the process of photosynthesis by which plants convert carbon dioxide and water into carbohydrates, is a complex of the nontransition metal magnesium (Figure 14.16). The carbohydrates synthesized are then used by plants for growth. Thus chlorophyll allows the plants to store solar energy in the form of chemical compounds that can be used by the plants when needed.

Vitamin B_{12} is a cobalt compound containing organic ligands of even greater complexity than chlorophyll (Figure 14.17). This vitamin is

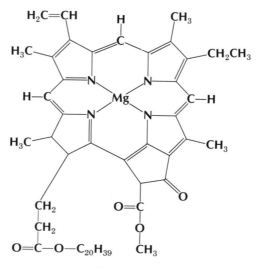

Figure 14.16
Structure of chlorophyll.

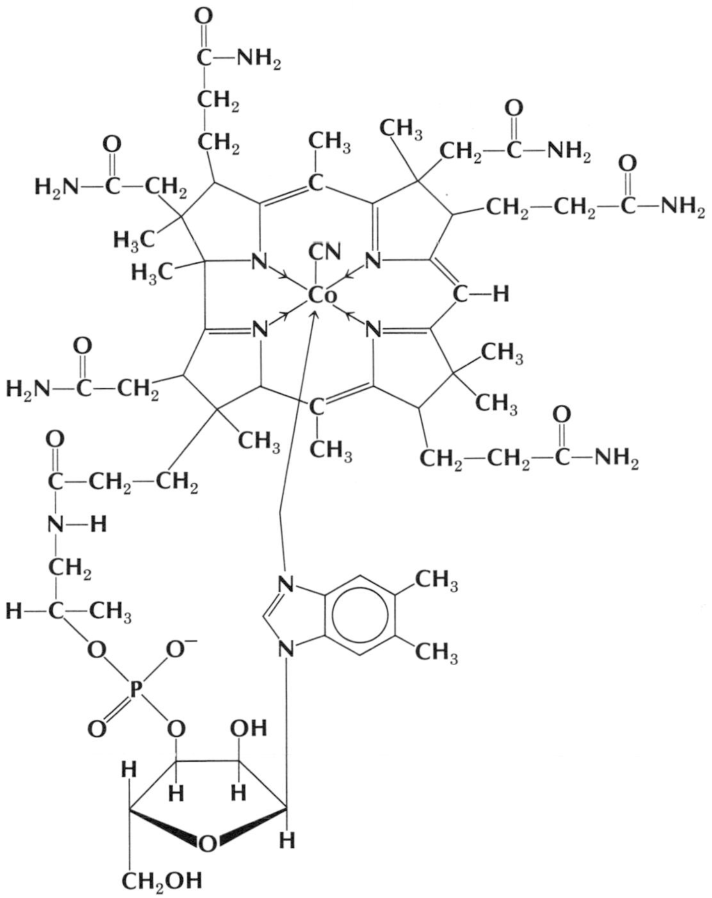

Figure 14.17
Structure of vitamin B_{12}.

necessary for the maintenance of healthy red blood cells and is used in the treatment of anemia. A vitamin B_{12} deficiency leads to stunted growth and metabolic disturbances.

Hemoglobin is similar to chlorophyll in structure but contains iron in place of magnesium at the center of the molecule (Figure 14.18). The iron complex is called *heme*. Association with the high molecular mass protein, globin (Chapter 24), produces hemoglobin. Four of the positions about the iron are associated with four nitrogen atoms, and other positions are occupied by donor atoms of globin. In biological processes oxygen is absorbed and released by the iron by replacing the globin ligands. Toxic gases such as hydrogen sulfide (H_2S) and carbon monoxide (CO) also can react with iron to form complexes so strongly bonded that they cannot be replaced by oxygen. Because they prevent the hemoglobin from performing its usual oxygen-carrying function, they produce oxygen starvation of the tissues.

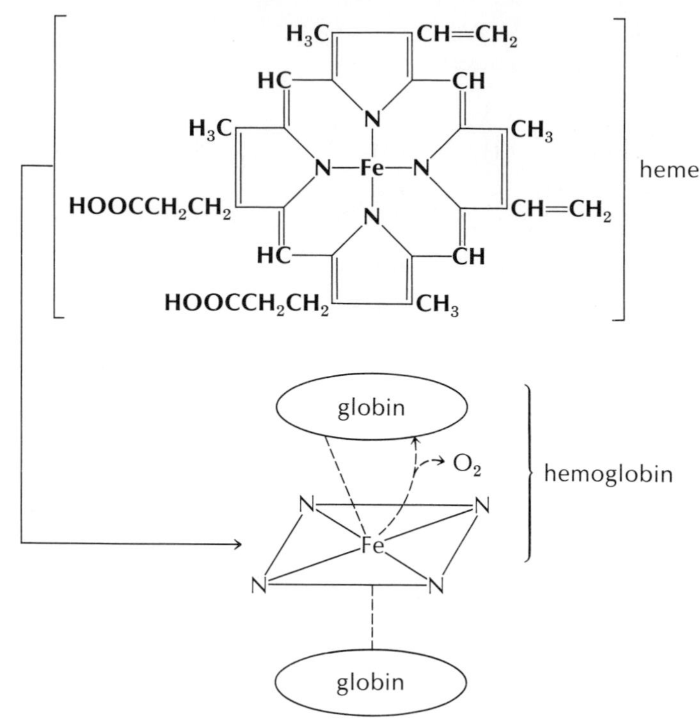

Figure 14.18
Structure of heme and hemoglobin.

Suggested further readings

Bailar, J. C., Jr., "The Numbers and Structures of Isomers of Hexacovalent Complexes," *J. Chem. Educ.,* **34,** 334 (1957).

Cotton, F. A., "Ligand Field Theory," *J. Chem. Educ.,* **41,** 467 (1964).

Johnson, R. C., "A Simple Approach to Crystal Field Theory," *J. Chem. Educ.,* **42,** 147 (1965).

Kauffman, G. B., "Alfred Werner's Coordination Theory," *Educ. Chem.,* **4,** 11 (1967).

Schubert, J., "Chelation in Medicine," *Sci. Amer.,* 40, (May 1966).

Sutton, L. E., "Some Recent Developments in the Theory of Bonding in Complex Compounds of the Transition Metals," *J. Chem. Educ.,* **37,** 498 (1960).

Terms and concepts

chlorophyll	isomerism
hemoglobin	ligand

ligand field theory
octahedral complex
optical isomerism
paramagnetism
polydentate ligands
square planar complex
tetrahedral complex
transition elements

Questions and problems

1. Write the electronic configurations for the *d* orbital electrons in Mn, Fe^{2+}, Ni^{2+}, Pt^{4+}, La^{3+}, and Cd^{2+}.
2. Calculate the magnetic moment expected for the following ions: Cr^{2+}, Sc^{3+}, Fe^{2+}, Ni^{3+}, and V^{3+}.
3. Which of the following ions would be expected to be colorless in aqueous solution: Sc^{3+}, V^{4+}, Ni^{2+}, Ag^{+}, and Hg^{2+}? Why?
4. Ammonia and water both are ligands whereas methane CH_4 is not. Explain.
5. Suggest the geometry of the eight-coordinate complex ion $Mo(CN)_8{}^{4-}$ on the basis of spatial considerations.
6. What geometric shapes would be expected for $Cr(H_2O)_6{}^{2+}$, $NiCl_4{}^{2-}$, and $HgCl_4{}^{2-}$? Draw the electronic configurations for these compounds. What types of orbitals are used in forming these complexes?
7. Reduction of $Co(CN)_6{}^{3-}$ to $Co(CN)_6{}^{4-}$ leads to an increase in the magnetic moment of 1.7 Bohr magnetons. Account for this change.
8. What types of orbitals are involved in the formation of the $Ni(CN)_4{}^{2-}$ square planar complex ion? Calculate the magnetic moment of the complex.
9. What orbitals are used in bonding $CdBr_4{}^{2-}$ and $Au(CN)_2{}^{-}$? What geometry would be expected in these compounds?
10. Cerium can exist in the +4 oxidation state. What electrons does cerium lose in the formation of this oxidation state? Would you expect Ce^{4+} to be colored?
11. In the square planar complex ion $PtCl_4{}^{2-}$ the d_{z^2} orbital is of lower energy than the other three occupied *d* orbitals. Explain.
12. In square planar complexes the d_{xy} orbital is of higher energy than the d_{xz} and d_{yz} orbitals, which are of equal energy. Explain this order.
13. The chromate ion $CrO_4{}^{2-}$ is tetrahedral. Draw the electronic arrangement for this ion.
14. There are two isomeric $Pt(NH_3)_2Cl_2$ compounds. One of these compounds has a dipole moment whereas the other does not. What are the structures for these two compounds? Account for the observed difference in the dipole moments.
15. Draw the geometric isomers having the molecular formula $Cr(H_2O)_3F_3$.
16. Draw the geometric isomers of square planar $PdCl_2BrI^{2-}$.
17. Ethylene diamine is abbreviated as en. Which of the following complex ions involving the bidentate ligand can exist as optical isomers: $Co(NH_3)_4en^{3+}$, *cis*-$Co(NH_3)_2(en)_2{}^{3+}$, and *trans*-$Co(NH_3)_2(en)_2{}^{3+}$?

15

NUCLEAR CHEMISTRY

Up to this point the chemistry of the elements and their properties have been discussed only in terms of the behavior of electrons in atoms and molecules; the nucleus has been mentioned in a purely cursory fashion. Although the mass of an atom is concentrated largely in the nucleus, the chemical reactivity of the electrons does not seem to be significantly altered by the mass of the nucleus. Isotopes of an element undergo the same types of chemical reactions at closely similar rates. Because of their charge, the number of protons present in the nucleus affects ordinary chemical reactions involving the electrons about the nucleus, but the only contribution of the uncharged neutrons is to increase the atomic mass. However, matter undergoes many important transformations that involve the nucleus, and the neutrons play an important role in these processes. In this chapter the chemistry of the nucleus will be examined.

15.1
THE NUCLEUS

The nuclei of atoms contain protons and neutrons that collectively are called *nucleons*. Although many more fundamental particles have been discovered in recent years, the fundamental chemistry of the nucleus can be adequately interpreted on the basis of these two elementary particles. The number of protons in a nucleus is its atomic number Z, and the total number of nucleons is its mass number A. The number of neutrons is $A - Z$. The chemical symbol of a given isotope consists of the elemental letter symbol with a superscript mass number and a subscript atomic number. For example, the two naturally occurring isotopes of hydrogen (Chapter 13), are written ${}^{1}_{1}H$ and ${}^{2}_{1}H$.

The volumes of nuclei vary directly with the mass number and, therefore, the density of all nuclei are approximately constant. The density of any nucleus is approximately 2.4×10^{14} g/cm^3. One liter of nuclear matter would weigh more than 5×10^{14} lb.

The most striking feature of the nucleus is its ability to maintain an aggregation of positively charged particles in a region that is about 10^{-13} cm in radius. The protons would be expected to fly apart as a result of electrostatic repulsion. Indeed, some nuclei are unstable and do undergo *radioactive* processes in which other nuclei and atomic particles are produced. The difference between stable and unstable nuclei is apparently a function of the number of neutrons present, which play an important role in binding the nucleus. Neutron–proton attractions are apparently stronger than neutron–neutron or proton–proton interactions. For example, the deuteron ${}^{2}_{1}H$ is a stable substance whereas particles containing two protons or two neutrons in the absence of other nucleons have never been observed.

15.2
ZONE OF STABILITY

An indication of the importance of neutrons in binding the nucleus is the zone of stability obtained by plotting the number of protons for naturally occurring nonradioactive *nuclides* vs the number of neutrons (Figure 15.1). This zone indicates the natural limits of nuclear stabilities. For any given element, a particular ratio of neutrons to protons is required for stability of the nucleus. If the number of neutrons is altered to either increase or decrease that ratio, the nucleus becomes less stable. For the elements of low atomic number, the ratio of neutrons to protons in a stable nuclide is approximately one to one as in ${}^{4}_{2}He$, ${}^{12}_{6}C$, ${}^{16}_{8}O$, and ${}^{20}_{10}Ne$. The zone of stability is narrow in the region of the elements of low atomic mass, and the addition or removal of one neutron from the nucleus of elements in this region usually leads to unstable elements. A straight line representing a neutron-to-proton ratio of one is shown in Figure 15.1, and it is evident that additional neutrons are needed in order

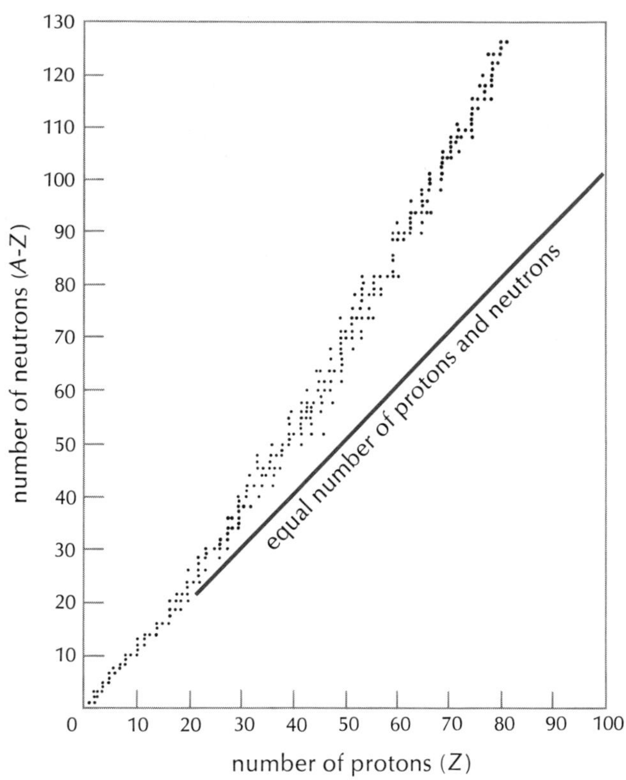

Figure 15.1
Zone of stability.

to increase the neutron-to-proton ratio in the elements of high atomic mass. In addition, the width of the zone of stability increases somewhat. The behavior represented by the zone of stability plot strongly suggests that neutrons are partly responsible for the binding of protons in the nucleus. As more protons are packed into the very small confines of the nucleus the electrostatic forces of repulsion between all protons increase sharply. Therefore, a larger excess of neutrons is required to counteract the increased number of combinations of repulsive forces.

The majority of stable nuclides (157) have an even number of protons and an even number of neutrons. Fifty-two nuclides have an even number of protons and an odd number of neutrons; 50 nuclides have an odd number of protons and an even number of neutrons. Only five stable nuclides have both an odd number of protons and an odd number of neutrons. For no element of odd atomic number are there more than two stable nuclides, whereas elements of even atomic numbers have many stable nuclides. These observations and the limitation of nuclear stability indicated by the zone of stability tend to suggest that a consistent structural concept of the nucleus similar to that proposed for electronic configurations may be developed in the future.

15.3 RADIOACTIVITY

A nucleus that lies outside the zone of stability is by definition unstable. Such a nucleus could achieve a stable nuclear configuration by any process that produces a nucleus with neutron-to-proton ratio inside the zone of stability. If the neutron-to-proton ratio of an unstable nucleus is too high, the process might involve a diminution of the number of neutrons, an increase in the number of protons, or a simultaneous alteration of the number of both particles. Conversely, if the neutron-to-proton ratio is too low, stabilization could be achieved by an increase in the number of neutrons, a decrease in the number of protons, or both. All processes involving conversion of unstable to more stable nuclei are said to be examples of *radioactive decay*. Radioactivity has been observed in nature and also has been induced by the production of unstable elements in the laboratory. Both naturally occurring and man-made elements undergo decay reactions that are independent of their origin. Natural radioactivity is found in all elements with atomic numbers higher than 83. Note that the zone of stability picture in Figure 15.1 stops at $Z = 83$, implying the existence of no stable isotopes above that atomic number.

The types of radioactive decay are few and are governed by the neutron-to-proton ratio. First the case of a high neutron-to-proton ratio and then the case of a low neutron-to-proton ratio will be discussed.

β Emission: If the nucleus has a high neutron-to-proton ratio it would appear that it could achieve stability by neutron ejection, but this process is rarely observed. The more common method of reducing the neutron-to-proton ratio involves *beta (β) emission.* A β particle is an electron and is represented for the purposes of balancing nuclear equations as ${}_{-1}^{0}e$. The subscript refers to the charge on the electron; the superscript indicates that the mass of the electron is negligible compared to that of an atom. When a nucleus undergoes β decay, the mass of the product is to a first approximation the same as that of the decaying nucleus. However, the number of protons has increased by one, and the number of neutrons has decreased by one. The net result of the reaction is conversion of a neutron into a proton and an electron. An example of β decay is the conversion of the unstable isotope carbon-14, ${}_{6}^{14}C$, to nitrogen-14, ${}_{7}^{14}N$, in which the neutron-to-proton ratio is decreased from 1.33 to 1.0:

$$ {}_{6}^{14}\mathbf{C} \longrightarrow {}_{7}^{14}\mathbf{N} + {}_{-1}^{0}e $$

Another example of β decay is the conversion of ${}_{11}^{24}Na$ to ${}_{12}^{24}Mg$, in which the neutron-to-proton ratio is decreased from 1.18 to 1. Both the charge and mass are conserved, as indicated in the balanced equation:

$$ {}_{11}^{24}\mathbf{Na} \longrightarrow {}_{12}^{24}\mathbf{Mg} + {}_{-1}^{0}e $$

When the unstable nucleus lies below the zone of stability the low neutron-to-proton ratio can be adjusted either by gaining neutrons, losing protons, or doing both simultaneously. Three processes that commonly occur are *K electron capture, positron emission,* and *alpha emission.*

K Capture: Absorption by the nucleus of one of the lowest energy level electrons converts one proton into a neutron and reduces the nuclear charge by one. Since the absorbed electron is from the K shell, the process is called *K electron capture.* The nuclear mass is unchanged, but the atomic number decreases by one, as indicated by the conversion of $^{40}_{19}K$ into $^{40}_{18}Ar$. The neutron-to-proton ratio increases from 1.10 to 1.22 in this process:

$$^{40}_{19}\mathbf{K} \longrightarrow {}^{40}_{18}\mathbf{Ar}$$

K capture

Mass as well as charge is conserved in K electron capture, but the latter fact is the less obvious. While the nuclear charge has decreased by one positive unit, the number of extranuclear electrons also has decreased by one and electrical neutrality is maintained.

Positron emission: Another process by which the neutron-to-proton ratio is increased involves *positron emission.* A positron is a particle whose mass is equivalent to an electron but which is positively charged. Using the symbol $^{0}_{1}e$ for the positron, nuclear reactions such as the conversion of $^{30}_{15}P$ to $^{30}_{14}Si$ can be balanced:

$$^{30}_{15}\mathbf{P} \longrightarrow {}^{30}_{14}\mathbf{Si} + {}^{0}_{1}e$$

Both the mass and charge are conserved in this reaction. A proton has disappeared and a neutron is generated in its place by positron emission. The neutron-to-proton ratio is increased from 1 to 1.14.

The positron is a member of the class of *antiparticles*—particles that correspond in mass to a common particle, but bear the opposite charge. Another member of this class is the *antiproton,* which has been produced in the laboratory. It is of the same mass as a proton but is negatively charged. Antimatter is destroyed upon collision with matter, producing large quantities of energy. Antiparticles can exist only for short periods of time because they quickly collide with members of the class of matter and are destroyed. It is interesting to speculate whether, in another part of the universe, an entire solar system of antimatter exists. In such an environment antielements could exist theoretically without danger of destruction, but the introduction of matter as we know it into this hypothetical antimatter world would lead to destruction of our supposedly stable matter as well as an equal quantity of antimatter.

α Emission: A third way in which the neutron-to-proton ratio can be increased is by *alpha (α) emission.* The α particle is the helium nucleus, which consists of two neutrons and two protons, $^{4}_{2}He$. Nuclei containing 84 or more protons commonly decay by α emission, as exemplified by the decay of $^{212}_{84}Po$ to $^{208}_{82}Pb$:

$$^{212}_{84}\mathbf{Po} \longrightarrow {}^{208}_{82}\mathbf{Pb} + {}^{4}_{2}\mathbf{He}$$

At first glance it does not appear that the loss of an equal number of protons and neutrons in the form of an α particle alters the neutron-to-proton ratio. All the elements that emit α particles have more neutrons than protons, and the loss of the two protons causes a larger percentage change in the number of protons than the loss of two neutrons. The neutron-to-proton ratio of $^{212}_{84}Po$ is 1.52, whereas in the product $^{208}_{82}Pb$ it is 1.54. Elements of mass number higher than 84 may require emission of α particles in several steps to achieve a stable nucleus.

15.4 TRANSURANIUM ELEMENTS

Elements of atomic number greater than 92 are rare in the earth's crust. These elements, which are called *transuranic* since they have atomic numbers greater than uranium, have been synthesized in the laboratory using reactors and accelerators. The first transuranic element, neptunium-239, $^{239}_{93}Np$, was synthesized in 1940 by an American group of scientists who bombarded the nucleus $^{238}_{92}U$ with deuterons $^{2}_{1}H$ of very high energy. The first product was an isotope, uranium-239, which then underwent β emission to produce $^{239}_{93}Np$:

$$^{238}_{92}\mathbf{U} + {}^{2}_{1}\mathbf{H} \longrightarrow {}^{239}_{92}\mathbf{U} + {}^{1}_{1}\mathbf{H}$$
$$^{239}_{92}\mathbf{U} \longrightarrow {}^{239}_{93}\mathbf{Np} + {}^{0}_{-1}e$$

In a similar fashion plutonium-239, $^{239}_{94}Pu$, has been produced in atomic reactors by neutron bombardment of $^{238}_{92}U$. The nuclear reactions are given below:

$$^{238}_{92}\mathbf{U} + {}^{1}_{0}n \longrightarrow {}^{239}_{92}\mathbf{U}$$
$$^{239}_{92}\mathbf{U} \longrightarrow {}^{239}_{93}\mathbf{Np} + {}^{0}_{-1}e$$
$$^{239}_{93}\mathbf{Np} \longrightarrow {}^{239}_{94}\mathbf{Pu} + {}^{0}_{-1}e$$

The preparation of elements of high atomic numbers is a difficult, expensive, and tedious task that requires great experimental skill. The instability of both the elements used for the targets and the resultant products is such that rapid analysis of very small numbers of atoms is required. Some of the transuranic elements have been prepared and characterized with less than 10 atoms present. Much of this masterful work has been carried out by the Nobel prize winner Glenn T. Seaborg and other members of the faculty at the University of California at Berkeley, where plutonium, americium, curium, berkelium, and californium have been synthesized and identified.

In order to effectively bombard nuclei with positively charged particles the electrostatic repulsion between projectile and nuclei must be overcome. The simplest way of accomplishing this is to accelerate the projectiles and thus increase their energy. The positive particles such as α particles, which are emitted in the radioactive decay of nuclei, are usually of insufficient energy to be used for nuclear transformations of other nuclei, so artificial means of accelerating atomic particles have been devised. The particle accelerators that are used for nuclear research are expensive but have yielded rich dividends in terms of the knowledge of fundamental forces that control nuclear stability and ultimately the whole of the physical universe.

Particle accelerators: Two types of particle accelerators are the *cyclotron* and the *linear accelerator.* In the cyclotron particles are accelerated in circular paths by proper application of magnetic and electric fields. The source of particles is at the center of the instrument between two hollow D-shaped plates called *dees* (Figure 15.2). The dees are separated by a gap and are enclosed in an evacuated chamber, which in turn is located between the poles of a powerful electromagnet. The dees are

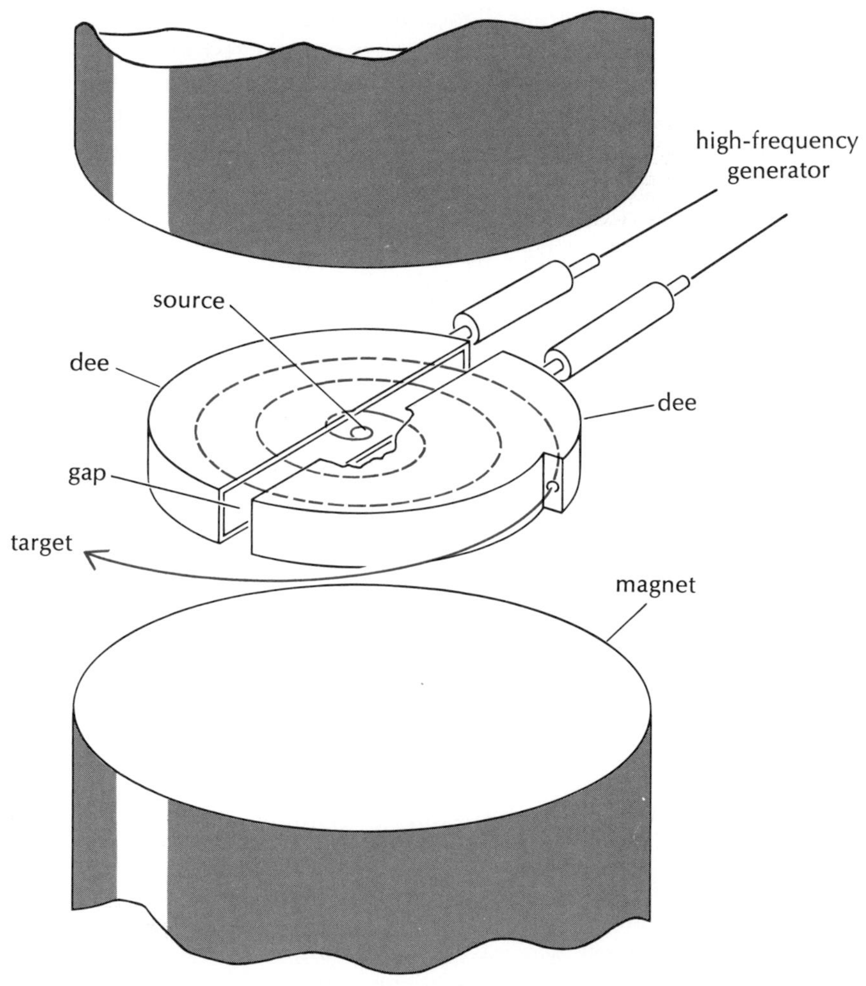

Figure 15.2
Path of a charged particle in a cyclotron.

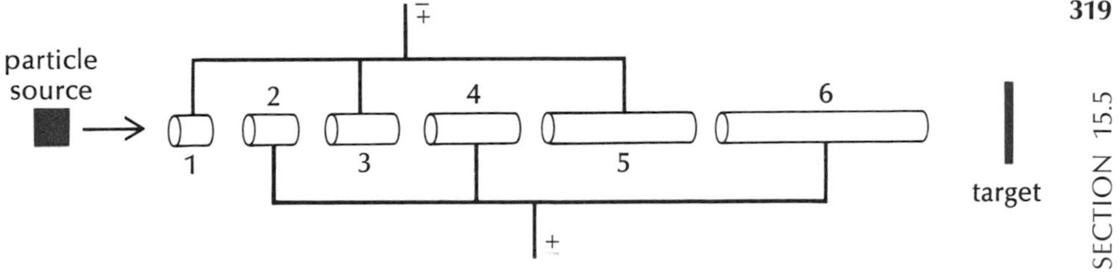

Figure 15.3
Representation of a linear accelerator.

kept oppositely charged by means of a high frequency generator. The charged particles move in a circular path controlled by the magnetic and electric fields. As the particles reach the gap between the dees, the charge of the dees is reversed such that the positive particles are repulsed out of a positive dee and attracted into a negative dee. Because of the acceleration of the particles as they traverse the gap, they travel in a spiral path of increasing radius. Eventually, after several spirals, the particles leave the instrument at high speed and collide with a target nucleus.

In a linear accelerator the particles are accelerated through a series of charged tubes in an evacuated chamber (Figure 15.3). In the absence of a magnetic field the particles travel in a straight line, causing the instrument necessary for acceleration to be very long as compared to the dimensions of the cyclotron. The positive particles are attracted into the first tube, which is at that moment negatively charged. At this time all odd-numbered tubes are negatively charged and the even numbered tubes are positively charged. Just as the particles leave tube 1, the polarity of the tubes is reversed. The particles are attracted to tube 2 and repelled by tube 1 and as a result are accelerated. Repetition of this process at constant time intervals leads to increased acceleration. Each tube must be successively larger in order to allow the accelerating particles the same residence time in each tube. At the end of the accelerator the projectiles meet the target and a nuclear reaction occurs.

15.5 BINDING ENERGY

The instability of a nucleus with a neutron-to-proton ratio that lies outside the zone of stability can be expressed in terms of energy. Consider the helium nucleus, which contains two neutrons and two protons. The masses of the neutron and proton are 1.00867 and 1.00728 amu, respectively. On this basis the mass of the helium nucleus would be expected to be 4.03190 amu. However, experimentally the mass of the helium nucleus has been determined as 4.0026 amu. The mass difference corresponds to an energy that would be liberated in forming the helium nucleus from two neutrons and two protons. Mass and energy are related by the Einstein equation $E = mc^2$, where E represents energy, m is mass, and c is the speed of light. The 0.0293 amu difference in mass corresponds to 6.3×10^8 kcal/mole of helium atoms and is called the *binding energy* of the helium nucleus. In order to disrupt 1 mole of helium atoms and form 2 moles of protons and 2 moles of neutrons, energy equivalent to the binding energy must be supplied.

In a similar manner the binding energies of other nuclei can be calculated. As the atomic mass increases, the binding energy also increases. However, if the binding energy per nucleon (neutrons and protons) is calculated and plotted as a function of mass number, the intriguing graph in Figure 15.4 is obtained. Elements of intermediate mass have the highest binding energy per nuclear particle and are, therefore, the most stable. The point of maximum stability occurs in the middle of the transition metals of the fourth period. All other elements are unstable with respect to those at the maximum of the graph. If a heavy element could be converted into elements of intermediate mass, energy would be liberated. Similarly, conversion of several low mass elements into an element of intermediate mass also would liberate energy. As sources of energy both of these conversions are of considerable potential benefit to man.

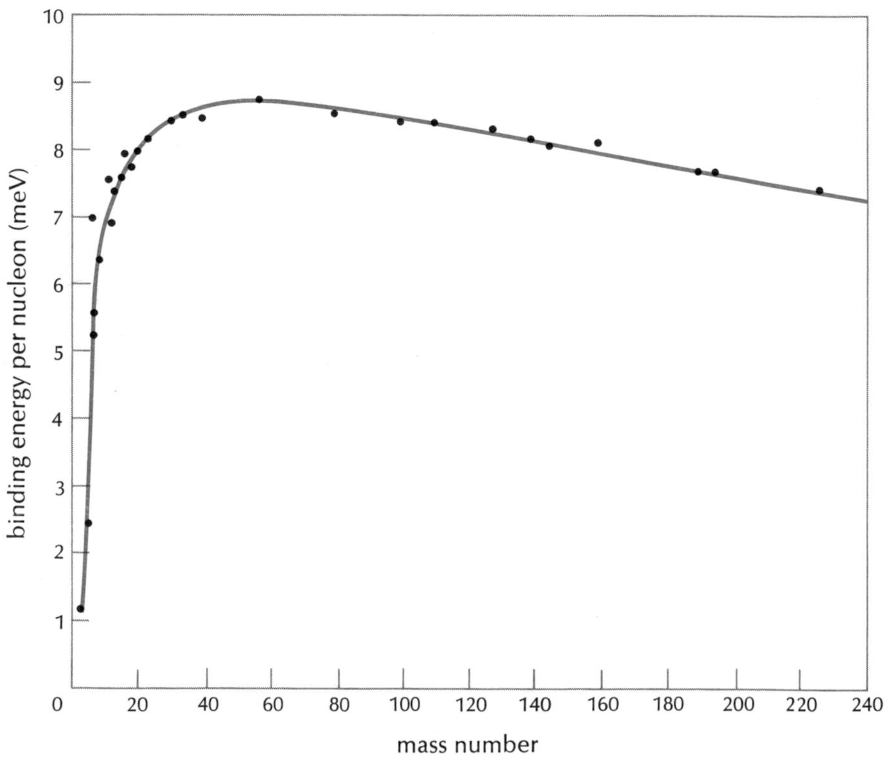

Figure 15.4
Binding energy per nucleon.

15.6
NUCLEAR FISSION

If nuclear energy is obtained by conversion of a heavy element into two or more lighter elements or nuclear particles, the process is called *nuclear fission.* An example of nuclear fission is shown by the bombardment of uranium-235 with neutrons. The unstable uranium-236 nucleus is formed, which rapidly splits into nuclei of lower atomic mass. Although many pairs of elements are actually produced in the process, $^{137}_{52}Te$ and $^{97}_{40}Zr$ are typical:

$$^{235}_{92}U + ^{1}_{0}n \longrightarrow ^{236}_{92}U$$
$$^{236}_{92}U \longrightarrow ^{137}_{52}Te + ^{97}_{40}Zr + 2^{1}_{0}n$$

In addition to technetium and zirconium, two neutrons are produced. The fission process then can be repeated since these neutrons are available for collisions. Since more neutrons are produced than are used in initiating the reaction, the fission process rapidly becomes a self-sustaining chain reaction that releases large amounts of energy in a short time. This principle is utilized in the construction of atomic bombs. Other decomposition modes for $^{236}_{92}U$ are listed below:

$$^{236}_{92}U \longrightarrow ^{140}_{56}Ba + ^{95}_{36}Kr + ^{1}_{0}n$$
$$^{236}_{92}U \longrightarrow ^{141}_{54}Xe + ^{93}_{38}Sr + 2^{1}_{0}n$$
$$^{236}_{92}U \longrightarrow ^{89}_{37}Rb + ^{144}_{55}Cs + 3^{1}_{0}n$$

Atomic energy as contained in the arsenals of numerous countries is potentially destructive and could seriously limit human progress. It should be emphasized that on the other side of the coin controlled nuclear energy can be of service in a world of limited dimensions and resources. Fossil-derived fuels (oil and coal) are being used at an ever-increasing rate, and in the future this consumption could increase in a dramatic manner. Nuclear fission is and will continue to be an important source of vast amounts of energy. The energy equivalent of 1 ton of coal can be produced by only $\frac{1}{2}$ g of uranium-235. To a large degree the fission process can be controlled in nuclear reactors, in which the rate of the chain reaction is kept in control by the use of materials that can absorb some of the neutrons. The energy produced in atomic reactors is removed as heat and converted into electrical energy. There are limitations in the general applicability of nuclear reactors because they are large installations that require extensive shielding and trained personnel to run. However, if they were used only for the generation of energy, this alone would relieve the demands being made on our natural resources.

15.7
NUCLEAR FUSION

The second method by which the nuclear binding energy can be liberated is *nuclear fusion.* The fusion of two light elements to form a more massive one is a source of even greater energy than nuclear fission because the curve of the binding energy per nuclear particle is very steep in the area of low mass elements. Unlike nuclear fission, fusion requires extremely high temperatures (1,000,000°C) for initiation.

Fusion occurs on the sun, where the temperature is high enough to convert hydrogen into helium. Only 45 mg of hydrogen are required to produce the energy equivalent of 1 ton of coal. At the present rate the sun will expend itself in approximately 100 billion years. This problem is not a cause for great concern. The steps proposed to account for the net transformation of ${}^{1}_{1}H$ into ${}^{4}_{2}He$ on the sun are as follows:

$$2\,{}^{1}_{1}\mathbf{H} \longrightarrow {}^{2}_{1}\mathbf{H} + {}^{0}_{1}e$$
$${}^{2}_{1}\mathbf{H} + {}^{1}_{1}\mathbf{H} \longrightarrow {}^{3}_{2}\mathbf{He}$$
$$2\,{}^{3}_{2}\mathbf{He} \longrightarrow {}^{4}_{2}\mathbf{He} + 2\,{}^{1}_{1}\mathbf{H}$$

On earth, nuclear fusion has been obtained by employing atomic bomb "triggers" to provide sufficient energy to initiate fusion. This is the principle upon which the hydrogen bomb is based. Since hydrogen is so readily available from water, controlled nuclear fusion for practical purposes would eliminate any concern about the source of energy for future human progress.

15.8 USES OF RADIOISOTOPES

MEDICINE

The use of radium in cancer treatment is well known. The radiation emitted by the decaying radium is focused on a malignant tissue in order to destroy or retard its growth. Great care must be used to limit concomitant destruction of healthy tissue. Radium is being replaced by cobalt-60 for this purpose because it is less expensive and easier to handle.

Hyperthyroidism and cancer of the thyroid gland have been treated with ${}^{131}_{53}I$. When iodine is introduced into the body, it is concentrated in the thyroid gland, which requires it to function properly. Administration of sodium iodide containing iodine-131 results in transmission of the radioisotope directly to the source of difficulty. Consequently, the danger of damage to healthy tissue is reduced. ${}^{24}_{11}Na$ is used to monitor and study the circulatory system. Injection of sodium chloride containing the radioisotope into the bloodstream allows a physician to follow the course of the sodium by using the proper radiation detector. In this way, abnormalities of the circulatory system may be examined and diagnosed.

The radioisotope of phosphorus, ${}^{32}_{15}P$, has been used to treat certain blood disorders in which red blood cells are overproduced. The phosphorus is taken up more rapidly by developing red blood cells than by mature cells. Breast cancers and brain tumors have been detected by the use of phosphorus.

INDUSTRY

The petroleum industry uses radioisotopes in pipelines in order to delineate the boundary between batches of different grades of oil being transported through a single pipeline. When the radioisotope is located at a switching station or storage site, the separation can be made. The radioactivity has been used to initiate automatically the switches necessary to separate each batch of oil into its proper pipeline.

The radioactivity of cobalt-60 is used to detect flaws in metal castings. A flaw in metal absorbs less radiation than sound metal.

The effectiveness of lubricants has been ascertained by exposing the metal object to be lubricated to a nuclear reactor. The radioactive metal, if not properly lubricated, will wear down and the oil becomes radioactive. In a similar manner the wear of rubber tires has been determined by incorporating radioisotopes in the tread.

Sterilization of food by radioisotopes allows the food to be kept fresh longer. Experiments have been carried out on both animal- and plant-derived food products. The radioisotope ${}^{137}_{55}Cs$ has been used to eliminate trichinosis in fresh pork at a very low cost.

CHEMISTRY

Since radioactive nuclei can be detected in minute amounts by devices such as a Geiger counter, small amounts of radioactive compounds can be mixed with nonradioactive compounds and the resultant mixture can be traced in a chemical or biological transformation. Examples of the elucidation of the mechanism of chemical reactions by means of radioisotopes are given in later chapters. The mechanism of the hydrolysis of esters will be described in Chapter 21. The process of photosynthesis has been determined by using radioactive carbon dioxide (Chapter 29). The pathways of metabolism of amino acids, carbohydrates, and lipids have been postulated on the basis of studies with carbon-14-labeled compounds (Chapter 28).

15.9 HALF-LIVES

It is a characteristic of all radioactive decay reactions that the rate at which nuclear disintegration occurs is governed by a statistical process in which one-half of the nuclei will decay in a given unit of time. The time required for the decay of one-half of a given number of nuclei is called a *half-life*. The half-life of a given nucleus is independent of the number of atoms considered. Thus 100 g of an element might decay at a rate such that its half-life is 5 days. At the end of 5 days 50 g of the element will remain. In the next 5-day period one-half of the remaining 50 g will decay. At the end of 50 days, or 10 half-lives, approximately 0.1 g will remain.

One of the best-known uses of the half-lives of radioisotopes is the carbon-14-dating method. The age of ancient materials made of plant or animal matter can be established on the basis of their $^{14}_{6}C$ content. Carbon dioxide in the atmosphere consists mainly of carbon-12 with trace amounts of carbon-14, which is radioactive and decays. However, the concentration of carbon-14 does not decrease because it is constantly being formed in the atmosphere from the action of cosmic rays on nitrogen, $^{14}_{7}N$. All plants absorb carbon dioxide from the atmosphere, and as long as the plant is living the amount of carbon-14 incorporated into the molecules it produces and uses will be a constant fraction of the amount of carbon present. When the plant dies and is utilized for purposes such as construction, the amount of carbon-14 diminishes with time. Since it is known that the half-life of carbon-14 is 5570 years, it is possible to obtain a good measure of the age of an object by determining the amount of radioactive carbon remaining in it. A wooden dish that contains only 25 percent of the carbon-14 that trees have today is, therefore, approximately 11,000 years old. The carbon-14-dating technique is limited for the dating of objects that are older than 50,000 years. After many half-lives have elapsed it is difficult to measure accurately the small amount of carbon-14 remaining.

Because of the time limitation of the carbon-14-dating method, another dating method has been developed involving krypton, $^{40}_{19}Kr$, and argon, $^{40}_{18}Ar$. The half-life of krypton-40, which undergoes K electron capture to produce argon-40, is 1.3×10^9 years. By determining the amount of argon-40 in a potassium-bearing mineral, the approximate data of origin of the mineral can be estimated. Objects as old as 2 million years have been dated by this method.

Suggested further readings

Asimov, I., "The Radioactivity of the Human Body," *J. Chem. Educ.*, **32,** 84 (1955).

Deevey, E. S., Jr., "Radiocarbon Dating," *Sci. Amer.*, 24 (February 1952).

Fowler, T. K., and Post, R. F., "Progress Toward Fusion Power," *Sci. Amer.*, 21 (December 1966).

Garrett, A. B., "The Flash of Genius: Radioactivity: Henri Becquerel," *J. Chem. Educ.*, **39,** 533 (1962).

Garrett, A. B., "The Flash of Genius: Carbon-14 Dating: Willard F. Libby," *J. Chem. Educ.*, **40,** 76 (1963).

Garrett, A. B., "The Flash of Genius: The Positron: C. D. Anderson," *J. Chem. Educ.*, **40,** 123 (1963).

Hahn, O., "Discovery of Fission," *Sci. Amer.*, 76 (February 1958).

Kamen, M. D., "The Early History of Carbon-14," *J. Chem. Educ.*, **40,** 234 (1963).

Leachman, R. B., "Nuclear Fission," *Sci. Amer.*, 49 (August 1965).

Post, R. F., "Fusion Power," *Sci. Amer.*, 73 (December 1957).
Wallmann, J. C., "The First Isolations of the Transuranium Elements—A Historical Survey," *J. Chem. Educ.*, **36,** 340 (1959).

Terms and concepts

alpha (α) particles
antiparticle
beta (β) particles
binding energy
cyclotron
fission
fusion
half-life
K electron capture
linear accelerator
nucleons
nucleus
nuclides
positron
radioactivity
zone of stability

Questions and problems

1. Write balanced nuclear equations for each of the following processes:

a. β emission of ${}^{21}_{9}F$
b. α emission of ${}^{220}_{86}Rn$
c. K electron capture by ${}^{56}_{28}Ni$
d. positron emission by ${}^{75}_{35}Br$
e. β emission by ${}^{115}_{48}Cd$
f. positron emission by ${}^{104}_{47}Ag$

2. The half-life of ${}^{32}_{15}P$ is 14 days. How many grams of a 2-g sample of this isotope would remain after 56 days?

3. The bombardment of ${}^{7}_{3}Li$ by an α particle produces a neutron and an element. What is the element?

4. Bombardment of ${}^{238}_{92}U$ by ${}^{14}_{7}N$ leads to five neutrons and a transuranium element. What is this element?

5. A nuclear particle when projected into ${}^{58}_{26}Fe$ produces ${}^{59}_{26}Fe$ and a proton. What is the nuclear particle?

6. Conversion of ${}^{234}_{92}U$ to ${}^{206}_{82}Pb$ occurs by multiple steps. The order in which the particles are emitted is α, α, α, α, α, β, α, β, β, β, and α. What elements are produced in this process?

7. An article of archaeological importance has 3.1 percent of the ${}^{14}_{6}C$ contained in the atmosphere. How old is it?

8. The mass of ${}^{58}_{26}Fe$ is 57.9333 amu. Using the values of 1.00867 and 1.00728 amu for the mass of the neutron and the proton, respectively, calculate the mass lost in forming ${}^{58}_{26}Fe$ from these elementary particles. How much energy per mole of iron would be liberated if it were formed from neutrons and protons?

PART FOUR

ORGANIC CHEMISTRY

In Chapter 13 the structural modifications of carbon and a few of its compounds were discussed and compared with the properties of the other members of Group IV and their compounds. Carbon is unique in its ability to form stable bonds between carbon atoms, resulting in large networks of connected atoms. Of all the elements in the periodic table, carbon forms the greatest number of stable and high molecular mass substances. Elements such as oxygen and nitrogen are capable of forming compounds such as hydrogen peroxide $H—O—O—H$ *and hydrazine* $H_2N—NH_2$*, but longer chains of connected oxygen or nitrogen atoms for the most part* *are highly unstable chemical curiosities. Aside from the almost unlimited variety of arrangements of carbon atoms that are possible in compounds, carbon also is capable of combining with virtually every other element at many sites within a given carbon compound. Although the exact number of carbon compounds is unknown, the number is estimated to be well over 3.5 million. More than 90 percent of all known compounds contain carbon, and the rate at which new carbon compounds are being produced is approximately 75,000 a year. A detailed study of the chemistry of carbon can be justified by statistics alone.*

The importance of carbon compounds to every facet of our lives is an even more compelling reason for studying the chemistry of carbon. In the present era of rapid scientific advances, the well-educated individual should understand how science is affecting his life. It is difficult to be unaware of the more dramatic strides that man has made in recent years in the fields of electronic and space technology. However, our life is intimately entwined with and dependent on the science of organic chemistry in ways not always obvious.

16

THE ALKANES

The simplest class of organic compounds is the *hydrocarbons*. Hydrocarbons contain only carbon and hydrogen. Substitution of other atoms or groups of atoms for the hydrogen of hydrocarbons yields other classes of compounds. The general class hydrocarbons is further divided into subgroups based on the types of reactions they undergo. The first subgroup is the *alkanes,* which contains all compounds with the general molecular formula C_nH_{2n+2}, where *n* may be any integer. All members of this subgroup have names ending in *-ane*.

16.1 VITAL FORCE

The term *organic* means pertaining to plant or animal organisms. It was introduced originally as a method of classifying substances derived from living sources that were thought to contain a mysterious "vital force."

Compounds containing carbon were obtained either directly from living sources by simple physical extraction or by chemical conversion of these products into other organic substances. Prior to the eighteenth century the rather crude and destructive method of pyrolysis was employed to obtain organic compounds. At the beginning of the nineteenth century plant and animal products were obtained essentially unaltered from their native state by extraction with solvents. The Swedish chemist C. W. Scheele obtained citric acid from lemons, malic acid from apples, tartaric acid from grapes, lactic acid from milk, uric acid from urine, and oxalic acid from wood sorrel. Other chemists isolated urea from urine (H. M. Rouelle, 1773), hippuric acid from horse urine (J. von Liebig, 1830), cholesterol from fats (M. E. Chevreul, 1815), and morphine from opium (F. W. Sertürner, 1805).

In 1828 Wohler demonstrated that urea could be obtained by heating ammonium cyanate:

$$NH_4CNO \longrightarrow NH_2CONH_2$$

Since the sources of ammonium cyanate now are considered to be mineral, it would appear that the vital force concept should have fallen into disrepute. However, at that time ammonium cyanate was obtained by roasting bones, and so it was argued that the vital force still might be contained in the ammonium cyanate (surviving the high temperatures at which all other forms of life would have been destroyed). It remained for chemists to obtain a variety of compounds from mineral sources before the idea of a vital force was finally abandoned. The preparation of organic compounds today is a relatively straightforward process that requires only appropriate modifications of many known experimental procedures. The student should not conclude from the preceeding statement that organic chemistry is now a collection of recipes. Considerable planning, ingenuity, and persistence is required to synthesize new substances of known classes of compounds. New classes of compounds and reactions still are being discovered.

Organic chemistry today not only involves the study of carbon-containing compounds similar in structure to those that are present in living organisms but in addition includes the preparation and examination of the reactions of compounds that do not occur in nature. The special role of carbon is the result of its ability to form numerous frameworks to which other elements may be attached.

16.2 STRUCTURE OF THE ALKANES

METHANE

The lightest alkane, *methane,* is a product of the decomposition of organic material given off in swamps. It is a major component of the fuel "natural

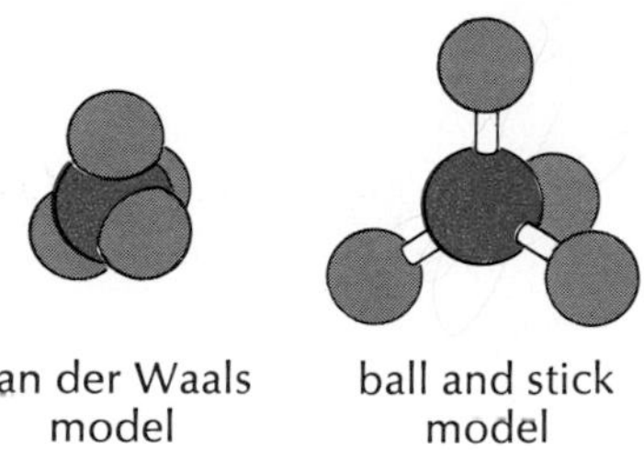

Figure 16.1
Methane.

gas." The molecular formula of methane is CH_4. The four hydrogen atoms surround the central carbon and are covalently bound to it. As in diamond the angle between any two bonds is 109° 28′, the tetrahedral angle. The molecule appears in Figure 16.1. The first form, called the *van der Waals model,* depicts the dimensions of the atoms. The second, a *ball and stick model* in which the bond lengths are distorted and the atoms are pictured as spheres, emphasizes the geometrical shape of the molecule but is unrealistic in the sense that bonded atoms are not spherical and are not separated in space by stick bonds.

Since the electron configuration of carbon is $1s^2 2s^2 2p^2$, it might be expected that only two single bonds would be formed by sharing the electrons in the $2p$ orbitals. These bonds would be approximately perpendicular to one another, and carbon would have six valence electrons rather than a full octet. In order to form four bonds it is necessary to utilize all four valence electrons. To account for the four identical bonds, valence electrons of carbon in methane are considered to be sp^3 hybridized. The sp^3 orbitals, each with one electron, are directed toward the corners of a regular tetrahedron where the hydrogen atoms with their single electrons in s orbitals can overlap to produce four covalent bonds. All four bonds are identical, and the hydrogens are indistinguishable. The electron density in each *sigma* (σ) *bond* is symmetrical about a line connecting the centers of the two bonded atoms.

It is convenient to devise a method of representing the three-dimensional molecules of the alkane series in two dimensions. Methane will be utilized to illustrate this method. Placing two hydrogens on the plane of a page and projecting the molecule onto the page produces a two-dimensional representation of methane as shown in Figure 16.2. Remember that all four hydrogens directed up and down in Figure 16.2 are the same as those directed left and right. This fact will become more important to remember when dealing with higher members of the alkane series.

ETHANE

The second member of the alkane series, *ethane,* has the molecular formula C_2H_6. The ethane molecule consists of two carbon atoms bound to

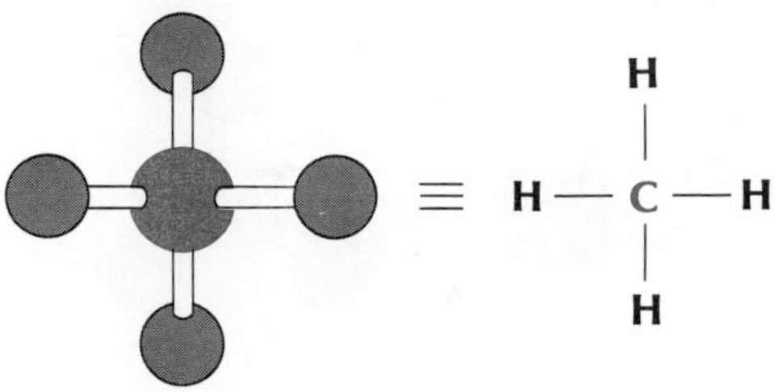

Figure 16.2
Planar projection of methane.

each other by overlap of sp^3 hybrid orbitals and separated by a distance of 1.54 Å. The hydrogens are bound by means of the remaining three sp^3 hybrid orbitals of each carbon atom. In Figure 16.3 the ethane molecule is represented using both the van der Waals and the ball and stick models. Projection into the plane of a page results in a planar structure that is used for simplicity. The three dimensionality of the molecule, however, must be held in mind in order to understand the reactions of organic chemistry. The three hydrogens attached to each carbon are identical and cannot be physically or chemically distinguished from one another. Furthermore, both carbon atoms are identical and, therefore, all six hydrogen atoms are equivalent. Both carbon atoms may rotate about the internuclear axis because the only requirement for bonding between the carbon atoms is that the two cylindrically symmetrical sp^3 orbitals be directed along the same line. There are no constraints placed on the locations of the hydrogen atoms on each carbon relative to those on the neighboring carbon. Again it is emphasized that each of the three hydrogens on each carbon atom in the planar projection are identical. The hydrogens directed up and down in the plane of the page are identical to those directed left and right in the same plane.

PROPANE

The third member of the alkane series is *propane* (C_3H_8). Its structure can be visualized as being formed by the attachment of one additional carbon atom to the two-carbon skeleton of ethane. Since all orbitals emanating

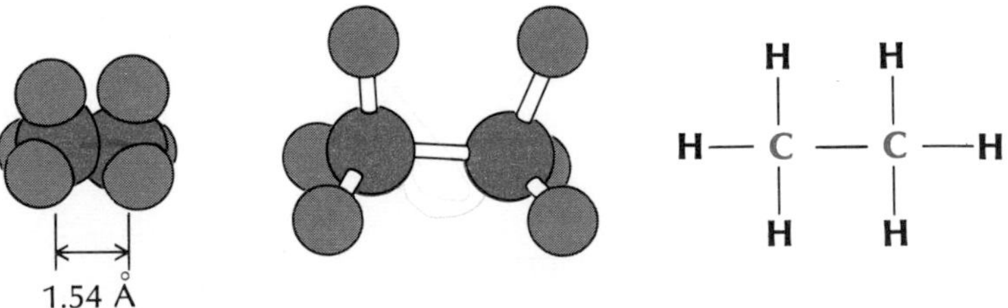

Figure 16.3
Structural representations of ethane.

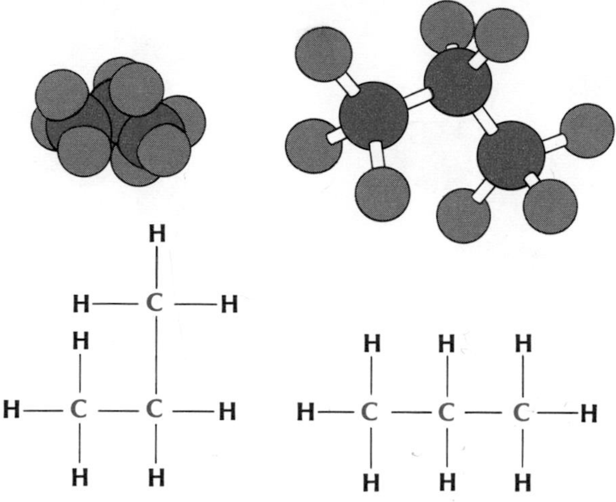

Figure 16.4
Structural representations of propane.

from each carbon in the ethane skeleton are identical, the replacement of a hydrogen atom by a carbon atom will yield only one possible structure. Propane and all the other high molecular mass alkanes are written in the planar form so that the carbon skeleton defines a straight line. Therefore, while the two planar representations for propane drawn in Figure 16.4 are equally correct, it is less confusing to write the linear arrangement. The angles between all bonds in propane are nearly 109° and not 90 and 180° as they appear in the two planar projections. The structure on the left with a carbon and its three hydrogens directed up from a two-carbon, ethane-like skeleton is really the same as the linear arrangement at the right.

Not all the hydrogen atoms in propane are equivalent. Since the carbon atoms that occupy the terminal positions are indistinguishable, the six hydrogens attached to them are equivalent. However, the hydrogens in the center of the molecule are obviously different in that they are attached to an interior carbon atom. It should then be expected that replacement of any of the six exterior hydrogens would lead to a different species than would replacement of the two interior hydrogens.

Propane, which is gaseous at standard conditions, can be liquified and stored under pressure. It serves as a fuel for heating and cooking in the home and is used by explorers and weekend campers in portable stoves and refrigerators.

BUTANES

The molecular formula C_4H_{10} corresponds to two isomeric compounds. In order to distinguish between the two compounds a structural representation is necessary. Replacing a hydrogen atom of the propane

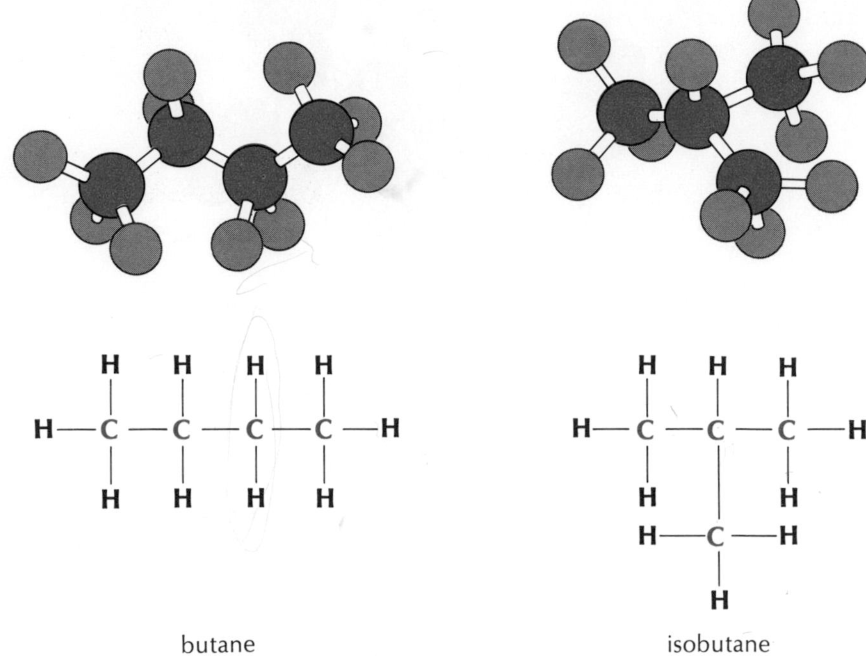

Figure 16.5
The isomeric butanes.

molecule by a carbon atom bearing three hydrogens can lead to two different compounds, because there are two different sets of hydrogens attached to propane. The two-dimensional representations of these compounds are given in Figure 16.5 along with the three-dimensional ball and stick pictures.

Butane, in which all carbons are connected in a continuous, nonbranching chain, is called a *normal alkane,* whereas *isobutane* is classified as a *branched alkane*. The branch involves a CH_3 unit on the second carbon of a chain of three carbon atoms. Butane and isobutane are often written in a more compressed form in order to save time:

$CH_3CH_2CH_2CH_3$	$CH_3CH(CH_3)CH_3$ **or** $CH_3CH(CH_3)_2$
butane	isobutane

Examination of butane reveals that there are two different sets of hydrogens. Six hydrogens are located on the terminal carbons and are identical because there is no way of distinguishing one end from the other. The two internal carbon atoms are identical and, therefore, the four hydrogens attached to them also are identical. In isobutane there also are two different sets of hydrogens. Attached to the central atom are three identical CH_3 units and a single hydrogen. The nine hydrogens of the three CH_3 units form one set, and the lone hydrogen on the central carbon is the only member of a second set.

Addition of one carbon to each of the two butane carbon skeletons at the various nonidentical positions leads to the structures of the three isomeric pentanes with molecular formula C_5H_{12}:

```
   H  H  H  H              H  H  CH3 H             H  CH3  H
   |  |  |  |              |  |  |   |             |  |    |
H—C—C—C—C—CH3           H—C—C—C———C—H           H—C—C———C—H
   |  |  |  |              |  |  |   |             |  |    |
   H  H  H  H              H  H  H   H             H  CH3  H

                                   CH3                  CH3
                                   |                    |
CH3CH2CH2CH2CH3           CH3CH2—C—CH3           CH3—C—CH3
                                   |                    |
                                   H                    CH3

     pentane                  isopentane            neopentane
```

Pentane, which is a normal alkane, can be drawn by replacing any of the exterior hydrogens of butane with a CH_3 unit. Replacing an interior hydrogen by a CH_3 unit yields *isopentane,* a branched alkane. Isopentane also can be drawn by replacing any of the set of nine equivalent hydrogens of isobutane by a CH_3 unit. The structures produced by each method are identical, although they have been derived in two different ways. *Neopentane* is derived by replacement of the single hydrogen on the central carbon atom of isobutane by a CH_3 unit.

At this time it should be reemphasized that the chains of carbon atoms are not straight. The actual molecules are free to rotate internally and assume many geometric shapes in space within the confines of the structural relationships represented by the bonds of the projection formulas.

16.3 NOMENCLATURE

Before additional members of the alkane series are considered, two things should be pointed out. First, the number of possible isomers increases rapidly as the number of carbon atoms involved increases. A listing of the isomers as a function of molecular formulas is given in Table 16.1. While only a few of the possible isomers of the high molecular mass alkanes have been isolated from nature or prepared in the laboratory, the knowledge necessary to prepare all of them is available if there were a reason to do so. Second, it is necessary to consider how to name the five isomers having the molecular formula C_6H_{14} and the even greater number of isomers for compounds of higher molecular mass. The names given to the isomeric C_4H_{10} and C_5H_{12} compounds can be committed to memory, but it would be difficult and indeed undesirable to give a trivial name to each compound.

The International Union of Pure and Applied Chemistry (IUPAC), has developed a systematic method for naming all organic compounds. Like

Table 16.1
Number of isomers of alkanes

molecular formula	*number of isomers*
CH_4	1
C_2H_6	1
C_3H_8	1
C_4H_{10}	2
C_5H_{12}	3
C_6H_{14}	5
C_7H_{16}	9
C_8H_{18}	18
C_9H_{20}	35
$C_{10}H_{22}$	75
$C_{20}H_{42}$	336,319
$C_{30}H_{62}$	4,111,846,763

many complex regulations it sometimes leads to cumbersome results, but the names derived by this system are unique for a given compound. According to the rules, a name can be assigned to a given structure or the correct structure can be written from a name. Rules for naming alkanes by the IUPAC system are listed below:

1. *The longest continuous chain of carbon atoms is selected as the basis for the name. This chain is named according to the stem name plus the suffix -ane as listed in the following series.* Beyond the fourth member of the series the stems are derived from Greek.

$\mathbf{CH_4}$	*methane*	$\mathbf{C_7H_{16}}$	*heptane*
$\mathbf{C_2H_6}$	*ethane*	$\mathbf{C_8H_{18}}$	*octane*
$\mathbf{C_3H_8}$	*propane*	$\mathbf{C_9H_{20}}$	*nonane*
$\mathbf{C_4H_{10}}$	*butane*	$\mathbf{C_{10}H_{22}}$	*decane*
$\mathbf{C_5H_{12}}$	*pentane*	$\mathbf{C_{11}H_{24}}$	*undecane*
$\mathbf{C_6H_{14}}$	*hexane*	$\mathbf{C_{12}H_{26}}$	*dodecane*

2. *The carbon atoms in the chain are numbered consecutively from that end of the chain nearest a branch.*
3. *Each branch is located by the number of the atom to which it is attached on the chain.* These branches are called *alkyl* groups and are named by the stem name plus the ending *-yl.*

$$\overset{4}{\mathbf{CH}}\ \overset{3}{\mathbf{CH_2}}{-}\overset{\overset{\mathbf{CH_3}}{|}}{\underset{\underset{\mathbf{H}}{|}}{\overset{2}{\mathbf{C}}}}{-}\overset{1}{\mathbf{CH_3}} \quad \text{is 2-methylbutane}$$

4. *If two or more of the same types of branches occur, the number of them is indicated by the prefixes di-, tri-, tetra-, etc. and the location of each on the main chain is indicated by a number.*

$$\begin{array}{c} CH_3 \\ {}^1CH_3-{}^2CH-{}^3CH-{}^4CH_3 \\ CH_3 \end{array}$$ is 2,3-dimethylbutane

5. *The numbers designating the positions of the alkyl groups are placed immediately before the names of the groups, and hyphens are placed before and after the numbers.* If two or more numbers occur together commas are placed between them.
6. *In the case of several alkyl groups being present, they are placed in alphabetical order and prefixed onto the name of the basic alkane.* The whole is written as a single word.

While the systematic names are more cumbersome than the trivial names, the structure corresponding to the IUPAC name can be written readily even if that particular name is unfamiliar. The name contains all the information needed to specify the complete structure. The rules can be applied to complex alkanes as well, as illustrated for the compound written below:

$$\begin{array}{c} HCH_3H \\ {}^1CH_3{}^2CH_2-{}^3C-{}^4C-{}^5CH_2{}^6CH_2-{}^7C-{}^8CH_2{}^9CH_2{}^{10}CH_3 \\ CH_2HCH_2 \\ CH_3CH_3 \end{array}$$

First: Because the longest chain has ten carbons this is a *decane.*

Second: The chain is numbered from left to right since the branch nearest the end is located on the left. This numbering will result in the branches being assigned low numbers.

Third: There are ethyl groups on carbon 3 and 7 and, therefore, these are named and numbered as *-3,7-diethyl.*

Fourth: A methyl group is located at position 4 and is named *-4-methyl.*

Fifth: Assembling all components of the name we have *3,7-diethyl-4-methyldecane.* The hyphen preceeding the 3 is dropped.

Two compounds with their correct names further illustrate the naming process:

$$\begin{array}{c} CH_3 \\ CH_2CH_3CH_3 \\ {}^9CH_3{}^8CH_2{}^7CH_2-{}^6C-{}^5CH_2-{}^4C-{}^3C-{}^2CH_2{}^1CH_3 \\ CH_3CH_3H \end{array}$$

6-ethyl-3,4,4,6-tetramethylnonane

```
                         CH3
                          |
             CH3   CH3   CH2
  1      2   3|    4|    5|   6   7   8
CH3—CH2—C——C——C—CH2CH2CH3
              |     |     |
             CH3   CH2   CH3
                    |
                   CH3
```

4,5-diethyl-3,3,4,5-tetramethyloctane

Care should be exercised to locate the longest possible straight chain and also to number the chain correctly. The correct name and a commonly encountered incorrect one are given for two examples. Only if the rules are carefully applied can each structure correspond to a unique name and each name correspond to a unique structure.

```
                 CH3
 5   4   3       2|  1
CH3CH2CH2—C—CH3
                  |
                 CH3
```

2,2-dimethylpentane
(*not* 4,4-dimethylpentane)

```
        H     CH3
       3|    4|    5   6
CH3—C——C—CH2CH3
       2|     |
       CH2    H
       1|
       CH3
```

3,4-dimethylhexane
(*not* 2-ethyl-3-methyl pentane)

A structural formula can be reconstructed from a name, such as 3,3-diethyl-4-propyl-2,4,5-trimethylheptane.

1. Draw a heptane skeleton and number it.

```
C—C—C—C—C—C—C
1  2  3  4  5  6  7
```

2. Place two ethyl groups on the chain at carbon atom 3.

```
      CH2CH3
      |
C—C—C—C—C—C—C
      |
      CH2CH3
```

3. Place a propyl group on carbon atom 4.

```
             CH3
              |
      CH3    CH2
       |      |
      CH2    CH2
       |      |
C—C—C——C—C—C—C
       |
      CH2CH3
```

4. Place three methyl groups, one each, on carbon atoms 2, 4, and 5.

```
                 CH3
                  |
            CH3  CH2
             |    |
      CH3   CH2  CH2
       |     |    |
   C—C———C———C———C—C—C
             |    |    |
            CH2  CH3  CH3
             |
            CH3
```

5. Fill in the necessary hydrogens so that each carbon atom has four bonds.

```
                 CH3
                  |
            CH3  CH2
             |    |
      CH3   CH2  CH2   H
       |     |    |    |
 CH3—C———C———C———C—CH2CH3
       |     |    |    |
       H    CH2  CH3  CH3
             |
            CH3
```

16.4 PHYSICAL PROPERTIES OF ALKANES

The normal alkanes form a set of compounds called a *homologous series.* Each member of the series differs from the preceding or following compound by one CH_2 unit. Such a series of compounds provides an interesting picture of the relationship between structure and physical properties. The melting points, boiling points, and densities of the liquids at 20° for the straight chain alkanes are listed in Table 16.2.

Table 16.2
Physical properties of alkanes

name	*melting point (°C)*	*boiling point (°C)*	*density at 20° (g/ml)*
methane	−182	−162	
ethane	−183	−89	
propane	−138	−42	
butane	−138	−1	
pentane	−130	36	0.626
hexane	−95	69	0.659
heptane	−91	98	0.685
octane	−57	126	0.702
nonane	−54	151	0.718
decane	−30	171	0.730

The properties of compounds are functions of the atomic composition, molecular mass, size, and shape of the molecules and the bond types. Since the alkanes are all composed of carbon and hydrogen, many possible variables are eliminated from consideration in discussing the observed trends in physical properties.

Boiling points reflect the size and shape of molecules and the polarity of the bonds in these molecules. Alkanes contain only nonpolar carbon-carbon and carbon-hydrogen bonds, and, therefore, the boiling points should be controlled by molecular mass and shape of the molecules. The expected increase in the boiling point between homologs is noted for compounds that are systematically of longer chain length (Figure 16.6). This increase is larger for the smaller members of the series where the additional CH_2 unit represents a larger fraction of the total molecule. The increase in the boiling point gradually becomes smaller and is nearly constant at 13° per $—CH_2—$ unit when approximately 20 carbons make up the chain.

In Table 16.3 the boiling points of the isomeric hexanes are listed. Since the molecular masses are identical, the differences must be due to molecular shape. With increasing branching the boiling point decreases. The branched compounds are more compact and do not become

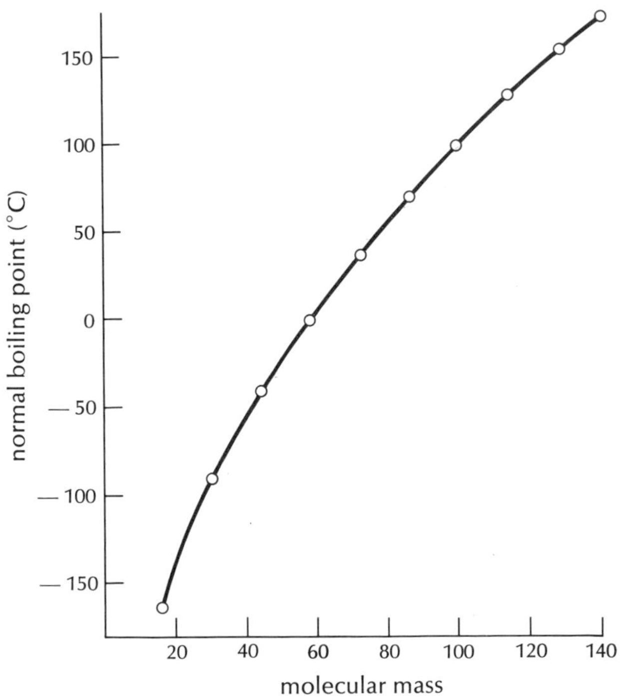

Figure 16.6
Effect of molecular mass on normal boiling point of alkanes.

Table 16.3
Boiling points of C_6H_{14} isomers

compound	*boiling point* (°C)
hexane	69
3-methylpentane	63
2-methylpentane	60
2,3-dimethylbutane	58
2,2-dimethylbutane	50

entangled with one another, resulting in a decrease in the van der Waals attractive forces that decreases in turn the boiling point.

The density of compounds in the liquid state is a function of the kinds of atoms contained in the molecules and the ratios among the different types. In the alkanes only two different types of atoms are present. However, the ratio of the light hydrogen atoms to the dense carbon atoms decreases with increasing chain length. In methane the ratio is 4:1. As the chain length increases the ratio gradually approaches 2:1. That the density of the alkanes increases and finally reaches a limiting constant value of 0.777 g/ml is a reflection of the limiting hydrogen-to-carbon ratio.

16.5 CHEMICAL PROPERTIES OF ALKANES

As a class, the alkanes are very unreactive compounds. The carbon–carbon and carbon–hydrogen bonds resist attack by both acids and bases and are insensitive to moderate heat. The few reactions that are discussed here occur with all alkanes. This general principle is the saving feature of attempting to cope with the 3.5 million organic compounds presently known. Classification of molecules according to structural types decreases the number of facts that must be memorized. In the remaining chapters this principle of classification of reactions will be used many times.

A second feature of organic chemistry that is frequently encountered is the concept of *minimum structural rearrangement.* Although the hydrocarbon skeleton does rearrange under some reaction conditions, most reactions occur with little or no change in the carbon framework. Therefore, throughout the remainder of the organic chemistry section of this text most of the reactions presented will be discussed in terms of the groups attached to the carbon framework, which itself does not undergo structural change.

COMBUSTION

In the presence of sufficient oxygen all alkanes will burn if ignited, yielding carbon dioxide and water. The number of calories liberated per mole

Table 16.4
Heats of combustion of alkanes

compound	*kcal/mole*	*increment*
pentane	833	
hexane	990	157
heptane	1150	160
octane	1303	153
nonane	1457	154
decane	1611	154

of alkane is called the *heat of combustion,* which for methane is 211 kcal/mole:

$$CH_4 + 2O_2 \longrightarrow CO_2 + 2H_2O \qquad \Delta H = -211 \text{ kcal/mole}$$

The heats of combustion of the liquids pentane through decane are tabulated in Table 16.4. A nearly constant increase per —CH_2— unit of 156 kcal/mole exists between homologs. As addition of a —CH_2— unit requires breaking more bonds and producing additional molecules of carbon dioxide and water, the constant increase is expected.

EXAMPLE 16.1

Estimate the heat of combustion of octadecane ($C_{18}H_{38}$).

Octadecane differs from decane by eight —CH_2— units. The heat of combustion of octadecane should be 8(156) kcal/mole larger than that of decane, or 2859 kcal/mole:

$$1611 + 8(156) = 2859$$

ISOMERIZATION

While the alkanes as a class of compounds are inert toward most acids, they react quite readily with hydrogen chloride in the presence of Lewis acids (Section 10.5) such as aluminum chloride. Starting from either butane or 2-methylpropane, the same equilibrium mixture of the two compounds is obtained:

$$\underset{\text{20 percent}}{CH_3CH_2CH_2CH_3} \xrightleftharpoons{HCl/AlCl_3} \underset{\text{80 percent}}{CH_3—\overset{\overset{CH_3}{|}}{\underset{\underset{H}{|}}{C}}—CH_3}$$

The larger amount of 2-methylpropane for this equilibrium process indicates that it is more stable than butane. Isomerization of higher alkanes produces complex equilibrium mixtures of alkanes in which the branched isomers predominate. Processes of this type are important to petroleum companies where it is necessary to produce the proper mixtures of compounds for use in industry.

CRACKING

Pyrolysis, the destructive heating of methane in the absence of oxygen, produces hydrogen and carbon. Higher molecular mass alkanes are cracked into smaller molecules. For example, tetradecane can produce hexane and an unsaturated hydrocarbon called 1-octene (Chapter 17).

$$\underset{\text{tetradecane}}{CH_3{-}(CH_2)_5{-}CH_2CH_2CH_2{-}(CH_2)_4{-}CH_3} \rightarrow$$

$$\underset{\text{1-octene}}{CH_3{-}(CH_2)_5{-}CH{=}CH_2} + CH_3{-}(CH_2)_4{-}CH_3$$

The indicated cleavage is not specific, and a large number of other alkanes and unsaturated compounds are produced as the result of cleavage between different carbon-carbon bonds. The petroleum industry has succeeded in devising procedures to produce the desired low molecular mass hydrocarbons from higher molecular mass materials not in commercial demand.

HALOGENATION

Alkanes react with halogens to produce hydrogen halides and organic products in which halogen atoms have been substituted for hydrogen atoms. The reaction is called a *substitution reaction* and is illustrated by the reaction of methane and chlorine to yield chloromethane:

$$CH_4 + Cl_2 \longrightarrow \underset{\text{chloromethane}}{CH_3Cl} + HCl$$

The reaction does not occur with the pure compounds in the absence of light, but in sunlight the process occurs readily.

Quantities of compounds containing more than one chlorine atom per molecule also are obtained in the chlorination of methane. The chloromethane formed in the early states of the reaction can continue to react with chlorine to yield CH_2Cl_2, $CHCl_3$, and CCl_4, which can be separated by distillation:

$$CH_3Cl + Cl_2 \rightarrow HCl + \underset{\text{dichloromethane}}{CH_2Cl_2}$$

$$CH_2Cl_2 + Cl_2 \rightarrow HCl + \underset{\substack{\text{trichloromethane} \\ \text{(chloroform)}}}{CHCl_3}$$

$$CHCl_3 + Cl_2 \rightarrow HCl + \underset{\substack{\text{tetrachloromethane} \\ \text{(carbon tetrachloride)}}}{CCl_4}$$

EXAMPLE 16.2

Propane reacts with chlorine to yield a mixture of two isomers having the molecular formula C_3H_7Cl. *Explain this observation.*

In propane there are two sets of nonequivalent hydrogen atoms. The six hydrogen atoms attached to the terminal carbon atoms are structurally different from those attached to the center carbon atom. Therefore, replacement of hydrogen by chlorine can lead to two possible products:

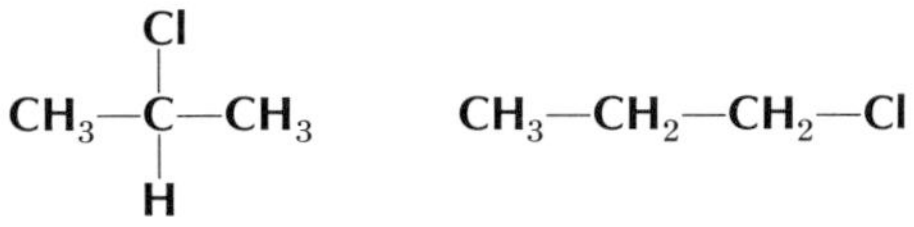

$$CH_3{-}C(Cl)(H){-}CH_3 \qquad CH_3{-}CH_2{-}CH_2{-}Cl$$

EXAMPLE 16.3

How many monochlorinated products can result from the chlorination of 2-methylbutane?

There are four sets of nonequivalent hydrogens in 2-methylbutane. Note that two methyl groups are equivalent in the following structure:

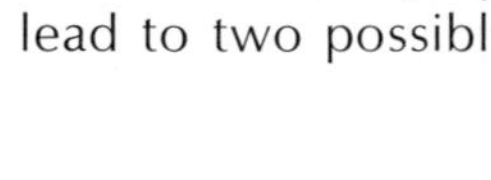

$$\underset{1}{CH_3}{-}\underset{2}{CH_2}{-}\underset{3}{CH}(\overset{4}{CH_3}){-}\underset{4}{CH_3}$$

Replacement of hydrogen by chlorine can result in four compounds of the molecular formula C_4H_9Cl:

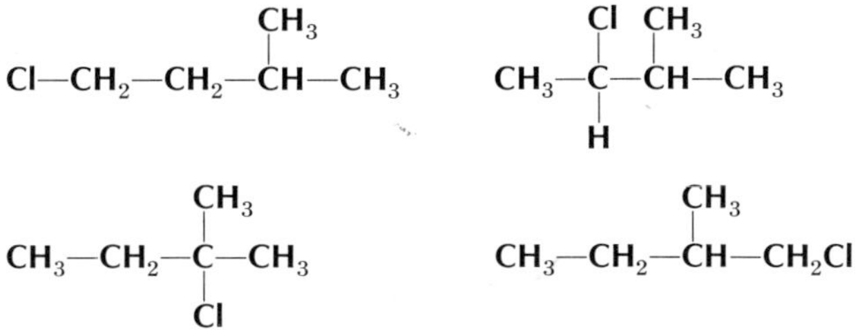

$$Cl{-}CH_2{-}CH_2{-}CH(CH_3){-}CH_3 \qquad CH_3{-}C(Cl)(H){-}CH(CH_3){-}CH_3$$

$$CH_3{-}CH_2{-}C(CH_3)(Cl){-}CH_3 \qquad CH_3{-}CH_2{-}CH(CH_3){-}CH_2Cl$$

16.6 MECHANISM OF CHLORINATION

The component of sunlight that initiates the chlorination reaction is a certain frequency of ultraviolet light whose energy corresponds to that required to cleave the chlorine molecule into chlorine atoms. The reaction with methane has been shown to proceed from the chlorine atoms. The detailed description of the reaction in terms of its separate steps (the reaction mechanism) is the following:

$$Cl_2 \longrightarrow 2\ \cdot\overline{\underline{Cl}}| \qquad \text{initiation step}$$

$$\left.\begin{aligned} |\overline{\underline{Cl}}\cdot + CH_4 &\longrightarrow \cdot CH_3 + H{-}Cl \\ \cdot CH_3 + Cl_2 &\longrightarrow CH_3Cl + \cdot\overline{\underline{Cl}}| \end{aligned}\right\} \quad \text{propagation steps}$$

The cleavage of the chlorine molecule into chlorine atoms is called the *initiation step* and is the prerequisite for the reaction to occur. The chlorine atom achieves a stable octet in the second step by abstracting a hydrogen atom from a methane molecule. In the third step the methyl radical $\cdot CH_3$ combines with a chlorine atom from the chlorine molecule, releasing another chlorine atom. The second and third steps are capable of sustaining the reaction if allowed to proceed unimpeded and are called *propagation steps.* Neither step requires much energy, since in each case the bond that is broken and the one that is formed are of comparable energy. In each step a species with an unpaired electron, a *radical,* and a stable molecule react to produce a radical and a neutral molecule. A small number of chlorine molecules rupturing to yield chlorine atoms thus leads to formation of a considerable amount of product. Without additional light the process will eventually slow down and stop as the result of *termination* steps in which radicals combine to form stable compounds:

$$\begin{aligned} 2|\overline{\underline{Cl}}\cdot &\longrightarrow Cl_2 \\ 2CH_3\cdot &\longrightarrow CH_3CH_3 \\ CH_3\cdot + |\overline{\underline{Cl}}\cdot &\longrightarrow CH_3Cl \end{aligned}$$

The likelihood of termination steps occurring is small compared to that of propagation steps. The concentration of free radicals is always low because formation of a radical requires destruction of a radical in the propagation steps. Since the concentration of methane and chlorine is high, reaction with these substances by radicals is more likely. The mechanism described for the chlorination of methane is one example of a *chain mechanism.*

The reaction of chlorine with propane illustrates some basic reactivity concepts that can be applied in later chapters. Propane possesses two sets of nonequivalent hydrogens. Consequently, abstraction of hydrogen atoms from propane by chlorine atoms can lead to two different products, the isomeric 1-chloropropane and 2-chloropropane, which are produced in equimolar amounts:

$$CH_3CH_2CH_3 + Cl_2 \longrightarrow \underset{\text{50 percent}}{CH_3CH_2CH_2Cl} + \underset{\text{50 percent}}{CH_3{-}\overset{\overset{\displaystyle Cl}{|}}{\underset{\underset{\displaystyle H}{|}}{C}}{-}CH_3}$$

Because there are six exterior hydrogens and only two interior ones, it is statistically three times more likely that an exterior hydrogen will be abstracted to yield 1-chloropropane. The higher than expected yield of 2-chloropropane compared to 1-chloropropane suggests that some factor in addition to statistics is controlling the reaction. It must be energetically more favorable to abstract a single interior hydrogen as compared to a single exterior hydrogen. If this is the case then the two following reactions must be considered to supply a rationale for this difference:

$$CH_3CH_2CH_3 + :\ddot{\underset{..}{Cl}}\cdot \longrightarrow HCl + CH_3CH_2CH_2\cdot$$

$$CH_3CH_2CH_3 + :\ddot{\underset{..}{Cl}}\cdot \longrightarrow HCl + CH_3-\underset{\substack{|\\H}}{\dot{C}}-CH_3$$

The sole difference between the two reactions is the structure of the two alkyl radicals. If the radical with the unpaired electron on the interior carbon atom is more stable than the radical with the odd electron at the exterior carbon atom, then the second of the two abstraction reactions will be energetically favored, although statistically less likely. In order to account for the observed experimental results the reactivity of the interior hydrogen must be three times that of a single exterior hydrogen. The relative amount of product derived from attack at an exterior carbon is a product of the relative unit reactivity multiplied by the number of equivalent hydrogens, or $6(1) = 6$. The relative amount of product derived from attack at an interior carbon is $2(3) = 6$. The ratio of the two products is therefore 1:1. Only with the assumption of a 3:1 ratio of reactivities of hydrogens at the two positions can the experimental data be rationalized.

The difference in stability between two radicals derived from propane can be rationalized in terms of the ability of atoms or groups of atoms attached to the carbon containing the unpaired electron to supply electrons to the electron deficient center. If each of the bonded pairs of electrons attached to the electron deficient carbon are on the average closer to the atom than in the hydrocarbon, its deficiency would be partially relieved. Thus any group that is able to release electrons will stabilize the radical. The observed stability of the radical in which the central atom is electron deficient indicates that the alkyl groups are better able to release electrons via the bonds than is hydrogen. We shall accept this conclusion as one that will be encountered several times in other reactions and is a basic determinant in organic chemistry.

The terminal carbon atom is connected to a single alkyl fragment and is called the *primary* carbon atom. Hydrogens attached to the primary carbon are called *primary hydrogens,* and the radical that results from abstraction of a primary hydrogen is called a *primary radical.* The interior carbon atom is a *secondary* carbon atom if it is attached to two alkyl groups. *Secondary hydrogens* are located on this carbon, and abstraction leads to the formation of a *secondary radical.* A carbon atom attached to three alkyl groups is called *tertiary,* and abstraction of a *tertiary hydrogen* leads to a *tertiary radical.*

Chlorination of 2-methylpropane produces 1-chloro-2-methylpropane and 2-chloro-2-methylpropane in the ratio of 2:1, whereas statistically a 9:1 ratio would be expected.

$$CH_3-\underset{H}{\overset{CH_3}{C}}-CH_3 + Cl_2 \longrightarrow \underset{\text{66 percent}}{CH_3-\underset{H}{\overset{CH_3}{C}}-CH_2Cl} \qquad \underset{\text{34 percent}}{CH_3-\underset{Cl}{\overset{CH_3}{C}}-CH_3}$$

Therefore, abstraction of a tertiary hydrogen to yield a tertiary radical is energetically more favorable than abstraction of a primary hydrogen by a reactivity factor of 4.5. The order of stability of radicals is tertiary > secondary > primary and is a reflection of the number of electron-releasing alkyl groups attached to the carbon atom containing the unpaired electron.

16.7 PREPARATION OF ALKANES

While alkanes are commonly obtained by fractional distillation of petroleum, it is often necessary to prepare for specific purposes compounds that are not available in large quantities in petroleum. The methods then employed involve classes of compounds that have not been examined as yet. Since the necessary background has not been laid, the discussion of methods of preparation will be brief. When we have discussed other classes of organic compounds, the number of experimental approaches to the synthesis of a given class of compounds, such as alkanes, will increase rapidly. The synthetic methods for the preparation of alkanes to be considered now utilize alkyl halides, which in turn can be prepared from alkanes.

THE WURTZ REACTION

The French chemist Charles Wurtz discovered in 1855 that iodomethane, when treated with sodium metal, produces ethane. The mechanism of this reaction is complex and not well understood:

$$2CH_3I + 2Na \longrightarrow CH_3CH_3 + 2NaI$$

By analogy it would be expected that the reaction of iodoethane or bromoethane with sodium would yield butane, which does in fact proceed in fair yield.

$$2CH_3CH_2Br + 2Na \longrightarrow CH_3CH_2CH_2CH_3 + 2NaBr$$

The success of reasoning by analogy permits the prediction of the products of specific combinations of reactants that we have not encountered previously. The organization of organic chemistry depends on the cataloging of reactions according to classes that have common characteristics and can be explained by common mechanisms.

It should be noted that in the formation of both ethane and butane symmetrical compounds containing twice as many carbons as the original organic molecules are obtained. The carbon atoms join at the point at which the halogen was attached originally. The *Wurtz reaction,* however, has serious limitations in the preparation of hydrocarbons containing an odd number of carbon atoms. If a mixture of iodomethane and iodoethane is treated with sodium, ethane and butane are produced as well as propane:

$$2Na + CH_3I + C_2H_5I \longrightarrow CH_3CH_2CH_3 + CH_3CH_3 + CH_3CH_2CH_2CH_3 + 2NaI$$

The problem of isolating the desired product limits the usefulness of the Wurtz reaction.

THE GRIGNARD REAGENT

The French chemist Victor Grignard discovered that alkyl halides react with magnesium in dry diethyl ether (Chapter 19) to yield a solution of a substance that has come to be known as the *Grignard reagent.* The exact structure of these reagents is still a point of controversy a half century after their discovery. The reaction of 1-bromobutane with magnesium is illustrated using the commonly accepted structural representation of the Grignard reagent:

$$CH_3CH_2CH_2CH_2Br + Mg \xrightarrow{\text{diethyl ether}} \underset{\text{Grignard reagent}}{CH_3CH_2CH_2CH_2\text{—}MgBr}$$

Grignard reagents are very reactive and serve as intermediates in the synthesis of many classes of organic compounds. Some of the reactions the reagents undergo are discussed in Chapter 20.

Water must be excluded from contact with the Grignard reagents as they readily react with it. The product of the reaction is the hydrocarbon derived by replacement of the MgBr fragment with hydrogen. Therefore, the synthesis of hydrocarbons can be accomplished by deliberately adding water to the solution containing the Grignard reagent:

$$CH_3CH_2CH_2CH_2\text{-}MgBr + H_2O \longrightarrow CH_3CH_2CH_2CH_3 + MgBr(OH)$$

16.8 CYCLOALKANES

STRUCTURE

In addition to arranging carbon atoms in a continuous or branched chain it is possible to envision rings of carbon atoms. Such compounds are

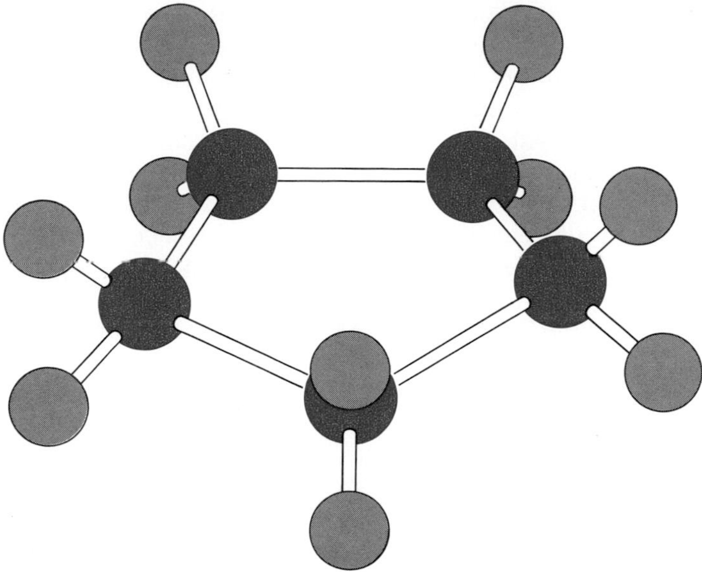

Figure 16.7
Cyclopentane.

known as *cycloalkanes,* and their general formula is C_nH_{2n}. The deficiency of two hydrogens, compared to the alkanes, is necessary because the formation of a ring requires that two carbons of an alkane be joined by bonds that are occupied by hydrogen atoms in the alkane.

Using ball and stick models the smallest cycloalkane that can be constructed is cyclopentane (Figure 16.7). A pentagon whose internal angle closely corresponds to the tetrahedral angle results. The ten hydrogens radiate outward both above and below the plane of the ring and are equivalent. Construction of larger rings results in flexible nonplanar structures that can assume numerous geometrical orientations.

STRAIN ENERGY

Cyclobutane and cyclopropane are known compounds that cannot be constructed from ball and stick models. In order to arrange the atoms in a ring it is necessary to alter the bond angles that carbon customarily shows when bonded to four other atoms. The internal angles of a square and an equilateral triangle deviate drastically from the tetrahedral angle of 109.5°. These molecules must be somewhat unstable as a result of the abnormal bonding required in these compounds and are said to be strained.

```
      H   H                  H     H
       \ /                   |     |
        C                 H—C-----C—H
 H     / \     H             |     |
  \   /   \   /           H—C-----C—H
   C-------C                 |     |
  /         \                H     H
 H           H
```

The energy that is expended in forming these strained ring systems is incorporated into the resultant molecules and can be released if the ring is ruptured. A measure of the *strain energy* is available from the heats of combustion of these compounds. An unstrained compound such as cyclohexane has a heat of combustion of 989 kcal/mole, a value that is close to that predicted for six CH_2 units. By contrast, the heat of combustion of cyclopropane is 496 kcal/mole, whereas the three CH_2 units of cyclopropane should produce 468 kcal/mole. The excess energy, 28 kcal/mole, is the strain energy of cyclopropane and is a reflection of the strain involved in three bond angles.

Because of their relative instability cyclobutane and cyclopropane undergo reactions that the other cycloalkanes do not. Cyclopropane reacts with hydrogen at 120° in the presence of a nickel catalyst; cyclobutane reacts with hydrogen at 200° The relative activation energies required to open cycloalkane rings are indicated by the temperatures necessary for the reaction:

$$\triangle + H_2 \xrightarrow[Ni]{120^\circ} CH_3CH_2CH_3$$

cyclopropane

$$\square + H_2 \xrightarrow[Ni]{120^\circ} CH_3CH_2CH_2CH_3$$

cyclobutane

In these equations an equilateral triangle and a square are used to represent cyclopropane and cyclobutane. Each corner of the figures represents a carbon atom. The hydrogens attached to the carbon atoms are not represented but are implied by the structures.

GEOMETRICAL ISOMERISM

A type of isomerism not observed in alkanes is possible in cycloalkanes. The two sides of the cycloalkane ring can be distinguished if the proper substituents are present. This phenomenon can be illustrated by considering the possible isomeric dimethylcyclopropanes. Since all the hydrogens in cyclopropane are identical, the replacement of one hydrogen by a methyl group leads to only one compound:

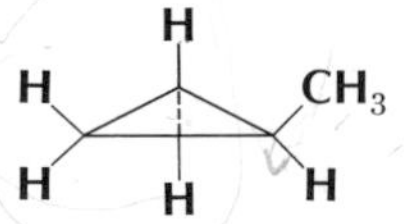

H H CH_3 H H H

The hydrogens in methylcyclopropane are not all identical. The ring carbon bearing the methyl group is tertiary and obviously different from the other carbons in the ring. Both secondary ring carbon atoms are identically situated with respect to the tertiary carbon, but the four secondary ring hydrogens form two sets of geometrically nonequivalent hydrogens. Two hydrogen atoms are on the same side of the ring as the methyl group and are called *cis;* hydrogens on the opposite side of the ring are called *trans.* There are three possible dimethylcyclopropanes because there are three different positions for the location of a second methyl group. These compounds are drawn below using a convention to indicate the geometry of the methyl groups relative to each other. The wedge-shaped bond is used to represent the top side of the ring and the dashed bond, the bottom side of the ring. The numbering system gives the substituents the lowest possible numbered positions:

CH_3 CH_3

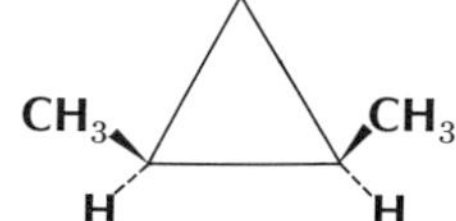

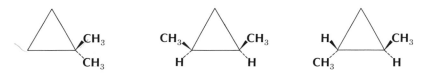

H CH_3 CH_3 H

1 1-dimethylcyclopropane *cis*-1,2-dimethylcyclopropane *trans*-1,2-dimethylcyclopropane

MULTIRING COMPOUNDS

The carbon skeleton of many biologically important substances contain multiring systems. For example, the *steroid* ring system that is characteristic of many biologically important compounds involves four rings that can be fused *cis* or *trans* with respect to each other:

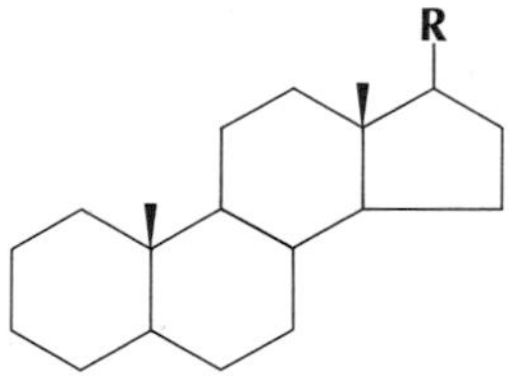

R

The lines at the juncture of two rings represent methyl groups; the R represents an alkyl group that may contain as many as eight carbon atoms. Cholesterol, for example, is a steroid substance that contains a hydroxyl group and a double bond. It is present in large amounts in the brain and spinal cord. Circulatory ailments such as hardening of the arteries are the result of deposition of excess blood cholesterol on the arterial walls. Other substances that contain the steroid ring system are ergosterol, which when irradiated by ultraviolet light produces vitamin D_2, commonly added to milk. The male and female sex hormones and substances controlling pregnancy are steroid derived materials (Chapter 26).

Many highly strained ring compounds have been synthesized in order to better understand the limitations of chemical bonding. A few of these are shown in Figure 16.8.

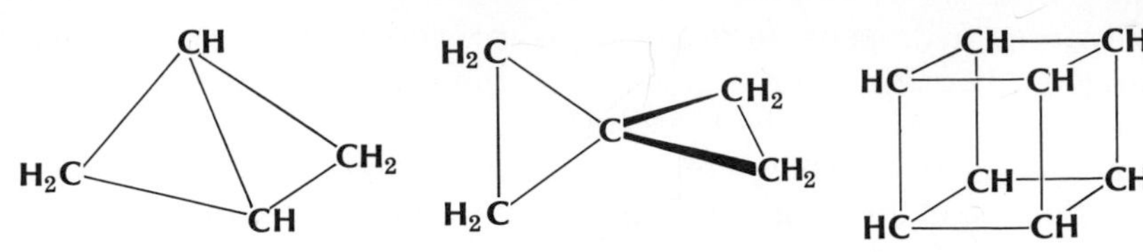

Figure 16.8
Strained multicyclic hydrocarbons.

Suggested further readings

Campaigne, E., "Wohler and the Overthrow of Vitalism," *J. Chem. Educ.*, **32,** 403 (1955).

Evieux, E. A., "The Geneva Congress on Organic Nomenclature, 1892," *J. Chem. Educ.*, **31,** 326 (1954).

Hartman, L., "Wohler and the Vital Force," *J. Chem. Educ.*, **34,** 141 (1957).

Hurd, C. D., "The General Philosophy of Organic Nomenclature," *J. Chem. Educ.*, **38,** 43 (1961).

Reinmuth, O., "Some Aspects of Organic Molecules and Their Behavior. II. Bond Energies," *J. Chem. Educ.*, **34,** 318 (1957).

Rheinboldt, H., "Fifty Years of the Grignard Reaction," *J. Chem. Educ.*, **27,** 476 (1950).

Rossini, F. D., "Hydrocarbons in Petroleum," *J. Chem. Educ.*, **37,** 554 (1960).

Schmerling, L., "The Mechanisms of the Reactions of Aliphatic Hydrocarbons," *J. Chem. Educ.*, **28,** 562 (1951).

Shoemaker, R. H., d'Ouville, E. L., and Marscher, R. F., "Recent Advances in Petroleum Refining," *J. Chem. Educ.*, **32,** 30 (1955).

Terms and concepts

alkanes
branched alkane
chain mechanism
cis
cycloalkane
geometrical isomerism
Grignard reagent
halogenation
heat of combustion
homologous series
initiation step
isomer
isomerization
IUPAC
mechanism
minimum structural rearrangement
normal alkane
primary

propagation step	substitution
radical	termination step
secondary	tertiary
sigma bond	*trans*
steroid	Wurtz
strain energy	

Questions and problems

1. Write the structural formula for each of the following compounds:
a. 3-methylpentane
b. 2,2,3-trimethylpentane
c. 1,1,2,2-tetramethylcyclobutane
d. 1,3-dibromo-2-methylpropane
e. 2,2-dimethyl-3-ethylheptane
f. 1,2-dibromo-3,3-dichloro-2-methylhexane

2. Indicate why each of the following names is incorrect and assign the correct name in each case:
a. 5-methylheptane
b. 1-iodo-1-methylpentane
c. 2,2,1-trichloroethane
d. 2-methylcyclopropane
e. 2-ethylbutane
f. 1,2-dimethylcyclobutane

3. Write the structural formulas for all of the isomers for each of the following formulas (the correct number of isomers is indicated in parentheses):
a. C_6H_{14} (5)
b. C_7H_{16} (9)
c. C_4H_9Br (4)
d. $C_2H_3Cl_3$ (2)
e. $C_3H_6Cl_2$ (4)
f. $C_3H_5Cl_3$ (5)

4. Draw and name all the possible isomeric dimethylcyclobutanes.

5. Draw and name all the possible cycloalkanes having the molecular formula C_5H_{10}.

6. A hydrocarbon of molecular formula C_4H_{10} may be either butane or isobutane. How could one of these substances be synthesized to provide a sample with which to compare and identify either of the two compounds.

7. Predict the percent composition of the monochlorinated materials derived from 2,3-dimethylbutane. Do the same thing with pentane.

8. Predict the heat of combustion of tetradecane ($C_{14}H_{30}$). Do the same thing for 1,2-dicyclopropylethane.

9. The heat of combustion of cyclobutane is 650 kcal/mole. What is the strain energy of cyclobutane?

10. What reactions might occur if 1,3-dibromopropane were reacted with sodium? How might these two reactions be affected by adding an inert solvent to dilute 1,3-dibromopropane?

11. A compound C_3H_7Br, when reacted with magnesium in ether followed by treatment with heavy water D_2O, yields CH_3CHDCH_3. What is the structure of C_3H_7Br? Write the equations for the reactions described.

12. What generalized formula can be used to represent a class of compounds containing two rings?

13. What is the molecular formula for cubane. (See Figure 16.8 for the structural formula.)

14. Draw as many structures having the molecular formula C_5H_8 as possible. (There are five structures possible containing only carbon–carbon single bonds.)

15. Suggest a synthesis of 3,4-dimethylhexane using compounds containing no more than four carbon atoms.

16. Propose a structure that would have the molecular formula C_4H_4.

17

ALKENES AND ALKYNES

In the alkanes and the chemically related cycloalkanes all of the valence electrons of carbon are engaged in single bonds to either hydrogen or carbon atoms, yielding molecules that are quite unreactive. In this chapter two additional types of hydrocarbons, *alkenes* and *alkynes,* are examined. These compounds contain multiple bonds between carbon atoms and are said to be *unsaturated* because they have fewer hydrogen atoms than they could have if the multiple bonds were absent. Because of the high chemical reactivity of multiple bonds, there is a wide array of chemical reactions from which to develop synthetic routes to many desired end products.

The multiple bonds of alkenes and alkynes are the first of several molecular features called *functional groups,* which are sites of reaction in a molecule. Because functional groups react by well-defined pathways a chemist can generalize and extrapolate his observations of a limited number of representative molecules to other molecules containing the

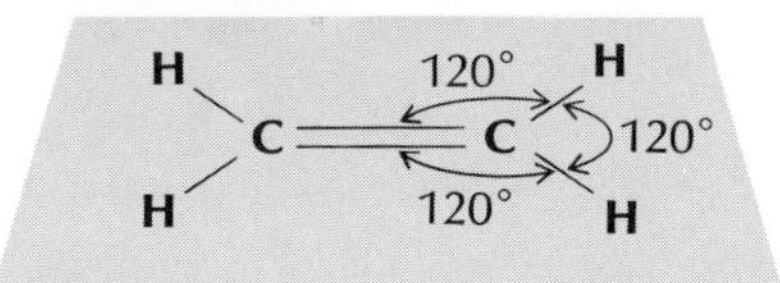

Figure 17.1
Ethylene.

same functional group. We shall see examples of this principle in this chapter and later ones.

17.1 STRUCTURE AND BONDING IN ETHYLENE

Ethylene (C_2H_4) is the simplest member of the homologous series of compounds called *alkenes.* The two carbon atoms and four hydrogen atoms of ethylene all lie in a plane with angles of 120° between bonds (Figure 17.1). The molecule is considered to consist of two carbon atoms, each with three sp^2 hybrid orbitals. The hybrid orbitals, each containing one electron, are bonded *via* a sigma (σ) bond to either carbon or hydrogen. The fourth electron of each carbon is in the remaining *p* orbital, whose axis is perpendicular to the plane of the three sp^2 hybrid orbitals. The bond formed by the side-to-side overlap of the parallel *p* orbitals is called a *pi* (π) *bond.* The carbon-to-carbon bond in ethylene consists of both a sigma bond and a pi bond, which together constitute a double bond (Figure 17.2).

Because the *p* orbitals that form the π bond are perpendicular to the sp^2 hybrid orbitals, π bonds can exist only when the sp^2 orbitals of both carbon atoms lie in the same plane. The need to maintain the π bond in order to satisfy the bonding requirements of carbon precludes free rotation about the carbon–carbon double bond. Rotation of one CH_2 unit of ethylene with respect to the other can occur only if enough energy is added to the molecule by heating it to a high temperature or by irradiating it with ultraviolet light.

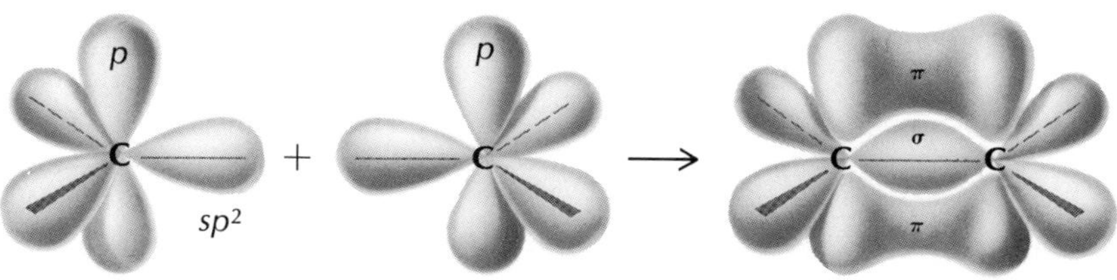

Figure 17.2
Bonding orbitals in a carbon–carbon double bond.

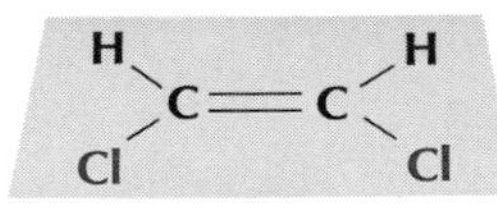

cis-dichloroethylene

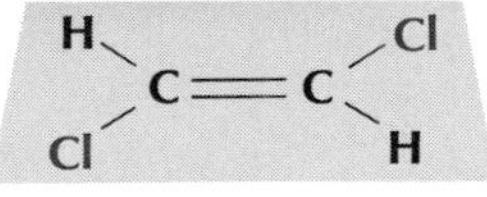

trans-dichloroethylene

Figure 17.3
Geometric isomers of dichloroethylene.

The restriction imposed by the double bond gives rise to geometric isomers such as those observed in transition metals and in cycloalkanes if the proper number of different substituents are attached to the two double-bonded carbon atoms. If each carbon atom is attached to two different groups, two isomers are possible. These isomers are called *cis* and *trans,* according to the location of the substituents on the same or opposite "side" of the double bond. The notation is like that employed for transition metal complex ions and cycloalkanes.

The two possible dichloroethylenes are illustrated in Figure 17.3. The physical properties of the isomers differ considerably. The *trans* isomer has no dipole moment and melts at $-50°$, whereas the *cis* isomer has a dipole moment and melts at $-80°$.

If two of the groups bound to one of the carbon atoms involved in the double bond are identical, then no geometrical isomerism exists. There is only one compound, 1,1-dichloroethylene, and there is only one chloroethylene:

$$\mathrm{Cl_2C{=}CH_2} \qquad \mathrm{H_2C{=}CHCl}$$

1,1-dichloroethylene   chloroethylene

17.2 NOMENCLATURE OF ALKENES

The general formula for alkenes is C_nH_{2n}, which reflects its lack of two hydrogens relative to alkanes. The lower molecular mass alkenes are known by their common names,

$$\mathrm{CH_2{=}CH_2} \qquad \mathrm{CH_2{=}CH{-}CH_3} \qquad \mathrm{CH_2{=}C(CH_3){-}CH_3}$$

ethylene   propylene   isobutylene

but the more complex alkenes are named according to the IUPAC system under rules similar to those for naming alkanes. The ending of the name of a member of the alkene class of compounds is *-ene*. The root name is derived from the longest continuous chain of carbon atoms containing the double bond. Numbering of the chain is such that the carbon atoms of the double bond have the lowest possible number. The position of the double bond is indicated in the name by placing the number of the lower-numbered carbon atom before the root name. Several examples of proper nomenclature follow:

$$\overset{1}{C}H_2{=}\overset{2}{C}H\overset{3}{C}H_2\overset{4}{C}H_3$$

1-butene

$$\underset{1}{CH_3}{-}\underset{2}{\overset{CH_3}{C}}{=}\underset{3}{\overset{CH_3}{C}}{-}\underset{4}{CH_3}$$

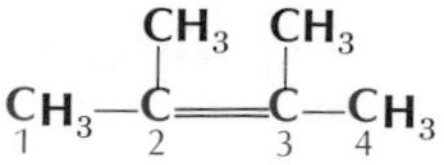

2,3-dimethyl-2-butene

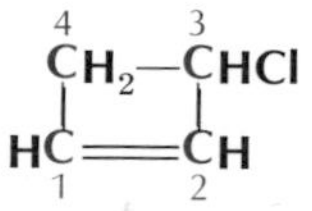

$$\begin{matrix} \overset{4}{C}H_2 & - & \overset{3}{C}HCl \\ | & & | \\ H\underset{1}{C} & = & \underset{2}{C}H \end{matrix}$$

3-chlorocyclobutene

$$\underset{1}{C}H_2{=}\underset{2}{C}H\underset{3}{C}H_2\underset{4}{C}H_2\underset{5}{C}H_2\underset{6}{\overset{Cl}{C}}H\underset{7}{C}H_3$$

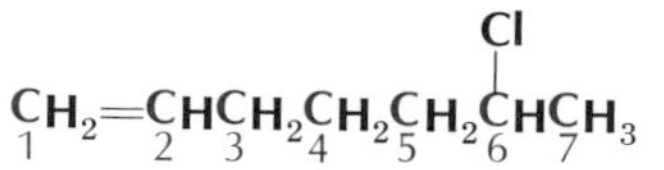

6-chloro-1-heptene

If the compound to be named is a *cis* or *trans* compound by virtue of the attachment of two nonequivalent groups on each of the two carbon atoms of the double bond, that term precedes the entire name of the substance. The *cis* or *trans* term refers to the orientation of the longest continuous chain in the molecule.

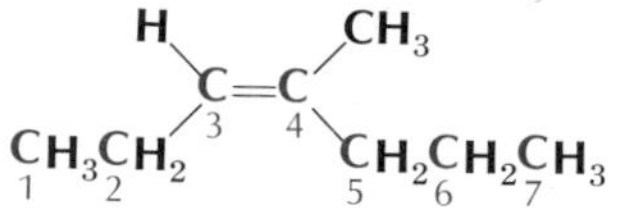

$$\begin{matrix} H & & CH_3 \\ & \overset{3}{C}{=}\overset{4}{C} & \\ \underset{1}{C}H_3\underset{2}{C}H_2 & & \underset{5}{C}H_2\underset{6}{C}H_2\underset{7}{C}H_3 \end{matrix}$$

cis-4-methyl-3-heptene

$$\begin{matrix} H & & \overset{4}{C}H_2\overset{5}{C}H_3 \\ & \overset{2}{C}{=}\overset{3}{C} & \\ \underset{1}{C}H_3 & & CH_3 \end{matrix}$$

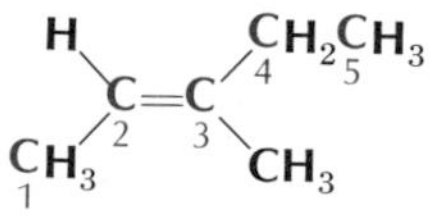

trans-3-methyl-2-pentene

17.3 ADDITION REACTIONS

In one of the most common reactions that alkenes undergo a reagent adds to the carbon-carbon double bond, forming an *addition product* in which the components of the reagent are attached by single bonds to each carbon that was originally part of the double bond:

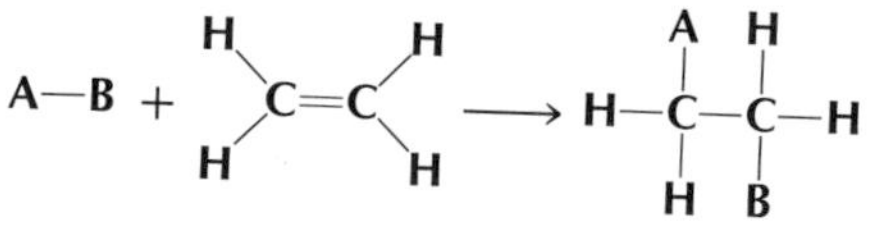

$$A{-}B + \begin{matrix} H & & H \\ & C{=}C & \\ H & & H \end{matrix} \longrightarrow H{-}\underset{H}{\overset{A}{C}}{-}\underset{B}{\overset{H}{C}}{-}H$$

As will be seen shortly the π electrons of the double bond are the causative factor in this general reaction because they are susceptible to reaction with electron-seeking species called electrophiles.

The halogens chlorine, bromine, and iodine, the related hydrogen halides, and water can add to a double bond, as illustrated for the case of ethylene:

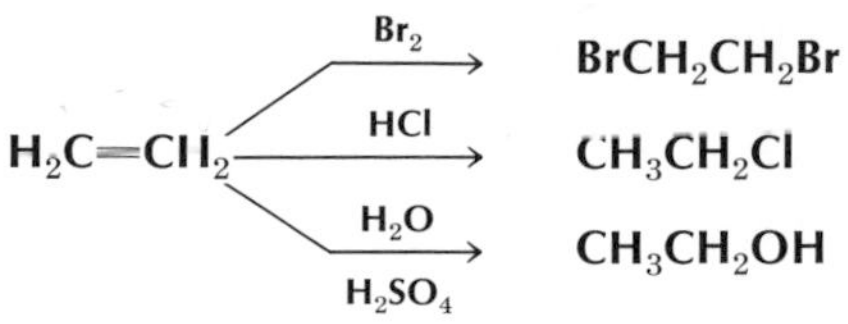

$$H_2C{=}CH_2 \begin{cases} \xrightarrow{Br_2} BrCH_2CH_2Br \\ \xrightarrow{HCl} CH_3CH_2Cl \\ \xrightarrow[H_2SO_4]{H_2O} CH_3CH_2OH \end{cases}$$

Bromine reacts with ethylene at room temperature in the absence of light. These conditions are distinctly different than those for the substitution reaction of alkanes in which HBr is formed. The facile reaction with bromine provides a simple laboratory test for the presence of unsaturation such as is present in alkenes. (Other unsaturated compounds are discussed later in this chapter.) The addition of bromine to an alkene is marked by the disappearance of the characteristic red-brown color of the bromine as the colorless dibromoalkane product is formed.

EXAMPLE 17.1

A compound C_5H_{10} reacts rapidly with bromine to yield $C_5H_{10}Br_2$. What is a possible structure for the compound?

The molecular formula fits the general formula C_nH_{2n}, which is indicative of either a cycloalkane or an alkene. Cycloalkanes such as cyclopentane do not contain any multiple bonds and do not react with Br_2. Therefore, the unknown compound must be an alkene. Any five-carbon structure containing one double bond will fit the data. Some examples are:

$$CH_2{=}CHCH_2CH_2CH_3 \qquad \begin{matrix} CH_3 & & CH_3 \\ & C{=}C & \\ CH_3 & & H \end{matrix}$$

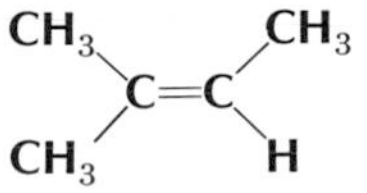

Since there are six isomeric alkenes of the formula C_5H_{10}, obviously much more information about the compound is necessary before a structure can be proposed.

MARKOWNIKOFF'S RULE

For more complex alkenes in which the carbon skeleton is not symmetric about the double bond two addition products are possible when unsymmetrical reagents such as HCl are employed. The addition of hydrogen chloride to a symmetrical alkene such as *cis*-2-butene can yield only 2-chlorobutane regardless of the point of attachment of the adding atoms:

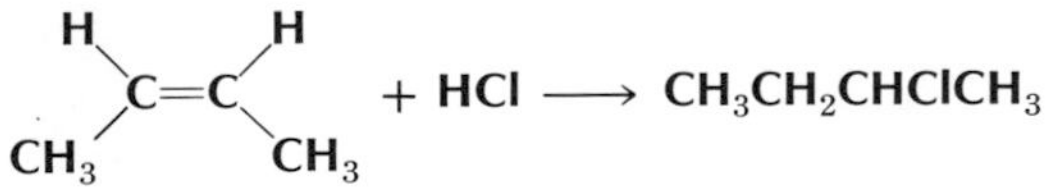

$$\begin{matrix} H & & H \\ & C{=}C & \\ CH_3 & & CH_3 \end{matrix} + HCl \longrightarrow CH_3CH_2CHClCH_3$$

In the case of an unsymmetrical alkene such as propene the carbon atoms involved in the π bond are not equivalent and, therefore, two possible addition products could result:

$$\underset{CH_3}{\overset{H}{\ }}\!\!\!>C{=}C<\!\!\!\underset{H}{\overset{H}{\ }} + HCl \begin{cases} \not\longrightarrow CH_3CH_2CH_2Cl \\ \longrightarrow CH_3CHClCH_3 \end{cases}$$

The experimental observation is that only 2-chloropropane is formed. Similarly, the addition of HBr to methylpropene results in the formation of only 2-bromo-2-methylpropane with no 1-bromo-2-methylpropane being formed:

$$\underset{CH_3}{\overset{CH_3}{\ }}\!\!\!>C{=}C<\!\!\!\underset{H}{\overset{H}{\ }} + HBr \begin{cases} \not\longrightarrow CH_3{-}\underset{H}{\overset{CH_3}{\overset{|}{\underset{|}{C}}}}{-}CH_2Br \\ \longrightarrow CH_3{-}\underset{Br}{\overset{CH_3}{\overset{|}{\underset{|}{C}}}}{-}CH_3 \end{cases}$$

After observing a large number of such addition reactions, the Russian chemist Vladimir Markownikoff formulated a general rule to account for the experimental facts. In modern language, and taking into account present knowledge of the mechanism of addition reaction (see following section), the rule can be stated as follows: When an unsymmetrical reagent adds to an unsymmetrical double bond the derived product will contain the positive part of the addend attached to the double-bonded carbon that has the greater number of directly attached hydrogen atoms.

MECHANISM OF ADDITION TO DOUBLE BONDS

The π electrons of a double bond are not as tightly bound to the carbon nuclei as are the electrons of σ bonds. Furthermore, because of the shape of the π bond, they are exposed to easy attack by electron-seeking species. The most common such species, called *electrophiles* or *electrophilic rea-*

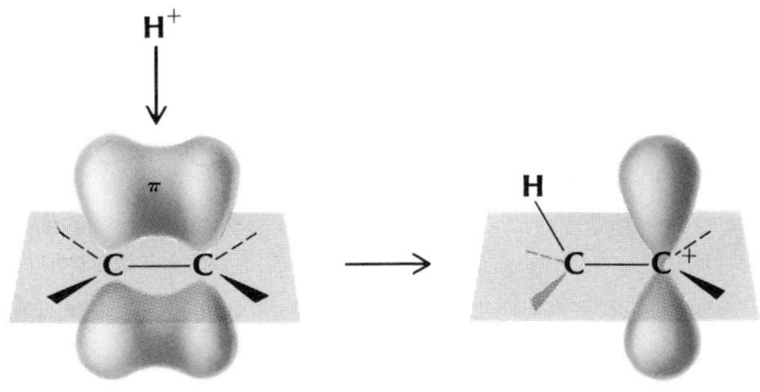

Figure 17.4a
Addition of an electrophile to a π system.

gents, are positive ions that can form a covalent bond with an available electron pair.

An electrophile such as H^+ can approach the π electrons in a direction perpendicular to the plane of the two carbon atoms and the four atoms directly attached to them (Figure 17.4a). As the H^+ moves sufficiently close to the π system, a covalent bond is formed, with the result that the adjacent carbon atom has only six electrons and an empty *p* orbital. The positively charges species is called a *carbonium ion.*

A reagent seeking a center of positive charge (*nucleophile*), such as the negative chloride ion, can then approach the carbonium ion from the side opposite the electrophile and form a bond to carbon (Figure 17.4b). Actually the course of addition reactions does not proceed in discrete steps as pictured above. Rather the attack of the electrophile and nucleophile are part of a closely timed, almost simultaneous sequence in which they operate on opposite sides of the molecules.

The observant student may have noticed that the description of an electrophile is similar to that of a Lewis acid and that the nucleophile, which must have an unshared electron pair, is a Lewis base (Chapter 10). The terms *Lewis acid* and *Lewis base* were proposed to describe equilibrium processes, that is, acid-base reactions. The terms *electrophile* and *nucleophile* are commonly used in describing the mechanism and rate of a reaction that is considered to proceed in one direction. In addition the electrophilicity or nucleophilicity of a reagent is a function not only of a molecule itself but of the type of system with which it reacts.

The addition of symmetrical reagents, such as bromine, is visualized as involving attack by a positive ion, the bromonium ion Br^+, which has only six valence electrons. The species is generated in low concentrations from Br_2 and can attack the π electrons of alkenes as well as other centers containing available electrons:

$$Br_2 \rightleftharpoons |\overline{\underline{Br}}^+ + |\overline{\underline{Br}}|^-$$

$$CH_2{=}CH_2 + Br^+ \longrightarrow BrCH_2{-}CH_2^+$$

$$Br{-}CH_2{-}CH_2^+ + Br^- \longrightarrow BrCH_2CH_2Br$$

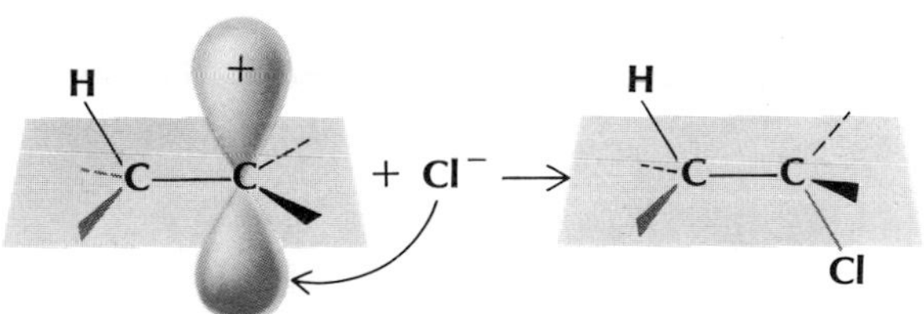

Figure 17.4b
Addition of a nucleophile to a carbonium ion.

Markownikoff's Rule, that the positive portion of the adding molecule is attached to the carbon with more hydrogens, is easily interpreted in the light of our knowledge of the mechanism of addition reactions. When a proton adds to the double bond of propene the position of addition determines the location of the positive charge in the carbonium ion:

$$H^+ + CH_3CH{=}CH_2 \begin{cases} \not\rightarrow CH_3CH_2CH_2^+ \\ \rightarrow CH_3\overset{}{\underset{+}{C}}HCH_3 \end{cases}$$

Experimentally it is observed that the product contains the electrophile H^+ attached to the terminal carbon atom in the ultimate product. Therefore, the intermediate carbonium ion, which has the positive charge on a secondary carbon atom, must be more stable than the one with a positive charge on a primary carbon atom.

The product from the reaction of methylpropene with HBr has the electrophile H^+ attached to a primary carbon atom rather than a tertiary carbon atom. Therefore, the intermediate carbonium ion with a positive charge on the tertiary carbon atom must be more stable than the one with the positive charge on a primary carbon atom:

$$H^+ + CH_3{-}\overset{CH_3}{\overset{|}{C}}{=}CH_2 \begin{cases} \not\rightarrow CH_3{-}\overset{CH_3}{\underset{H}{C}}{-}CH_2^+ \\ \rightarrow CH_3{-}\overset{CH_3}{\underset{+}{C}}{-}CH_3 \end{cases}$$

Like radicals, carbonium ions are classed as *tertiary, secondary,* or *primary,* depending on the number of alkyl groups attached to the carbon atom bearing the charge. Their stabilities decrease in the order tertiary > secondary > primary. The difference in stability of carbonium ions can be rationalized as it was for radicals (Section 16.4): The attached alkyl groups are better able than hydrogen to release the electrons of their bonds to the electron deficient center. In the case of radicals the electron deficiency is one, whereas with carbonium ions it is two. The increased deficiency causes a greater difference in the stability of carbonium ions than of radicals. While various potential radicals compete in their rates of formation in reactions such as chlorination of alkanes, in addition reactions one carbonium ion is formed to the virtual exclusion of other less stable ones.

What is the product of the addition of HBr to 2-methyl-2-butene?

The structure of the alkene is:

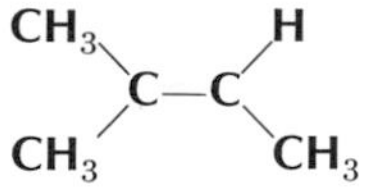

$$\begin{array}{ccc} CH_3 & & H \\ & C{=}C & \\ CH_3 & & CH_3 \end{array}$$

According to Markownikoff's rule the HBr should add to yield:

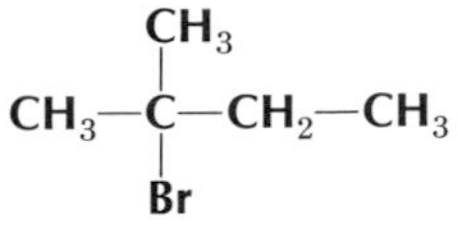

$$CH_3-\underset{\displaystyle Br}{\overset{\displaystyle CH_3}{\underset{|}{\overset{|}{C}}}}-CH_2-CH_3$$

This product is derived from the addition of a proton to yield a tertiary carbonium ion rather than a secondary carbonium ion.

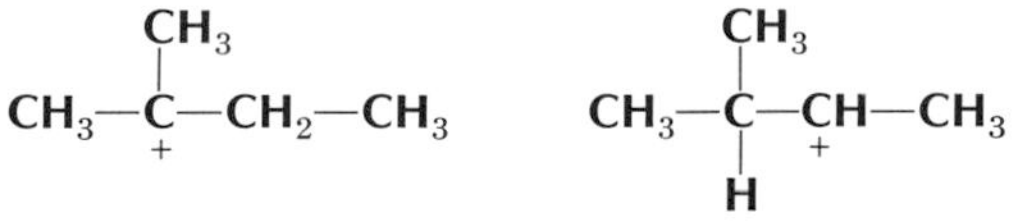

$$CH_3-\underset{+}{\overset{\displaystyle CH_3}{\overset{|}{C}}}-CH_2-CH_3 \qquad CH_3-\underset{\displaystyle H}{\overset{\displaystyle CH_3}{\underset{|}{\overset{|}{C}}}}-\underset{+}{CH}-CH_3$$

EXAMPLE 17.3

Hypochlorous acid (HOCl) adds to alkenes to yield addition products that contain an OH group and a Cl on adjacent atoms. What product would be expected for the reaction with propene?

Oxygen is more electronegative than chlorine (Chapter 6), and therefore the electrophile would be expected to be Cl^+ and the nucleophile HO^-. The adduct should have the following structure that results from typical Markownikoff addition:

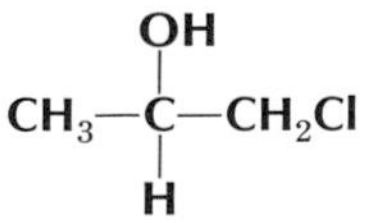

$$CH_3-\underset{\displaystyle H}{\overset{\displaystyle OH}{\underset{|}{\overset{|}{C}}}}-CH_2Cl$$

17.4 REDOX REACTIONS

REDUCTION

In the presence of certain metal catalysts such as palladium, platinum, or nickel an alkene will react with hydrogen gas to form an alkane:

$$CH_2{=}CH_2 + H_2 \xrightarrow{Pd} CH_3CH_3$$

This process is called *hydrogenation* or *reduction of an alkene*. (Note that the oxidation state of carbon in ethylene is −2, whereas in ethane it is −3.)

The hydrogenation reaction is of considerable utility in ascertaining the general class of compounds to which a substance of unknown structure belongs. A compound of molecular formula C_6H_{10} could be any substance containing (1) two double bonds, (2) a double bond and a ring, (3) two rings, or (4) a triple bond (an alkyne, which will be described in a later section). If the substance when hydrogenated yields C_6H_{12} by the uptake of only 1 mole of H_2, it can be concluded that the compound originally contained a double bond and a ring.

Hydrogenation eliminates the structural feature responsible for *cis* and *trans* isomerism. Thus, both *cis*-2-butene and *trans*-2-butene yield butane when hydrogenated. Of course 1-butene also will yield butane:

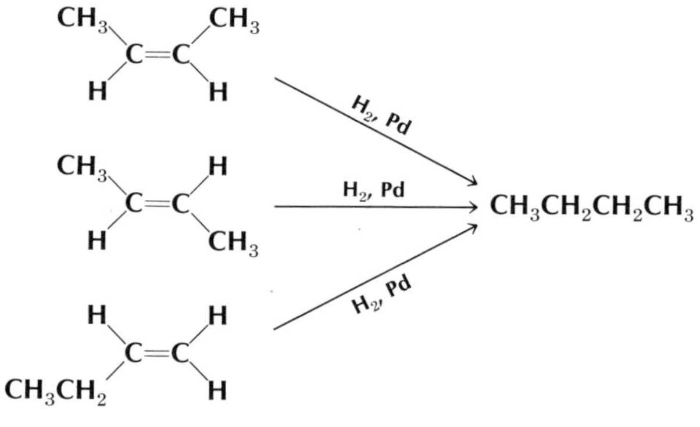

The hydrogenation reaction is of commercial importance. Many oils such as cottonseed oil or corn oil contain double bonds, and their hydrogenation yields the semisolid "shortenings" such as Crisco, Spry, or Fluffo. (Oils and fats are discussed in Chapters 21 and 26.)

EXAMPLE 17.4

How could one distinguish between cyclohexene and 1,4-hexadiene:

$$CH_2{=}CH{-}CH_2{-}CH{=}CH{-}CH_3$$

1,4-hexadiene

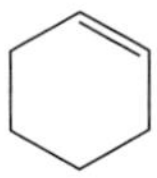

cyclohexene

There are two double bonds contained in 1,4-hexadiene, whereas cyclohexene contains only one double bond. If the same molar quantities of each substance are compared, the 1,4-hexadiene will react with twice the number of moles or twice the volume of H_2 under the same conditions of temperature and pressure.

OXIDATION

The π electrons of alkenes are susceptible to reaction with oxidizing agents. This simple fact allows ready distinction between alkenes and alkanes. Potassium permanganate ($KMnO_4$) is reduced to manganese dioxide by olefins; the color change from purple to brown is easily detectable and is called the *Baeyer test* for alkenes. The organic products of this reaction are seldom isolated, but they include *aldehydes, ketones* (Chapter 20), and *acids* (Chapter 21).

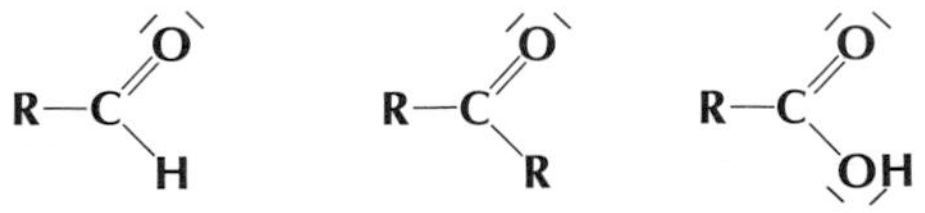

$$R{-}C({=}O){-}H \qquad R{-}C({=}O){-}R \qquad R{-}C({=}O){-}OH$$

an aldehyde   a ketone   an acid

Ozone reacts rapidly with alkenes to yield short-lived materials called *ozonides,* which usually are not isolated because of their explosive character. In the presence of metallic zinc and acid the ozonide is cleaved into two fragments, which may be either aldehydes or ketones:

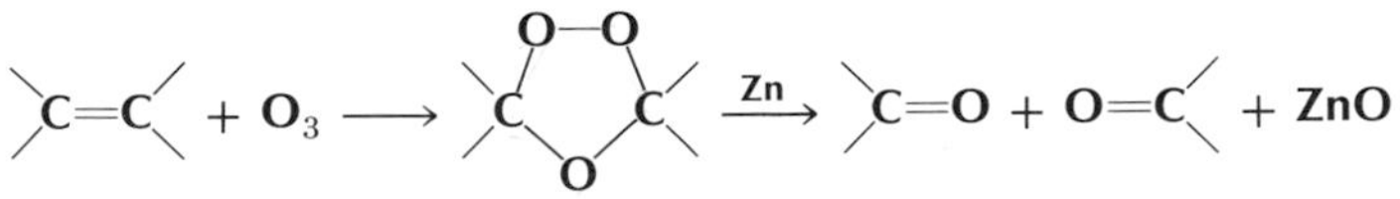

$$>C{=}C< + O_3 \longrightarrow \underset{\text{an ozonide}}{>C(O{-}O)(O)C<} \xrightarrow{Zn} >C{=}O + O{=}C< + ZnO$$

Identification of the fragments resulting from *ozonolysis* determines the structure of the original alkene. A compound of molecular formula C_5H_{10} that upon ozonolysis yields the products shown in the following equation must be 2-methyl-2-butene.

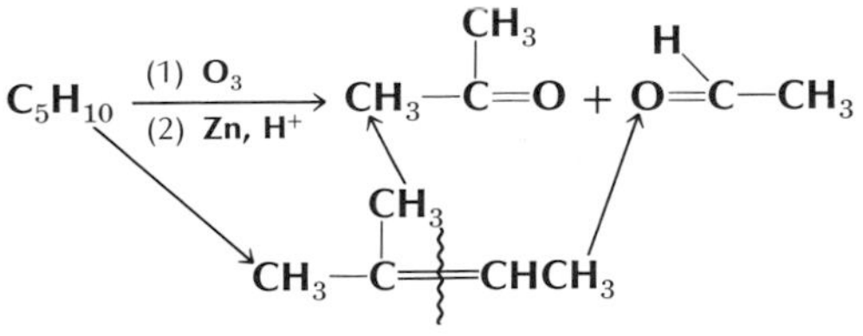

$$C_5H_{10} \xrightarrow[(2)\ Zn,\ H^+]{(1)\ O_3} CH_3{-}\overset{CH_3}{\overset{|}{C}}{=}O + O{=}\overset{H}{C}{-}CH_3$$

$$CH_3{-}\overset{CH_3}{\overset{|}{C}}{=}CHCH_3$$

Can the ozonolysis of an alkene yield only one product?

Yes. This situation will arise whenever a symmetrical alkene is ozonized. Since the alkyl groups attached to the carbon atoms of the double bond are identical, the product from each half of the molecule must be one and the same. An example is given below:

$$CH_3(H)C{=}C(H)CH_3 \xrightarrow[(2)\ Zn]{(1)\ O_3} 2\ CH_3(H)C{=}O$$

17.5 POLYMERIZATION OF ALKENES

A *polymer* (from the Greek *poly,* many, and *meros,* parts) is a high molecular mass substance consisting of many identical units or molecules linked to each other in a chain. The molecules that yield polymers are called *monomers,* and the reaction by which hundreds and even thousands of these monomers join together is called *polymerization.* Specifically polymerization of alkenes is termed *addition polymerization* since it involves addition of one monomer to the next.

Polymerization of alkenes can be initiated by radicals, carbonium ions, or carbanions (R_3C^-). All of the reactions have in common the addition of a reactive species to a double bond to produce an intermediate that then adds to another molecule of alkene. This process is illustrated for the polymerization of ethylene in which an electrophilic catalyst R^+ is used to initiate the chain-building process:

$$R^+ + CH_2{=}CH_2 \longrightarrow RCH_2CH_2^+$$
$$RCH_2CH_2^+ + CH_2{=}CH_2 \longrightarrow RCH_2CH_2CH_2CH_2^+$$
$$RCH_2CH_2CH_2CH_2^+ + CH_2{=}CH_2 \longrightarrow RCH_2CH_2CH_2CH_2CH_2CH_2^+$$

Eventually the growing carbonium ion reacts with some negative species that terminates the chain, and the polyethylene molecule is produced.

Polyethylene is not a homogeneous substance but rather consists of molecules of different chain lengths. The exact mixture of molecules in industrial processes is controlled by reaction conditions. Various polyethylene mixtures are used in the fashioning of electrical insulators, hospital equipment, and innumerable household articles. Among the many useful polymers for mass consumption are Saran (polydichloroethylene), Teflon (polytetrafluoroethylene), and Orlon (polyacrylonitrile).

$$CH_2{=}CCl_2 \longrightarrow \left(CH_2{-}\overset{Cl}{\underset{Cl}{C}} \right)_n \quad \text{Saran}$$

monomer unit

$$CF_2{=}CF_2 \longrightarrow (CF_2CF_2)_n \quad \text{Teflon}$$

$$CH_2{=}CHCN \longrightarrow \left(CH_2\overset{CN}{CH} \right)_n \quad \text{Orlon}$$

17.6 PREPARATION OF ALKENES

In general a double bond can be introduced in a saturated molecule by the elimination of two atoms or groups of atoms located on adjacent carbon atoms. One of the atoms or groups leaves without its bonding pair of electrons while the other carries with it its bonding pair of electrons. The pair of electrons left by one of the groups provides the necessary electrons for the π bond. The process is called an *elimination reaction* and may be regarded as the reverse of an addition reaction. The general process is illustrated using E and N as the groups to be eliminated, where E represents the group leaving without its bonding electron pair and N represents the group leaving with its electron pair. In the reverse of the elimination reaction—that is, the addition reaction—the E would be the electrophile and N the nucleophile. The arrows serve to represent the changes that occur both with respect to atomic and electronic redistribution in the course of the reaction. They are models for the progress of the reaction and do not necessarily correspond to reality.

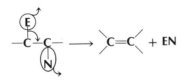

$$-\overset{E}{\underset{|}{C}}-\underset{N}{\overset{|}{C}}- \longrightarrow \;>C{=}C<\; + \text{EN}$$

DEHYDRATION OF AN ALCOHOL

The removal of water from an alcohol (Chapter 19) is called *dehydration* and may be effected by means of an acid catalyst, such as sulfuric or phosphoric acid. The mechanism of dehydration is illustrated for the formation of ethylene from ethyl alcohol:

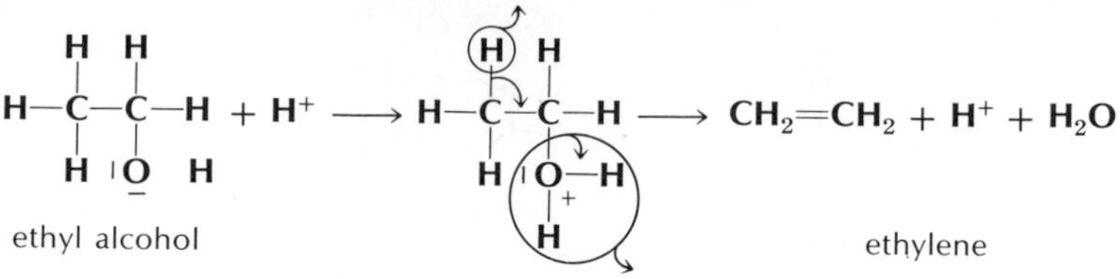

$$\text{H-}\underset{\text{H}}{\overset{\text{H}}{\text{C}}}\text{-}\underset{\text{OH}}{\overset{\text{H}}{\text{C}}}\text{-H} + H^+ \longrightarrow \text{H-}\underset{\text{H}}{\overset{\text{H}}{\text{C}}}\text{-}\underset{\text{H-}\overset{+}{\text{O}}\text{-H}}{\overset{\text{H}}{\text{C}}}\text{-H} \longrightarrow CH_2{=}CH_2 + H^+ + H_2O$$

ethyl alcohol ethylene

For more complex alcohols such as 2-butanol, the dehydration produces a mixture of products, because there are two neighboring carbon atoms adjacent to the carbon atom bearing the hydroxyl group. In such cases the isomers that contain the greatest number of alkyl groups attached to the double bond predominate in the mixture:

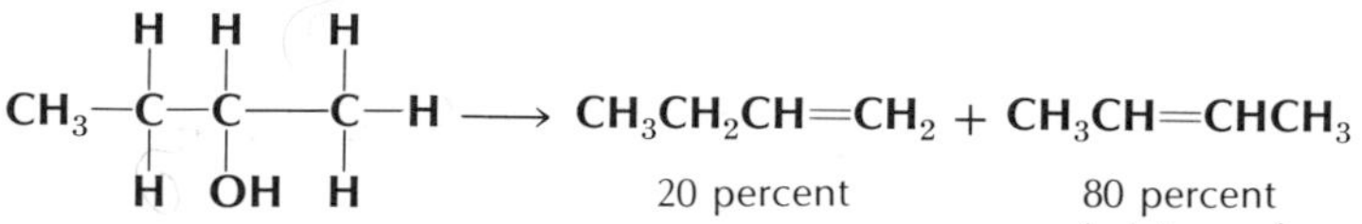

$$CH_3\text{-}\underset{\text{H}}{\overset{\text{H}}{\text{C}}}\text{-}\underset{\text{OH}}{\overset{\text{H}}{\text{C}}}\text{-}\underset{\text{H}}{\overset{\text{H}}{\text{C}}}\text{-H} \longrightarrow CH_3CH_2CH{=}CH_2 + CH_3CH{=}CHCH_3$$

20 percent; 80 percent (mixture of *cis* and *trans*)

DEHYDROHALOGENATION

Dehydrohalogenation, the elimination of a molecule of a hydrogen halide from a haloalkane can be accomplished with a solution of hydroxide ions in alcohol as the solvent. The mechanism of the reaction is shown for bromoethane:

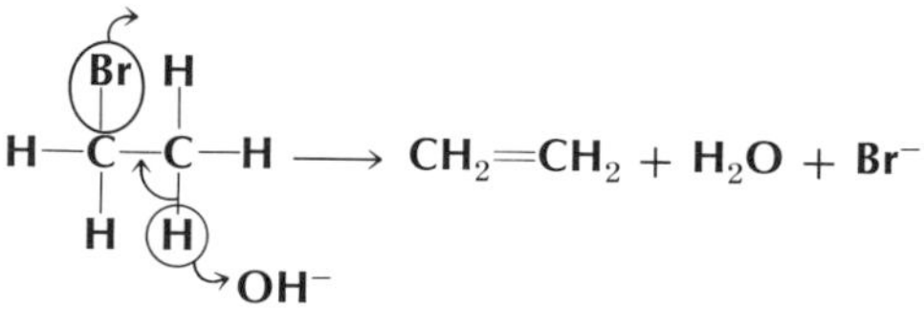

$$\text{H-}\underset{\text{H}}{\overset{\text{Br}}{\text{C}}}\text{-}\underset{\text{H}}{\overset{\text{H}}{\text{C}}}\text{-H} \longrightarrow CH_2{=}CH_2 + H_2O + Br^-$$

OH^-

As in the case of the dehydration of alcohols, elimination can occur in several ways and a mixture of alkenes results:

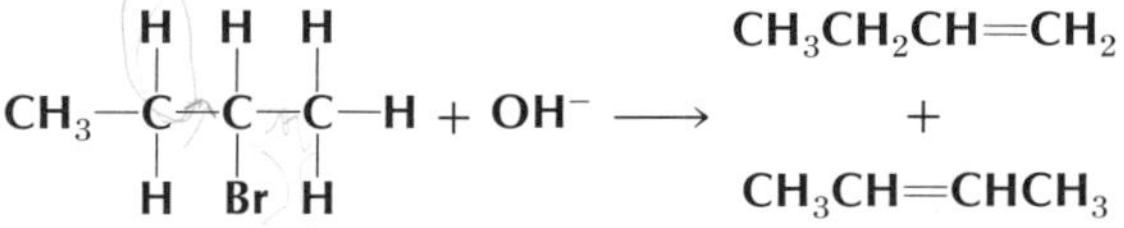

$$CH_3\text{-}\underset{\text{H}}{\overset{\text{H}}{\text{C}}}\text{-}\underset{\text{Br}}{\overset{\text{H}}{\text{C}}}\text{-}\underset{\text{H}}{\overset{\text{H}}{\text{C}}}\text{-H} + OH^- \longrightarrow CH_3CH_2CH{=}CH_2 + CH_3CH{=}CHCH_3$$

(mixture of *cis* and *trans* isomers)

17.7 ALKYNES

STRUCTURE

The compounds of molecular formula C_nH_{2n-2} that contain a triple bond are called *alkynes* or acetylenes after the simplest member of the homol-

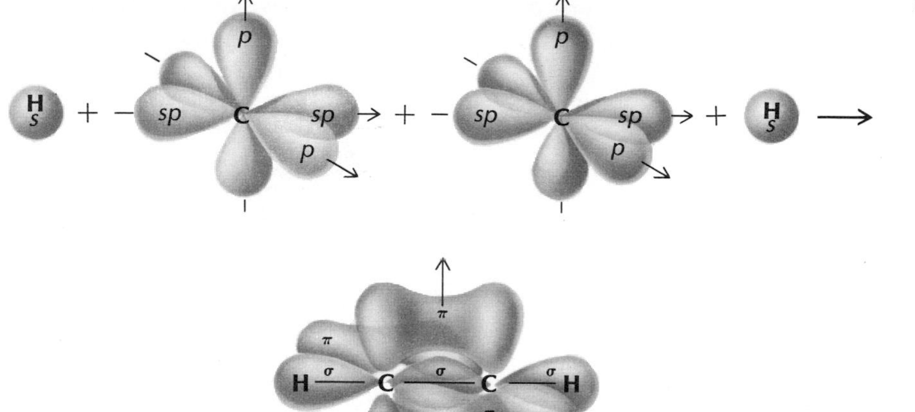

Figure 17.5
Schematic representation of bonding orbitals in a carbon–carbon triple bond.

ogous series, C_2H_2 acetylene or ethyne. The acetylene molecule is linear and has a triple bond that is constructed from one σ bond and two π bonds: H—C≡C—H. The two carbon atoms are joined by a σ bond formed from two overlapping *sp* hybrid orbitals. Each carbon is bound to its respective hydrogen atom by a σ bond formed by an *sp* hybrid orbital of carbon and the *s* orbital of hydrogen. The two remaining electrons of carbon are in *p* orbitals that are mutually perpendicular and also are perpendicular to the axis of the *sp* hybrid orbitals. Two π bonds are formed by electrons in overlapping *p* orbitals of each carbon atom, as shown for the acetylene molecule in Figure 17.5.

NOMENCLATURE

The general rules for nomenclature of alkynes are similar to those of alkenes except that the ending *-yne* is used. Since the two carbon atoms involved in the triple bond and the immediately adjacent atoms are linearly arranged there are no geometrical isomers possible in the alkynes. A few examples should suffice to illustrate the nomenclature of alkynes:

$$\underset{1}{CH_3}-\underset{2}{C}\equiv\underset{3}{C}-\underset{4}{CH_3}$$

2-butyne

$$\underset{5}{CH_3}-\underset{4}{\overset{\overset{\large CH_3}{|}}{CH}}-\underset{3}{C}\equiv\underset{2}{C}-\underset{1}{CH_3}$$

4-methyl-2-pentyne

$$\underset{6}{CH_3}-\underset{5}{CH_2}-\underset{4}{C}\equiv\underset{3}{C}-\underset{2}{\overset{\overset{\large CH_3}{|}}{CH}}-\underset{1}{CH_3}$$

2-methyl-3-hexyne

REACTIONS

The π electrons of alkynes are susceptible to attack by electrophiles in the same manner as alkenes. Addition of reagents may occur in a stepwise manner to produce an alkene that then reacts further to produce a saturated compound. Under some circumstances the reaction may be terminated at the alkene stage, but it is usually difficult to do so. The addition of bromine to acetylene ultimately yields 1,1,2,2-tetrabromoethane:

$$H{-}C{\equiv}C{-}H \xrightarrow{Br_2} CHBr{=}CHBr \xrightarrow{Br_2} CHBr_2CHBr_2$$

In the case of unsymmetrical reagents, such as HCl, the addition proceeds according to Markownikoff's rule:

$$H{-}C{\equiv}C{-}H \xrightarrow{HCl} CH_2{=}CHCl \xrightarrow{HCl} CH_3CHCl_2$$

Hydrogenation of alkynes ultimately leads to the consumption of 2 moles of hydrogen per mole of alkyne. By controlling the amount of hydrogen available and using platinum as catalyst the reaction can be terminated at the alkene:

$$H{-}C{\equiv}C{-}H \xrightarrow[Pt]{H_2} \begin{matrix} H & & H \\ & C{=}C & \\ H & & H \end{matrix} \xrightarrow[Pt]{H_2} CH_3CH_3$$

ACIDITY

Acetylenes containing a hydrogen atom attached directly to the carbon atom of a triple bond are weak acids. Active metals such as sodium will react with terminal acetylenes to liberate hydrogen gas in the same way as does the weak acid water:

$$2R{-}C{\equiv}C{-}H + 2Na \longrightarrow 2R{-}C{\equiv}C^-Na^+ + H_2$$
$$2H_2O + 2Na \longrightarrow 2Na^+OH^- + H_2$$

This reaction, which is characteristic of terminal acetylenes, can be used as a test to determine the general structure of alkynes and to distinguish some of the alkynes from the alkenes that do not react with sodium.

Certain higher atomic mass metal ions also react to form *acetylide* salts that are insoluble in water. These compounds are seldom isolated because they are explosive when dry. The copper salt and the silver salt are red and white, respectively.

$$2R{-}C{\equiv}C{-}H + Cu_2Cl_2 + 2NH_3 \longrightarrow (R{-}C{\equiv}C)_2Cu_2 + 2NH_4Cl$$
$$R{-}C{\equiv}C{-}H + Ag(NH_3)_2^+ \longrightarrow R{-}C{\equiv}CAg + NH_3 + NH_4^+$$

EXAMPLE 17.6

A compound C_4H_6 reacts with $Ag(NH_3)_2^+$ to yield a white precipitate. What is the structure of the compound?

The C_4H_6 compound is deficient by four hydrogen atoms from that expected for a saturated hydrocarbon. This deficiency could be due to the presence of (1) two double bonds, (2) two rings, (3) a ring and a double bond, or (4) a triple bond. The reaction with $Ag(NH_3)_2^+$ indicates that the compound contains a triple bond that is in a terminal position. A partial structure is:

$$H{-}C{\equiv}C{-}?$$

The remaining two carbon atoms must be attached to the carbon atom that is triply bonded. There is only one structural possibility for two carbon atoms attached at a single position, that is, an ethyl group: 1-butyne.

$$H{-}C{\equiv}C{-}CH_2{-}CH_3$$

PREPARATION

A triple bond can be introduced into a molecule by a double elimination reaction. For example, two successive dehydrobromination of a compound containing bromine atoms on adjacent carbon atoms lead to the formation of an alkyne:

$$R{-}\underset{\displaystyle Br}{\overset{\displaystyle H}{\underset{|}{\overset{|}{C}}}}{-}\underset{\displaystyle H}{\overset{\displaystyle Br}{\underset{|}{\overset{|}{C}}}}{-}H + 2OH^- \xrightarrow{\text{alcohol}} R{-}C{\equiv}C{-}H + 2H_2O + 2Br^-$$

Note that the necessary dibromo compound can be made from the addition of bromine to the double bond of an alkene. Therefore, in theory an alkene could be used to produce an alkyne by a two-step sequence of reactions:

$$CH_3CH{=}CH_2 + Br_2 \rightarrow CH_3CHBrCH_2Br \xrightarrow[\text{alcohol}]{OH^-} CH_3C{\equiv}CH$$

Multiple reactions are often required to obtain compounds of interest from available starting materials. The above two-step sequence is a simple representation of the techniques that often are necessary to achieve the synthesis of a desired compound.

Alkynes can be synthesized by joining organic compounds containing the requisite number of carbon atoms. The sodium salt of an alkyne reacts with haloalkanes to displace a halide ion and form a carbon–carbon bond:

$$CH_3CH_2C{\equiv}C^-Na^+ + CH_3Br \longrightarrow CH_3CH_2C{\equiv}C{-}CH_3 + Na^+Br^-$$

The above reaction is the second synthetic method of joining lower molecular mass fragments introduced in this book. The first was the Wurtz reaction (Section 16.5). An advantage of the synthesis involving salts of alkynes is that the carbon fragments joined need not contain the same number of carbon atoms.

17.8
ISOPRENE AND TERPENES

The *isoprene* (2-methyl-1,3-butadiene) molecule is one of the basic building blocks of chemicals found in plants and animals:

$$CH_2{=}\underset{}{\overset{CH_3}{\overset{|}{C}}}{-}CH{=}CH_2$$

isoprene

Terpenes are compounds that contain two or more of the basic isoprene units joined in a head-to-tail fashion. The functional groups present in terpenes depend on the biochemistry of the organism that produces them. Citronellol, which is found in oil of geranium, consists of two isoprene units. The dotted line separates the structure into its two component isoprene units. This type of representation is employed with the remaining structures in this section.

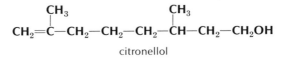

$$CH_2{=}\overset{CH_3}{\overset{|}{C}}{-}CH_2{-}CH_2{-}CH_2{-}\overset{CH_3}{\overset{|}{CH}}{-}CH_2{-}CH_2OH$$

citronellol

The yellow coloring material found in carrots, tomatoes, and spinach is due to β-carotene, which consists of eight isoprene units. The following line structure, which does not indicate hydrogen atoms, represents β-carotene:

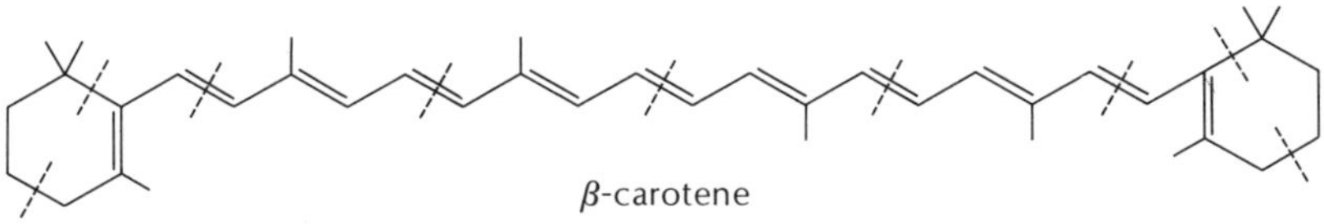

β-carotene

Vitamin A is produced by oxidation of β-carotene at the midpoint of the molecule:

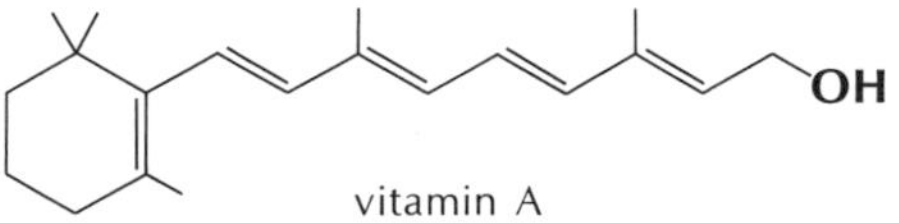

OH

vitamin A

Squalene is a terpene that can be converted into steroids (Chapters 16 and 26) in living systems:

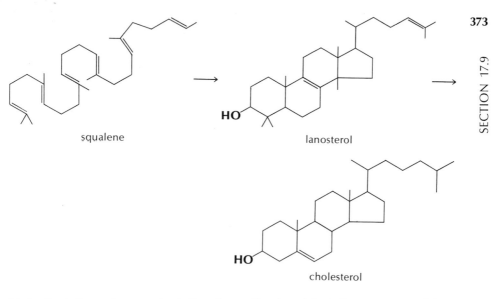

squalene → lanosterol → cholesterol

Note that there is one deviation from the head-to-tail sequence of isoprene units in squalene.

17.9 ULTRAVIOLET SPECTROSCOPY

Many molecules containing unsaturated centers absorb ultraviolet light in the range of wavelengths from 200 to 400 *nanometers* (1 nm = 10^{-9} m). Upon absorption of the energy associated with a specific wavelength of light the electrons of multiple bonds are promoted to higher energy level orbitals. Thus the wavelength of the absorbed light indicates the energy difference between the normal (ground state) electronic arrangement of the molecule and its excited state. Molecules containing only σ bonds do not absorb light in the ultraviolet region because the electrons are in a low energy (relatively stable) state.

Alkenes and alkynes containing only one multiple bond do not absorb light in the ultraviolet region because the separation of energy levels is too large. However, if multiple bonds alternate with single ones, as in 1,3-butadiene, the molecule does absorb in the ultraviolet region:

$$CH_2{=}CH{-}CH{=}CH_2$$

1,3-butadiene

Such molecules are said to be *conjugated* by virtue of the interaction that results from the proximate double bonds. If the double bonds are separated by a methylene unit (—CH_2—) the molecule is not conjugated and does not absorb in the ultraviolet region:

$$CH_2{=}CH{-}CH_2{-}CH{=}CH_2$$

Ultraviolet spectroscopy is a convenient means of determining the bonding characteristics of a molecule without destroying it in a chemical reaction. Furthermore, analysis can be accomplished with amounts of material as small as 1 mg. The occurrence of the wavelength of absorption for molecules containing conjugated π bonds is a well-documented subject. Experience in the area allows the chemist to make predictions of the wavelength of the absorption of a compound that has been newly synthesized. Alternatively an unknown material from a natural source can be assigned a partial structure on the basis of its ultraviolet spectrum.

A few empirical rules illustrate the value of ultraviolet spectroscopy. A 1,3-butadiene unit absorbs at 215 nm. Alkyl groups attached directly to the π system shift the position of the absorption by 5 nm to longer wavelength. Thus 2,4-hexadiene will absorb at 225 nm, since there are two methyl groups attached to the 1,3-butadiene-like structure:

$$CH_3{-}CH{=}CH{-}CH{=}CH{-}CH_3$$

2,4-hexadiene

The isomeric 1,3-hexadiene absorbs at 220 nm since there is one ethyl group attached to the 1,3-butadiene-like structure:

$$CH_2{=}CH{-}CH{=}CH{-}CH_2CH_3$$

1,3-hexadiene

If the double bond is attached directly to the exterior of a ring, a shift of 5 nm to a longer wavelength results. The absorption of

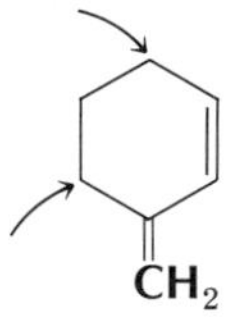

CH_2

occurs at 230 nm, which is consistent with the presence of alkyl groups attached at two positions shown by arrows in the above figure and with a double bond attached to the exterior of the ring.

Additional conjugate double bonds produce a 30-nm shift to longer wavelength, as in 1,3,5-hexatriene, which absorbs at 245 nm.

Ascertaining the complete structure of organic molecules requires a much wider body of knowledge than is possible to describe in this short section. Indeed, this knowledge in general is useful only to the professional chemist.

Suggested further readings

Jones, G., "The Markovnikov Rule," *J. Chem. Educ.*, **38,** 297 (1961).

Juster, N. J., "Color and Chemical Constitution," *J. Chem. Educ.*, **39,** 596 (1962).

Leicaster, H. M., "Vladimir Vasilevich Markovnikov," *J. Chem. Educ.*, **18,** 53 (1941).
Natta, G., "Polymerization," *Sci. Amer.* (September 1957).
Traynham, J. G., "The Bromonium Ion," *J. Chem. Educ.*, **40,** 392 (1963).

Terms and concepts

acetylide
addition polymerization
addition reaction
aldehyde
alkene
alkyne
Baeyer test
carbanion
carbonium ion
dehydration
dehydrohalogenation
electrophile
elimination reaction
hydrogenation
isoprene
ketone
Markownikoff rule
monomer
nucleophile
ozonide
ozonolysis
(π) bond
polymer
terpene
ultraviolet spectroscopy

Questions and problems

1. Name the following compounds by the IUPAC system:

a. $CH_3CH{=}CHCH_2CH_3$
b. $(CH_3)_2C{=}CH_2$
c. $(CH_3)_2C{=}C(CH_3)_2$
d. $CH_2{=}CHCH_2CH{=}CH_2$
e. $CH_3C{\equiv}CCH_2CH_2CH_3$
f. $CH_3CHBrC{\equiv}CCH_2CH_3$
g. 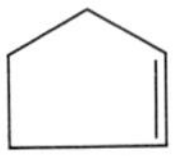
h. $CH_3C{\equiv}C{-}C{\equiv}CH$

2. Why is each of the following names incorrect?

a. 2-ethyl-1-propene
b. 3,3-dimethyl-1-propyne
c. 3-butyne
d. 3,3-dimethyl-4-pentene

3. Write structural formulas for all the possible isomers of the indicated molecular formula (the correct number is given in parentheses):

a. C_3H_6 (2)
b. C_3H_4 (3)
c. C_4H_8 (6)
d. C_4H_6 (9)
e. C_5H_{10} (11)

4. Using reactions discussed in Chapters 16 and 17, write equations indicating how each of the following compounds can be prepared from 2-bromopropane:

a. propene
b. 1,2-dibromopropane
c. propyne
d. 2,2-dichloropropane
e. 2,3-dimethylbutane
f. 4-methyl-1-pentene
g. 2-methylpentane

5. What laboratory test can be used to distinguish between each member of the following pairs of compounds?

   a. propane and propene
   b. 1-butyne and 2-butyne
   c. 1-hexene and cyclohexane

6. A compound, C_5H_8, reacts with only 1 mole of bromine to yield $C_5H_8Br_2$. Suggest a structure for C_5H_8 that is in accord with this fact.

7. A compound, C_5H_8, reacts with bromine to yield $C_5H_8Br_4$. The C_5H_8 substance reacts with $Ag(NH_3)_2^+$ to yield a white precipitate. Suggest two structures for C_5H_8 that are in accord with these facts.

8. A compound, C_5H_{10}, does not react with bromine. Suggest one structure for this compound that is in accord with this fact.

9. What alkene will yield the following compound(s) upon ozonolysis?

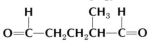

   a. $HC_2{=}O$ and CH_3CHO
   c. $CH_3\overset{H}{\overset{|}{C}}{=}O$ and $(CH_3)_2C{=}O$
   b. only $(CH_3)_2C{=}O$
   d. only $CH_3{-}\overset{H}{\overset{|}{C}}{=}O$

10. What bromoalkane will give only 2-pentene upon dehydrobromination?

11. A compound, C_6H_{10}, reacts with only 1 mole of hydrogen in the presence of a metal catalyst. Upon ozonolysis a single product shown below is obtained. What is the structure of C_6H_{10}?

$$O{=}\overset{H}{\overset{|}{C}}{-}CH_2CH_2\overset{CH_3}{\overset{|}{C}}H{-}\overset{H}{\overset{|}{C}}{=}O$$

12. Predict the position of absorption of each of the following compounds:

   a. 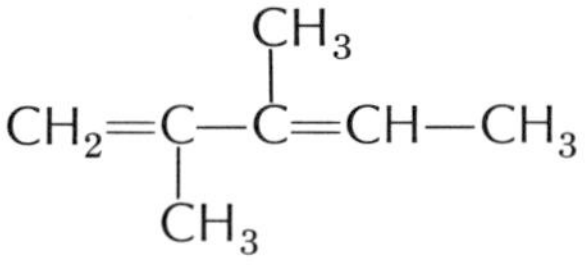 $CH_2{=}\underset{CH_3}{\underset{|}{C}}{-}\overset{CH_3}{\overset{|}{C}}{=}CH{-}CH_3$

   b. [cyclohexene ring bearing CH_3 on the ring double-bond carbon, with an exocyclic $=C(CH_3)_2$ group]
   CH₃ CH₃ CH₃

   c. $CH_3{-}CH{=}CH{-}\underset{CH_3}{\underset{|}{C}}{=}\overset{CH_3}{\overset{|}{C}}{-}CH{=}CH{-}CH_3$

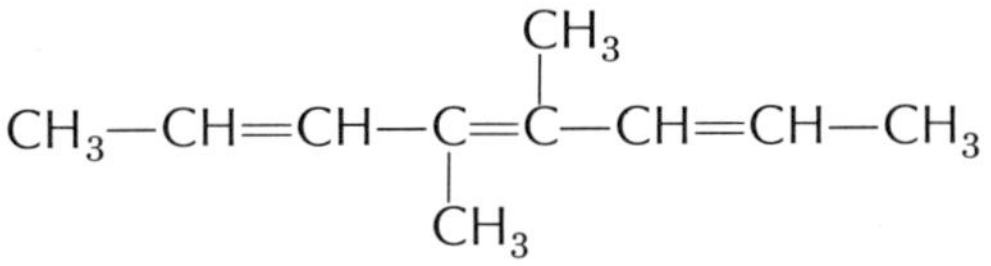

13. How can the following pair of compounds be distinguished from each other?

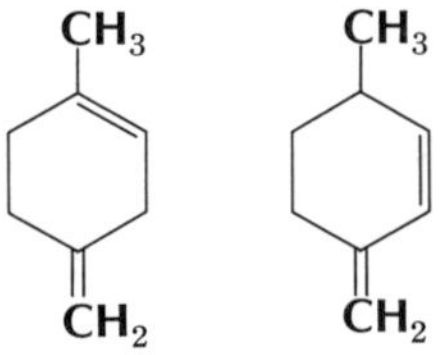

[two six-membered rings, each with CH_3 at top and exocyclic $=CH_2$ at bottom, differing in ring double-bond position]
CH_3 CH_3
CH_2 CH_2

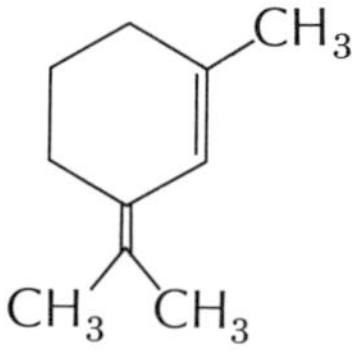

18

AROMATIC HYDROCARBONS

The fragrant oils of wintergreen, clove, and almond belong to a class of compounds called *aromatic* because of their odor. Chemically these and other members of the class are distinguished by a low hydrogen-to-carbon ratio, which indicates a high degree of unsaturation. They are, however, unreactive toward many reagents of the type discussed in Chapter 17 that typically react with carbon–carbon multiple bonds. This baffling conflict of evidence has intrigued chemists of several generations.

This chapter studies the structure and properties of aromatic compounds in general through the simplest member of the class, *benzene*. Benzene and other aromatics are obtained in quantity from coal tar, the tarry distillate that results when coal is heated in the absence of air to form coke, a form of carbon. Coal tar constitutes less than 5 percent of the weight of the coal but is available in large quantities as a result of the demand for coke by industries such as steel. Today benzene and other aromatic hydrocarbons are produced by altering the structure of petroleum hydrocarbons. The study of their unique structures and reactivities constitutes a large branch of organic chemistry.

18.1 THE BENZENE PROBLEM

EXPERIMENTAL OBSERVATIONS

The formula for benzene was established in the mid-nineteenth century as C_6H_6 on the basis of chemical analysis and molecular mass determination. The hydrogen-to-carbon ratio is quite low and suggests the presence of several unsaturated carbon atoms. However, benzene does not decolorize bromine, nor is it oxidized readily by permanganate, in sharp contrast to unsaturated compounds such as alkenes and alkynes. Benzene normally does not undergo addition reactions, but in a manner reminiscent of the chemistry of alkanes it can be made to undergo substitution reactions. The reaction conditions and the mechanisms, however, are distinctly different from those involving alkanes.

An example of the type of substitution reaction that benzene commonly undergoes is bromination. Bromine will substitute for a hydrogen atom in the presence of an iron catalyst, and only one organic compound, C_6H_5Br (bromobenzene), is obtained.

$$C_6H_6 + Br_2 \xrightarrow{Fe} C_6H_5Br + HBr$$

This observation could indicate that only one particular hydrogen is selectively replaced or that all hydrogen atoms are equivalent. It is now known that the latter alternative is correct because no other compounds isomeric with bromobenzene ever have been detected even in trace amounts in this reaction. A further indication of the structure of benzene is derived from its reaction with 3 moles of hydrogen in the presence of a catalyst, such as rhodium, with the formation of cyclohexane:

$$C_6H_6 + 3H_2 \xrightarrow{\text{catalyst}} \text{cyclo-}(CH_2)_6$$

This reaction is much more difficult than the hydrogenation of alkenes and alkynes. On the basis of its bromination and hydrogenation benzene now is considered a cyclic, six-membered system containing one hydrogen attached to each carbon atom. The problem now is to arrive at a final representation for benzene, accounting for the low hydrogen-to-carbon ratio without employing double bonds.

F. A. Kekulé suggested in 1865 that benzene could be represented with alternating single and double bonds. There are two equivalent ways in which the bonds may be arranged:

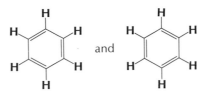

and

Kekulé suggested that the two forms could not be isolated because the bonds alternated positions so rapidly that the resultant molecule was different from unsaturated compounds. It was thought that this special cyclic arrangement of alternating double bonds somehow accounted for the chemistry of benzene and other aromatic compounds.

In more modern terms the two Kekulé forms, which differ only in

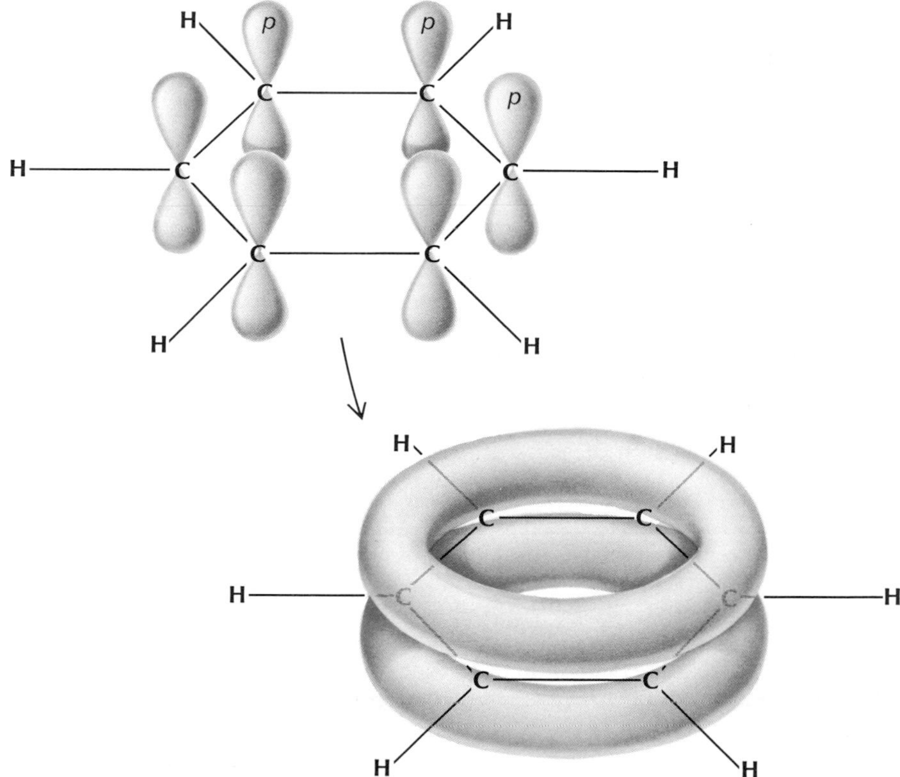

Figure 18.1
Structural representation of a π electron cloud in benzene.

their arrangement of π electrons, are considered to be resonance forms of a molecule that cannot be described adequately with a single structural representation. Each carbon in benzene is considered to be bound to two neighboring carbons and to one hydrogen by sp^2 hybrid orbitals. In addition a *p* orbital perpendicular to the plane of the ring and containing one electron is available for overlap with neighboring *p* orbitals to form π bonds. Because the two neighboring carbon atoms are equidistant, there is no reason why overlap should occur with one specific neighboring *p* orbital to yield ultimately one of the Kekulé forms. Rather, each *p* orbital overlaps with both adjacent *p* orbitals to produce a *delocalized* (that is, over many atoms) cloud of electrons above and below the plane of the ring (Figure 18.1). The bonding between any two carbon atoms is neither single nor double but rather an intermediate bond type. Experimentally it has been shown that all carbon–carbon bond distances in benzene are 1.39 Å. This value is intermediate between a single bond (1.54 Å) and a double bond (1.34 Å).

RESONANCE ENERGY

Benzene contains bonds that resemble double bonds, but they do not react in the same manner as the localized π bonds in alkenes and alkynes. The difference in reactivity reflects the delocalization of the π electrons, which results in a stabilization of the molecule. The π network is not easily destroyed by chemical reagents, and the reactions of benzene proceed with the preservation of the π system.

As is the case in many other chemical reactions it is possible to evaluate the energy terms associated with a reaction and to apply these to furthering our understanding of molecular structure. When unsaturated molecules react with hydrogen, heat is evolved. The quantity of energy evolved per mole of compound (an enthalpy change) is called its *heat of hydrogenation*. The heat of hydrogenation of alkenes averages 28 kcal/mole. As a model for benzene, cyclohexene (whose heat of hydrogenation is 28.6 kcal/mole) can be used as a reference compound:

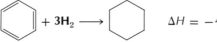

$$\text{cyclohexene} + \mathbf{H_2} \longrightarrow \text{cyclohexane} \qquad \Delta H = -28.6\ kcal/mole$$

If benzene contained three double bonds its heat of hydrogenation would be expected to be three times that of cyclohexene, or 85.8 kcal/mole. This value is for a hypothetical molecule that could be called cyclohexatriene. The experimental heat of hydrogenation of benzene is 49.8 kcal/mole, or 36 kcal/mole less energy than predicted by the model.

$$\text{benzene} + \mathbf{3H_2} \longrightarrow \text{cyclohexane} \qquad \Delta H = -49.8\ kcal/mole$$

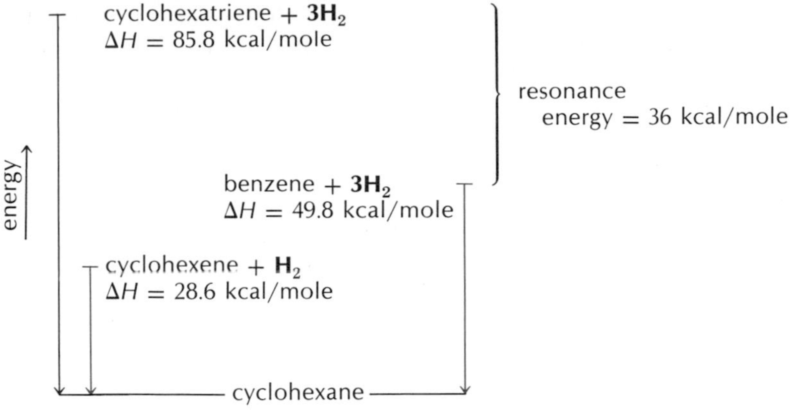

Figure 18.2
Resonance energy of benzene.

Evidently benzene is different in its behavior toward hydrogen than hypothetical cyclohexatriene with its three normal double bonds. This difference indicates that the π system of benzene is more stable than predicted from the model compound (Figure 18.2). The difference in kilocalories per mole, which represents the stability of benzene, is called the *resonance energy*. While the term *resonance energy* of aromatic compounds and its specific value of 36 kcal/mole for benzene is indicative of the stability of the delocalized π system, it should be emphasized that it is more accurately a measure of our inability to correctly represent benzene in simple terms.

18.2 NOMENCLATURE FOR AROMATIC COMPOUNDS

As has been the case for a variety of other compounds, chemical research on aromatics developed more rapidly than did interest in developing a systematic method of nomenclature. Consequently, many aromatic compounds are known by their common or trivial names. The general class name of aromatic compounds is *arenes,* and the removal of a hydrogen atom from an arene produces a fragment called an *aryl* group. The benzene ring with one hydrogen removed is called a *phenyl* group:

H H H H H

phenyl group

Thus for compounds in which a benzene ring is attached to a substantially larger collection of carbon atoms the term *phenyl* is used as a prefix similar to the use of methyl.

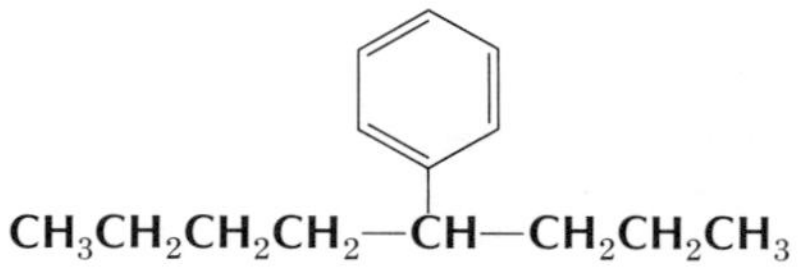

$CH_3CH_2CH_2CH_2—CH—CH_2CH_2CH_3$

4-phenyloctane

In the case of replacement of a hydrogen of benzene by one or more simple atoms or collections of atoms, it is preferable to name the compounds as substituted benzenes. Many simple benzene derivatives are known by their common names:

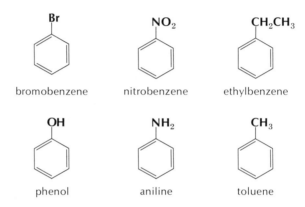

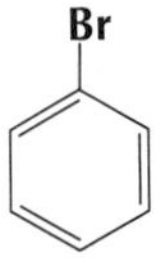

Br — bromobenzene

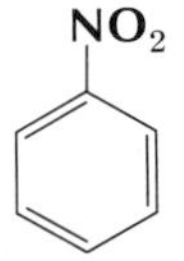

NO_2 — nitrobenzene

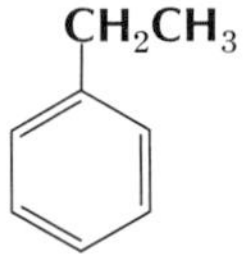

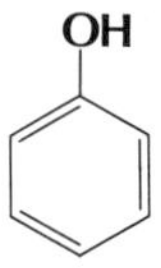

CH_2CH_3 — ethylbenzene

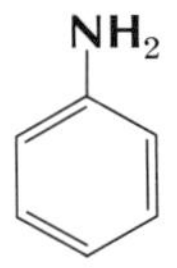

OH — phenol

NH_2 — aniline

CH_3 — toluene

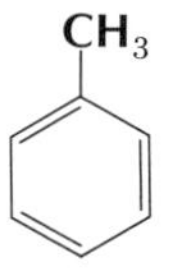

When two or more groups are substituted on benzene a method for designating their relative position on the molecule is required. Substitution of two groups leads to three possible isomers. Prefixes *ortho*- (*o*), *meta*- (*m*), and *para*- (*p*) are used for disubstituted benzenes.

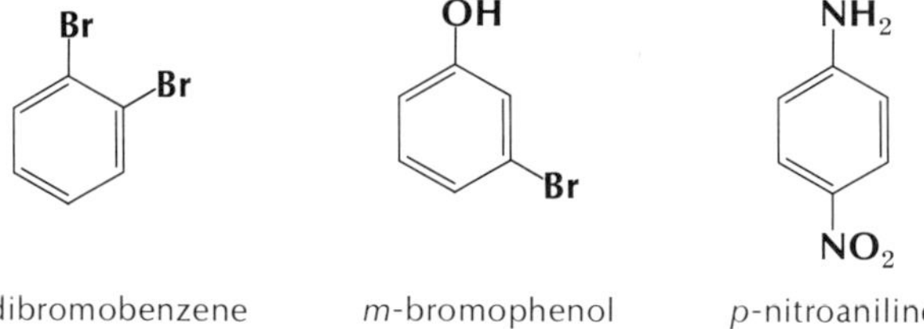

o-dibromobenzene *m*-bromophenol *p*-nitroaniline

If three or more substituents are present a numbering system is used. A carbon atom bearing a substituent is designated as number 1, and the ring is numbered so that substituents are located on low-numbered atoms. The choice of the carbon to be designated as 1 is not easily or unambiguously determined in all cases. Occasionally it is done alphabetically; sometimes the most reactive or important substituent is chosen, and this may be a function of the interest of the individual chemist. In addition, since the combination of certain groups with a benzene ring often is known by its common name, the choice may be dictated by a desire to employ the common name and thus shorten the final name. In this case the atom bearing the substituent that is the basis for the common name is labeled 1.

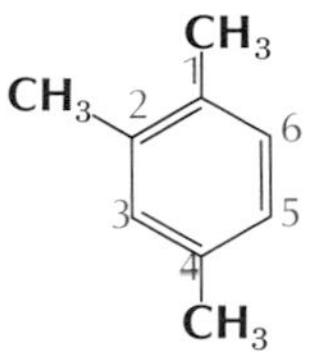

1,2,4-trimethylbenzene

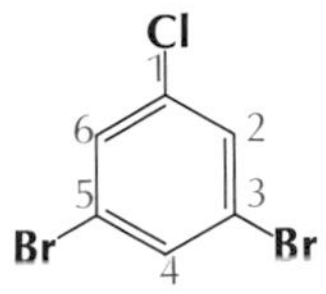

1-chloro-3,5-dibromobenzene

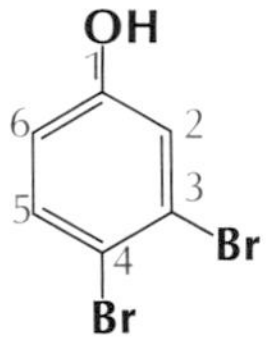

3,4-dibromophenol

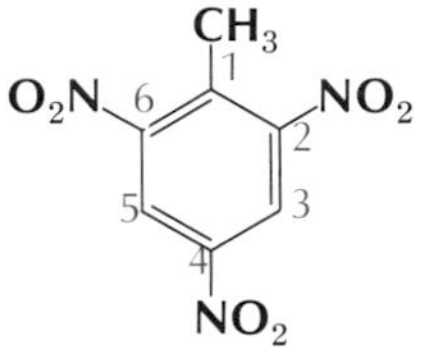

2,4,6-trinitrotoluene

18.3 MECHANISM OF AROMATIC SUBSTITUTION

The special stability of the π electron system of benzene decreases its susceptibility to attack by electrophilic reagents. Thus weak electrophiles that will react with alkenes will not react with benzene under comparable conditions. It is possible to increase the electrophilicity of a potential electrophilic reagent by use of Lewis acid catalysts, which react with the nucleophilic portion of the reagent and remove it with an electron pair. The electrophilic portion of the molecule is left without the pair of electrons used in bonding it to the nucleophilic portion of the molecule and, therefore, its electrophilicity is greater than that of the original molecule in the absence of a catalyst.

In the reaction of benzene with halogens the attacking electrophile is a halonium ion or, in the specific case of bromination, a bromonium ion. The electron-deficient bromonium ion is produced by the reaction of bromine with the Lewis acid $FeBr_3$, which is generated from the reaction of bromine with the iron catalyst:

$$|\overline{\underline{Br}}—\overline{\underline{Br}}| + FeBr_3 \longrightarrow |\overline{\underline{Br}}^+ + FeBr_4^-$$

The electrophilic Br^+ attacks the π electron system of benzene, causing the withdrawal of an electron pair to form a σ bond with the electrophile. This results in an intermediate in which the special stability of the aromatic system has been altered. The positive charge of the intermediate carbonium ion is distributed over the ring, as indicated by the three resonance forms in the following equation:

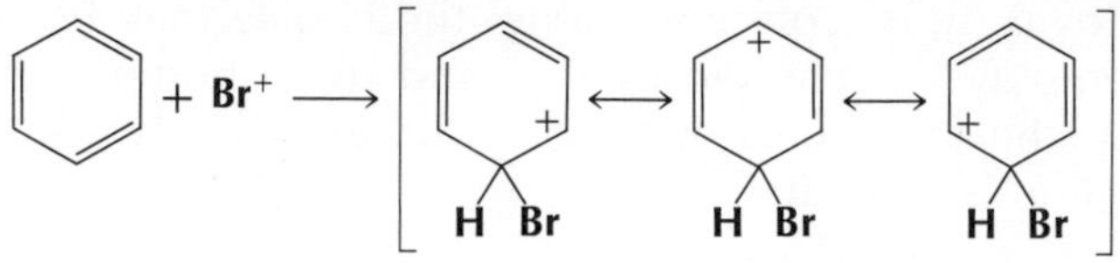

In contrast to the reaction of alkenes with electrophiles, a second step involving neutralization of the positive charge of the carbonium ion by a nucleophile does not occur in benzene and other aromatic compounds. Hypothetically, addition of a bromide ion Br^- to the intermediate carbonium ion would result in two dibromides. These products are not observed.

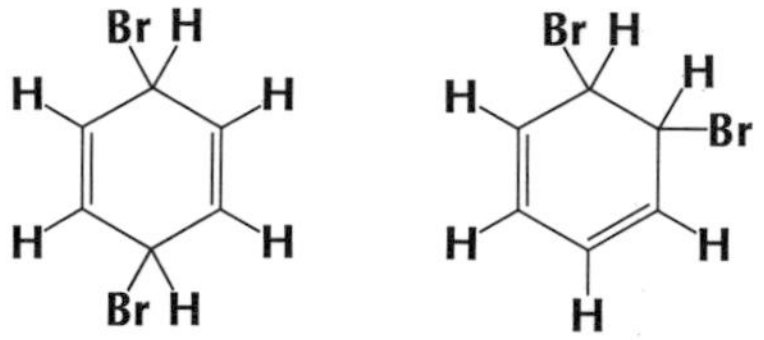

Addition of bromide ion to the carbonium ion would result in a nonaromatic product. On the other hand, if a proton is ejected from the intermediate carbonium ion the resonance stabilized aromatic ring is regenerated.

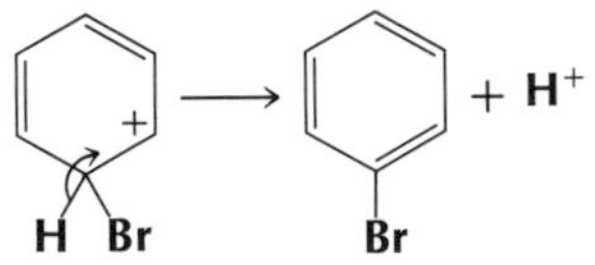

This second alternative is energetically more feasible, as it will form a product approximately 36 kcal/mole lower in energy, which is the resonance energy of the parent substance benzene.

18.4 SUBSTITUTION REACTIONS

HALOGENATION

Bromination or chlorination of aromatic compounds occurs in the presence of iron or an iron halide catalyst. The attacking electrophiles are Br^+ and Cl^+, respectively. Iodination of aromatic compounds can be carried out with some experimental modifications, but the resultant compounds are not as widely used. Direct fluorination to produce monofluoroaromatic compounds cannot be accomplished selectively. An indirect route involving the replacement of an amino group ($—NH_2$) from substituted compounds, such as aniline, is used. While few halogenated aromatic compounds are of commercial importance, *p*-dichlorobenzene is used as moth crystals. The chlorination of chlorobenzene produces a mixture of *ortho* and *para* isomers:

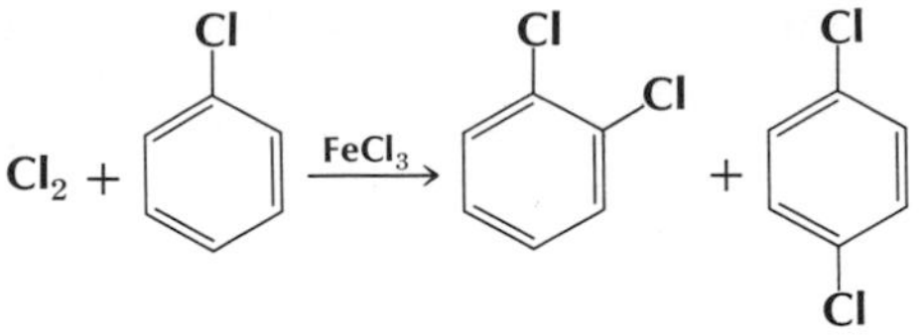

$$Cl_2 + C_6H_5Cl \xrightarrow{FeCl_3} o\text{-}C_6H_4Cl_2 + p\text{-}C_6H_4Cl_2$$

If the aromatic compound has an attached alkyl side chain, the reaction conditions can be altered to produce products resulting from a free radical pathway like that for halogenation of alkanes. The products derived from this free radical reaction are halogen substituted at the carbon directly attached to the benzene ring rather than at any other position.

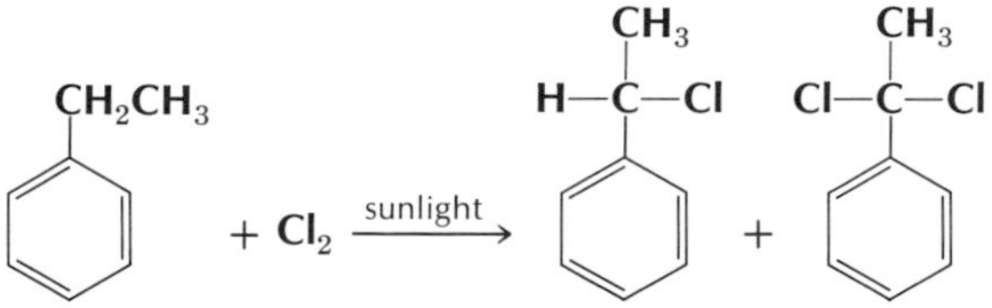

$$C_6H_5CH_2CH_3 + Cl_2 \xrightarrow{\text{sunlight}} C_6H_5\text{—CH(Cl)—}CH_3 + C_6H_5\text{—CCl}_2\text{—}CH_3$$

This indicates that the radical produced by hydrogen abstraction from the carbon atom attached directly to the ring is more stable than any alternate hydrocarbon radical. The unpaired electron in a *p* orbital adjacent to the π system is said to be resonance stabilized since the electron deficiency of the radical is shared by several carbon atoms.

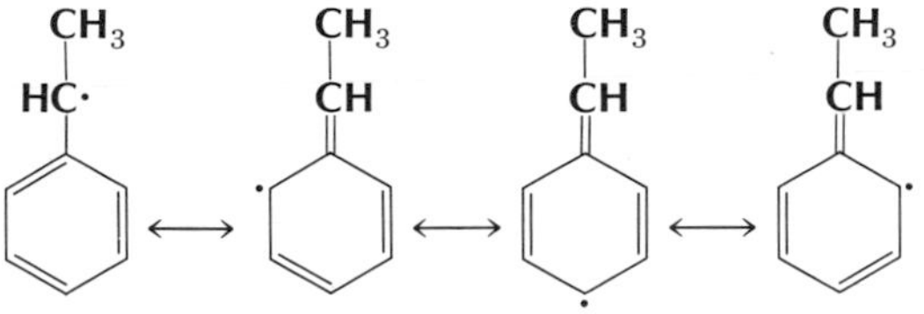

$$C_6H_5\text{—}\dot{C}H\text{—}CH_3 \longleftrightarrow \longleftrightarrow \longleftrightarrow$$

The specific halogenation of the carbon atom attached directly to the ring is just one of many examples of the special reactivity of this carbon center. Other positions of a side chain behave like the carbons of alkanes and their derivatives. In later chapters this special reactivity of carbon directly attached to an aryl group will be encountered again.

NITRATION

Aromatic compounds can be nitrated by nitric acid in the presence of sulfuric acid. The use of sulfuric acid is another example of the need to use an acid catalyst to assist in the reaction of aromatic compounds. The active electrophile is the *nitronium ion* (NO_2^+):

$$H_2SO_4 + HNO_3 \longrightarrow HSO_4^- + H_2NO_3^+$$
$$H_2NO_3^+ \longrightarrow H_2O + NO_2^+$$

As is the case with all electrophilic aromatic substitution reactions, the type of catalyst and temperature that must be employed are functions of the reactivity of the aromatic compound. Benzene can be nitrated at 50° using concentrated nitric acid and sulfuric acid. Some very active aromatic compounds will react with nitric acid directly without the need for sulfuric acid. Aromatic molecules containing several nitro groups are unstable and frequently explosive, the best-known example being 2,4,6-trinitrotoluene (TNT), which can be obtained by multiple nitration of toluene.

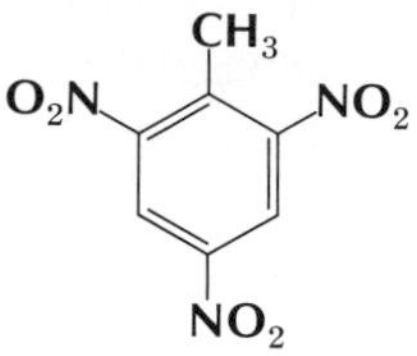

CH_3 O_2N NO_2 NO_2

SULFONATION

Sulfuric acid at high temperatures can sulfonate benzene and other arenes to produce *aryl sulfonic acids.* The attacking electrophile is a protonated sulfur trioxide ion, which results from a proton transfer in which sulfuric acid behaves as both an acid and a base:

$$2H_2SO_4 \longrightarrow H_3SO_4^+ + HSO_4^-$$

$$H_3SO_4^+ \longrightarrow H_2O + HSO_3^+$$

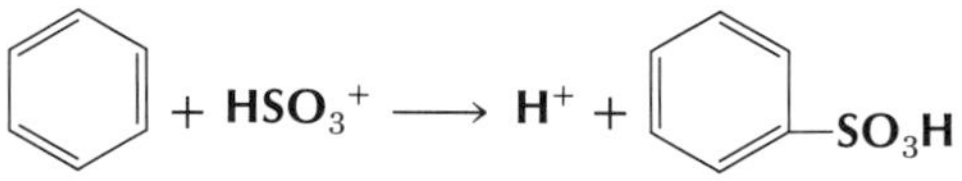

$$C_6H_6 + HSO_3^+ \longrightarrow H^+ + C_6H_5\text{—}SO_3H$$

ALKYLATION

Alkylation, the substitution of an alkyl group in benzene derivatives, can be accomplished by the combined action of aluminum halides and alkyl halides. The aluminum halide serves as a Lewis acid catalyst and allows the formation of a powerful electrophile, the alkyl carbonium ion. The carbonium ion is probably closely associated with $AlCl_4^-$ as a complex aluminum salt.

$$CH_3CH_2Cl + AlCl_3 \rightleftharpoons CH_3CH_2^+AlCl_4^-$$

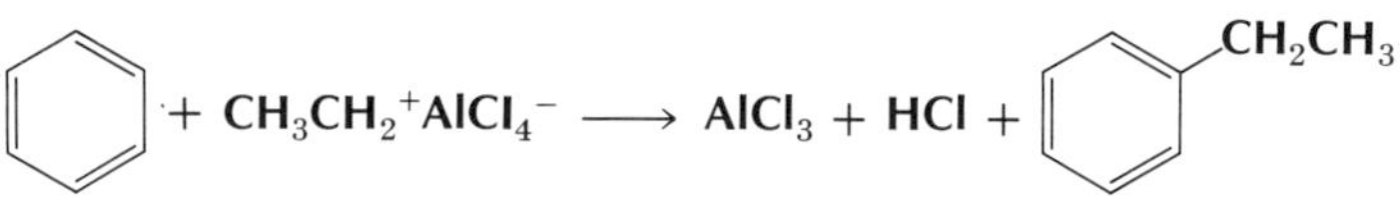

$$C_6H_6 + CH_3CH_2^+AlCl_4^- \longrightarrow AlCl_3 + HCl + C_6H_5CH_2CH_3$$

The use of aluminum chloride as a catalyst for the alkylation reaction was discovered in 1878 by Charles Friedel and James Crafts. Like many important reactions that are referred to by the names of the chemists who first pioneered their use, this alkylation reaction is called the Friedel-Crafts reaction.

A closely related reaction involves the use of alkenes instead of alkyl halides. Alkenes are produced in large quantities from petroleum, and alkylation is cheaper when these items of commerce are used. Ethylbenzene is produced from benzene and ethylene using an aluminum chloride-hydrogen chloride catalyst. The catalyst reacts with the alkene to produce a carbonium ion that then alkylates the aromatic ring:

$$CH_2{=}CH_2 + HCl + AlCl_3 \longrightarrow CH_3CH_2{}^+AlCl_4{}^-$$

With higher molecular mass alkenes, the alkyl aromatic compound produced contains a branched chain, as would be predicted from Markownikoff's rule:

$$CH_3{-}CH{=}CH_2 + HCl + AlCl_3 \longrightarrow CH_3{-}\overset{+}{C}H{-}CH_3 + AlCl_4{}^-$$

$$CH_3{-}\overset{+}{C}H{-}CH_3\ AlCl_4{}^- + C_6H_6 \longrightarrow C_6H_5{-}\underset{H}{\overset{CH_3}{C}}{-}CH_3 + HCl + AlCl_3$$

A variety of detergents has been made utilizing the alkylation reaction in conjunction with the sulfonation reaction previously discussed. One example involves the use of a *tetramer* (four monomer units) of propylene, which is obtained by a partial polymerization process. The tetramer still contains a double bond that can be used to alkylate benzene. Sulfonation of the resultant alkylbenzene produces the sulfonic acid that in the form of the sodium salt is a detergent (Chapter 21):

$$C_{12}H_{24} + C_6H_6 \xrightarrow[AlCl_3]{HCl} C_6H_5{-}C_{12}H_{25}$$

$$H_2SO_4 + C_6H_5{-}C_{12}H_{25} \longrightarrow o\text{-}C_{12}H_{25}C_6H_4SO_3H + p\text{-}HO_3S{-}C_6H_4{-}C_{12}H_{25}$$

Another example of the commercial use of alkylbenzenes is conversion of ethylbenzene to the very useful polymer polystyrene. Styrene can be produced from ethylbenzene by a commercial dehydrogenation process and polymerizes even more readily than ethylene:

$$C_6H_5{-}CH_2CH_3 \longrightarrow C_6H_5{-}CH{=}CH_2 + H_2$$

$$n\ C_6H_5{-}CH{=}CH_2 \longrightarrow {-}\underset{C_6H_5}{CH}{-}CH_2{-}\underset{C_6H_5}{CH}{-}CH_2{-}$$

18.5 OXIDATION OF SIDE CHAINS

A striking example of the stability of the benzene ring is provided by the behavior of alkylbenzenes subjected to strong oxidizing agents, such as potassium permanganate. Unsaturated compounds are oxidized with ease with this reagent (Section 17.4), while saturated compounds are attacked only under vigorous conditions. Even under conditions where the saturated hydrocarbon side chain is attacked, the aromatic ring remains unscathed. Toluene is oxidized to a compound called benzoic acid. (The CO_2H group and its properties are discussed in Chapter 21.)

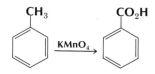

$$C_6H_5CH_3 \xrightarrow{KMnO_4} C_6H_5CO_2H$$

Longer chains also are attacked and produce the same product. All the carbon atoms in the chain, save the one attached directly to the ring, are lost as carbon dioxide.

$$C_6H_5CH_2CH_2CH_3 \xrightarrow{KMnO_4} C_6H_5CO_2H + 2\ CO_2$$

The oxidation of side chains provides a means of introducing the carboxyl function ($—CO_2H$) into an aromatic compound. Other methods can be used, but a discussion of them will be postponed until Chapter 21. More important than the synthetic possibilities of the oxidation reaction is the ability to determine the position of the side chain in a complicated aromatic molecule of unknown substitution pattern. The position of the final carboxyl group ($—CO_2H$) and the determination of the number of carbon atoms lost in the oxidation provide information about the structure of the unknown compound. Most of the simple substituted benzoic acids are known, and the product of the reaction of an unknown compound can be identified by comparing its physical properties with those of the known benzoic acids. As an example of this technique, it can be readily seen that the two isomers, ethylbenzene and *p*-dimethylbenzene (*p*-xylene), can be distinguished by the number of carbon atoms retained in the oxidation product.

$$C_6H_5CH_2CH_3 \xrightarrow{KMnO_4} C_6H_5CO_2H + CO_2$$

ethylbenzene

$$p\text{-}CH_3C_6H_4CH_3 \xrightarrow{KMnO_4} p\text{-}HO_2CC_6H_4CO_2H$$

p-dimethylbenzene

A compound C_9H_{10} reacts with potassium permanganate to yield the following diacid:

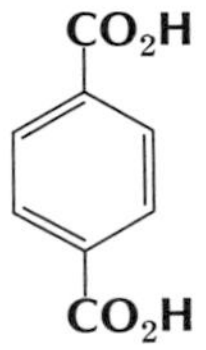

What is the structure of C_9H_{10}?

The structure of the diacid indicates that there are only two points of attachment of side chains to the benzene ring and that these are *para* to each other. Since six of the carbons of C_9H_{10} are contained in the ring, the remaining three carbon atoms must be contained in the side chains.

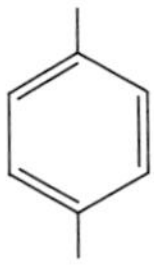

+ 3 carbon fragment(s)

There is only one possible way to arrange three carbon atoms at two positions with a *para* relationship to each other. A one-carbon fragment must be at one position and a two-carbon fragment at the other. *p*-Ethylmethylbenzene looks like a reasonable structure at first glance:

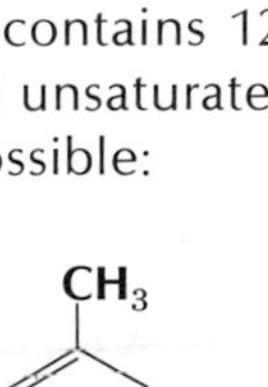

However, this structure contains 12 hydrogens and cannot be correct. An additional unsaturated bond must be present. Only one structure is possible:

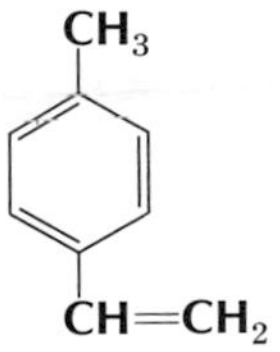

18.6 REACTIVITY AND ORIENTATION IN SUBSTITUTION REACTIONS

EXPERIMENTAL OBSERVATIONS

The reaction of benzene with an electrophile produces only one monosubstituted product because all hydrogens on benzene are identical. If a substituent is already on the benzene ring and a second substituent is to be added, two questions are immediately prompted. First, will the second substituent become attached to the aromatic ring more or less readily than to benzene itself? Second, since the various hydrogens are no longer all equivalent in a monosubstituted benzene derivative, which of the three possible disubstituted products will be produced and in what amounts? The first question deals with *reactivity;* the second, with *orientation.*

Substituents attached to the ring can be divided into two classes according to their effects on reactivity and orientation. One group of substituents increases the rate of further substitution and tends to produce *ortho* and *para* products. The second group of substituents deactivates the ring toward further substitution and tends to produce *meta* isomers. These observations can be arrived at by examining the following two sequences of reactions to form chloronitrobenzenes:

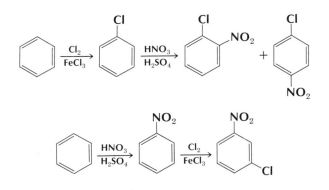

$$C_6H_6 \xrightarrow[FeCl_3]{Cl_2} C_6H_5Cl \xrightarrow[H_2SO_4]{HNO_3} o\text{-}ClC_6H_4NO_2 + p\text{-}ClC_6H_4NO_2$$

$$C_6H_6 \xrightarrow[H_2SO_4]{HNO_3} C_6H_5NO_2 \xrightarrow[FeCl_3]{Cl_2} m\text{-}ClC_6H_4NO_2$$

The product is determined by the order in which the substitutions are carried out. Chlorination followed by nitration produces the *ortho* and *para* isomers. Nitration followed by chlorination yields essentially only the *meta* isomer. What effect does the group already attached to the ring have on the orientation that the second group assumes? What preference for a particular position does the second entering group possess? Answers to these two questions are obtained by examining two additional reactions:

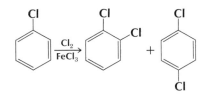

$$C_6H_5Cl \xrightarrow[FeCl_3]{Cl_2} o\text{-}C_6H_4Cl_2 + p\text{-}C_6H_4Cl_2$$

$$\text{C}_6\text{H}_5\text{NO}_2 \xrightarrow[\text{H}_2\text{SO}_4]{\text{HNO}_3} m\text{-C}_6\text{H}_4(\text{NO}_2)_2$$

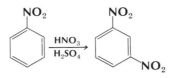

Further chlorination of chlorobenzene leads to *ortho* and *para* products, whereas nitration of nitrobenzene produces *meta* product. It is evident that the orientation in the resultant product is not a function of the attacking electrophile but is rather a result of the groups already attached to the ring. All substituents attached to benzene fall into one of two classes, called *ortho-para directors* or *meta directors.* Chlorine is an *ortho-para* director, and the nitro group is a *meta* director.

In addition to observations of the directing properties of substituents it has also been noted that *ortho-para* directing groups usually activate or increase the reactivity of the aromatic ring to which they are attached. The *meta* directors deactivate the aromatic ring and make further substitution difficult. Of the groups mentioned in this chapter the halogens, hydroxyl, amino ($—NH_2$), and alkyl groups are *ortho-para* directors. The nitro, carboxyl ($—CO_2H$), and sulfonic acid groups are *meta* directors.

EXAMPLE 18.2

Indicate how the following compound can be prepared from benzene:

$$p\text{-NO}_2\text{C}_6\text{H}_4\text{CO}_2\text{H}$$

The compound contains two *meta*-directing groups in a *para* relationship to each other. Clearly the introduction of one of the groups onto the benzene ring will direct the second group into the *meta* position. Therefore, an *ortho-para* directing group must be used to establish the proper isomer position, and then one of the groups somehow must be changed by a chemical reaction. Toluene can be produced from benzene by the Friedel–Crafts reaction.

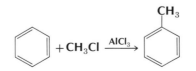

$$\text{C}_6\text{H}_6 + \text{CH}_3\text{Cl} \xrightarrow{\text{AlCl}_3} \text{C}_6\text{H}_5\text{CH}_3$$

Nitration of toluene will yield a mixture of *ortho*- and *para*-nitrotoluene:

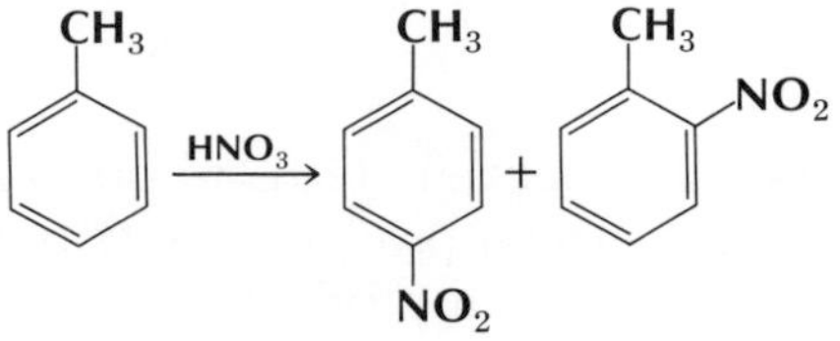

$$\text{C}_6\text{H}_5\text{CH}_3 \xrightarrow{\text{HNO}_3} p\text{-CH}_3\text{C}_6\text{H}_4\text{NO}_2 + o\text{-CH}_3\text{C}_6\text{H}_4\text{NO}_2$$

Isolation of the *para* isomer from the reaction mixture followed by oxidation with potassium permanganate yields the desired compound:

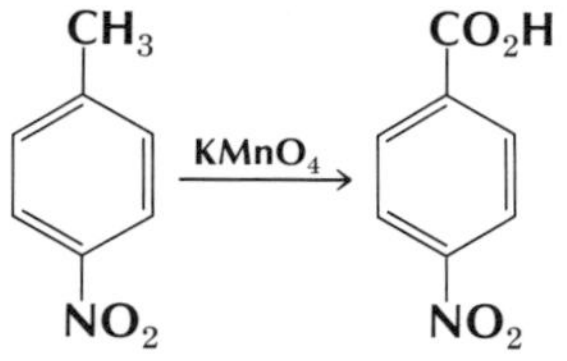

$$p\text{-CH}_3\text{C}_6\text{H}_4\text{NO}_2 \xrightarrow{\text{KMnO}_4} p\text{-HO}_2\text{CC}_6\text{H}_4\text{NO}_2$$

A RATIONALE FOR ORIENTATION RULES

The chemistry of aromatic compounds developed splendidly for many years without an understanding of reaction mechanisms and of electronic structure of molecules. If a particular compound was needed it was obtained by taking care to introduce the substituents in the proper order into the proper aromatic derivative. The orientation rules could be used without understanding why they worked.

It is now possible to rationalize why the orientation rules do work. This rationalization is more than an intellectual exercise. As is the case with other reaction mechanisms presented in this text, it is possible to obtain information about the controlling features of this reaction that applies to other reactions in which the same features are operative.

The attack of an electrophile on an aromatic ring produces an intermediate carbonium ion. Any substituent attached to the ring that can supply electrons to the π system or help disperse the positive charge developed should stabilize the intermediate. For example, the hydroxyl group of phenol contains unshared electrons that can stabilize the intermediate that results from attack of an electrophile (E^+) at the *para* position.

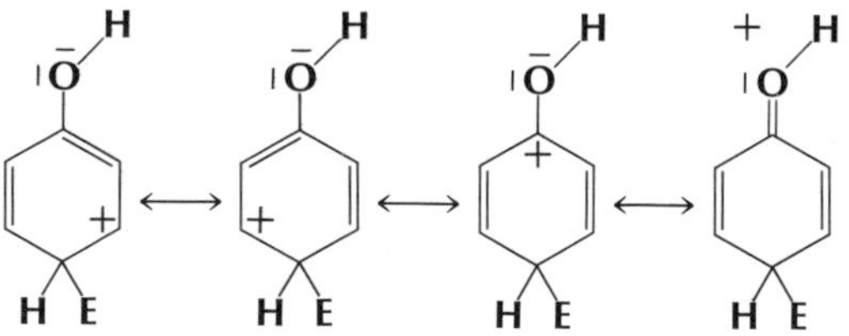

[Resonance structures of the para-attack intermediate of phenol: four structures connected by ↔, each with OH on top and H, E on the bottom carbon; the positive charge on ring carbons in the first three and on oxygen (C=O⁺–H) in the fourth.]

In a similar manner attack at the *ortho* position by the electrophile also enables the oxygen atom to donate its electrons and stabilize the intermediate carbonium ion. However, attack at the *meta* position leads to an intermediate in which the hydroxyl group cannot contribute its electrons to relieve the deficiency:

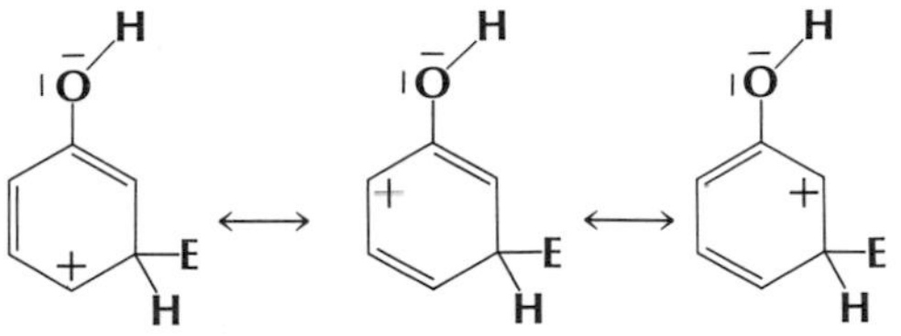

The resultant product is, therefore, a mixture of *ortho* and *para* isomers.

Both the nitro and sulfonic acid groups contain atoms in high positive oxidation states which are attached directly to the ring in benzene derivatives. If a positive charge were placed on the carbon to which either of these groups were attached, the resultant species would be less stable than alternate structures in which the positive charge does not reside on the carbon next to the nitro or sulfonic acid group. One of the resonance forms of the intermediate resulting from attack of an electrophile at either the *ortho* or *para* position places the positive charge on the carbon bearing the *meta*-directing group.

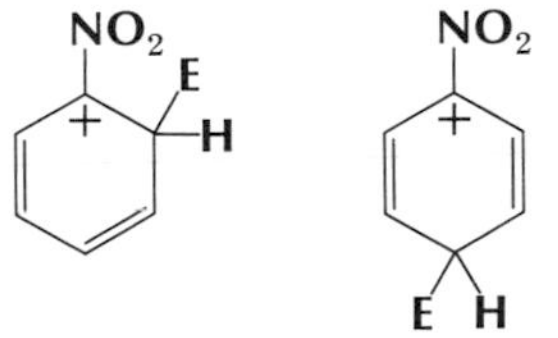

The contribution of these resonance forms destabilizes the intermediates resulting from *ortho* or *para* attack relative to the intermediate resulting from *meta* attack. In the *meta* intermediate the positive charge does not appear on the carbon bearing the *meta*-directing group in any of the resonance forms.

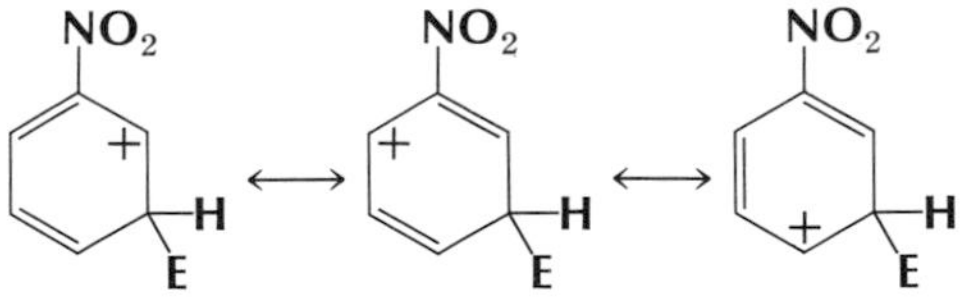

The electronic properties of the *ortho–para* directing groups also provide an explanation for the reactivity of the benzene ring as a function of attached substituents. *Ortho–para* directors are capable of donating electrons to increase the electron density at the *ortho* and *para* positions and thereby to facilitate the reaction of an electrophile with the π system. *Meta*-directing groups withdraw electrons from the aromatic ring and decrease the potential of the ring to supply electrons to the electrophile.

One final word of explanation about orientation rules is necessary. In most aromatic substitutions all possible isomers are obtained. The orientation rules only indicate the major products of the reaction.

18.7
DETERMINATION OF THE STRUCTURE OF ISOMERS

In this chapter it has been tacitly assumed that the orientation of the substituents in a particular compound is known. How is this assignment determined? In practice it is possible at this stage of the development of chemistry to convert compounds of unknown structure by chemical manipulation of the attached groups into groups that are present in compounds of known structure. For example, a given chlorotoluene is established as the *meta* isomer if upon oxidation with potassium permanganate the known *m*-chlorobenzoic acid is obtained:

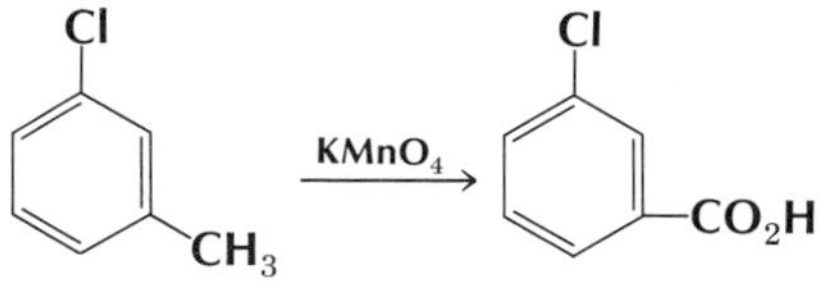

Cl, CH_3 $\xrightarrow{KMnO_4}$ Cl, CO_2H

Physical methods can be employed as well. The three isomeric dibromobenzenes can be assigned structures on the basis of dipole moments. The *para* isomer does not have a dipole moment (Chapter 6) because the molecule is symmetrical and the bond moments of the two carbon–bromine bonds cancel each other. Of the two remaining isomers, the *ortho* isomer is the one that has the larger dipole because the two bond moments are more nearly coincident, and their sum is larger than for the *meta* isomer.

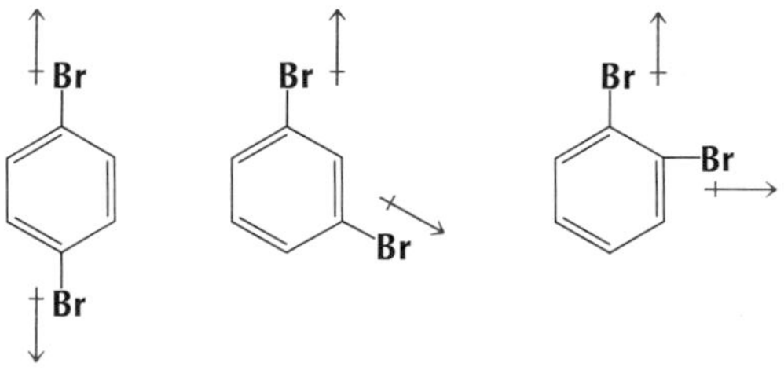

Br Br | Br Br | Br Br

Ultraviolet spectroscopy can be used to confirm or establish the structure of aromatic hydrocarbons. The necessary information is available to the organic chemist in texts devoted to this subject. In addition absorption spectroscopy using light in the infrared region (Section 20.8) is useful in the determination of substitution patterns of aromatic compounds.

18.8
POLYNUCLEAR AROMATIC HYDROCARBONS

The aromatic hydrocarbons all have in common a ring system in which every carbon contributes an electron in a *p* orbital to a cyclic system.

Naphthalene ($C_{10}H_8$) contains two rings in which two carbon atoms are common to both rings. Three Kekulé forms can be written for naphthalene. The compound is regarded as a resonance hybrid of these three forms:

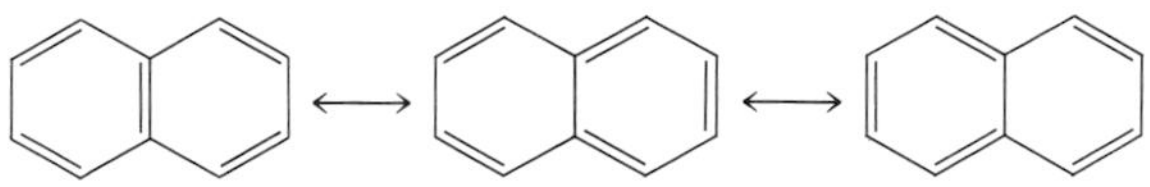

Anthracene ($C_{14}H_{10}$) contains three condensed rings, allowing four Kekulé forms to be written. One of these is drawn below. Phenanthrene is isomeric with anthracene but contains an angular arrangement of rings. Three Kekulé forms can be drawn for phenanthrene:

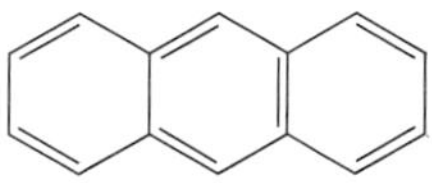

anthracene 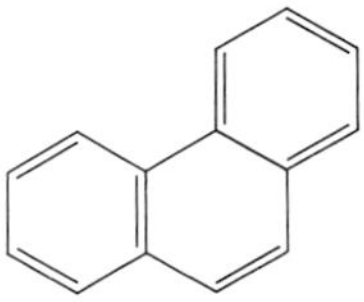 phenanthrene 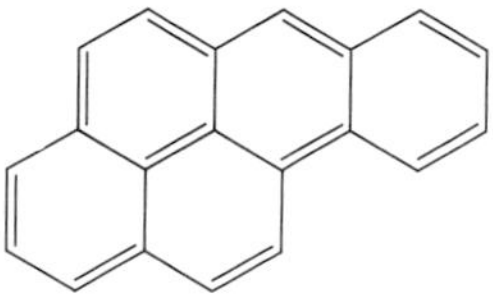 benzpyrene

The five-ring aromatic hydrocarbon benzpyrene is a cancer-producing substance formed in the combustion of many diverse substances, such as cigarettes and the fat that drips onto hot charcoal.

Suggested further readings

Ferguson, L. N., "The Orientation and Mechanism of Electrophilic Aromatic Substitution," *J. Chem. Educ.*, **32,** 42 (1955).

Gero, A., "Kekulé's Theory of Aromaticity," *J. Chem. Educ.*, **31,** 201 (1954).

Lambert, F. L., "Substituent Effects in the Benzene Ring. A Demonstration," *J. Chem. Educ.*, **35,** 342 (1958).

Newell, L. C., "Faraday's Discovery of Benzene," *J. Chem. Educ.*, **3,** 1248 (1926).

Varshni, Y. P., "Directive Influence of Substituents in the Benzene Ring," *J. Chem. Educ.*, **30,** 342 (1953).

Willemart, A., "Charles Friedel (1932–1899)," *J. Chem. Educ.*, **26,** 2 (1949).

Terms and concepts

alkylation
arenes
aromatic
aromatic substitution
aryl group
bromonium ion
carboxyl group
delocalization
heat of hydrogenation
Kekulé
meta
meta director

orientation
ortho
ortho-para director
para
phenyl
reactivity
resonance energy
sulfonation

Questions and problems

1. Name the following compounds:

a. 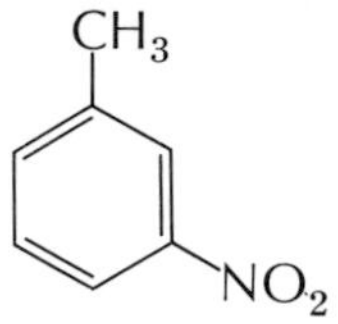 CH_3, NO_2

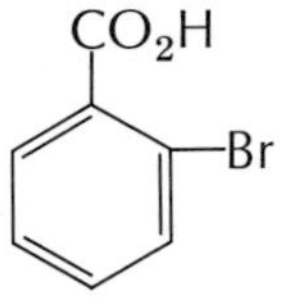

b. 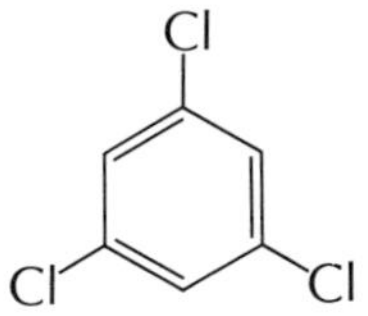 Cl, Cl, Cl

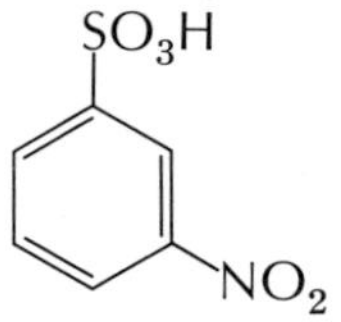

c.  OH, CH_3

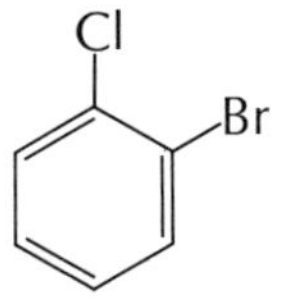

d. 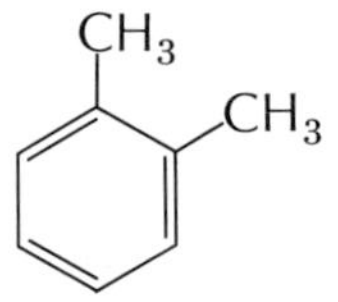 CH_3, CH_3

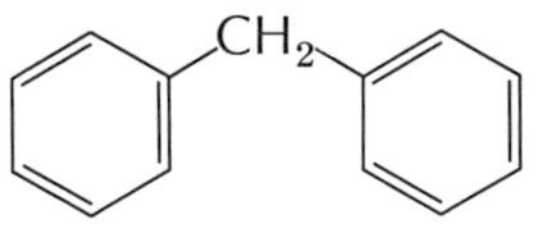

e. 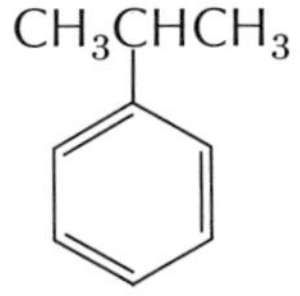 CH_3CHCH_3

f. CO_2H, Br

g. SO_3H, NO_2

h. Cl, Br

i. CH_2

j. 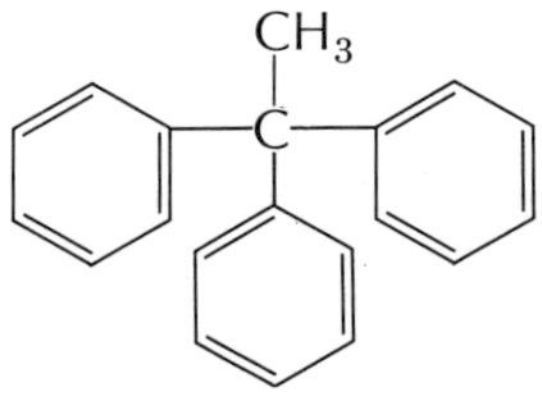 CH_3, C

2. Write structural formulas for the following compounds:

a. 3-phenylpentane
b. dichlorodiphenylmethane
c. *m*-bromotoluene
d. *o*-nitrobenzene sulfonic acid
e. 2,4-dibromo-5-nitrotoluene

3. Write all the structural formulas for the aromatic compounds with the following molecular formulas: **a.** C_8H_{10} **b.** $C_6H_3Br_3$ **c.** $C_{10}H_{14}$

4. How many different monobrominated compounds of naphthalene are there? of anthracene? Draw the structural formulas for the compounds.

5. How many possible dibrominated naphthalenes are there? Draw them.

6. The three isomeric tribromobenzenes have melting points of 120, 87, and 44°. Mononitration of these compounds produces one, two, and three compounds, respectively. Assign the structures to the three tribromobenzenes and their nitrated derivatives.

7. The compound having the following structure can be obtained by oxidation of any of three compounds having the formula $C_{10}H_{14}$. What compounds are these?

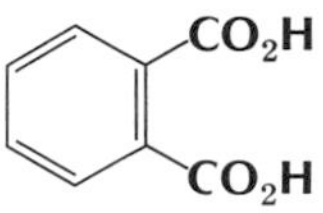

CO_2H

CO_2H

8. A compound $C_{10}H_{12}$ yields upon oxidation the same compound obtained in Problem 7. What is the structure of $C_{10}H_{12}$?

9. Write equations to show how each of the following conversions can be accomplished with a minimum of undesired isomers; several steps may be required:

a. benzene to *p*-chloronitrobenzene
b. benzene to isopropylbenzene
c. benzene to benzoic acid
d. toluene to methylcyclohexane
e. toluene to *p*-nitrobenzoic acid
f. bromobenzene to 3,4-dibromonitrobenzene

10. The heat of hydrogenation of 1,3-cyclohexadiene to cyclohexane is 55.4 kcal/mole of cyclohexadiene. What is the heat of hydrogenation of benzene to cyclohexadiene? Is this process energetically favorable?

11. Suggest chemical tests that would enable you to distinguish visually between members of the following pairs of compounds:

a. cyclohexene and benzene
b. benzene and benzene sulfonic acid
c. aniline and bromobenzene
d. 1-phenyl-1-propyne and 3-phenyl-1-propyne

12. Draw all of the Kekulé representations of anthracene. Do the same for phenanthrene.

13. There are four isomeric aromatic compounds of the molecular formula $C_{18}H_{12}$ that contain four rings. Draw their structures.

19

ALCOHOLS, PHENOLS, AND ETHERS

Virtually all primitive societies have discovered that fermentation of various fruits yields an intoxicating liquid. The distillation of the liquid produces a concentrated solution of a substance known as *alcohol.* Alcohol contains oxygen in addition to the elements carbon and hydrogen usually found in organic materials. The many compounds that are structurally related to the intoxicating liquid are all classed as alcohols. The specific names of the many alcohols are discussed in this chapter along with their physical properties, methods of preparation, and chemical reactions. Two other classes of substances, *phenols* and *ethers,* that also contain oxygen and are structurally related to alcohols are discussed briefly.

19.1 PROPERTIES OF ALCOHOLS, PHENOLS, AND ETHERS

STRUCTURE

Alcohols, phenols, and ethers can be thought of as organic analogs of water. In alcohols a hydroxyl group (—OH) is attached to a saturated carbon atom. In phenols the hydroxyl group is bonded directly to an aromatic ring. The presence of the hydroxyl group in alcohols and phenols leads to physical properties and chemical reactivities distinctly different from those of hydrocarbons. Two distinguishing structural features of alcohols serve to rationalize the physical and chemical properties of alcohols and phenols. These features are (1) the unshared electron pairs on the oxygen atom, lone pairs that serve as a Lewis base and nucleophilic center, and (2) the proton attached to the electronegative oxygen atom that makes the compounds acidic and forms hydrogen bonds. It should be anticipated that the presence of an alkyl or aryl group should attenuate the reactivity of the hydroxyl group, but in general the reactivity is not markedly different from that of water.

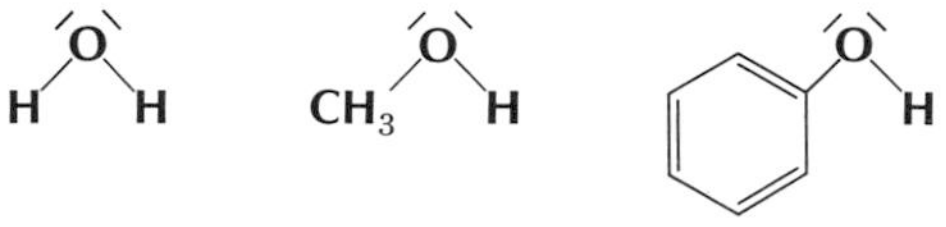

water   methanol   phenol

Ethers differ from alcohols in that they do not contain hydrogen atoms bonded to oxygen. Both hydrogens of water are replaced in a formal sense by carbon-containing fragments such as alkyl or aryl groups. Ethers may be aliphatic, aromatic, or mixed aliphatic-aromatic.

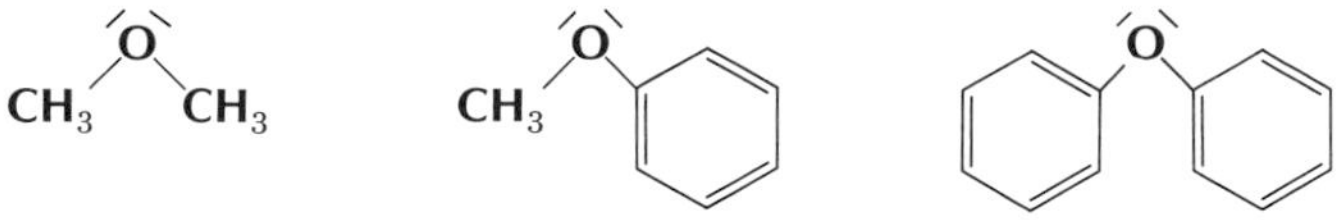

dimethyl ether   methyl phenyl ether   diphenyl ether

The chemical and physical properties of ethers are different from those of alcohols because of the absence of a proton bonded to oxygen. Unlike alcohols, ethers cannot function as acids and cannot hydrogen bond to neighboring ether molecules. However, the presence of the oxygen atom distinguishes ethers from hydrocarbons; ethers can serve as bases and are more polar than the hydrocarbons.

HYDROGEN BONDING IN ALCOHOLS AND PHENOLS

As a class of compounds, the alcohols boil at higher temperatures than alkanes of comparable molecular mass. For example, ethanol (CH_3CH_2—OH) of molecular mass 46 boils at 78° whereas propane of molecular mass 44 boils at −44.5°. This dramatic difference in boiling points is directly attributable to association of alcohols in the liquid state. This association, which involves an attraction between a partially positive hydrogen and the oxygen atom of a neighboring alcohol molecule, is an example of hydrogen bonding not unlike the interactions discussed in earlier chapters in the cases of hydrogen fluoride, water, and ammonia.

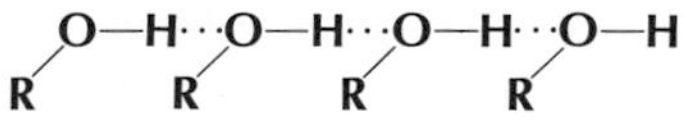

Table 19.1
Properties of normal alcohols

name	*formula*	*boiling point (°C)*	*solubility in H_2O (g/100 g at 25°)*
methanol	CH_3OH	65	completely miscible
ethanol	CH_3CH_2OH	78	completely miscible
propanol	$CH_3CH_2CH_2OH$	97	completely miscible
butanol	$CH_3CH_2CH_2CH_2OH$	118	7.9
pentanol	$CH_3CH_2CH_2CH_2CH_2OH$	138	2.7
hexanol	$CH_3CH_2CH_2CH_2CH_2CH_2OH$	157	0.6
heptanol	$CH_3CH_2CH_2CH_2CH_2CH_2CH_2OH$	176	0.1
octanol	$CH_3CH_2CH_2CH_2CH_2CH_2CH_2CH_2OH$	195	slight

The boiling points of a series of alcohols containing normal alkyl groups are listed in Table 19.1.

Phenols also associate strongly in the liquid phase. The boiling points of phenol and toluene, which are of similar molecular mass, are 182 and 110°, respectively.

The solubility of alcohols and phenols in water distinguishes them from aliphatic and aromatic hydrocarbons. Table 19.1 lists the solubilities of some alcohols containing normal alkyl groups. The lower molecular mass alcohols are miscible in all proportions with water. Introduction of an alcohol into the liquid structure of water apparently can be accomplished with ease if the alkyl group is not large. However, with increasing size of the alkyl group the alcohol more closely resembles an alkane, and the polar hydroxyl group becomes an increasingly smaller contributor to the overall physical properties of the alcohols. As a result, the solubility of alcohols decreases with the increasing size of the alkyl group.

In a similar fashion phenols are soluble in water. Phenol is soluble to the extent of 6.8 g/100 g of water at 25°, whereas toluene is immiscible with water.

ETHERS

Dimethyl ether is a gas that boils at −24°. Diethyl ether is a volatile liquid with a boiling point of 35°. The higher homologs and aromatic ethers are either liquids or solids, depending on molecular mass. Ethers have boiling points similar to those of alkanes of comparable molecular mass. The lower boiling points of ethers as compared to the isomeric alcohols are a reflection of the absence of hydrogen bonds in ethers.

The low molecular mass ethers are only slightly soluble in water. This fact makes them unusually good solvents for the separation of organic compounds from inorganic substances. If ether and water are shaken together with a mixture of inorganic and organic substances, the organic material will dissolve primarily in the ether layer, which separates and rises to the top of the container because it is less dense than water. After separation the ether layer can be removed by evaporation or distillation to yield the organic material in pure form. Diethyl ether ($CH_3CH_2—O—CH_2CH_3$) is most commonly used for this purpose. Its low boiling point is advantageous, but it is extremely flammable.

Although ethers are not soluble in water, they are soluble in strong acidic media since the ether oxygen can be protonated to form an oxonium salt in much the same way that water can be protonated to form the hydronium ion H_3O^+.

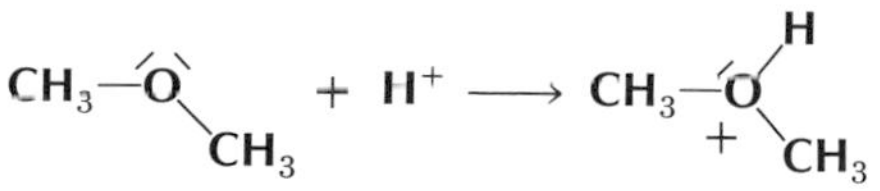

$$CH_3—\ddot{O}—CH_3 + H^+ \longrightarrow CH_3—\overset{+}{O}(H)—CH_3$$

Water is not sufficiently acidic to protonate ethers, and as a result the following equilibrium lies far to the right:

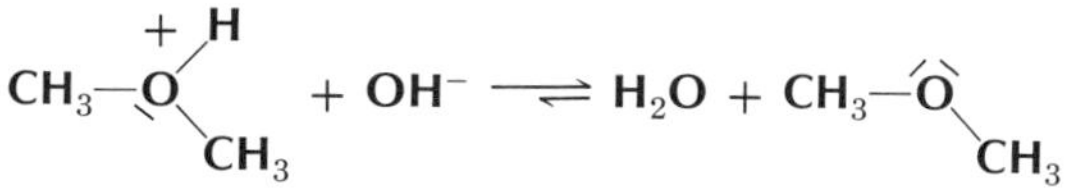

$$CH_3—\overset{+}{O}(H)—CH_3 + OH^- \rightleftharpoons H_2O + CH_3—\ddot{O}—CH_3$$

Sulfuric acid is sufficiently strong to protonate the ether oxygen. Ethers are soluble in concentrated sulfuric acid, and this property may be used to distinguish them from alkanes, which are not soluble in cold acid.

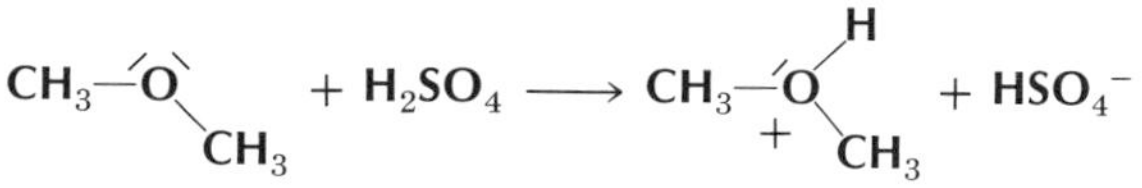

$$CH_3—\ddot{O}—CH_3 + H_2SO_4 \longrightarrow CH_3—\overset{+}{O}(H)—CH_3 + HSO_4^-$$

ACIDITY OF ALCOHOLS AND PHENOLS

The proton attached to the electronegative oxygen atom in alcohols and phenols is acidic. The relative acidities of these two classes of compounds can be explained using the concepts of the electron-releasing ability (*inductive effect*) of alkyl groups and the resonance stabilization of an aromatic ring. Because alkyl groups are electron releasing, the electron density at the oxygen is already higher than in water, making it more difficult to dissociate into an *alkoxide* ion and a proton. Another way of stating this fact is that the increase in the electron density on oxygen due to the effect of the alkyl group makes the alkoxide ion a stronger base. The following equilibrium lies to the left:

$$R—O—H + HO^- \rightleftharpoons R—O^- + H_2O$$

Phenols are stronger acids than water because the conjugate base (*phenoxide* ion) is resonance stabilized. That is, it is of lower energy than a structure in which the negative charge resides exclusively on the oxygen atom.

phenoxide ion: [resonance structures] $\longleftrightarrow$ $\longleftrightarrow$ $\longleftrightarrow$

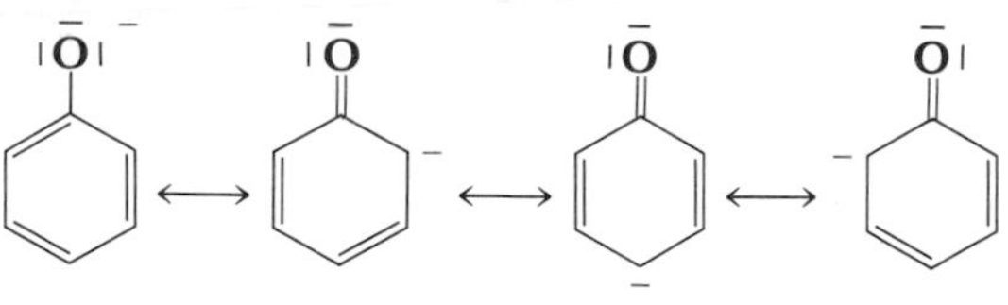

The equilibrium resulting from mixing sodium hydroxide and phenol lies far to the right:

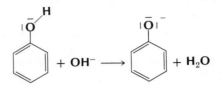

$$C_6H_5OH + OH^- \longrightarrow C_6H_5O^- + H_2O$$

The presence of electron-donating groups on the aromatic ring decreases the acidity of phenols by destabilizing the phenoxide ion. Electron-withdrawing groups increase the acidity by distributing the negative charge over the molecule. The compound 2,4,6-trinitrophenol (picric acid) is 1 billion times more acidic than phenol.

A qualitative test for the type of hydroxyl group in an organic molecule is suggested by the difference between the positions of the equilibrium involving an alcohol and sodium hydroxide and the equilibrium between phenol and sodium hydroxide. Because the sodium salts of phenols are soluble in water, phenols can be extracted from other organic material with an aqueous base. However, high molecular mass alcohols will not dissolve to any appreciable extent.

The reaction of sodium or any other active metal, such as those of the alkali metal family, with alcohol or phenols is similar to the reaction of water with active metals. Hydrogen gas is liberated, and sodium is oxidized to the sodium ion. The sodium alkoxide or phenoxide salts are formed in 100 percent yield from this reaction.

$$2R{-}O{-}H + 2Na \longrightarrow 2R{-}O^-\ Na^+ + H_2$$

19.2 NOMENCLATURE

The lower molecular mass alcohols have common names that frequently are used. The names consist of the alkyl group name plus the term alcohol:

CH_3OH	methyl alcohol
CH_3CH_2OH	ethyl alcohol
$CH_3CH_2CH_2OH$	*n*-propyl alcohol
$CH_3CHOHCH_3$	isopropyl alcohol
$CH_3CH_2CH_2CH_2OH$	*n*-butyl alcohol
$CH_3CH(CH_3)CH_2OH$	isobutyl alcohol
$CH_3{-}C(CH_3)_2{-}OH$	tertiary butyl alcohol

In naming more complex alcohols the IUPAC system is used. The longest chain of carbon atoms that includes the hydroxyl group is chosen as the basis for the name. The ending *–ol* is substituted for the e of the alkane. The position of the hydroxyl group is indicated by the number of the carbon atom to which it is attached, with the numbering arranged such that the hydroxyl group receives the lowest possible number. This deference to the hydroxyl group instead of to the side chain is an indication of the chemical importance of such functional groups. Proper nomenclature is illustrated by the examples below:

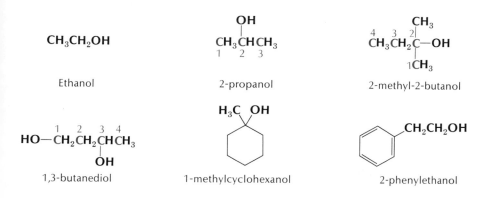

CH_3CH_2OH

Ethanol

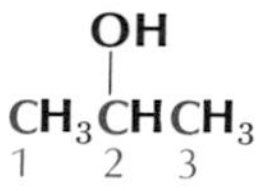

OH
|
CH_3CHCH_3
1 2 3

2-propanol

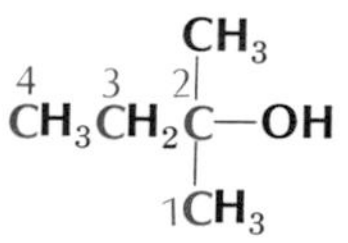

CH₃ above C2; 4 3 2 numbering
$CH_3CH_2C{-}OH$ with CH_3 (1) below

2-methyl-2-butanol

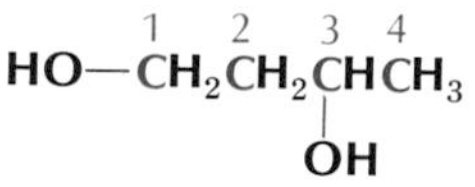

1 2 3 4
$HO{-}CH_2CH_2CHCH_3$
OH (on C3)

1,3-butanediol

H_3C OH

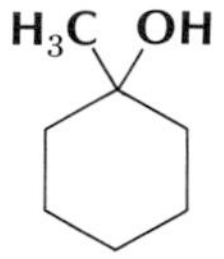

1-methylcyclohexanol

CH_2CH_2OH

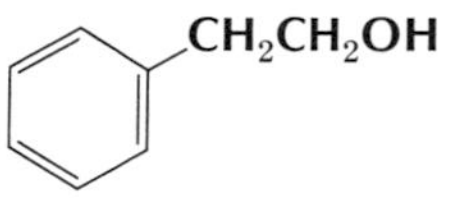

2-phenylethanol

The chemistry of alcohols can be classified according to the type of alkyl group to which the hydroxyl group is attached. An alcohol in which the hydroxyl group is located on a carbon atom to which are bonded one, two, or three alkyl groups is referred to as a *primary, secondary,* or *tertiary* alcohol, respectively. This type of distinction between alkyl groups also was encountered in discussing the free radical reactions of alkanes.

$CH_3CH_2CH_2CH_2OH$

a primary alcohol

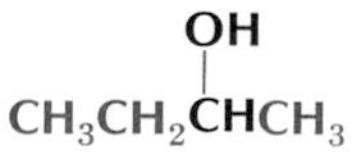

OH
|
$CH_3CH_2CHCH_3$

a secondary alcohol

OH
|
$CH_3{-}C{-}CH_3$
|
CH_3

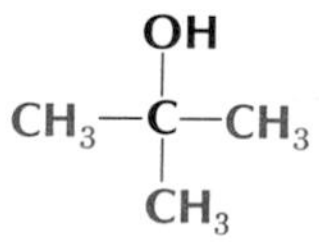

a tertiary alcohol

Phenols are named as derivatives of the parent compound. The term *hydroxyl* is not used in naming phenols. Certain phenols are known by common names, and as a result complex phenols are usually named utilizing the common name as a base.

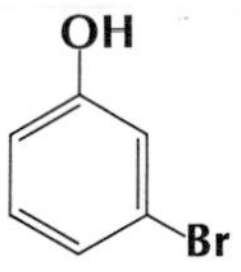

OH, Br

m-bromophenol

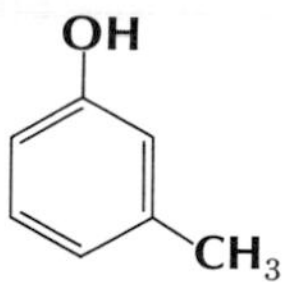

OH, CH_3

m-cresol

OH, OH

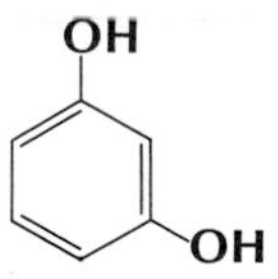

resorcinol

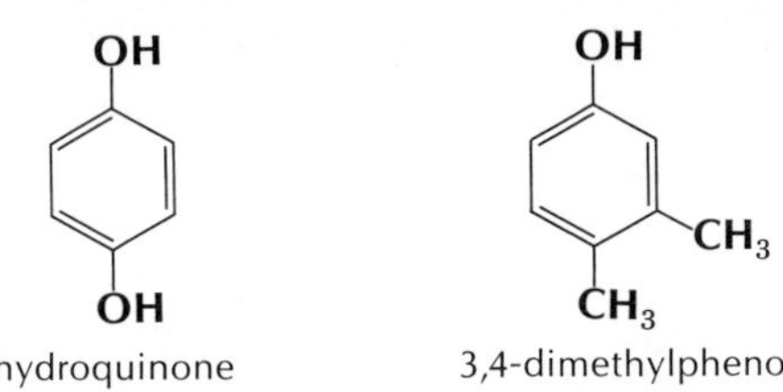

hydroquinone 3,4-dimethylphenol

19.3 COMMON ALCOHOLS, PHENOLS, AND ETHERS

ALCOHOLS

Methanol sometimes is called wood alcohol because it was originally obtained by heating wood to a very high temperature in the absence of air. Under these conditions decomposition occurs, and methanol is one of the volatile substances produced. Temporary blindness, permanent blindness, or death can result from the consumption of methanol present as an impurity in illegal sources of ethanol. The prolonged breathing of methanol vapors also can be a serious health hazard.

Ethanol is the substance popularly known as alcohol, and its effects on animals are well known. In sharp contrast to methanol, ethanol is poisonous only if huge amounts are consumed, which is usually difficult because of physiological reactions. Fermentation of almost any substance containing sugar will produce ethanol but only in concentrations up to 14 percent (28 proof). Higher concentrations must be achieved by distillation. The term *proof* is numerically equal to twice the alcohol percentage in the beverage. A 100-proof liquor contains 50 percent alcohol; the remainder is water and substances that impart flavor and odor.

Ethanol, which is used for commercial purposes, is denatured by the presence of a poisonous substance such as methanol that imparts a disagreeable odor and taste to provide sufficient warning against human consumption. Undenatured ethanol is heavily taxed. A "fifth" of pure alcohol costs approximately 15 cents to produce; the tax on it is $4.

Isopropyl alcohol is the substance used in the production of rubbing alcohol. It also is used in some paints, inks, and cosmetic preparations.

Ethylene glycol is a compound containing two hydroxyl groups—one each on the adjacent carbons of an ethane-like structure:

$$\begin{array}{cc} CH_2 - & CH_2 \\ | & | \\ HO & OH \end{array}$$

It is widely used for automobile antifreeze and for the production of Dacron fibers (Chapter 21) and Mylar film.

Glycerol, also known as glycerine, contains three hydroxyl groups, one each on the three carbons of a propane-like structure:

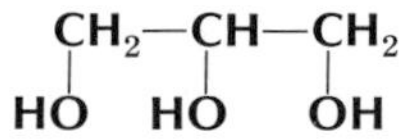

$$\underset{\text{HO}}{CH_2}-\underset{\text{HO}}{CH}-\underset{\text{OH}}{CH_2}$$

Obtained originally as a by-product in the formation of soap (Chapter 21), glycerol is now in such great demand that a synthetic process utilizing propene has been developed. The moisture-retaining properties of glycerol make it useful in the production of candy, skin lotions, inks, and pharmaceuticals.

PHENOLS

The phenols as a class of compounds are active bactericides. Phenol itself is commonly called *carbolic acid* and is used as an antiseptic in hospitals. The efficiency of other antiseptics is measured in arbitrary units called the *phenol coefficient.* A 1-percent solution of a germicide that is as effective as a 10-percent solution of phenol is assigned a phenol coefficient of 10. If alkyl groups are attached to the aromatic ring the germicidal action of the compound is greater than that of phenol itself. Mouthwash solutions invariably contain phenolic substances. The antiseptic hexyl resorcinol is used in throat lozenges. Hexachlorophene is used in some toothpastes, deodorants, and soaps.

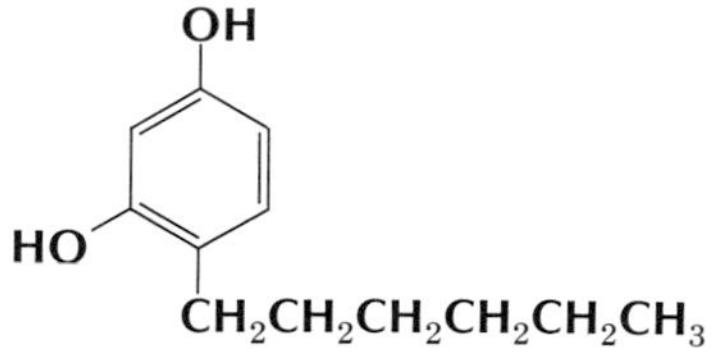

OH, HO, $CH_2CH_2CH_2CH_2CH_2CH_3$

hexylresorcinol

OH, OH, Cl, Cl, CH_2, Cl, Cl, Cl, Cl

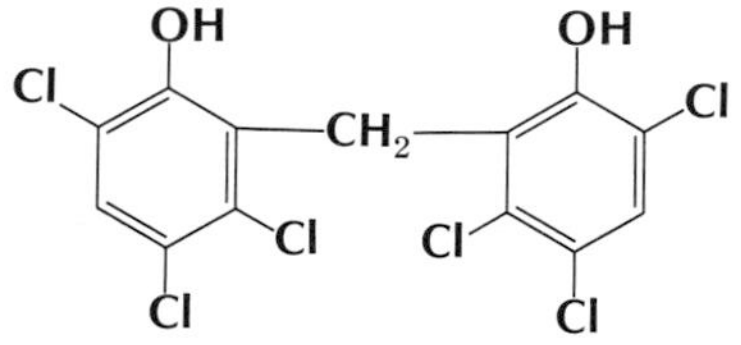

hexachlorophene

ETHERS

Diethyl ether is a familiar general anesthetic in use since 1842. Because its high flammability and volatility are hazards in the operating room, it must be administered by highly trained personnel. Other precautions that must be observed to reduce the possibility of static electricity include wearing conducting shoes and banning synthetic fiber uniforms that might generate sparks of static electricity.The wide use of diethyl ether by organic chemists attests to its excellent solvent characteristics. There are relatively few organic compounds that are not soluble in diethyl ether.

Methyl ethers of phenolic substances are common in naturally occurring compounds. Eugenol from oil of cloves and anethole from oil of anise are examples of such substances.

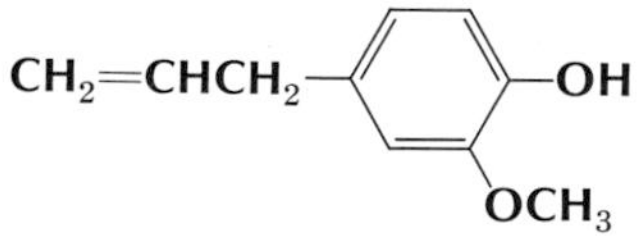

$CH_2{=}CHCH_2-$ —OH, OCH_3

eugenol

$CH_3CH{=}CH-$ —OCH_3

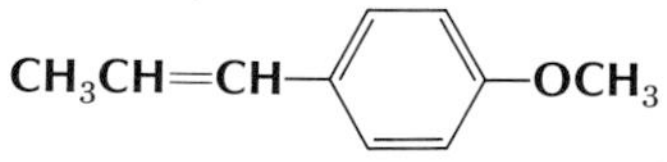

anethole

19.4 PREPARATION OF ALCOHOLS

HYDRATION OF ALKENES

The hydration of alkenes in the presence of an acid catalyst was noted in Chapter 17. The net result of the addition of water is a product in accord with Markownikoff's rule. The ready availability of propene from petroleum sources makes the production of 2-propanol (isopropyl alcohol) an inexpensive process.

$$CH_3CH{=}CH_2 + H_2O \xrightarrow{H^+} CH_3{-}\underset{\displaystyle H}{\overset{\displaystyle OH}{\overset{|}{\underset{|}{C}}}}{-}CH_3$$

For many years the preparation of anti-Markownikoff alcohols such as 1-propanol from the hydration of alkenes was an unfulfilled goal of organic chemists. The compound diborane, generated from sodium borohydride ($NaBH_4$) and boron trifluoride reacts with alkenes to form trialkylboranes. These can be oxidized with hydrogen peroxide to the trialkylborates, which upon treatment with base produce an alcohol. The overall conversion produces an alcohol from an alkene by means of anti-Markownikoff addition:

$$6CH_3CH{=}CH_2 + B_2H_6 \longrightarrow 2(CH_3CH_2CH_2)_3B$$

$$(CH_3CH_2CH_2)_3B \xrightarrow{H_2O_2} (CH_3CH_2CH_2O)_3B \xrightarrow{OH^-} CH_3CH_2CH_2OH$$

EXAMPLE 19.1

How can isobutyl alcohol be prepared from methylpropene?

Isobutyl alcohol consists of the same carbon framework as methylpropene:

$$\underset{\text{methylpropene}}{CH_3{-}\underset{\displaystyle CH_3}{\underset{|}{C}}{=}CH_2} \qquad \underset{\text{isobutyl alcohol}}{CH_3{-}\underset{\displaystyle CH_3}{\overset{\displaystyle H}{\overset{|}{\underset{|}{C}}}}{-}CH_2{-}OH}$$

The synthesis must involve conversion of a double bond into a single bond and the addition of the elements of H_2O to the alkene. Hydration in the presence of an acid will yield a Markownikoff-type product.

$$CH_3{-}\underset{\displaystyle CH_3}{\underset{|}{C}}{=}CH_2 + H_2O \xrightarrow{H^+} CH_3{-}\underset{\displaystyle CH_3}{\overset{\displaystyle OH}{\overset{|}{\underset{|}{C}}}}{-}CH_3$$

The Markownikoff-type product is tertiary butyl alcohol. In order to produce the desired isobutyl alcohol an anti-Markownikoff addition of water must be achieved. The diborane method described above will yield the desired product.

DISPLACEMENT OF HALIDE IONS

Alcohols can be generated by displacement of a halide ion from a haloalkane by treatment with hot aqueous solutions of hydroxide ions.

$$CH_3CH_2Br + OH^- \longrightarrow CH_3CH_2OH + Br^-$$

This reaction is accompanied by an interfering side reaction—the dehydrohalogenation to yield alkenes. The elimination reaction is very competitive for halides attached to a tertiary carbon atom.

19.5
PREPARATION OF PHENOLS

In general the preparative reactions for alcohols cannot be used for the preparation of phenols. The common laboratory and commercial preparation involves the fusion of an aromatic sulfonic acid (Chapter 18) with sodium hydroxide at temperatures above 250°. The sodium salt of the phenol generated reacts with acid to liberate the phenol.

$$C_6H_5SO_3H \xrightarrow[\text{room temperature}]{Na^+OH^-} C_6H_5SO_3^-Na^+ \xrightarrow[250°]{2NaOH} C_6H_5O^-Na^+ + Na_2SO_3 + H_2O$$

$$C_6H_5O^-Na^+ + HCl \longrightarrow C_6H_5OH + NaCl$$

The Dow Chemical Company converts chlorobenezene into phenol by means of sodium hydroxide at high temperature and pressure. The sodium salt is formed and is treated with carbon dioxide to generate the phenol.

$$C_6H_5Cl + 2NaOH \xrightarrow[150\ \text{atm}]{300°} C_6H_5O^-Na^+ + NaCl + H_2O$$

$$2\,C_6H_5O^-Na^+ + CO_2 + H_2O \longrightarrow 2\,C_6H_5OH + Na_2CO_3$$

19.6 PREPARATION OF ETHERS

The British chemist Alexander Williamson devised a procedure for preparing ethers in 1851. The method, which involves the displacement of a halide ion from a haloalkane by the sodium salt of an alcohol or phenol, is analogous to the displacement of a halide ion from a haloalkane by a hydroxide ion to produce an alcohol:

$$RO^-Na^+ + R'{-}X \longrightarrow R{-}O{-}R' + NaX$$
$$HO^-Na^+ + R'{-}X \longrightarrow R'{-}OH + NaX$$

The Williamson method is still the best way to prepare ethers. The preparation of mixed ethers can be accomplished in two ways. For example, methyl ethyl ether can be prepared from bromomethane and the sodium salt of ethanol or from bromoethane and the sodium salt of methanol.

$$CH_3CH_2O^-Na^+ + CH_3Br \longrightarrow CH_3CH_2OCH_3 + NaBr$$
$$CH_3O^-Na^+ + CH_3CH_2Br \longrightarrow CH_3OCH_2CH_3 + NaBr$$

19.7 REACTIONS OF ALCOHOLS

ACIDITY

The acidity of the hydroxyl proton of alcohols was discussed earlier in this chapter. The reaction of alcohols with sodium is a feature that distinguishes them from ethers that are unreactive.

EXAMPLE 19.2

A compound $C_5H_{10}O$ does not react with either sodium or bromine. Suggest a structure for this compound.

The compound contains one oxygen atom per molecule and could be either an alcohol or an ether. Both classes of compounds have the same general molecular formula. The lack of a reaction with sodium indicates that the compound is not an alcohol and therefore an ether. (It should be noted that other classes of compounds contain oxygen. These substances will be discussed in subsequent chapters.) The general molecular formula for a saturated alcohol or ether is $C_nH_{2n+2}O$. Therefore the compound described must either contain a double bond or a ring. The lack of a reaction with Br_2 precludes the presence of a double bond. Only those compounds which are ethers and contain a ring can satisfy the structural requirements suggested by the reactions described. One possibility is the following cyclic ether; there are many other isomeric substances that could be suggested:

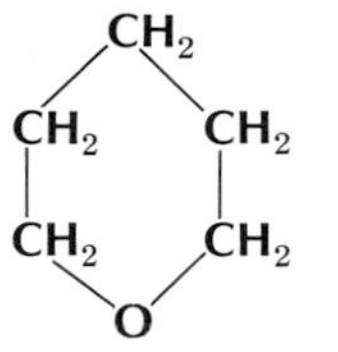

$$\begin{array}{ccc} & CH_2 & \\ CH_2 & & CH_2 \\ | & & | \\ CH_2 & & CH_2 \\ & O & \end{array}$$

DEHYDRATION

Dehydration of alcohols yields alkenes. This reaction was presented in Chapter 17.

EXAMPLE 19.3

A compound $C_6H_{14}O$ reacts with sulfuric acid to yield C_6H_{12}, which when ozonized yields only acetone:

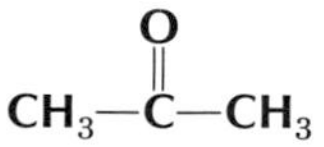

$$CH_3-\overset{\overset{\large O}{\|}}{C}-CH_3$$

What are the structures of the compounds C_6H_{12} and $C_6H_{14}O$?

The ozonolysis reaction yields a single product and, therefore, the alkene must be symmetrical:

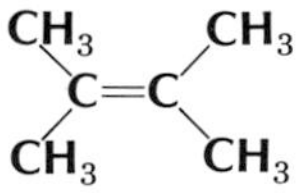

$$(CH_3)_2C{=}C(CH_3)_2$$

The dehydration reaction with sulfuric acid eliminated a molecule of water from an alcohol. Since the alkene is symmetrical the fragments H and OH could only be arranged in one way in the original alcohol:

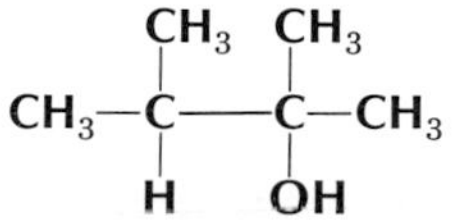

$$CH_3-\underset{\underset{\large H}{|}}{\overset{\overset{\large CH_3}{|}}{C}}-\underset{\underset{\large OH}{|}}{\overset{\overset{\large CH_3}{|}}{C}}-CH_3$$

REPLACEMENT OF HYDROXYL BY HALIDE

The hydroxyl group of alcohols can be replaced by halogens in several different ways:

$$CH_3CH_2CH_2OH + HCl \xrightarrow{ZnCl_2} CH_3CH_2CH_2Cl + H_2O$$
$$3CH_3CH_2CH_2OH + PBr_3 \longrightarrow CH_3CH_2CH_2Br + H_3PO_3$$
$$CH_3CH_2CH_2OH + SOCl_2 \longrightarrow CH_3CH_2CH_2Cl + SO_2 + HCl$$

All these reactions occur to yield specific products in which the halogen replaces the hydroxyl group from a particular carbon atom. These reactions are a more efficient way of obtaining haloalkanes than by halogenating alkanes (Chapter 16), which produces a mixture of products. The rate for the reaction with HCl, with $ZnCl_2$ as a catalyst, decreases in the order tertiary > secondary > primary alcohols. For tertiary butyl alcohol the reaction occurs in a few minutes at room temperature. The reaction with *n*-butyl alcohol requires hours at higher temperatures. Therefore, this combination of reagents can be used qualitatively to establish the class to which an alcohol belongs. The reaction for classification purposes is called the *Lucas test.*

EXAMPLE 19.4

How could 1-pentanol and 2-pentanol be distinguished?

The reaction of 2-pentanol with HCl and $ZnCl_2$ should occur readily at room temperature although slower than for a tertiary alcohol. The reaction with 1-pentanol should be extremely slow or not occur at all at room temperature.

$CH_3CH_2CH_2CH_2CH_2OH$	$CH_3\underset{\displaystyle OH}{\underset{\vert}{C}}HCH_2CH_2CH_3$
1-pentanol	2-pentanol

ESTERIFICATION

Alcohols react with acids to form a product in which the hydroxyl group of the alcohol is replaced by the conjugate base of the acid. These products are called *esters* and are discussed more extensively in Chapter 21. When cold sulfuric acid reacts with primary alcohols an ester called an *alkyl hydrogen sulfate* is produced. When an alcohol containing 12 carbons in a continuous chain (lauryl alcohol or 1-dodecanol) is used the resultant hydrogen sulfate compound is acidic and can be reacted with a base to yield a sodium salt. The product, sodium lauryl sulfate, is the commercial detergent Dreft:

$$CH_3(CH_2)_{10}CH_2{-}OSO_3^-Na^+$$

sodium lauryl sulfate

Nitric acid forms nitrate esters with alcohols. Glyceryl trinitrate or nitroglycerine, the explosive component in dynamite, is produced from glycerol. It also serves to dilate blood vessels in patients who suffer arterial tension in heart disease.

$$\begin{array}{l} CH_3OH \\ | \\ CHOH \\ | \\ CH_2OH \end{array} + 3HNO_3 \longrightarrow \begin{array}{l} CH_2ONO_2 \\ | \\ CHONO_2 \\ | \\ CH_2ONO_2 \end{array} + 3H_2O$$

Organic acids (Chapter 21) also react with acids to produce esters.

OXIDATION

The primary, secondary, and tertiary alcohols behave in individually characteristic ways toward oxidizing agents, such as sodium dichromate ($Na_2Cr_2O_7$). Primary alcohols are oxidized to aldehydes (Chapter 20), which in turn are oxidized to acids if they are not removed from contact with the oxidizing agent.

$$\underset{\text{primary alcohol}}{R{-}CH_2OH} \xrightarrow[H^+]{Na_2Cr_2O_7} \underset{\text{an aldehyde}}{R{-}C{\overset{O}{\underset{H}{\lessgtr}}}} \xrightarrow[H^+]{Na_2Cr_2O_7} \underset{\text{an acid}}{R{-}C{\overset{O}{\underset{OH}{\lessgtr}}}}$$

Secondary alcohols are oxidized to ketones, which are insensitive to further oxidation unless drastic reaction conditions are employed. The difference in behavior lies in the absence of a hydrogen atom on the carbon atom bearing oxygen in the ketones as compared to the aldehydes.

$$\underset{\text{secondary alcohol}}{R{-}\overset{\overset{OH}{|}}{\underset{\underset{H}{|}}{C}}{-}R'} \xrightarrow[H^+]{Na_2Cr_2O_7} \underset{\text{a ketone}}{R{-}\overset{\overset{O}{\|}}{C}{-}R'}$$

In the oxidation to an aldehyde or ketone two hydrogen atoms are eliminated from the alcohol. An increase in the oxidation number of carbon results. Tertiary alcohols are not oxidized under mild reaction conditions. Under severe reaction conditions the compounds are oxidized but are cleaved into numerous lower molecular mass fragments. The general formula for an alcohol is $C_nH_{2n+2}O$, as is the case for the isomeric ethers. The general formula for an aldehyde or ketone is $C_nH_{2n}O$. Of course, other structural features alter these general representations. A compound containing a double bond and a hydroxyl group has the formula $C_nH_{2n}O$, and its corresponding aldehyde or ketone is $C_nH_{2n-2}O$.

EXAMPLE 19.5

A compound of molecular formula $C_5H_{10}O$ reacts with sodium to evolve hydrogen gas. The compound is oxidized by sodium dichromate to yield C_5H_8O. No reaction occurs with $C_5H_{10}O$ and bromine. Suggest a structure for the compound.

The molecular formula implies that the compound is either an alcohol, which contains a double bond or a ring, or is an aldehyde or ketone. The reaction with sodium indicates that the compound is an alcohol. Furthermore, the oxidation to yield a compound with two less hydrogen atoms per molecule is also indicative that the original compound is an alcohol. Since an acid is not formed in the oxidation step the alcohol must be secondary. Finally, the absence of a reaction with bromine indicates that the alcohol does not contain a double bond. Therefore, the shortage of two hydrogens per molecule must be due to the presence of a ring in the molecule. Any compound which is a secondary alcohol and contains a ring of carbon atoms will be in agreement with the data provided. One example is:

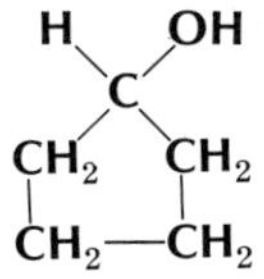

```
H   OH
 \ /
  C
 / \
CH2  CH2
|    |
CH2—CH2
```

THE IODOFORM REACTION

Alcohols of the general structure $CH_3CH(OH)R$ react in a characteristic manner with basic solutions of iodine. A mixture of sodium carbonate and iodine is an oxidizing agent that will oxidize the alcohol to the ketone:

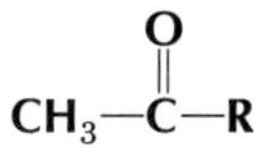

$$CH_3—\overset{\overset{\displaystyle O}{\|}}{C}—R$$

The halogen then reacts with the ketone to yield a halogenated derivative that is unstable in basic solutions. One of the ultimate products is iodoform (CHI_3) a yellow solid with a characteristic medicinal odor:

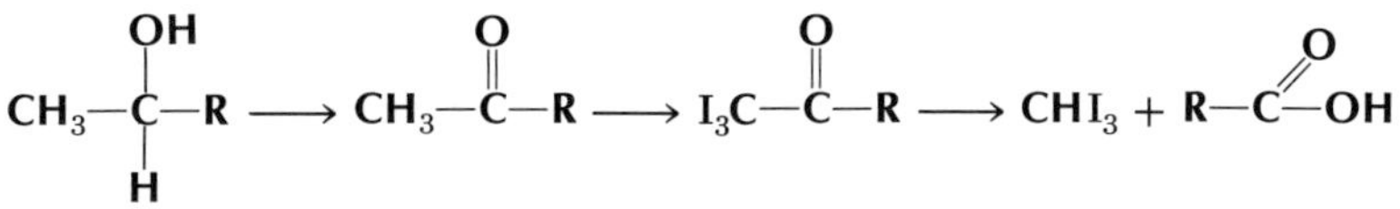

$$CH_3—\underset{\underset{\displaystyle H}{|}}{\overset{\overset{\displaystyle OH}{|}}{C}}—R \longrightarrow CH_3—\overset{\overset{\displaystyle O}{\|}}{C}—R \longrightarrow I_3C—\overset{\overset{\displaystyle O}{\|}}{C}—R \longrightarrow CHI_3 + R—\overset{\overset{\displaystyle O}{/\!/}}{C}—OH$$

Only compounds of structure $CH_3CH(OH)R$ can yield iodoform. If the hydrogen attached to carbon-bearing oxygen is absent, the compound cannot be oxidized by the reaction medium. If the methyl group is replaced by a larger alkyl group iodoform will not be produced. The iodoform reaction is therefore a specific test for alcohols of one general structure.

EXAMPLE 19.6

One of the isomeric alcohols of the molecular formula $C_4H_{10}O$ reacts with a mixture of sodium carbonate and iodine to yield iodoform. What is the structure of the alcohol?

The iodoform reaction occurs for the general structure:

$$CH_3-\underset{\large H}{\overset{\large OH}{\underset{|}{\overset{|}{C}}}}-R$$

Since the compound contains four carbon atoms per molecule, the alkyl group R must consist of two carbon atoms. There is only one way to arrange two carbon atoms, that is, in the form of an ethyl group. The alcohol must be 2-butanol:

$$CH_3-\underset{\large H}{\overset{\large OH}{\underset{|}{\overset{|}{C}}}}-CH_2-CH_3$$

19.8 MECHANISM OF DISPLACEMENT REACTIONS

In this chapter are several reactions in which one atom or group of atoms displaces another atom or group of atoms from a saturated carbon atom. In each case the displacing group is a nucleophile or Lewis base, which causes the second group with the original bonding pair of electrons to leave from the site of the carbon atom. Such substitution reactions can occur via either of two reaction mechanisms. In one the rate of the reaction depends only on the concentration of the carbon-containing molecule and is independent of the concentration of the nucleophilic reagent. This process is called *substitution, nucleophilic, unimolecular,* or simply S_N1. In the second reaction mechanism the reaction rate is dependent on both the concentrations of the organic molecule and the attacking nucleophile. This process is called *substitution, nucleophilic, bimolecular,* or S_N2.

In S_N1 reactions the rate-determining step is the formation of a carbonium ion in a dissociation process. In the case of 2-chloro-2-methylpropane the dissociation leads to a tertiary carbonium ion:

$$CH_3-\underset{CH_3}{\overset{CH_3}{\underset{|}{\overset{|}{C}}}}-Cl \longrightarrow CH_3-\underset{CH_3}{\overset{CH_3}{\underset{|}{\overset{|}{C^+}}}} + Cl^-$$

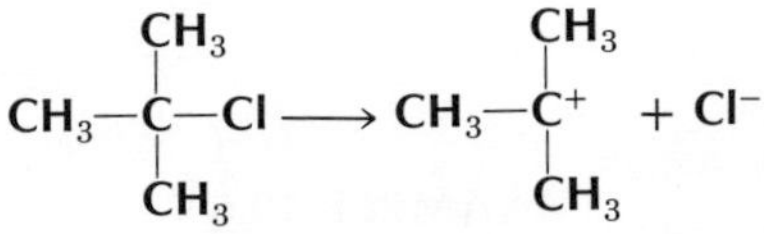

The carbonium ion then reacts with a nucleophile such as the hydroxide ion in a fast subsequent step. Since the second step is rapid with respect to the first step, the concentration of the nucleophile does not appear in the rate expression:

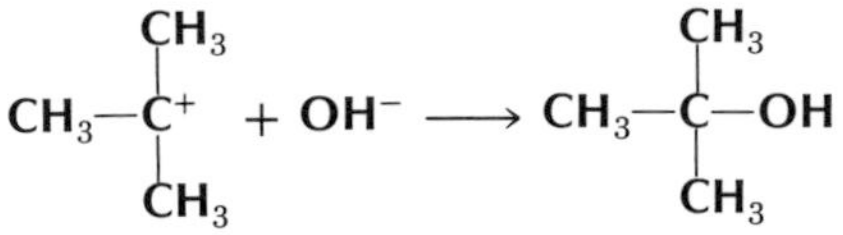

$$CH_3-\underset{CH_3}{\overset{CH_3}{\underset{|}{\overset{|}{C^+}}}} + OH^- \longrightarrow CH_3-\underset{CH_3}{\overset{CH_3}{\underset{|}{\overset{|}{C}}}}-OH$$

In the S_N2 reaction the rate of reaction is dependent on the concentrations of both the organic molecule and the nucleophile. In the case of 1-chlorobutane the transition state for the reaction in which the hydroxide ion displaces a chloride ion must incorporate both substances in an activated complex. In this transition state it is postulated that the hydroxide ion approaches the carbon atom from the direction opposite the carbon chlorine bond. As a new bond is formed between the carbon and oxygen, the bond between the carbon and chlorine is broken. The transition state is pictured below:

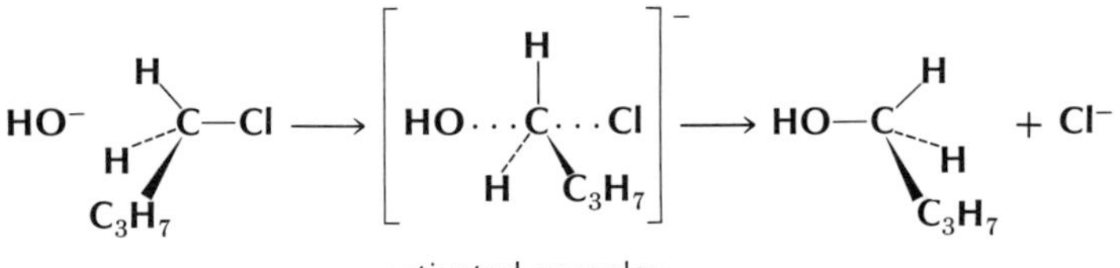

$$HO^- + \underset{C_3H_7}{\overset{H}{\underset{H}{C}}}-Cl \longrightarrow \left[HO\cdots\underset{H\ \ C_3H_7}{\overset{H}{\underset{}{\overset{|}{C}}}}\cdots Cl \right]^- \longrightarrow HO-\underset{C_3H_7}{\overset{H}{\underset{H}{C}}} + Cl^-$$

activated complex

The displacement reaction corresponds to a physical process similar to the inversion or turning inside out of an umbrella. The chemical consequences of this process will become evident in Chapter 23 when optical activity is presented.

Why does the reaction of 2-chloro-2-methylpropane occur via an S_N1 process whereas 1-chlorobutane proceeds via an S_N2 process? One of the reasons is that the tertiary carbonium ion derived from the tertiary chloro compound is more stable (lower energy) than that derived from the primary chloro compound. Therefore the achievement of the transition state corresponding to an S_N1 process is easier (of lower activation energy) in the transition state for the tertiary chloro compound than for the primary chloro compound. Another reason is that the S_N2 transition state for the tertiary chloro compound is of higher energy than that for the primary chloro compound because of the spatial bulk of the three methyl groups through which the hydroxide ion would have to pass in order to approach the back side of the carbon atom.

Suggested further readings

Beecher, H. K., "Anesthesia," *Sci. Amer.* (January 1957).
Deasy, C. L., "The Walden Inversion in Nucleophilic Aliphatic Substitution Reactions," *J. Chem. Educ.*, **22,** 455 (1962).
Ferguson, L. N., "Hydrogen Bonding and the Physical Properties of Substances," *J. Chem. Educ.*, **33,** 267 (1956).
Greenberg, L. A., "Alcohol in the Body," *Sci. Amer.* (December 1953).
Lesser, M. A., "Glycerin—Man's Most Versatile Chemical Servant," *J. Chem. Educ.*, **26,** 327 (1949).

Terms and concepts

alcohol
aldehyde
Dacron
denatured
displacement
ether
hydration
hydrogen bonds
iodoform
ketone
Lucas test
nitroglycerine
nucleophilic substitution
phenol
Williamson ether synthesis

Questions and problems

1. *n*-Butyl alcohol is moderately soluble in water whereas *t*-butyl alcohol is infinitely soluble. Suggest a reason for this difference.
2. Write structural formulas for each of the following compounds:
- **a.** 2-methyl-2-pentanol
- **b.** 2,3-dimethyl-1-butanol
- **c.** 2,3-butanediol
- **d.** *o*-bromophenol
- **e.** diphenyl ether
- **f.** cyclopentanol
- **g.** 2-methyl-1-butanol
- **h.** methyl cyclopentyl ether

3. What catalysts and conditions are necessary for each of the following reactions?
- **a.** 1-butanol from 1-butene
- **b.** 2-butanol from 1-butene
- **c.** isopropyl alcohol from 2-bromopropane
- **d.** 1-chlorobutane from 1-butanol
- **e.** diethyl ether from ethanol

4. 1,2-Hexanediol is very soluble in water. Why?

5. How can 1-propanol be converted into 2-propanol? How can 2-propanol be converted into 1-propanol?

6. What chemical conversions are needed in order to accomplish the following transformations?

- **a.** 1-pentanol into pentane
- **b.** 1-pentanol into 1-pentyne
- **c.** ethanol into butane
- **d.** 1-pentanol into 2-bromopentane

7. What general formula will describe all compounds containing the following?

- **a.** a hydroxyl group and a ring
- **b.** a hydroxyl group and two double bonds
- **c.** a hydroxyl group and two rings
- **d.** a hydroxyl group and a triple bond
- **e.** an ether oxygen and a double bond
- **f.** an ether oxygen and a hydroxyl group

8. What chemical test will serve to distinguish between the members of each of the following pairs?

- **a.** diethyl ether and 1-butanol
- **b.** phenol and 1-hexanol
- **c.** 2-butanol and 1-butanol
- **d.** 2-pentanol and 3-pentanol
- **e.** 3-methyl-3-pentanol and 2-methyl-3-pentanol

9. A compound, C_3H_6O, does not react with sodium. It readily reacts with H_2 in the presence of platinum to yield C_3H_8O. What is the structure for C_3H_6O?

10. A compound, $C_4H_{10}O$, reacts with sodium, and hydrogen gas is evolved. The substance does not react with sodium dichromate in acidic solution. What is the structure of $C_4H_{10}O$?

11. A compound, $C_5H_{12}O$, reacts with sodium carbonate and iodine to yield CHI_3. What two compounds can react as indicated?

12. A compound, $C_5H_{10}O$, does not react with hydrogen in the presence of a platinum catalyst. Upon treatment with sodium dichromate only C_5H_8O is produced. What compound could react as indicated?

13. Which of the pentanols when dehydrated will yield only 2-pentene? Which pentanol will yield only 1-pentene?

14. What reactions might occur when 1-bromopentane is reacted with sodium amide ($NaNH_2$)?

15. Consider and discuss the Williamson ether synthesis in terms of its probable mechanism using bromoethane and the sodium salt of ethanol as the reagents.

16. Suggest a structure for the tertiary carbonium ion derived from 2-chloro-2-methylpropane. Compare its shape to that of an inorganic compound. (A review of Chapter 6 would aid in answering this question.)

20

ALDEHYDES AND KETONES

The carbon-to-oxygen double bond, $-\overset{O}{\overset{\|}{C}}-$, occurs in a wide variety of natural products. When an alkyl group and a hydrogen are attached to this functional group called the *carbonyl group,* the compounds are termed *aldehydes.* When two alkyl groups are attached to the carbonyl carbon, the compounds are called *ketones.* In the next chapter other combinations of atoms attached to the carbonyl group will be considered.

20.1 NOMENCLATURE

Aldehydes are assigned common names on the basis of the common names of the acids to which they can be oxidized:

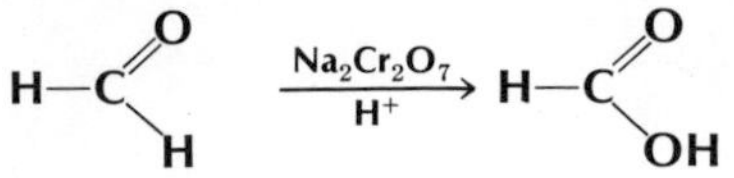

$$\underset{\text{formaldehyde}}{H{-}C(=O)H} \xrightarrow[H^+]{Na_2Cr_2O_7} \underset{\text{formic acid}}{H{-}C(=O)OH}$$

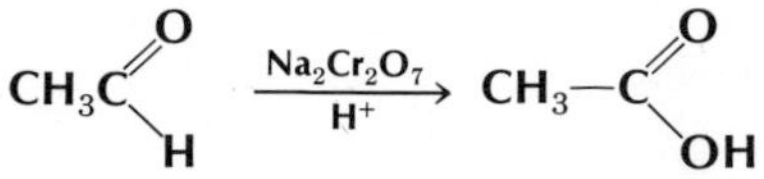

$$\underset{\text{acetaldehyde}}{CH_3C(=O)H} \xrightarrow[H^+]{Na_2Cr_2O_7} \underset{\text{acetic acid}}{CH_3{-}C(=O)OH}$$

Ketones are named by adding the word *ketone* to the groups attached to the carbonyl carbon atom.

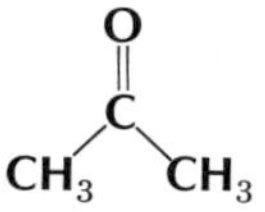

$CH_3C(=O)CH_3$	$CH_3C(=O)CH_2CH_3$	$CH_3CH_2C(=O)CH_2CH_3$
dimethyl ketone (acetone)	methyl ethyl ketone	diethyl ketone

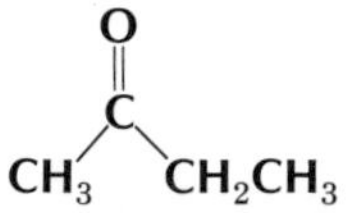

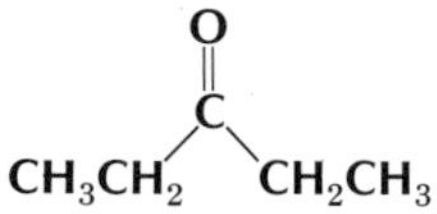

The IUPAC system uses the root name of the longest continuous chain containing the functional group and the ending *-al* for aldehydes and *-one* for ketones. The following examples illustrate the proper use of systematic nomenclature. Note that a number is not used to designate the position of the carbonyl oxygen of an aldehyde because it must of necessity be located on the 1 position:

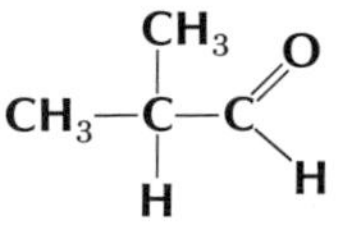

$CH_3{-}CH(CH_3){-}C(=O)H$	$\overset{4}{C}H_3\overset{3}{C}H_2\overset{2}{C}H(CH_3)\overset{1}{C}(=O)H$	$Cl\overset{4}{C}H_2\overset{3}{C}H_2\overset{2}{C}H_2\overset{1}{C}(=O)H$
methylpropanal	2-methylbutanal	4-chlorobutanal

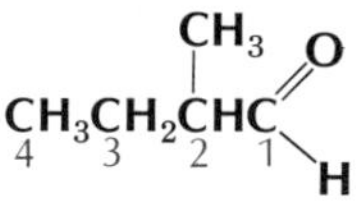

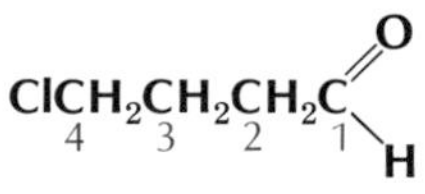

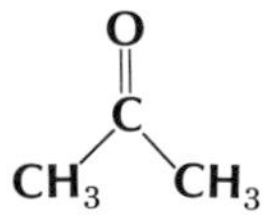

$CH_3C(=O)CH_3$	$\overset{1}{C}H_3C(=O)\overset{3}{C}H_2\overset{4}{C}H_2\overset{5}{C}H_3$	cyclohexanone ring (C=O at 1, ring positions 2–6), CH_3 at 3
propanone	2-pentanone	3-methylcyclohexanone

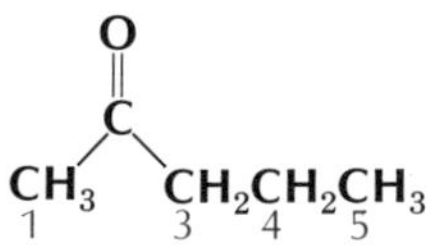

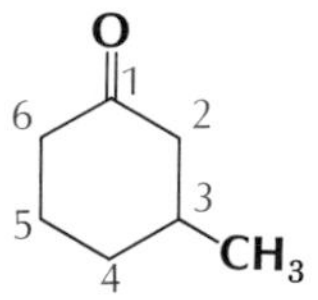

20.2 THE CARBONYL GROUP

BONDING

The trigonal carbon atom of the carbonyl group utilizes three sp^2 hybrid orbitals to form σ bonds with its three attached atoms. The remaining electron of carbon in a p orbital forms a π bond with an electron in a p orbital of oxygen (Figure 20.1). The geometry of the carbonyl group is thus

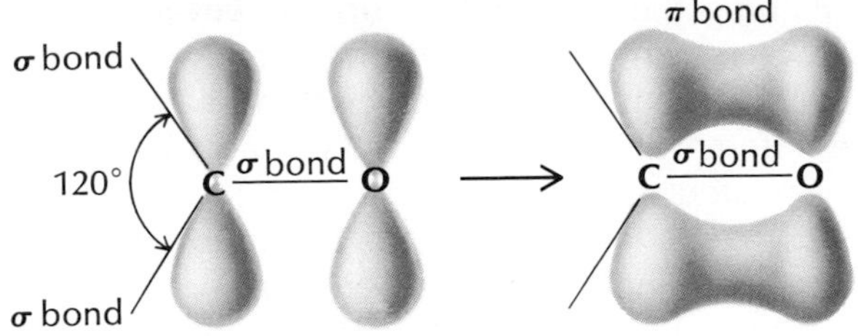

Figure 20.1
The carbonyl group.

similar to that of an alkene. However, the electron distribution of the carbonyl group is not symmetrical as it is for an alkene. Because oxygen is more electronegative than carbon it attracts both the σ and π electrons toward it. This fact coupled with the reactivity of the carbonyl group suggests that two resonance contributors are important in the bonding.

$$\text{C}{=}\ddot{\text{O}} \longleftrightarrow \overset{+}{\text{C}}{-}\overset{-}{\ddot{\text{O}}}$$

As will be demonstrated shortly, the carbonyl group reacts with nucleophilic reagents that supply an electron pair to the carbon atom.

PROPERTIES

The aldehydes and ketones have higher boiling points than hydrocarbons of similar molecular mass (Table 20.1), which reflects the polarity of the

Table 20.1
Comparison of boiling points of carbonyl compounds with those of alcohols and alkanes

compound	*molecular mass (g/mole)*	*boiling point (°C)*
ethane	30	−172
methanol	32	64.6
formaldehyde	30	−21
propane	44	−187
ethanol	46	78.3
acetaldehyde	44	20.2
butane	58	−135
1-propanol	60	97.8
propionaldehyde	58	48.8
isobutane	58	−12
2-propanol	60	82.5
acetone	58	56.1

molecules containing the carbonyl group. However, the aldehydes and ketones have lower boiling points than the related alcohols because they cannot form hydrogen bonds without a proton bound to an electronegative atom. The differences in boiling points of a carbonyl compound and the structurally related alcohol become smaller as the molecular mass increases. This change reflects the diminished importance of the hydrogen bond to the total molecule as the mass increases.

Except for those aldehydes and ketones containing four or fewer carbon atoms, these two classes of organic compounds are not very soluble in water. The low molecular mass compounds can form hydrogen bonds with water, as illustrated for acetone, which is completely miscible in water:

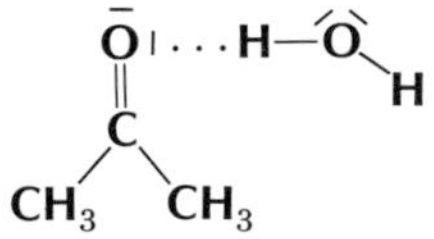

$$(CH_3)_2C{=}\bar{O}| \cdots H{-}O{-}H$$

Low molecular mass aldehydes have sharp irritating odors, but the high molecular mass compounds and the ketones are fragrant substances. The odor of the fragrant substances makes them useful as flavoring agents and perfumes. For example, the cyclic ketone civetone, which is a glandular secretion of the civet cat, is used in perfumes.

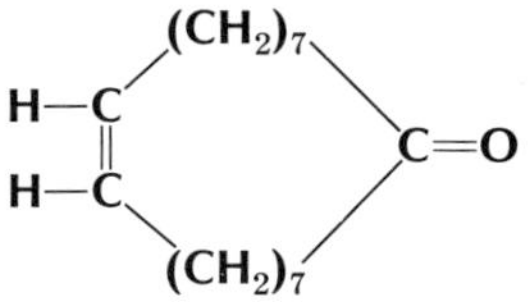

$$\text{H}{-}\text{C}{=}\text{C}{-}\text{H} \text{ joined via } (CH_2)_7 \text{ and } (CH_2)_7 \text{ to } C{=}O$$

civetone

NATURALLY OCCURRING CARBONYL COMPOUNDS

Acetone (propanone) is a product of the metabolism of fat. Normally it is oxidized in the body to carbon dioxide and water. However, in a person with diabetes mellitus the excessively fast metabolism of fat leads to the accumulation of acetone, which can be detected in the urine and in severe cases gives a characteristic odor to the breath.

The aldehyde citral is found in the oils of citrus fruits. Its structure is related to citronellol (Chapter 17) and is another representative of the class of terpene compounds.

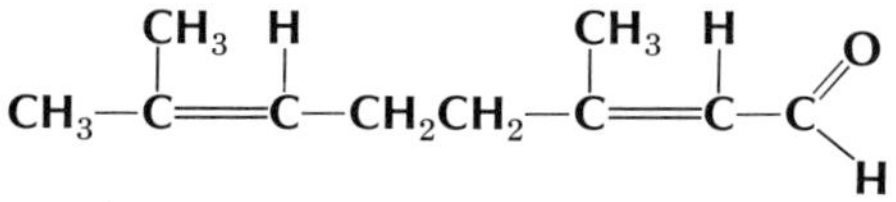

$$CH_3{-}\underset{}{C}(CH_3){=}C(H){-}CH_2CH_2{-}C(CH_3){=}C(H){-}C(=O)H$$

citral

The aromatic aldehydes benzaldehyde, vanillin, and cinnamaldehyde smell like almonds, vanilla, and cinnamon, respectively.

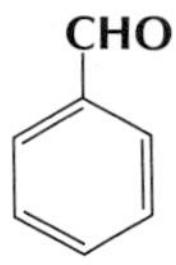

CHO OH OCH$_3$ CH=CHC(=O)H CHO

benzaldehyde 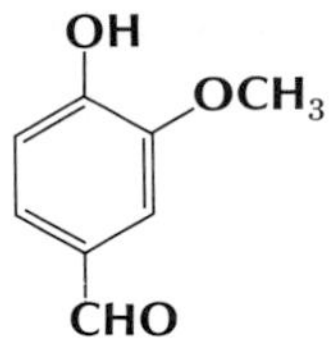 vanillin 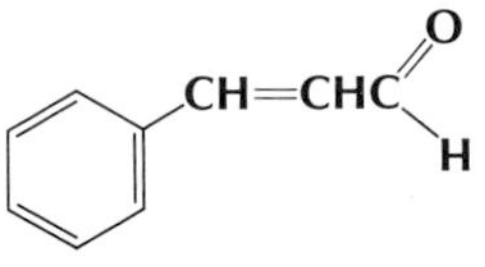 cinnamaldehyde

Among the more interesting ketones are estrone, one of the female sex hormones, and androsterone, one of the male sex hormones. The chemical difference is obviously subtle for such an important physiological difference.

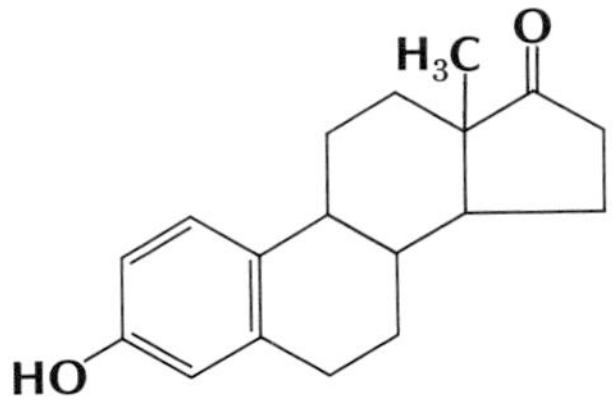

H$_3$C O HO H$_3$C O CH$_3$ HO

estrone 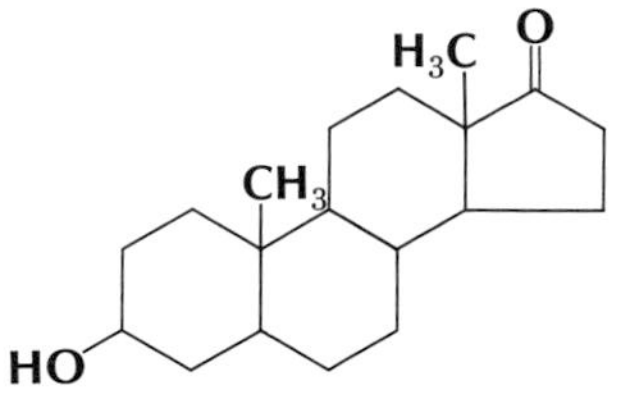 androsterone

20.3
PREPARATION OF ALDEHYDES AND KETONES

OXIDATION OF ALCOHOLS

The oxidation of primary alcohols yields aldehydes, and the oxidation of secondary alcohols yields ketones. While ketones are resistant to further oxidation under mild experimental conditions, aldehydes are not. If the aldehyde is not removed from the presence of the oxidizing agent it will become oxidized to an acid.

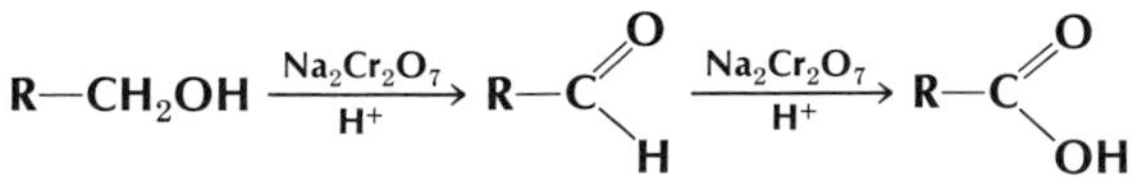

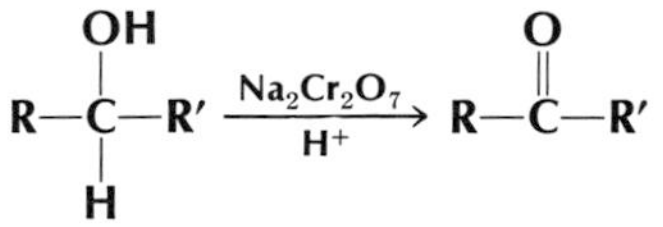

$$\mathrm{R{-}CH_2OH \xrightarrow[H^+]{Na_2Cr_2O_7} R{-}C(=O)H \xrightarrow[H^+]{Na_2Cr_2O_7} R{-}C(=O)OH}$$

$$\mathrm{R{-}CH(OH){-}R' \xrightarrow[H^+]{Na_2Cr_2O_7} R{-}C(=O){-}R'}$$

Aldehydes have lower boiling points than the corresponding primary alcohols and often can be removed from the oxidizing medium by distillation. If the reaction is carried out above the boiling point of the aldehyde but below that of the alcohol the aldehyde can be distilled out of the reaction mixture as it is formed.

The displacement by base of two halogen atoms from the same carbon atom of a dihalide leads to the formation of a carbonyl compound. By analogy with the displacement reaction of haloalkane with hydroxide ions, it would be expected that a *diol* in which both hydroxyl groups are on the same carbon atom would result. However, the intermediate halohydroxy compound, which results from displacement of one halide by hydroxide, loses hydrogen halide to form the carbonyl compound.

$$R{-}CCl_2{-}R + \xrightarrow{OH^-} \left[R{-}C(OH)(Cl){-}R \right] \overset{OH^-}{\not\longrightarrow} R{-}C(OH)_2{-}R$$

$$\downarrow$$

$$R{-}\overset{O}{\overset{\|}{C}}{-}R + HCl$$

INDUSTRIAL PROCESSES

In one industrial process employing the hydrolysis reaction described in the previous section, benzal chloride produced by the free radical chlorination of toluene is converted to benzaldehyde.

$$C_6H_5CH_3 \xrightarrow[\text{ultraviolet light}]{2Cl_2} C_6H_5CHCl_2 \xrightarrow[OH^-]{H_2O} \left[C_6H_5CH(OH)Cl \right] + Cl^- \longrightarrow C_6H_5CHO + HCl$$

toluene — benzal chloride — benzaldehyde

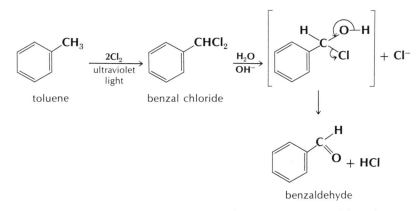

The Oxo process is used in industrial reactions to add carbon monoxide and hydrogen to alkenes in the presence of the catalyst dicobalt octacarbonyl, $[Co(CO)_4]_2$.

$$R{-}CH{=}CH_2 + CO + H_2 \xrightarrow{[Co(CO)_4]_2} RCH_2CH_2CHO + \underset{\displaystyle CHO}{\underset{|}{RCHCH_3}}$$

The reaction requires both high pressures and temperatures. Of the two possible aldehydes formed, the straight-chain compound usually predominates. The utility of this process is very high because the starting materials are readily available and cheap. Note that in this reaction one additional carbon atom per molecule is added to the starting material.

Acetaldehyde is prepared commercially from acetylene by a hydration step in the presence of mercuric sulfate and sulfuric acid. The unsaturated alcohol intermediate is not stable and undergoes a shift of a hydrogen atom to yield the aldehyde:

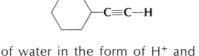

$$H{-}C{\equiv}C{-}H + H_2O \xrightarrow[H_2SO_4]{HgSO_4} \left[\begin{matrix} H & & OH \\ & C{=}C & \\ H & & H \end{matrix} \right] \longrightarrow CH_3CHO$$

The hydration of other alkynes can be carried out in the laboratory. The addition of water proceeds to yield the Markownikoff product, which then rearranges.

EXAMPLE 20.1

What product would result from the treatment of the following compound with water in the presence of $HgSO_4$ and H_2SO_4?

$$C_6H_{11}{-}C{\equiv}C{-}H$$

The addition of water in the form of H^+ and OH^-, according to Markownikoff's rule, must proceed to place the hydrogen on the carbon already bearing the most hydrogens.

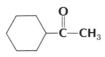

$$C_6H_{11}(HO)C{=}CH_2$$

Subsequent shift of a hydrogen will yield the following ketone:

$$C_6H_{11}{-}\overset{O}{\overset{\|}{C}}{-}CH_3$$

20.4 REDOX REACTIONS

OXIDATION

Both aldehydes and ketones that do not contain any other function groups and are noncyclic have the general formula $C_nH_{2n}O$. Note that this formula is deficient by two hydrogen atoms with respect to the general formula for saturated alcohols. The behavior of aldehydes and ketones toward oxidizing agents provides a useful distinction between them. Aldehydes are easily oxidized to acids by even mild oxidizing agents, whereas ketones are unreactive (Section 20.3).

The *Tollens' test reagent* is an oxidizing solution consisting of ammoniacal silver nitrate. The colorless silver ammonia complex ion is reduced to metallic silver by aldehydes at room temperature. The silver deposits on the interior surface of the vessel used for the test and forms a mirror. Ketones do not react with Tollens' reagent.

$$R{-}C(=O){-}H + 2Ag(NH_3)_2^+ + 3OH^- \longrightarrow R{-}C(=O){-}O^- + 2Ag + 4NH_3 + 2H_2O$$

EXAMPLE 20.2

Can the isomeric carbonyl-containing compounds of the molecular formula C_4H_8O be distinguished from each other?

There are three isomeric carbonyl-containing compounds, two aldehydes and one ketone:

$$CH_3CH(CH_3)CHO \qquad CH_3CH_2CH_2CHO \qquad CH_3CH_2C(=O)CH_3$$

Both of the aldehydes will react with Tollens' reagent to produce a silver mirror. Therefore, these two compounds cannot be distinguished from each other by the use of the reagent. However, the ketone will not react with Tollens' reagent and, therefore, the compound that does not yield a silver mirror is butanone.

REDUCTION

Aldehydes and ketones can be reduced to alcohols. Until recently the reduction was accomplished by hydrogen in the presence of a catalyst such as nickel, but now lithium aluminum hydride in ether solution is used. The reaction can be carried out at room temperature and normal atmospheric pressure. The use of the potentially explosive hydrogen gas is avoided. In addition to convenience the use of lithium aluminum hydride allows the reduction of the carbonyl group without the reduction of a carbon-carbon double bond in the same molecule.

$$CH_2{=}CHCH_2{-}C(=O){-}H \xrightarrow[\text{followed by hydrolysis}]{LiAlH_4} CH_2{=}CHCH_2CH_2OH$$

A compound C_4H_8O reacts with lithium aluminum hydride to yield $C_4H_{10}O$. What are the possible structures for the C_4H_8O compound?

The original compound is deficient by two hydrogens relative to that expected for a saturated structure, that is, an alkane. The presence of one oxygen atom per molecule indicates that this compound may be an alcohol, ether, aldehyde, or ketone. If it is an alcohol or ether there would have to be a ring or double bond in the molecule to account for the number of hydrogen atoms present. If the compound is an aldehyde or ketone then there can be no other structural features such as a double bond or a ring present in the molecule. The reaction with lithium aluminum hydride to yield a compound containing two additional hydrogen atoms per molecule suggests that the compound is an aldehyde or ketone that is reduced to an alcohol. The compound may be any of the three structures given in Example 20.2.

20.5 ADDITION REACTIONS

MECHANISM

The unsaturated carbonyl group undergoes numerous addition reactions. The attacking group is usually a supplier of an electron pair, a nucleophile. The nucleophile N| approaches the carbonyl carbon in a direction perpendicular to the plane of the carbonyl group. An electrophile then becomes attached to the oxygen atom:

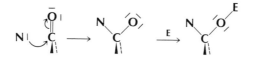

N| C=O ⟶ N–C–O ⟶(E) N–C–O–E

In general ketones are less reactive toward addition reactions than are aldehydes.

ADDITION OF WATER

Water adds reversibly to aldehydes to form hydrates. The fraction of hydrate in aqueous solutions of formaldehyde is high, but most higher molecular mass aldehydes are not appreciably hydrated.

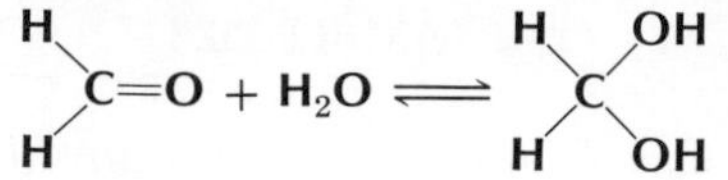

$$\mathrm{H_2C{=}O + H_2O \rightleftharpoons H_2C(OH)_2}$$

formaldehyde

In compounds in which the carbonyl group is attached to electron-withdrawing groups it is sometimes possible to isolate the hydrate. Trichloroacetaldehyde, or chloral, forms a stable hydrate, a soporific called "knockout drops."

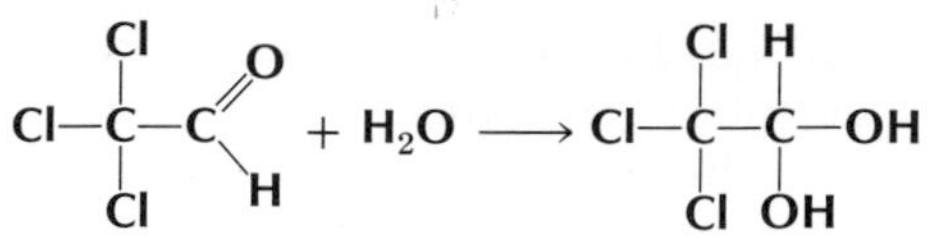

$$\mathrm{Cl_3C{-}CHO + H_2O \longrightarrow Cl_3C{-}CH(OH)_2}$$

trichloroacetaldehyde

ADDITION OF ALCOHOLS

Alcohols add to aldehydes in the presence of acid catalysts in much the same way as does water to yield unstable compounds called *hemiacetals*. The hemiacetals can react with a second molecule of alcohol to yield an isolable *acetal* that is actually a diether.

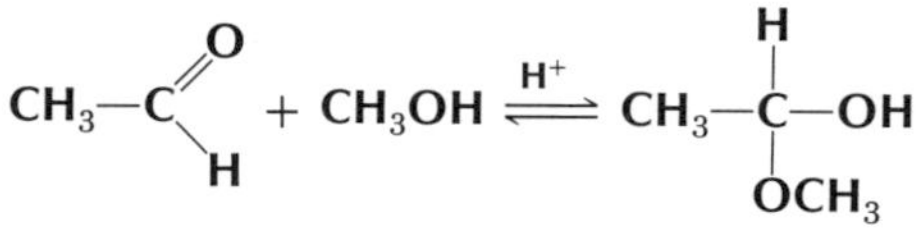

$$\mathrm{CH_3{-}CHO + CH_3OH \overset{H^+}{\rightleftharpoons} CH_3{-}CH(OH)(OCH_3)}$$

a hemiacetal

$$\mathrm{CH_3{-}CH(OH)(OCH_3) + CH_3OH \overset{H^+}{\rightleftharpoons} CH_3{-}CH(OCH_3)(OCH_3) + H_2O}$$

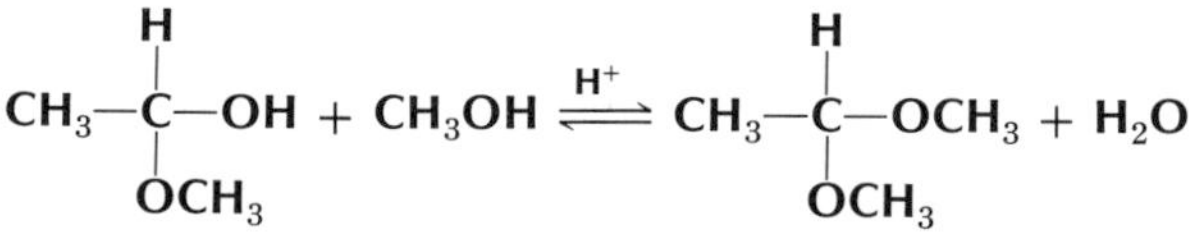

an acetal

Acetals resemble ethers in their reactivity. They are stable to bases but react with acids to regenerate the alcohol and aldehyde from which they were prepared. Therefore, acetal formation is a valuable way to protect an aldehyde functional group in a reaction in which the aldehyde might otherwise become oxidized. The acetal is first formed and the reaction is then carried out, after which the reversible acetal formation in the presence of excess water leads to the formation of the free aldehyde again. Hemiacetals and acetals are important in the chemistry of carbohydrates (Chapter 25).

ADDITION OF NITROGEN COMPOUNDS

Certain derivatives of ammonia that contain the amino group ($—NH_2$) add to carbonyl compounds to form intermediates that then lose water to form stable compounds containing a carbon–nitrogen double bond.

$$\rangle C{=}O + H_2\bar{N}{-}R \overset{H^+}{\rightleftharpoons} R_2C(OH){-}N(H)R \rightleftharpoons R_2C{=}NR + H_2O$$

These compounds are usually solids that can be used to characterize the carbonyl compound from which it was formed. These derivatives are useful in confirming the structure of carbonyl compounds isolated from both natural sources and chemical reactions. The carbonyl derivatives of hydroxylamine and phenylhydrazine are called *oximes* and *phenylhydrazones,* respectively.

$$CH_3{-}C(=O){-}H + HN_2OH \xrightarrow{H^+} CH_3{-}C(H){=}N{-}OH + H_2O$$

(bp 20°) acetaldoxime (mp 47°)

$$CH_3{-}C(=O){-}H + H_2NNHC_6H_5 \xrightarrow{H^+} CH_3{-}C(H){=}N{-}NHC_6H_5 + H_2O$$

acetaldehyde phenylhydrazone (mp 63°)

ADDITION OF GRIGNARD REAGENTS

The very reactive Grignard reagents (Section 16.7) can be represented as a carbanion (negatively charged carbon species) associated with a positive magnesium halide: R^-MgX^+. The alkyl portion of the Grignard reagent, which is negatively charged, adds to the carbonyl carbon while the positive magnesium joins the negative carbonyl oxygen:

$$\rangle C{=}\bar{O} \; R^-MgX^+ \longrightarrow XMgO{-}C{-}R \xrightarrow{H_2O} HO{-}C{-}R + Mg(OH)X$$

Hydrolysis converts the adduct to an alcohol. The importance of the reaction is that a compound of higher molecular mass can be obtained by joining together two lower molecular mass materials.

Formaldehyde reacts with Grignard reagents to give primary alcohols. Other aldehydes give secondary alcohols, and ketones yield tertiary alcohols.

$$R^-MgX^+ + H_2CO \xrightarrow[\text{hydrolysis}]{\text{followed by}} RCH_2OH + Mg(OH)X$$

formaldehyde primary alcohol

$$R^-MgX^+ + R'-\overset{O}{\underset{H}{C}} \xrightarrow[\text{hydrolysis}]{\text{followed by}} R-\overset{H}{\underset{R'}{C}}-OH + Mg(OH)X$$

an aldehyde — secondary alcohol

$$R^-MgX^+ + R'-\overset{O}{\underset{R''}{C}} \xrightarrow[\text{hydrolysis}]{\text{followed by}} R-\overset{R'}{\underset{R''}{C}}-OH + Mg(OH)X$$

a ketone — tertiary alcohol

Inspection of the structure of an alcohol is necessary in order to decide the route to be used to prepare it from a Grignard reagent and a carbonyl compound. 2,3-Dimethyl-3-pentanol can be made from

(1) the Grignard of bromomethane and 2-methyl-3-pentanone,

$$CH_3^-MgBr^+ + CH_3\overset{CH_3}{CH}\underset{O}{\overset{\|}{C}}CH_2CH_3 \xrightarrow[\text{hydrolysis}]{\text{followed by}} CH_3-\overset{CH_3}{CH}-\overset{CH_3}{\underset{OH}{C}}-CH_2CH_3$$

(2) the Grignard of bromoethane and methylbutanone,

$$CH_3CH_2^-MgBr^+ + CH_3\overset{CH_3}{CH}\underset{O}{\overset{\|}{C}}CH_3 \xrightarrow[\text{hydrolysis}]{\text{followed by}} CH_3-\overset{CH_3}{CH}-\overset{CH_3}{\underset{OH}{C}}-CH_2CH_3$$

and (3) the Grignard of 2-bromopropane and butanone.

$$CH_3\overset{CH_3}{CH}{}^-MgBr^+ + CH_3CH_2\underset{O}{\overset{\|}{C}}CH_3 \xrightarrow[\text{hydrolysis}]{\text{followed by}} CH_3-\overset{CH_3}{CH}-\overset{CH_3}{\underset{OH}{C}}-CH_2CH_3$$

EXAMPLE 20.4

Suggest two possible synthetic routes to the following compound:

$$C_6H_5-\overset{H}{\underset{OH}{C}}-CH_3$$

The compound is a secondary alcohol and can be produced by the reaction of a Grignard reagent with an aldehyde. Therefore, only two possibilities exist:

$$C_6H_5-\overset{O}{\underset{H}{C}} + CH_3^-MgBr^+ \quad \text{or} \quad CH_3-\overset{O}{\underset{H}{C}} + C_6H_5^-\ MgBr^+$$

20.6 REACTIVITY OF THE α-CARBON ATOM

The attraction of electrons by the oxygen atom of the carbonyl group is relayed to the immediately adjacent carbon atoms (α-carbon atoms).

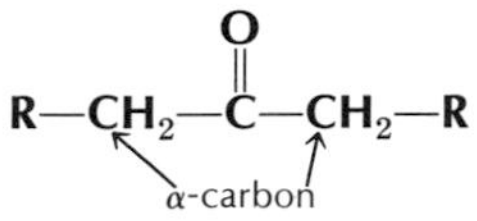

$$R{-}\underset{\alpha\text{-carbon}}{CH_2}{-}\overset{O}{\overset{\|}{C}}{-}\underset{\alpha\text{-carbon}}{CH_2}{-}R$$

This attraction is called an *inductive effect* and is responsible for an increase in the acidity of the attached hydrogen atoms. The carbanion that results after removal of a proton by a base can be represented by two contributing resonance forms:

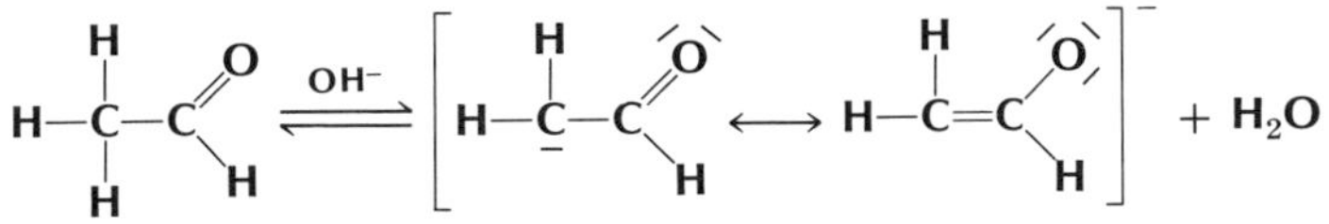

$$H{-}\underset{H}{\overset{H}{C}}{-}C(=O)H \xrightleftharpoons{OH^-} \left[H{-}\overset{H}{\underset{-}{C}}{-}C(=O)H \longleftrightarrow H{-}\overset{H}{C}{=}C(O^-)H \right]^- + H_2O$$

The conjugate bases of carbonyl compounds behave as nucleophilic reagents and undergo a large number of reactions. The only reaction considered here is the basis of the haloform reaction. The carbanions derived from carbonyl compounds react with halogens to yield halogenated carbonyl compounds. This reaction is illustrated below for the conjugate base of acetaldehyde and chlorine:

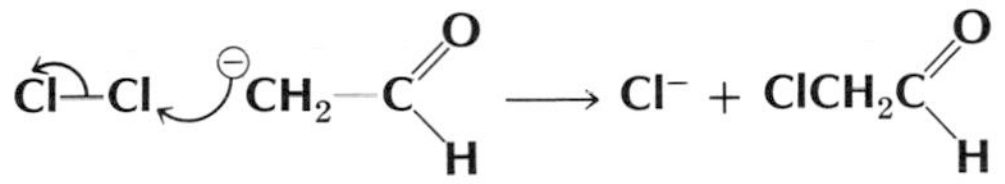

$$Cl{-}Cl \;\; {}^{\ominus}CH_2{-}C(=O)H \longrightarrow Cl^- + ClCH_2C(=O)H$$

For compounds of the general structure $CH_3{-}\overset{O}{\overset{\|}{C}}{-}R$ the replacement of hydrogen by halogen may occur successively to yield a trihalogenated compound:

$$X_3C{-}\overset{O}{\overset{\|}{C}}{-}R$$

Under the reaction conditions the bond between the α-carbon and the carbonyl carbon is cleaved by a hydroxide ion to form trihalogenated methane (haloform) and the salt of an acid:

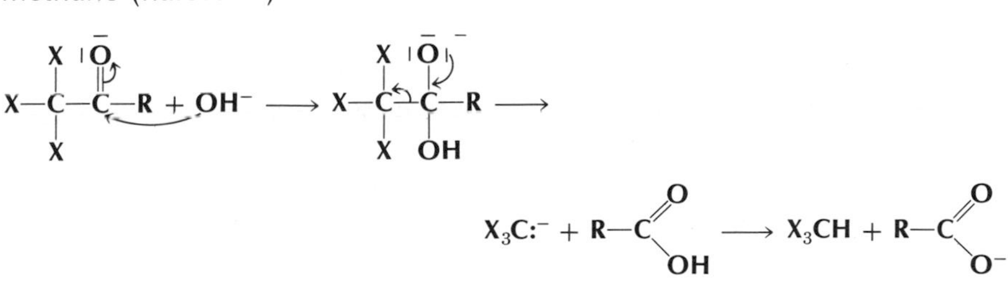

$$X{-}\underset{X}{\overset{X}{C}}{-}\overset{O}{\overset{\|}{C}}{-}R + OH^- \longrightarrow X{-}\underset{X}{\overset{X}{C}}{-}\underset{OH}{\overset{O^-}{C}}{-}R \longrightarrow$$

$$X_3C{:}^- + R{-}C(=O)OH \longrightarrow X_3CH + R{-}C(=O)O^-$$

When iodine in basic solution is used, solid iodoform is produced. The formation of this yellow substance is indicative of the structure

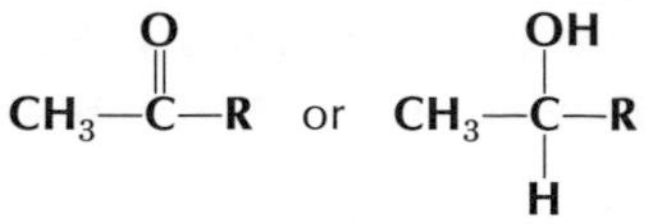

$$CH_3-\overset{\overset{\displaystyle O}{\|}}{C}-R \quad \text{or} \quad CH_3-\underset{\underset{\displaystyle H}{|}}{\overset{\overset{\displaystyle OH}{|}}{C}}-R$$

which is oxidized to the ketone under the reaction conditions (Section 19.7).

EXAMPLE 20.5

A compound C_4H_8O reacts with phenylhydrazine to yield a solid compound. The compound C_4H_8O reacts with a solution of sodium carbonate and iodine to yield iodoform. What is the structure of the unknown substance?

The fact that the substance reacts with phenylhydrazine is indicative that the compound is either an aldehyde or a ketone (see Section 20.5). Since aldehydes and ketones contain two fewer hydrogen atoms per molecule of compound the structure cannot contain any other structural features such as double bonds or rings. The iodoform product indicates that the compound is a ketone of the structure:

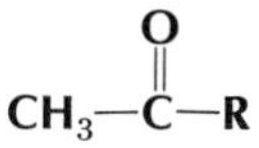

$$CH_3-\overset{\overset{\displaystyle O}{\|}}{C}-R$$

Since the compound contains only four carbon atoms the structure must be butanone because there is only one way to arrange the two carbon atoms unaccounted for by the above general structure:

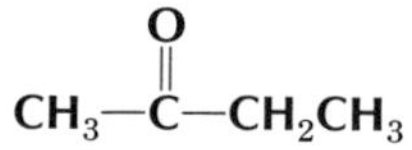

$$CH_3-\overset{\overset{\displaystyle O}{\|}}{C}-CH_2CH_3$$

20.7 POLYMERIZATION OF FORMALDEHYDE

Gaseous formaldehyde may be polymerized to yield paraformaldehyde, a white solid. The polymerization is carried out in aqueous solution and probably proceeds via the addition of the hydrate of formaldehyde to another formaldehyde molecule to yield a growing hemiacetal chain:

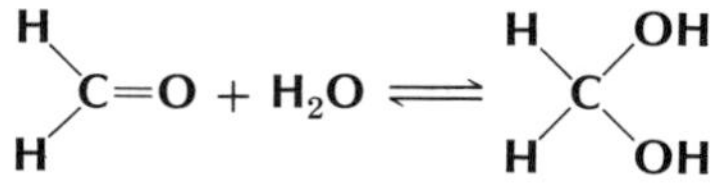

$$(H)_2C=O + H_2O \rightleftharpoons (H)_2C(OH)_2$$

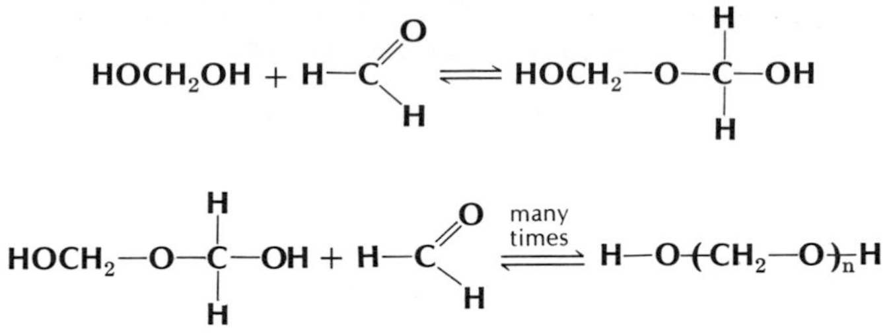

$$HOCH_2OH + H{-}\overset{\displaystyle O}{\overset{\|}{C}}{-}H \rightleftharpoons HOCH_2{-}O{-}\underset{H}{\overset{H}{C}}{-}OH$$

$$HOCH_2{-}O{-}\underset{H}{\overset{H}{C}}{-}OH + H{-}\overset{\displaystyle O}{\overset{\|}{C}}{-}H \xrightleftharpoons{\text{many times}} H{-}O{+}CH_2{-}O{+}_nH$$

paraformaldehyde

Paraformaldehyde, which contains approximately 30 monomer units, is depolymerized by heating and regenerates formaldehyde. This process is a convenient way of generating formaldehyde, which is a gas, at the time that it is needed in the laboratory.

If the terminal hydroxyl groups in a higher molecular mass polymer of formaldehyde are reacted with a carboxylic acid to form an ester (Chapter 21) the polymer is stable. This polymer is marketed under the trade name of Delrin by the E. I. Dupont Company and has found use as a structural material with which to mold many machine components for industry. Many gears in light industrial machines are now made of Delrin instead of metal.

Formaldehyde can be *copolymerized* with phenol to yield the hard, electrically nonconducting Bakelite polymer. *Copolymerization* is the term used in describing the reaction between two or more unlike monomers.

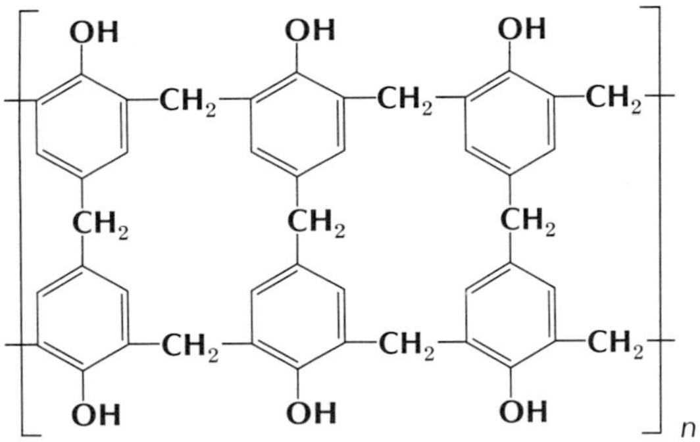

OH OH OH
CH_2 CH_2 CH_2
CH_2 CH_2 CH_2
CH_2 CH_2 CH_2
OH OH OH
n

Formaldehyde copolymerizes with melamine to form the polymer Melmac, which is used extensively in the production of unbreakable dinnerware.

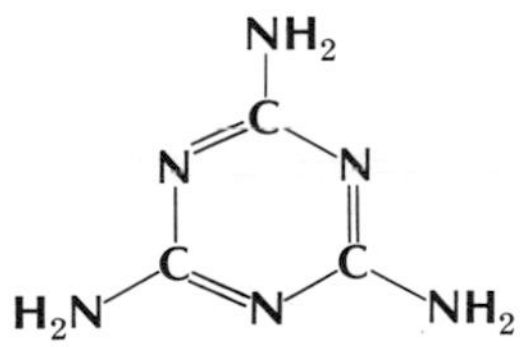

NH_2
C
N N
C C
H_2N N NH_2

melamine

20.8
INFRARED SPECTROSCOPY

Most functional groups absorb the energy associated with light in the infrared range, that is, 2.5–16 microns (1 $\mu = 10^{-6}$ m). This energy corresponds to the energy associated with molecular vibrations resulting from the stretching and bending of bonds. Since the bond types of groups such as the hydroxyl and the carbonyl differ considerably, the energies required to stretch these bonds are different. Therefore the respective regions of absorption for various functional groups can be easily distinguished.

In general the energy required to stretch or increase the distance between two atoms increases as the number of bonds between them increases. For example, the energy required to stretch a triple bond of an alkyne is greater than that of the double bond of an alkene, which in turn is greater than that of a single bond of an alkane. The average wavelengths of absorption of triple, double, and single carbon–carbon bonds are 4.5, 6.1, and 8.7 μ, respectively. Similarly, the average wavelengths of absorption of a carbon-oxygen bond in an alcohol- and a carbonyl-containing compound are 9.0 and 5.7 μ, respectively.

One of the great advantages of infrared spectroscopy is its sensitivity to variations in the wavelength of absorption within a particular class of compounds. For example, the carbonyl group of a cyclohexanone, cyclopentanone, and cyclobutanone absorb at 5.83, 5.73, and 5.62 μ, respectively. Therefore, the chemist can not only ascertain that a molecule contains a carbonyl group but can state its general structural environment.

The infrared spectrum of an organic molecule is usually quite complicated because it is a composite of all the possible stretching and bending vibrations of the bonds contained in a molecule. However, with experience and familiarity with the infrared spectra of many compounds the chemist can limit the types of function groups he must consider as possibilities in his search for the structure of an unknown compound.

Suggested further readings

Glasstone, S., "Oxidation Numbers and Valence," *J. Chem. Educ.*, **25,** 278 (1948).

Gregg, D. C. "A Simplified Electronic Interpretation of Oxidation-Reduction," *J. Chem. Educ.*, **22,** 548 (1945).

Seelye, R. N., and Turney, T. A., "The Iodoform Reaction," *J. Chem. Educ.*, **36,** 572 (1959).

Swinehart, D. F., "More on Oxidation Numbers," *J. Chem. Educ.*, **29,** 284 (1952).

Van der Werf, C. A., "A Consistent Treatment of Oxidation-Reduction," *J. Chem. Educ.*, **25,** 547 (1948).

Ward, C. H., "Keto-Enol Tautomerism of Ethyl Acetoacetate," *J. Chem. Educ.*, **39,** 95 (1962).
Zuffanti, S., "Enolization: An Electronic Interpretation," *J. Chem. Educ.*, **22,** 230 (1945).

Terms and concepts

acetal
acetaldehyde
acetone
aldehyde
carbonyl group
copolymerization
formaldehyde
Grignard reagent
hemiacetal
infrared spectroscopy
ketone
lithium aluminum hydride
sex hormones
Tollens' test

Questions and problems

1. Name the following compounds:

a. $(CH_3)_2CHCHO$

b. $CH_3—\overset{O}{\overset{\|}{C}}—C_2H_5$

c. 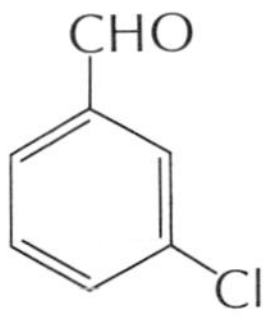 CHO Cl

d. 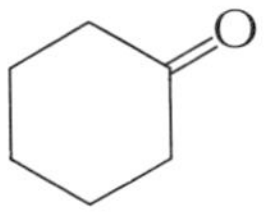 O

2. Write structural formulas for the following names:

- **a.** 2-octanone
- **b.** 2-bromopentanal
- **c.** 3-phenyl-2-butanone
- **d.** *m*-chlorocinnamaldehyde
- **e.** 2-methylcyclohexanone
- **f.** 2,2,2-trichloroethanol

3. What reagents are necessary to obtain the stated compound from the indicated starting material?

- **a.** butanone from 2-butanol
- **b.** 2-butanol from butanone
- **c.** 2-pentanone from 2,2-dichloropentane
- **d.** butanal from propene
- **e.** acetone from propene
- **f.** *p*-chlorobenzaldehyde from toluene

4. What single chemical test can be used to distinguish between the following pairs of compounds?

a. 3-pentanol and 3-pentanone
b. ethanal and butanal
c. 1-pentanal and 3-pentanone
d. 2-pentanone and 3-pentanone
e. methanal and ethanal
f. cyclopentanol and 3-pentanone

5. A compound, $C_5H_{12}O$, can be oxidized to yield $C_5H_{10}O$, which reacts with phenylhydrazine and gives an iodoform test. What two substances will react as indicated?

6. A compound, $C_5H_{10}O$, reacts with phenylhydrazine but does not react with Tollens' reagent. It does not give an iodoform test. What is the structure of the substance?

7. A compound, $C_5H_{10}O$, does not react with phenylhydrazine. It does not react with hydrogen in the presence of a catalyst, nor does it react with bromine. Suggest one possible structure that is in accord with these facts.

8. A compound, C_5H_8O, reacts with phenylhydrazone but does not react with Tollens' reagent. Treatment with lithium aluminum hydride produces a compound, $C_5H_{10}O$, which can be dehydrated to yield C_5H_8. Ozonolysis of C_5H_8 and treatment with zinc and acid yields the structure below. What is the structure of the C_5H_8O compound?

$$\mathrm{H{-}C({=}O){-}CH_2CH_2CH_2{-}C({=}O){-}H}$$

9. What product is formed after hydrolysis when the Grignard reagent of bromoethane is reacted with each of the following compounds?

a. acetone **b.** formaldehyde **c.** acetaldehyde

10. How can each of the following alcohols be prepared from an aldehyde or ketone and a Grignard reagent?

a. 2-butanol **b.** 1-pentanol **c.** 2-methyl-2-butanol

11. Suggest a structure for the polymer Melmac.

12. When methylpropanal is treated with heavy water (D_2O) and the base OD^- for a length of time the recovered aldehyde has a molecular mass of 73 amu. What is the structure of the recovered aldehyde?

13. When formaldehyde is dissolved in water containing some of the isotope $^{18}_{8}O$ and then is recovered from solution the aldehyde has the isotope incorporated in the molecule as the carbonyl oxygen. Why?

14. List as many reactions as possible that civetone will undergo.

21

CARBOXYLIC ACIDS AND ESTERS

Compounds that contain the *carboxyl group* $—C\!\begin{smallmatrix}\nearrow O\\ \searrow OH\end{smallmatrix}$ constitute a family of widely occurring natural products. Acetic acid, one of the simplest members of this class of compounds, has been known since biblical times. It is the substance to which ethyl alcohol is oxidized by bacteria and accounts for the sour taste that develops in wine with less than 14 percent alcohol content when it is left exposed to air. Vinegar is a 5 percent solution of acetic acid in water.

The carboxyl group cannot be regarded as simply a combination of a carbonyl group and a hydroxyl group in the same molecule. As shown in this chapter the whole carboxyl group has specific properties that reflect the electronic interaction of the component atoms.

Esters are products of the reaction of acids with alcohols. Many of the common fruits owe their odor to low molecular mass esters; *fats, oils,* and *waxes* are high molecular mass esters. The structures of these naturally occurring substances are examined in this chapter.

21.1 NOMENCLATURE OF CARBOXYLIC ACIDS

Most of the carboxylic acids containing continuous unbranched carbon chains have been known for a long time. Consequently they are known by common names that have their origins in Greek and Latin and indicate the natural source of the acid. The IUPAC nomenclature employs the suffix *-oic* acid added to the hydrocarbon stem minus its final e. The names, formulas, and derivation of the common names of some acids are listed in Table 21.1.

The carbon atom of the carboxyl group is the carbon 1 in the IUPAC nomenclature. Greek letters, (α, β, γ, δ) are used to designate substituent positions in common names. The alpha (α) position is the carbon atom adjacent to the carboxyl group, the beta (β) is the next on the chain, etc. The two systems of nomenclature should not be used interchangeably.

	$CH_3CH(Cl)CO_2H$	$CH_3-CH(CH_3)CH_2CO_2H$	$BrCH_2CH_2CH_2CO_2H$
common:	α-chloropropionic acid	β-methylbutyric acid	γ-bromobutyric acid
IUPAC:	2-chloropropanoic acid	3-methylbutanoic acid	4-bromobutanoic acid

21.2 PROPERTIES OF CARBOXYLIC ACIDS

PHYSICAL PROPERTIES

Carboxylic acids are responsible for the odors associated with goats, rancid butter, certain cheeses, and locker rooms. The common names caproic (hexanoic), caprylic (octanoic), and capric (decanoic) acid are all derived from the Latin *caper* for goat. Butyric acid occurs in rancid butter, aged cheese, and human perspiration.

Table 21.1
Names of the common acids

name of acid			
IUPAC	*common*	*formula*	*derivation*
methanoic	formic	HCO_2H	Latin, *formica,* ant
ethanoic	acetic	CH_3CO_2H	Latin, *acetum,* vinegar
propanoic	propionic	$CH_3CH_2CO_2H$	Greek, *protos, pion,* first fat
butanoic	butyric	$CH_3(CH_2)_2CO_2H$	Latin, *butyrum,* butter
pentanoic	valeric	$CH_3(CH_2)_3CO_2H$	Latin, *valere,* powerful
dodecanoic	lauric	$CH_3(CH_2)_{10}CO_2H$	laurel
hexadecanoic	palmitic	$CH_3(CH_2)_{14}CO_2H$	palm
octadecanoic	stearic	$CH_3(CH_2)_{16}CO_2H$	Greek, *stear,* tallow

Table 21.2
Physical properties of some acids

common name	*boiling point* (°C)	*solubility* (*g/100 g of water*)
formic acid	101	infinite
acetic acid	119	infinite
propionic acid	141	infinite
butyric acid	163	5.6
valeric acid	186	3.7
palmitic acid	268	insoluble
stearic acid	287	insoluble

The boiling points of the acids are abnormally high for their molecular masses (Table 21.2). Hydrogen bonding between acid molecules to form dimers that persist even in the vapor phase accounts for the elevated boiling points.

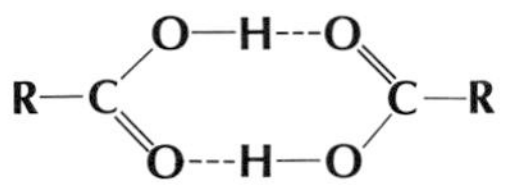

$$\begin{array}{c} \text{O—H---O} \\ \text{R—C} \qquad\qquad \text{C—R} \\ \text{O---H—O} \end{array}$$

The lower molecular mass acids are water soluble. Formic and acetic acids are completely miscible with water. The solubility of the acids decreases with increasing chain length (Table 21.2).

ACIDITY

Although carboxylic acids are weak acids, they are more acidic than alcohols and phenols. The discussion of the acidity of carboxylic acids will utilize the concept of the equilibrium constant developed for acid-base reactions in Chapters 9 and 10.

The equilibrium between the unionized and ionized forms of acetic acid lies far to the left. A 0.1*M* solution of acetic acid is only 1 percent ionized.

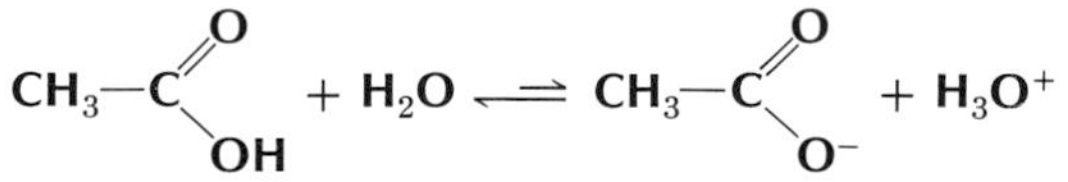

$$CH_3-C(=O)OH + H_2O \rightleftharpoons CH_3-C(=O)O^- + H_3O^+ \qquad K_a = 1.8 \times 10^{-5}$$

The equilibrium constant is 1.8×10^{-5}, or 1.8×10^9 times larger than the ionization constant of water. Two factors are important in determining the enhanced acidity of acids relative to water or alcohols. Resonance stabilization of the conjugate base, the acetate ion in the case of acetic acid, is important. The negative charge is delocalized over a π system consisting of orbitals of carbon and both oxygen atoms. Stabilization of a product through resonance assists its formation in an equilibrium reaction, which is controlled by energy considerations.

$$CH_3{-}C(=\ddot{O})\ddot{O}^- \longleftrightarrow CH_3{-}C(\ddot{O}^-)=\ddot{O}$$

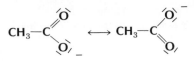

The second factor is the inductive effect of the positive carbon of the carbonyl group on the electron density of the O—H bond. Electron-withdrawing groups attached to the oxygen atom facilitate the departure of a proton.

A list of acids and their K_d values are given in Table 21.3. Acetic acid is a weaker acid than formic acid, a fact that can be rationalized on the basis of the electron-donating ability of a methyl group relative to hydrogen. The donation of electrons decreases the ease of ionization of the proton in the acid and increases the basicity of the conjugate base. The electronegative chlorine atom in chloroacetic acid increases the acidity relative to acetic acid. Additional chlorine atoms increase the acidity still further.

EXAMPLE 21.1

Compare the acidities of benzoic acid and p-nitrobenzoic acid.

The nitro group is an electron-withdrawing group, as evidenced by the classification of substituents in the chapter on aromatic hydrocarbons. The nitro group deactivates the aromatic ring toward reaction with electrophiles and is a *meta* director. Therefore, the nitro group in *p*-nitrobenzoic acid should increase the acidity of the compound above that of benzoic acid. (The K_d values of benzoic and *p*-nitrobenzoic acid are 6.4×10^{-5} and 3.8×10^{-4}, respectively.)

Table 21.3
Acid strengths of carboxylic acids

acid	*structure*	K_d
formic	HCO_2H	2×10^{-4}
acetic	CH_3CO_2H	1.8×10^{-5}
chloroacetic	$ClCH_2CO_2H$	1.6×10^{-3}
dichloroacetic	Cl_2CHCO_2H	5×10^{-2}
trichoroacetic	Cl_3CCO_2H	3×10^{-1}
fluoroacetic	FCH_2CO_2H	2.2×10^{-3}
bromoacetic	$BrCH_2CO_2H$	1.4×10^{-3}
iodoacetic	ICH_2CO_2H	7.5×10^{-4}

21.3
PREPARATION OF CARBOXYLIC ACIDS

Most straight chain acids and some aromatic acids are available from natural sources. The acids with an even number of carbon atoms are much more abundant than those with an odd number of atoms. The reason for this phenomenon is illustrated in Chapter 28. Some of the general synthetic methods that may be used to prepare acids follow.

OXIDATION

Oxidation of primary alcohols is the best route to carboxylic acids if the proper alcohol is available.

$$RCH_2OH \xrightarrow[H^+]{Na_2Cr_2O_7} R{-}C(=O)OH$$

However, the normal carbon chain alcohols with an odd number of carbon atoms are as scarce as the acids in natural sources.

The aromatic acids can be obtained from the oxidation of side chains (Chapter 18). The aromatic ring is resistant to oxidation and remains intact.

$$C_6H_5CH_3 \xrightarrow[H^+]{Na_2Cr_2O_7} C_6H_5CO_2H$$

$$C_6H_5CH_2CH_2CH_3 \xrightarrow[H^+]{Na_2Cr_2O_7} C_6H_5CO_2H + 2CO_2$$

CHAIN EXTENSION VIA NITRILES

Compounds containing the —C≡N group, *nitriles,* can be hydrolyzed to acids with either an acid or base catalyst.

$$CH_3CN + 2H_2O + HCl \longrightarrow CH_3CO_2H + NH_4Cl$$
$$CH_3CN + H_2O + NaOH \longrightarrow CH_3CO_2^-Na^+ + NH_3$$

The hydrolysis of nitriles is a very useful reaction when coupled with the formation of the nitrile itself. Potassium cyanide (KCN) reacts with haloalkanes to displace the halide ion, thereby lengthening the carbon chain.

$$R{-}Br + KCN \longrightarrow R{-}CN + KBr$$

Since even-numbered carbon chain compounds are abundant in natural sources, this reaction may be used to synthesize odd-numbered carbon compounds.

Another useful way to produce acids with one additional carbon atom than available starting materials is the addition of Grignard reagents to carbon dioxide. The mechanism of the reaction is similar to the addition to carbonyl compounds:

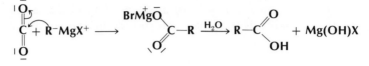

$$O{=}C{=}O + R^{-}MgX^{+} \longrightarrow BrMg^{+}\,{}^{-}O{-}C({=}O){-}R \xrightarrow{H_2O} R{-}C({=}O)OH + Mg(OH)X$$

EXAMPLE 21.2

How can phenylacetic acid be prepared from toluene?

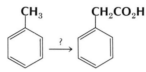

$$C_6H_5CH_3 \xrightarrow{?} C_6H_5CH_2CO_2H$$

The problem involves extension of the carbon chain by one carbon atom. However, the methyl group itself first must be converted to a functional group. Both the chain extension with cyanide displacement and the Grignard reaction require a haloalkane. Therefore, halogenation of the methyl group with a reagent such as bromine must be accomplished first.

$$C_6H_5CH_3 + Br_2 \longrightarrow C_6H_5CH_2Br + HBr$$

Note than an iron catalyst is not used in this reaction because that would lead to ring bromination. The chain then could be extended by reaction with potassium cyanide followed by hydrolysis.

$$C_6H_5CH_2Br \xrightarrow{KCN} C_6H_5CH_2CN \xrightarrow[HCl]{H_2O} C_6H_5CH_2CO_2H$$

Alternatively the Grignard reaction with CO_2 could be used.

$$C_6H_5CH_2Br \xrightarrow[Mg]{} C_6H_5CH_2MgBr \xrightarrow[(2)\ H_2O]{(1)\ CO_2} C_6H_5CH_2CO_2H$$

21.4 REACTIONS OF CARBOXYLIC ACIDS

REACTIONS WITH BASE

Carboxylic acids react with base to produce salts of the acids. The names of salts are derived from the acid by changing the *-ic* ending to *-ate* and preceeding this name by the name of the metal.

$$CH_3{-}C({=}O){-}OH + Na^+\ OH^- \longrightarrow CH_3C({=}O){-}O^-\ Na^+ + H_2O$$

sodium acetate
sodium ethanoate

The sodium and potassium salts of long chain unbranched acids are *soaps* (see Section 21.9 on saponification). The lithium salts of the same acids are blended with oils to form lubricating greases. Calcium propionate is added to bread to retard molding. Zinc undecylenate is used in the treatment of athlete's foot.

$(CH_3CH_2CO_2)_2Ca$
calcium propionate

$(CH_2{=}CH(CH_2)_8CO_2)_2Zn$
zinc undecylenate

REACTION WITH ALCOHOLS

A carboxylic acid and an alcohol react in the presence of an acid catalyst to produce an equilibrium mixture containing an ester and water:

$$R{-}C({=}O){-}OH + R'{-}OH \overset{H^+}{\rightleftharpoons} R{-}C({=}O){-}O{-}R' + H_2O$$

The equilibrium can be displaced to increase the yield of the ester by utilizing Le Châtelier's principle. If the water is distilled out of the reaction, as it can be for all but the very low molecular mass esters, a high yield of the ester can be obtained. If one of the starting reagents is rare or expensive, the yield of ester can be enhanced by using an excess of the cheaper component of the reaction. Ethyl esters of acids can be obtained by using the inexpensive ethanol as a solvent. Under such conditions the high concentration of ethanol favors a high conversion of the acid to the ester.

FORMATION OF ACYL HALIDES

Carboxylic acids yield *acyl chlorides* when treated with phosphorus trichloride or thionyl chloride:

$$3CH_3{-}\overset{\overset{O}{\|}}{C}{-}OH + PCl_3 \longrightarrow 3CH_3\overset{\overset{O}{\|}}{C}{-}Cl + H_3PO_3$$

acetic acid — acetyl chloride

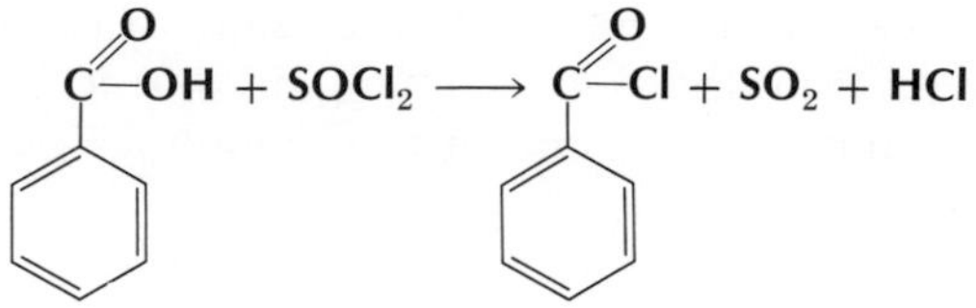

$$C_6H_5{-}\overset{\overset{O}{\|}}{C}{-}OH + SOCl_2 \longrightarrow C_6H_5{-}\overset{\overset{O}{\|}}{C}{-}Cl + SO_2 + HCl$$

benzoic acid — benzoyl chloride

These reactions parallel the reactions described to prepare chloroalkanes from alcohols (Chapter 19). Unlike alcohols, acids do not react with hydrogen chloride to yield acyl chlorides. The equilibrium overwhelmingly favors the reverse reaction, the hydrolysis of acyl chlorides.

$$CH_3{-}\overset{\overset{O}{\|}}{C}{-}Cl + H_2O \longrightarrow CH_3{-}\overset{\overset{O}{\|}}{C}{-}OH + HCl$$

This hydrolysis reaction accounts for the irritating action of these compounds upon the human system.

FORMATION OF AMIDES

The reaction of ammonia with carboxylic acids produces the ammonium salts of the acids, which in turn yield *amides* when heated.

$$CH_3{-}\overset{\overset{O}{\|}}{C}{-}O^-NH_4^+ \xrightarrow{\text{heat}} CH_3{-}\overset{\overset{O}{\|}}{C}{-}NH_2 + H_2O$$

ammonium acetate — acetamide

Amides can be made conveniently from the reaction of acyl halides with ammonia. This reaction is called *ammonolysis* and is similar to the reaction illustrated in the previous section in which an acyl chloride was hydrolyzed.

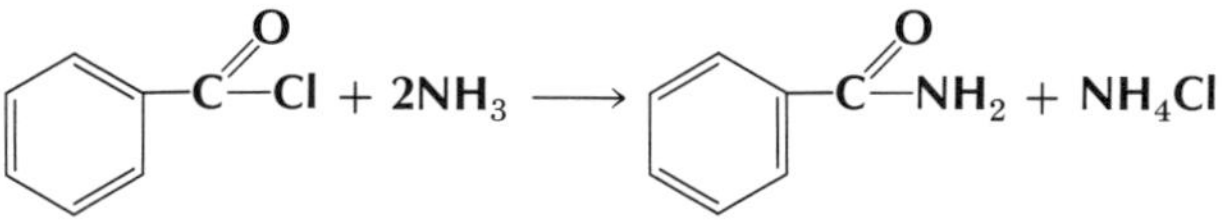

$$C_6H_5{-}\overset{\overset{O}{\|}}{C}{-}Cl + 2NH_3 \longrightarrow C_6H_5{-}\overset{\overset{O}{\|}}{C}{-}NH_2 + NH_4Cl$$

benzoyl chloride — benzamide

One of the many important synthetic fibers is a polyamide known commercially as Nylon. Nylon is produced by the reaction of a diamine (Chapter 22) and a diacid.

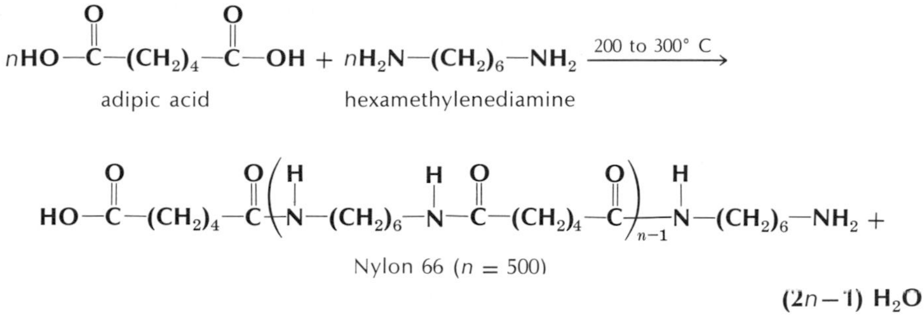

$$n\text{HO}-\overset{\overset{\text{O}}{\|}}{\text{C}}-(\text{CH}_2)_4-\overset{\overset{\text{O}}{\|}}{\text{C}}-\text{OH} + n\text{H}_2\text{N}-(\text{CH}_2)_6-\text{NH}_2 \xrightarrow{200 \text{ to } 300^\circ \text{ C}}$$

adipic acid hexamethylenediamine

$$\text{HO}-\overset{\overset{\text{O}}{\|}}{\text{C}}-(\text{CH}_2)_4-\overset{\overset{\text{O}}{\|}}{\text{C}}\left(\overset{\overset{\text{H}}{|}}{\text{N}}-(\text{CH}_2)_6-\overset{\overset{\text{H}}{|}}{\text{N}}-\overset{\overset{\text{O}}{\|}}{\text{C}}-(\text{CH}_2)_4-\overset{\overset{\text{O}}{\|}}{\text{C}}\right)_{n-1}\overset{\overset{\text{H}}{|}}{\text{N}}-(\text{CH}_2)_6-\text{NH}_2 + (2n-1)\,\text{H}_2\text{O}$$

Nylon 66 ($n = 500$)

The reaction involves the end of a particular diacid molecule reacting at both ends with a diamine. The amine ends of the resultant larger molecule can then react with additional diacid molecules, and the chain continues to grow. As the amine reacts with the acid a molecule of water is formed. The polymers formed with the simultaneous elimination of small molecules such as water are called *condensation polymers.* The student should contrast this type of polymerization process with that of the formation of addition polymers (Chapter 17) in which alkenes react to form polymers without the loss of component atoms or molecules.

Proteins (Chapter 24) also contain amide groupings in large molecular mass molecules. However, in these molecules the hydrocarbon fragments between the amide groupings are varied.

21.5 NOMENCLATURE OF ACID DERIVATIVES

The esters are named like the salts of acids, with the name of the alkyl group of the alcohol replacing the metal. Since there are two names of aliphatic carboxylic acids in common usage there are two names for the derived esters. A few examples illustrate this point:

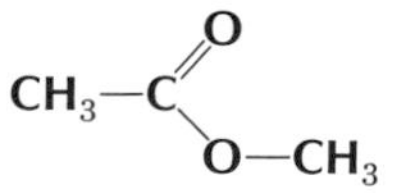

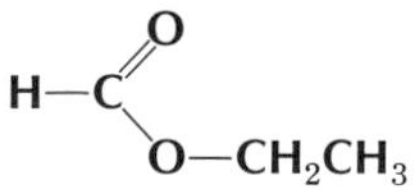

	$CH_3-C(=O)-O-CH_3$	$H-C(=O)-O-CH_2CH_3$
common:	methyl acetate	ethyl formate
IUPAC:	methyl ethanoate	ethyl methanoate

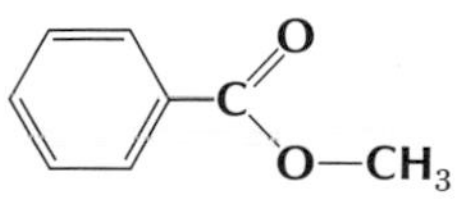

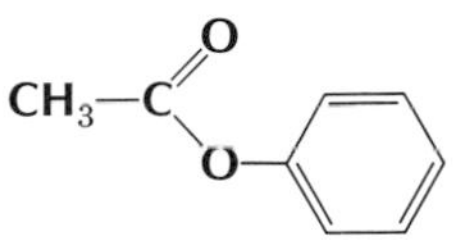

	$C_6H_5-C(=O)-O-CH_3$	$CH_3-C(=O)-O-C_6H_5$
common:	methyl benzoate	phenyl acetate
IUPAC:	methyl benzoate	phenyl ethanoate

The acyl halides are named by dropping the *-ic* of the acid name and adding *-yl* followed by the name of the halide. A few examples are given below:

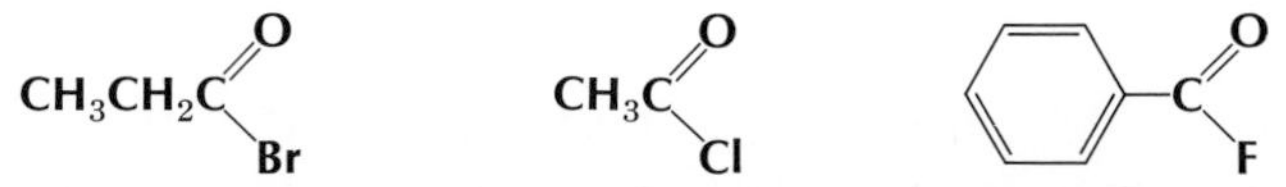

$$CH_3CH_2C(=O)Br \qquad CH_3C(=O)Cl \qquad C_6H_5C(=O)F$$

common:	propionyl bromide	acetyl chloride	benzoyl fluoride
IUPAC:	propanoyl bromide	ethanoyl chloride	benzoyl fluoride

The amides are named from the corresponding acids.

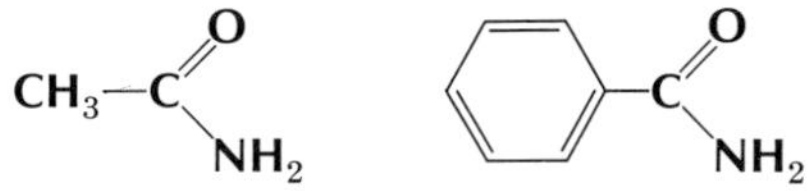

$$CH_3{-}C(=O)NH_2 \qquad C_6H_5{-}C(=O)NH_2$$

common:	acetamide	benzamide
IUPAC:	ethanamide	benzamide

21.6 REACTIONS OF ESTERS

HYDROLYSIS

Esters can be hydrolyzed in the presence of a mineral acid to produce an equilibrium mixture containing the component carboxylic acid and alcohol.

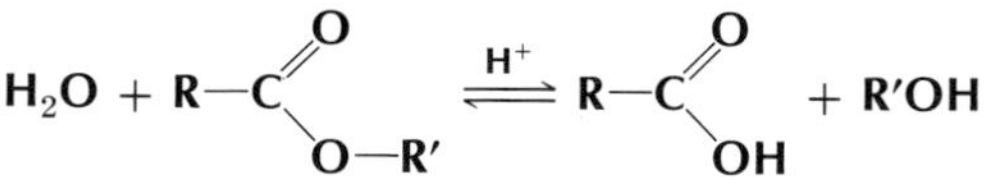

$$H_2O + R{-}C(=O){-}O{-}R' \overset{H^+}{\rightleftharpoons} R{-}C(=O){-}OH + R'OH$$

In order to hydrolyze an ester quantitatively the hydrolysis reaction can be made essentially irreversible by utilizing sodium hydroxide. The salt of the carboxylic acid is formed in this reaction, which is called a *saponification* reaction (Section 21.9). The acid itself may be obtained by treating the salt with mineral acid after isolating it from the reaction mixture.

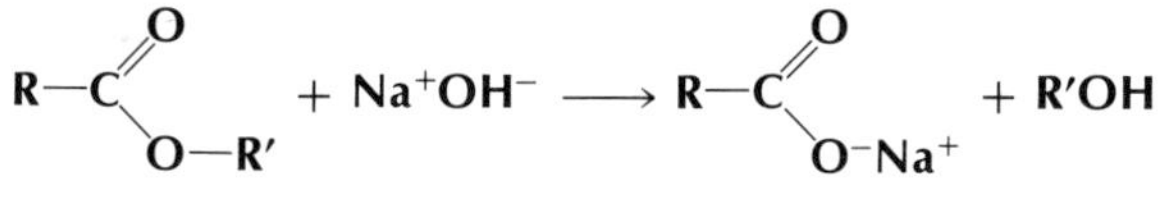

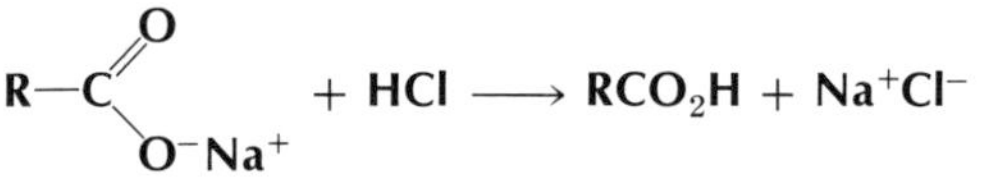

$$R{-}C(=O){-}O{-}R' + Na^+OH^- \longrightarrow R{-}C(=O){-}O^-Na^+ + R'OH$$

$$R{-}C(=O){-}O^-Na^+ + HCl \longrightarrow RCO_2H + Na^+Cl^-$$

REDUCTION

Esters may be reduced to alcohols by the use of lithium aluminum hydride. The intermediate oxidation state, aldehyde, is not produced in this reaction.

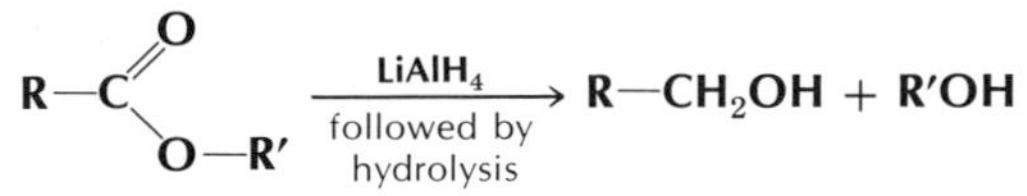

$$R{-}C(=O){-}O{-}R' \xrightarrow[\text{followed by hydrolysis}]{LiAlH_4} R{-}CH_2OH + R'OH$$

ADDITION OF A GRIGNARD REAGENT

Esters react with 2 moles of a Grignard reagent to form tertiary alcohols. Two of the three alkyl groups of the tertiary alcohol are identical and are derived from the Grignard reagent. The third alkyl group originates from the acid portion of the ester.

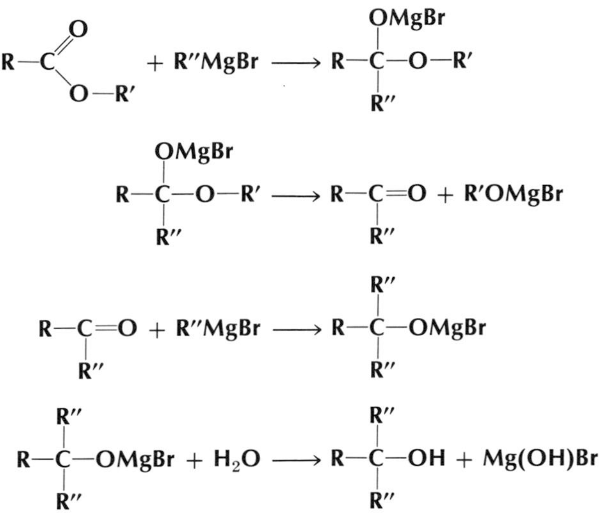

$$R{-}C(=O){-}O{-}R' + R''MgBr \longrightarrow R{-}C(OMgBr)(R''){-}O{-}R'$$

$$R{-}C(OMgBr)(R''){-}O{-}R' \longrightarrow R{-}C(R''){=}O + R'OMgBr$$

$$R{-}C(R''){=}O + R''MgBr \longrightarrow R{-}C(R'')_2{-}OMgBr$$

$$R{-}C(R'')_2{-}OMgBr + H_2O \longrightarrow R{-}C(R'')_2{-}OH + Mg(OH)Br$$

EXAMPLE 21.3

What reagents could be used to prepare 3-methyl-3-pentanol by using the above reaction?

Consideration of the structure of 3-methyl-3-pentanol reveals that two of the alkyl groups attached to the carbon atom bearing the hydroxyl group are identical.

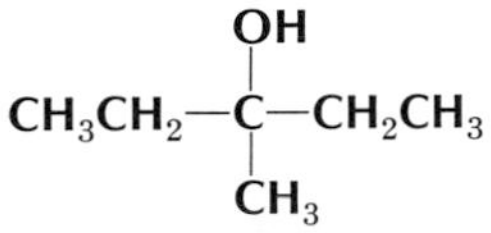

$$CH_3CH_2{-}C(OH)(CH_3){-}CH_2CH_3$$

The two identical ethyl groups could be introduced by the use of ethyl Grignard reagent. The ester that is necessary has to contain the remaining two carbon atoms of the final molecule in the acid portion. Therefore, an acetate ester is needed. For example, methyl acetate and ethyl Grignard reagent will provide the desired compound.

$$CH_3-\overset{\overset{\displaystyle O}{\|}}{C}-O-CH_3 + 2CH_3CH_2^{-}MgBr^{+} \xrightarrow[\text{hydrolysis}]{\text{after}} CH_3CH_2-\underset{\underset{\displaystyle CH_3}{|}}{\overset{\overset{\displaystyle OH}{|}}{C}}-CH_2CH_3 + CH_3OH$$

21.7 COMMON ESTERS

Many of the odors of fruits and flowers are esters. Ethyl acetate is found in pineapples and is a constituent of wines. Isoamyl acetate is contained in apples and bananas. Apples also contain isoamyl isovalerate. Octyl acetate is present in oranges.

$$CH_3-\overset{\overset{\displaystyle CH_3}{|}}{C}HCH_2CH_2O-\overset{\overset{\displaystyle O}{\|}}{C}-CH_3$$

isoamyl acetate

$$CH_3-\overset{\overset{\displaystyle CH_3}{|}}{C}HCH_2CH_2-O-\overset{\overset{\displaystyle O}{\|}}{C}-CH_2\overset{\overset{\displaystyle CH_3}{|}}{C}HCH_3$$

isoamyl isovalerate

In recent years there has been an increased demand for substances to enhance the flavor and aroma of processed foods. Most jams and jellies contain artificial flavoring, as designated on their labels. The esters used are not necessarily those contained in the natural fruits but may be others that produce the same odor or taste. Butyl butyrate is used as a pineapple essence, methyl butyrate resembles the scent of apples, and ethyl butyrate is used as a flavoring additive for peach products.

Salicylic acid can act as a phenol or as an acid. The acetate ester of the phenolic group is acetylsalicylic acid (aspirin), which is used to depress pain originating in the thalamus region of the brain. Approximately 12,000 tons of aspirin are produced annually in the United States. The methyl ester of the carboxyl group is methyl salicylate (oil of wintergreen) and is used in liniments.

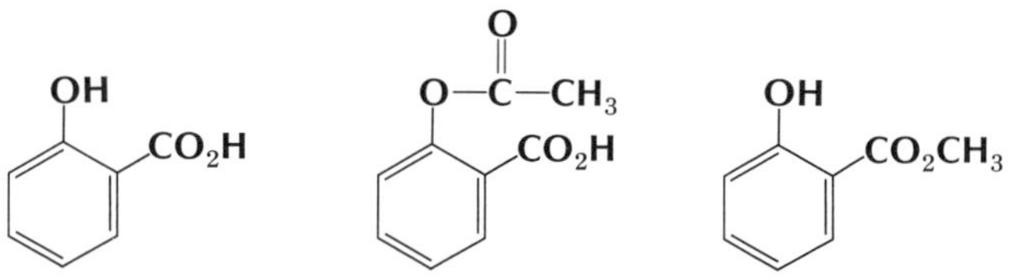

OH, CO_2H (benzene ring) — salicylic acid

$O-\overset{\overset{\displaystyle O}{\|}}{C}-CH_3$, CO_2H (benzene ring) — acetylsalicylic acid

OH, CO_2CH_3 (benzene ring) — methyl salicylate

Dacron is a polymeric ester fiber that can be set in permanent creases in cloths. It is made by copolymerization (Chapter 20) of ethylene glycol and terephthalic acid.

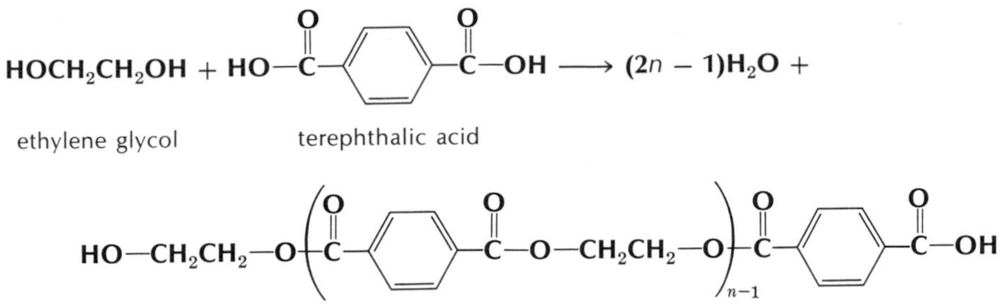

$$HOCH_2CH_2OH + HO{-}\overset{\overset{\displaystyle O}{\|}}{C}{-}C_6H_4{-}\overset{\overset{\displaystyle O}{\|}}{C}{-}OH \longrightarrow (2n-1)H_2O +$$

ethylene glycol terephthalic acid

$$HO{-}CH_2CH_2{-}O{-}\left(\overset{\overset{\displaystyle O}{\|}}{C}{-}C_6H_4{-}\overset{\overset{\displaystyle O}{\|}}{C}{-}O{-}CH_2CH_2{-}O\right)_{n-1}\overset{\overset{\displaystyle O}{\|}}{C}{-}C_6H_4{-}\overset{\overset{\displaystyle O}{\|}}{C}{-}OH$$

Dacron is classified as a condensation polymer.

21.8 WAXES

Waxes are esters of high molecular mass unbranched carboxylic acids and high molecular mass alcohols. They are found in both plants and animals, usually as mixtures of the ester with some of the acid and alcohol from which the esters are made. The waxes melt over a wide temperature range.

Carnauba wax, which is a principle component of floor and automobile wax and of the coating of mimeograph stencils, is a plant wax found on the leaves of certain palms. It is predominately myricyl cerotate and melts between 80 and 90°:

$$CH_3(CH_2)_{24}\overset{\overset{\displaystyle O}{\|}}{C}{-}O{-}CH_2(CH_2)_{29}CH_3$$

Spermaceti wax, used in cosmetics, is largely cetyl palmitate, and is obtained from the head of the sperm whale. It melts over the range of 40 to 50°:

$$CH_3(CH_2)_{14}{-}\overset{\displaystyle O}{\overset{/\!/}{C}}{-}OCH_2(CH_2)_{14}CH_3$$

Beeswax consists of cetyl myristate. It is used in shoe polishes and candles, and melts between 62 and 65°:

$$CH_3(CH_2)_{12}{-}\overset{\displaystyle O}{\overset{/\!/}{C}}{-}OCH_2(CH_2)_{24}CH_3$$

21.9 FATS AND OILS

STRUCTURE

Fats and oils are esters of high molecular mass acids and glycerol (Chapter 19). The component acids are unbranched compounds containing an even number of carbon atoms.

$$
\begin{array}{l}
\quad\; H \qquad\quad O \\
\quad\; | \qquad\quad \| \\
H-C-O-C-R_1 \\
\quad\; | \qquad\quad O \\
\quad\; | \qquad\quad \| \\
H-C-O-C-R_2 \\
\quad\; | \qquad\quad O \\
\quad\; | \qquad\quad \| \\
H-C-O-C-R_3 \\
\quad\; | \\
\quad\; H
\end{array}
$$

The naturally occurring fats and oils are not simple compounds but rather mixtures in which the acid fraction of the molecule may be of varying chain length and degree of unsaturation. In fact, a single molecule of a fat or oil may contain three different acid residues.

Fats are compounds in which the saturated acids predominate over unsaturated acids. They are solids or semisolids and are usually obtained from animals. The important acids found in these sources are lauric, palmitic, and stearic acid (Table 21.1). An example of a complex fat molecule containing all three of these acids is given below. There are two other isomeric fats corresponding to the same combination of acids.

$$
\begin{array}{l}
\quad\; H \qquad\quad O \\
\quad\; | \qquad\quad \| \\
H-C-O-C-(CH_2)_{16}CH_3 \\
\quad\; | \qquad\quad O \\
\quad\; | \qquad\quad \| \\
H-C-O-C-(CH_2)_{14}CH_3 \\
\quad\; | \qquad\quad O \\
\quad\; | \qquad\quad \| \\
H-C-O-C-(CH_2)_{10}CH_3 \\
\quad\; | \\
\quad\; H
\end{array}
$$

Oils are similar to fats except that the acid components are unsaturated. They are derived from vegetable sources such as olives, corn, and soybeans. The presence of unsaturated acids in the molecule of oils lowers their melting points, and they are usually liquids. Among the unsaturated acids found in oils are linoleic, linolenic, and oleic acid. These acids all contain 18 carbon atoms but differ in their degree of unsaturation.

$$CH_3(CH_2)_4CH{=}CHCH_2CH{=}CH(CH_2)_7CO_2H$$

linoleic acid

$$CH_3CH_2CH{=}CHCH_2CH{=}CHCH_2CH{=}CH(CH_2)_7CO_2H$$

linolenic acid

$$CH_3(CH_2)_7CH{=}CH(CH_2)_7CO_2H$$

oleic acid

Note that the positions of unsaturation are related. Oleic acid has a double bond at the 9-carbon atom, linoleic acid at the 9- and 12-carbon atoms, and linolenic acid at the 9-, 12-, and 15-carbon atoms.

The degree of unsaturation of fats and oils, which are mixtures of compounds, is expressed in terms of an *iodine value*—which is equal to the number of grams of iodine that will add to 100 g of a fat or oil. If the fat contains no unsaturated acids its iodine value is 0. An oil that is the glyceryl ester of linoleic acid will react with 6 moles of iodine per mole of the oil.

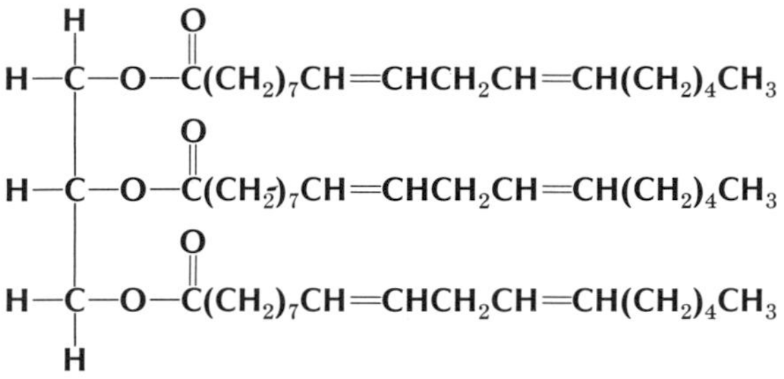

$$\begin{array}{l} \text{H—C—O—}\overset{\text{O}}{\overset{\|}{\text{C}}}(CH_2)_7CH{=}CHCH_2CH{=}CH(CH_2)_4CH_3 \\ \text{H—C—O—}\overset{\text{O}}{\overset{\|}{\text{C}}}(CH_2)_7CH{=}CHCH_2CH{=}CH(CH_2)_4CH_3 \\ \text{H—C—O—}\overset{\text{O}}{\overset{\|}{\text{C}}}(CH_2)_7CH{=}CHCH_2CH{=}CH(CH_2)_4CH_3 \end{array}$$

The molecular mass of the oil is 878. The amount of iodine (6 moles) that will react is 1523 g. Therefore, 100 g of the oil will react with 1523/878 $\times$ 100 g of iodine. The iodine value is 172.

The iodine values of the fats of butter and lard are approximately 25 and 60, respectively. The iodine values of olive oil, soybean oil, and safflower oil are 85, 135, and 150, respectively.

SAPONIFICATION

A solution of sodium or potassium hydroxide will hydrolyze fats or oils and yield glycerol and the salts of the constituent acids. The acid salts are collectively called soap, and the hydrolysis by base is called saponification. The balanced equation for the formation of sodium stearate from a saturated fat is given below:

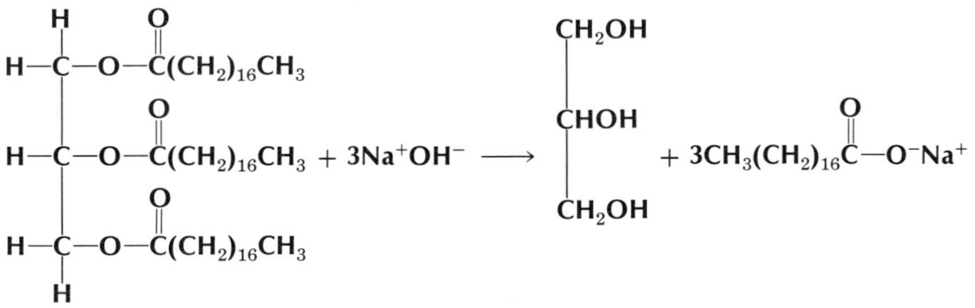

$$\begin{array}{l} \text{H—C—O—}\overset{\text{O}}{\overset{\|}{\text{C}}}(CH_2)_{16}CH_3 \\ \text{H—C—O—}\overset{\text{O}}{\overset{\|}{\text{C}}}(CH_2)_{16}CH_3 \\ \text{H—C—O—}\overset{\text{O}}{\overset{\|}{\text{C}}}(CH_2)_{16}CH_3 \end{array} + 3Na^+OH^- \longrightarrow \begin{array}{l} CH_2OH \\ CHOH \\ CH_2OH \end{array} + 3CH_3(CH_2)_{16}\overset{\text{O}}{\overset{\|}{\text{C}}}\text{—O}^-Na^+$$

The best soaps are made from fats having a high content of saturated acids, which are from 12 to 18 carbon atoms long. The sodium salts are commonly used in cake soaps, whereas the softer potassium salts are used in shaving creams. In order to achieve a high content of saturated acids, the fats are often hydrogenated prior to saponification. Floating soaps are produced by blowing air into the molten soap as it is produced. The density of the resultant solid products is in the 0.8 g/cm^3 range.

The soaps function by utilizing both the polar carboxylate *hydrophilic* (water-attracting) portion of the molecule and the nonpolar hydrocarbon *hydrophobic* (water-repelling) chain. The hydrocarbon segment of the molecule dissolves in the oily or greasy dirt that is dislodged from the fabric of clothes or skin. The tiny globule of a solution of hydrocarbon and dirt then becomes suspended in the water as the polar "head" of the soap is attracted to water molecules.

For many uses soaps have gradually been supplanted by detergents. Unlike soaps, detergents do not form insoluble precipitates with calcium and magnesium ions in hard water (Chapter 12). The detergents that are hydrocarbon derivatives of sulfuric acid are not precipitated by the Group II ions. Use of detergents is not without problems, however. Some of the early detergents contained an aromatic ring attached to an alkyl benzene sulfonate with a variety of branches (Chapter 18). Since the branched chain is resistant to oxidation, the detergent could not be degraded by the microorganisms present in water. These microorganisms degrade straight chains with ease but are less able to degrade branched material. Large-scale use of these detergents on the North American and European continents caused a serious pollution problem. With demand for action the chemical industry produced detergents that contain biodegradable molecules such as sodium lauryl sulfate:

$$\mathbf{CH_3(CH_2)_{10}CH_2{-}OSO_3^{-}Na^{+}}$$

21.10 MECHANISM OF ESTERIFICATION

The esterification reaction presented earlier in this chapter will be examined in this section in terms of its mechanism. The mechanisms of many other reactions presented in this chapter are related to the single example given below.

The esterification reaction might appear to be a simple acid–base reaction in which the proton of the acid and the hydroxyl group of the alcohol react to form water. If this were the case the oxygen of the alcohol molecule would eventually be found exclusively in the water. Actually the oxygen in the water has its origin in the acid molecule. The oxygen atom of the alcohol is located exclusively in the ester. These facts have been obtained by using the radioactive $^{18}_{8}O$ isotope. Alcohol containing this isotope was esterified, and the ester was found to contain the isotope. The following sequence of reactions has been proposed to explain how the reaction occurs:

$$\text{(1)}\quad CH_3{-}C({=}O){-}OH + H^+ \rightleftharpoons CH_3{-}\overset{+}{C}(OH)_2$$

$$\text{(2)}\quad CH_3{-}\overset{+}{C}(OH)_2 + H{-}\overset{18}{O}{-}C_2H_5 \rightleftharpoons CH_3{-}C(OH)_2{-}\overset{18}{O}^{+}(H){-}C_2H_5$$

$$\text{(3)}\quad CH_3{-}C(OH)_2{-}\overset{18}{O}^{+}(H){-}C_2H_5 \rightleftharpoons CH_3{-}C(OH)(\overset{+}{O}H_2){-}\overset{18}{O}{-}C_2H_5$$

$$\text{(4)}\quad CH_3{-}C(OH)(\overset{+}{O}H_2){-}\overset{18}{O}{-}C_2H_5 \rightleftharpoons H_2O + CH_3{-}\overset{+}{C}(OH){-}\overset{18}{O}{-}C_2H_5$$

$$\text{(5)}\quad CH_3{-}\overset{+}{C}(OH){-}\overset{18}{O}{-}C_2H_5 \rightleftharpoons H^+ + CH_3{-}C({=}O){-}\overset{18}{O}{-}C_2H_5$$

Step 1 is an acid-base reaction in which a carbonium ion is produced. The resultant carbonium ion reacts with the nucleophilic alcohol molecule in step 2. In step 3 proton transfer occurs to form a positively charged oxygen species, which then eliminates a water molecule in step 4. The carbonium ion formed in step 4 can then lose a proton in step 5 to yield the ester.

21.11 NUCLEAR MAGNETIC RESONANCE SPECTROSCOPY

Nuclear magnetic resonance spectroscopy involves the interaction of energy, provided by radiation in the radio frequency region, with the nuclei of atoms. Certain nuclei such as ${}^1_1H$ can spin in either of two directions similar to the spin of an electron described in Chapter 5. Therefore, in the presence of a magnetic field the energies of the two spin states are unequal because they generate magnetic fields in opposite directions. In the nuclear magnetic resonance experiment the hydrogen-containing sample is placed in a magnetic field and the energy difference between the favorable and unfavorable spin orientation is determined by an absorption of energy.

The difference in energy between the two spin states of a nucleus is a function of the molecular environment of the nucleus. This fact makes nuclear magnetic resonance of considerable value to the chemist because it enables him to identify the number of structurally nonequivalent hydrogen atoms in an organic molecule. As a result of studies with numerous organic compounds it is possible to estimate the conditions under which a given hydrogen undergoes transitions between spin states. These are recorded on an arbitrary scale with values of τ (tau) ranging from 0 to ten. Some examples of ranges of τ values for various types of structurally bound hydrogen atoms are given in Table 21.4.

In Figure 21.1 a representation of the nuclear magnetic resonance spectrum of methyl acetate is shown. The two absorption peaks have equal areas and illustrate the phenomenon that the strength of absorption is directly proportional to the number of protons responsible for the absorption. The ratio of the three methyl protons of the alcohol portion of the ester to the three methyl protons of the acetate group is 1:1, as is the ratio of the peak areas. The structures of some compounds can be established by accounting for the position of the absorptions and the relative areas of the absorptions.

Table 21.4
Some ranges of τ values for protons

structural type	*range of τ values*
H—C—C— (both carbons saturated)	8.6–9.2
H—C—O—C—	6.3–6.7
H—C—C(=O)—C—	7.7–8.0
H—C—C(=O)—O	7.8–8.0
H—C(=O)—C—	0.1–0.3
H—C(=O)—O—	1.9–2.1
H—C—O—C(=O)—	5.8–6.2

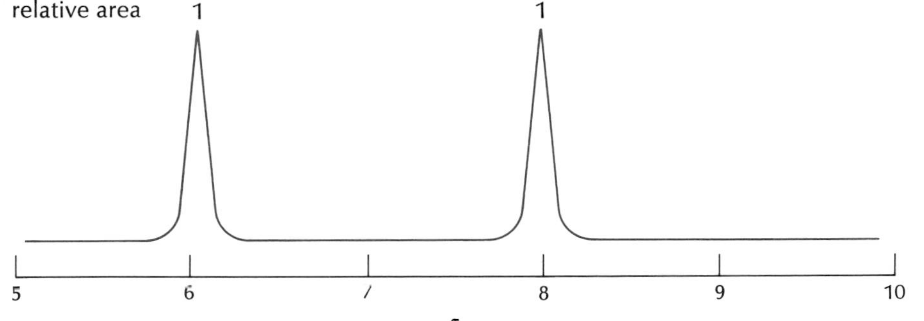

Figure 21.1
Nuclear magnetic resonance spectrum of methyl acetate.

EXAMPLE 21.4

What should the absorption spectrum of methyl formate look like?

$$H-\overset{\overset{\large O}{\|}}{C}-O-CH_3$$

There are two types of hydrogen atoms in the molecule. The proton of the formic acid portion of the ester will absorb at $\tau 2$, and the methyl protons will absorb at $\tau 6$. The ratio of the former to the latter will be 1 : 3.

The chemist can not only deduce how many types of hydrogen atoms are present in a molecule but also can establish how many hydrogen atoms of another structural type are immediately adjacent to a given hydrogen atom. The basis for this determination is the fact that non-equivalent hydrogen atoms in a magnetic field interact with each other so as to split their individual absorptions. In the case of methyl acetate the individual methyl protons are attached to atoms that do not contain attached hydrogen atoms. The signals observed are unsplit, that is, they are singlets. In ethyl acetate (Figure 21.2) the methyl proton absorption of the acetate group is unsplit. However, the methylene group absorption at $\tau 5.8$, which corresponds to two hydrogen atoms, is split into a quartet. The methyl hydrogen atom absorption at $\tau 8.7$ corresponding to three protons is split into a triplet. The *multiplicity*—that is, the number of subcomponents into which an absorption is split—is equal to $n + 1$ where n is the number of equivalent hydrogen atoms bound to an adjacent carbon atom. The methyl hydrogen atoms in the ethyl group have two neighboring protons on the adjacent carbon atom or $n = 2$. The methy-

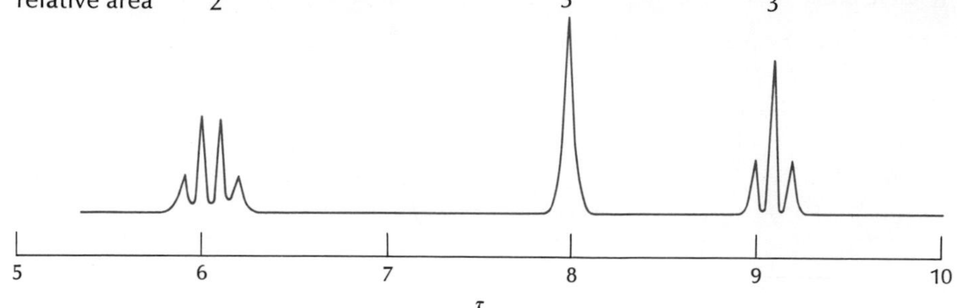

Figure 21.2
Nuclear magnetic resonance spectrum of ethyl acetate.

lene hydrogen atoms have three neighboring hydrogen atoms on the methyl carbon atom or $n = 3$.

In establishing a structure by nuclear magnetic resonance spectroscopy the chemist must check to see that the position, the intensity, and the multiplicity of all absorptions in the spectrum are correct for the compound suggested.

EXAMPLE 21.5

What features for methyl propanoate should be observed in the nuclear magnetic resonance spectrum of the compound?

$$\underset{1}{CH_3}-\underset{2}{CH_2}-\overset{O}{\overset{\|}{C}}-O-\underset{3}{CH_3}$$

The structure of the compound is given above, and the three types of hydrogen atoms are indicated by the numbers 1, 2, and 3. The absorptions of hydrogen atoms 1, 2, and 3 must be at $\tau 9$, $\tau 8$ and $\tau 6$, respectively. The intensities of these absorptions must be in the order 3 to 2 to 3, respectively. The signal at $\tau 6$ must be a singlet. The signals at $\tau 9$ and $\tau 8$ must be a triplet and a quartet, respectively.

Suggested further readings

Collier, H. O. J., "Aspirin," *Sci. Amer.*, offprint 169 (November 1963).

Davidson, D., "Acids and Bases in Organic Chemistry," *J. Chem. Educ.*, **19,** 154 (1942).

Heckert, W. W., "Synthetic Fibers," *J. Chem. Educ.*, **30,** 166 (1953).
Levey, M., "The Early History of Detergent Substances," *J. Chem. Educ.*, **31,** 521 (1954).
Shor, M., "Nylon," *J. Chem. Educ.*, **21,** 88 (1944).
Snell, F. D., "Soap and Glycerol," *J. Chem. Educ.*, **19,** 172 (1942).
Snell, F. D., and Snell, C. T., "Syndets and Surfactants," *J. Chem. Educ.*, **35,** 271 (1958).

Terms and concepts

acetic acid
acyl halide
amides
carboxyl group
condensation polymer
ester
fat
formic acid
iodine value
nitrile
nuclear magnetic resonance spectroscopy
Nylon
oil
saponification
soap
wax

Questions and problems

1. Name each of the following compounds:

a. $CH_3CH_2CH(CH_3)CO_2H$
b. CF_3CO_2H
c. $CH_3(CH_2)_{16}CO_2H$
d. 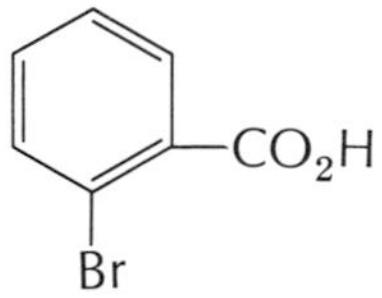 (benzene ring bearing CO_2H and, ortho to it, Br)
e. $CH_3CO_2^-Na^+$
f. CH_3CH_2COCl
g. $CH_3CH_2CO_2CH_2CH_3$
h. 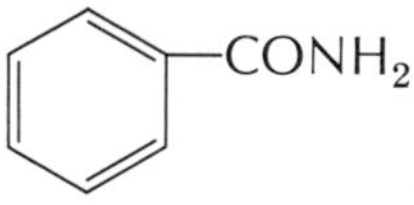 (benzene ring bearing $CONH_2$)

2. How many isomeric compounds of the formula $C_4H_8O_2$ are there?
3. The boiling points of methanol, acetic acid, and methyl acetate are 65, 118, and 57°, respectively. Why is the ester of lower boiling point than either of the two components that can produce it?
4. Which would be expected to be the stronger acid, CF_3CO_2H or CCl_3CO_2H? Why?
5. The K_a values of α-chloropropionic acid and β-chloropropionic acid are 1.6×10^{-3} and 8×10^{-5}, respectively. Explain the reason for this difference.
6. Devise two ways to synthesize 2-methylbutanoic acid from 2-bromobutane.
7. What reaction should occur between propanoyl chloride and methanol?
8. A compound, $C_6H_{12}O_2$, when reacted with lithium aluminum hydride followed by hydrolysis yields 1-propanol. What is the compound?

9. How can 2-methyl-2-butanol be prepared from ethyl propanoate and bromomethane?

10. What is the structural difference between an oil and a fat?

11. What is the calculated iodine value of the ester of oleic acid and glycerol?

12. A fat consists of 2 moles of palmitic acid and 1 mole of stearic acid combined with glycerol. How many isomeric fats are possible with this combination of components?

13. What product will be obtained by the reaction of 1,3-propanediol and $HO_2CCH_2CH_2CO_2H$ in the presence of an acid catalyst?

14. Oxalic acid, HO_2CCO_2H, is soluble in water but not in ether. The diethyl ester of oxalic acid is insoluble in water but soluble in ether. Explain these solubility differences.

15. How do soaps function?

16. Suggest a mechanism for the reaction of acetyl chloride with water.

17. Why was the ${}^{18}_{8}O$ not placed in the hydroxyl group of the acid in order to ascertain the mechanism of the esterification reaction?

18. Acetone exhibits a single absorption at $\tau 8$ in the nuclear magnetic resonance spectrum. Explain why.

19. The single proton attached to the interior carbon atom of 2-bromopropane is split into a septet in the nuclear magnetic resonance spectrum. Why?

20. Describe the expected spectrum for ethyl formate.

21. Could *p*-xylene and ethylbenzene be distinguished on the basis of their nuclear magnetic resonance spectra?

22

NITROGEN COMPOUNDS

Up to this point only the organic compounds of carbon, hydrogen, oxygen, and the halogens have been considered. In Chapter 16 it was stated that carbon can bond to most of the elements in the periodic table. It is now time to provide at least a glimpse of the vastness of the field of organic chemistry by considering in a general way the compounds of the element nitrogen. A great deal of knowledge has accumulated concerning the reactions and the mechanisms of nitrogen compounds. What we will discuss here is only a small indication of the amount that is known about the organic compounds of nitrogen. The organic compounds of phosphorus and sulfur, which also are of vital importance to life, cannot be discussed in even a limited way in this text. However, some of the chemistry of phosphorus parallels that of nitrogen, and some of the chemistry of sulfur parallels that of oxygen.

22.1 AMINES

NOMENCLATURE

The amines bear the same relationship to ammonia that the alcohols and ethers do to water. They are classified as *primary, secondary,* or *tertiary* amines, depending on the number of hydrogen atoms of ammonia that have been replaced by alkyl or aryl groups.

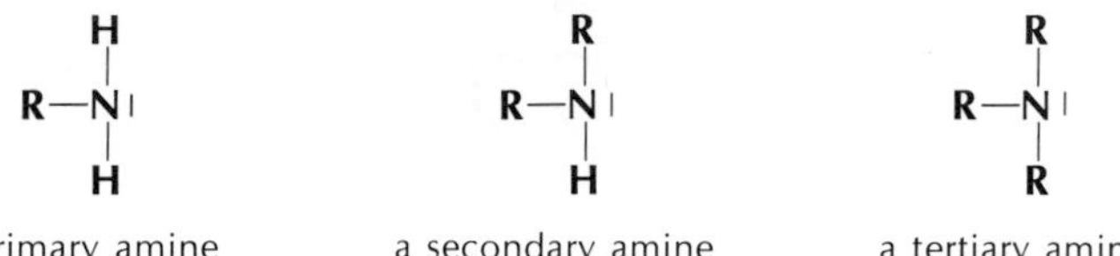

a primary amine　　a secondary amine　　a tertiary amine

Aliphatic amines are named by prefixing the names of the attached alkyl groups to the word *amine.* If identical substituents are present the prefixes *di-* and *tri-* are employed.

Many compounds found in nature contain nitrogen incorporated in cyclic systems. Several of the ring compounds are given below:

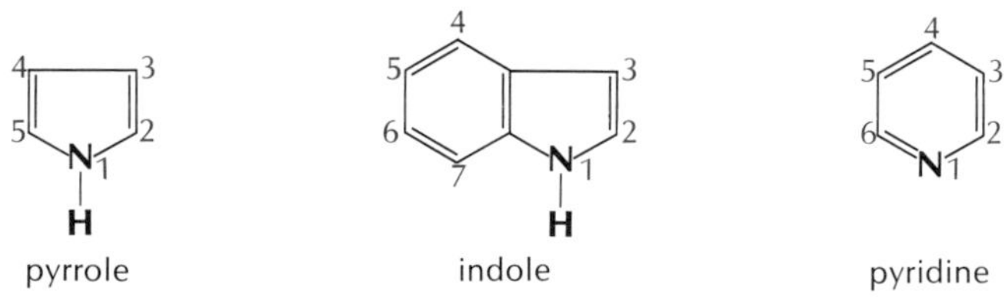

pyrrole　　indole　　pyridine

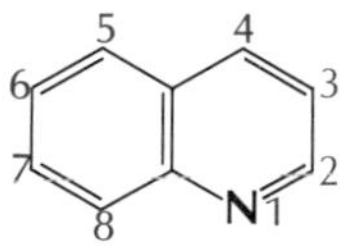

quinoline 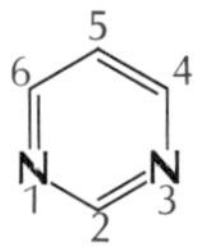　　pyrimidine 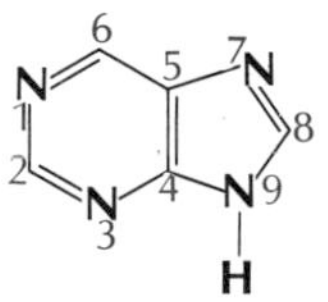　　purine

PHYSICAL PROPERTIES

The low molecular mass amines are gaseous, very soluble in water, and have odors similar to ammonia. Primary and secondary amines can form intermolecular hydrogen bonds similar to those formed by alcohols. However, the N—H···N bond is not as strong as the O—H···O bond because nitrogen is less electronegative than oxygen. This difference in hydrogen bond strength is responsible for the difference between the boiling points (Table 22.1) of methylamine (−6°) and methyl alcohol (65°). The fact that all three classes of amines can form hydrogen bonds with water accounts for their high solubility.

BASICITY

The unshared electron pair of amines allows them to behave as bases in a reaction similar to that of ammonia when it reacts with a proton to form a positive ammonium ion.

$$H_2O + NH_3 \rightleftharpoons NH_4^+ + OH^-$$
$$H_2O + CH_3NH_2 \rightleftharpoons CH_3NH_3^+ + OH^-$$

The equilibrium constant for the reaction of an amine with water is given by the symbol K_b. The larger the value of K_b, the stronger is the base. The base strengths of amines increase with the number of attached alkyl groups (Table 22.1). The electron-donating power of alkyl groups relative to hydrogen has been noted several times in this text (Chapters 16, 17, and 19). The electron density of the nitrogen atom is increased by the presence of alkyl groups, and this makes the amine a better acceptor of a proton.

EXAMPLE 22.1

Which should be the stronger base, aniline or p-nitroaniline?

The nitro group is electron-withdrawing and should decrease the electron availability of the electron pair on the amino group of *p*-nitroaniline. Therefore, the basicity of the compound will be reduced below that of aniline.

Table 22.1
Properties of amines

name	*formula*	*boiling point (°C)*	K_b
ammonia	NH_3	−33	2×10^{-5}
methylamine	CH_3NH_2	−6.5	44×10^{-5}
dimethylamine	$(CH_3)_2NH$	7.4	51×10^{-5}
ethylamine	$CH_3CH_2NH_2$	17	47×10^{-5}
propylamine	$CH_3CH_2CH_2NH_2$	49	38×10^{-5}
aniline	$C_6H_5NH_2$	184	4×10^{-10}
ethylenediamine	$H_2NCH_2CH_2NH_2$	116	8.5×10^{-5}
hexamethylenediamine	$H_2N(CH_2)_6NH_2$	204	85×10^{-5}
pyridine	C_6H_5N	115	23×10^{-10}

22.2 REACTIONS OF AMINES

FORMATION OF QUATERNARY AMMONIUM COMPOUNDS

All amines can react with haloalkanes to displace the halide ion and form an ammonium ion in which an alkyl group has been added to the original amine:

$$RNH_2 + R{-}X \longrightarrow R_2NH_2^+ + X^-$$
$$R_2NH + R{-}X \longrightarrow R_3NH^+ + X^-$$

When the amine is tertiary, the product is called a *quaternary ammonium* compound:

$$R_3N + R{-}X \longrightarrow R_4N^+ + X^-$$

Note that the products of the reaction of primary and secondary amines with haloalkanes can function as acids by donation of a proton to an appropriate base. However, the quaternary ammonium compound cannot function as an acid. These compounds are water soluble salts that can conduct electricity. Some detergents known as *invert soaps* contain quaternary ammonium salts. The ionic carbon-containing portion of the soap is positively charged, as contrasted with the negative charge of soaps and detergents (Chapter 21). An example of an invert soap is cetyltrimethylammonium chloride, which is used to sterilize some equipment used in hospitals:

$$CH_3(CH_2)_{15}{-}\overset{+}{N}(CH_3)_2{-}CH_3\ Cl^-$$

The invert soap can function in the same way as soap in that it contains a hydrocarbon chain that can interact with dirt and grease. The major difference is that the resultant globule of soap and dirt is attracted to water via a positive charge rather than a negative charge, as is the case in soap.

REACTION WITH ACID

The unshared electron pair on the nitrogen atom in amines allows it to function as a base toward a variety of materials. In the previous section the reaction described with haloalkanes illustrates that amines can serve as nucleophiles in displacement reactions.

When the electron pair of amines reacts with a proton the resultant cation tends to be soluble in water, that is, if the hydrocarbon portion of the amine is not too large. Therefore, amines that may be insoluble in water because of their relatively large molecular mass may become soluble in an acid solution. The amine-containing compounds in naturally occurring material may be removed by treatment with an organic solvent such as ether and an acid solution. The amines dissolve in the acid solution, leaving the other organic materials in the ether:

$$C_6H_5{-}NH_2 + H_3O^+ \longrightarrow C_6H_5{-}NH_3^+ + H_2O$$

The free amine can then be obtained by making the solution basic:

$$C_6H_5{-}NH_3^+ + HO^- \longrightarrow C_6H_5{-}NH_2 + H_2O$$

FORMATION OF AMIDES

As indicated in Chapter 21, amines react with carboxylic acids or acyl halides to yield amides. This reaction is useful in protecting amine groups in complex molecules while reactions at other functional groups are carried out:

$$R{-}NH_2 + R{-}C(=O){-}Cl \longrightarrow R{-}N(H){-}C(=O){-}R + HCl$$

The amine may be regenerated by hydrolysis in concentrated solutions of base:

$$R{-}N(H){-}C(=O){-}R + OH^- \longrightarrow R{-}NH_2 + R{-}C(=O){-}O^-$$

In Chapter 24 the protection of an amine group in the synthesis of peptides is illustrated.

REACTION WITH NITROUS ACID

Nitrous acid can be prepared by the reaction of a strong acid and sodium nitrite:

$$H_3O^+ + NO_2^- \longrightarrow H_2O + HNO_2$$

If the nitrous acid is generated in the presence of a primary amine a reaction occurs to liberate gaseous nitrogen. The primary amine is converted to a mixture of compounds of which one is the related alcohol. Of considerable importance to the field of protein research is the quantitative nature of this reaction. It is so complete that 1 mole of nitrogen is produced for every mole of amino groups. Measurement of the amount of nitrogen generated gives an accurate value for the quantity of free primary amine.

$$RNH_2 + HNO_2 \longrightarrow ROH + N_2 + H_2O$$

EXAMPLE 22.2

What volume of nitrogen at standard temperature and pressure will be liberated from 0.73 g of n-butyl amine?

The molecular mass of *n*-butyl amine $CH_3CH_2CH_2CH_2NH_2$ is 73 g/mole. Therefore, 0.73 g is 0.01 mole of compound:

$$\frac{0.73 \text{ g}}{73 \text{ g/mole}} = 0.01 \text{ mole}$$

One mole of nitrogen at STP occupies 22.4 liters. Therefore the volume of nitrogen liberated from a 0.01-mole sample is 0.224 liter or 224 ml.

$$(22.4 \text{ liters/mole})0.01 \text{ mole} = 0.224 \text{ liter}$$

Secondary amines react with nitrous acid to form nitrosamines that are insoluble in the aqueous solution and separate as a yellow oily layer.

$$\mathbf{R_2NH + HNO_2 \longrightarrow R_2N\text{—}NO + H_2O}$$

Tertiary amines simply dissolve in the acidic solution and form an ammonium salt. Therefore, the evolution of nitrogen gas, the formation of an oily layer, or dissolution without apparent reaction constitute tests by which amines may be classified.

22.3 DIAMINOALKANES

The simplest diaminoalkane is ethylenediamine, which was previously encountered as a ligand in complex ion formation (Chapter 14). It is a water soluble, basic liquid:

$$\mathbf{H_2N\text{—}CH_2\text{—}CH_2\text{—}NH_2}$$

The diamines putrescine and cadaverine are formed when meat or fish decays. The sources of these unpleasant-smelling compounds are the nitrogen-containing compounds called proteins (Chapter 24).

$$\mathbf{H_2N\text{—}CH_2\text{—}CH_2\text{—}CH_2\text{—}CH_2\text{—}NH_2}$$
putrescine

$$\mathbf{H_2N\text{—}CH_2\text{—}CH_2\text{—}CH_2\text{—}CH_2\text{—}CH_2\text{—}NH_2}$$
cadaverine

The next higher homologous diamine, hexamethylenediamine, is produced and used in carload quantities for the manufacture of Nylon (Chapter 21).

$$CH_2{=}CH{-}CH{=}CH_2 \xrightarrow{Cl_2} ClCH_2{-}CH{=}CH{-}CH_2Cl \xrightarrow{NaCN}$$

$$NC{-}CH_2{-}CH{=}CH{-}CH_2{-}CN \xrightarrow[Ni]{5H_2} H_2N{-}(CH_2)_6{-}NH_2$$

22.4
PHYSIOLOGICALLY ACTIVE AMINES

Choline is a quaternary ammonium salt of the simple amino alcohol β-hydroxyethylamine.

$$H_2N{-}CH_2CH_2{-}OH \qquad [(CH_3)_3\overset{+}{N}{-}CH_2CH_2OH]OH^-$$

β-hydroxyethylamine   choline

$$[(CH_3)_3\overset{+}{N}CH_2CH_2{-}O{-}\overset{O}{\overset{\|}{C}}CH_3]OH^-$$

acetylcholine

Acetylcholine, an ester of choline, plays a vital role in the transmission of nerve impulses in the body. Nerve gases, such as diisopropylfluorophosphate, developed but not used in World War II, are lethal by virtue of their action on the enzyme cholinesterase, which hydrolyzes acetylcholine to choline. They thereby prevent the transmission of the impulses that control vital functions.

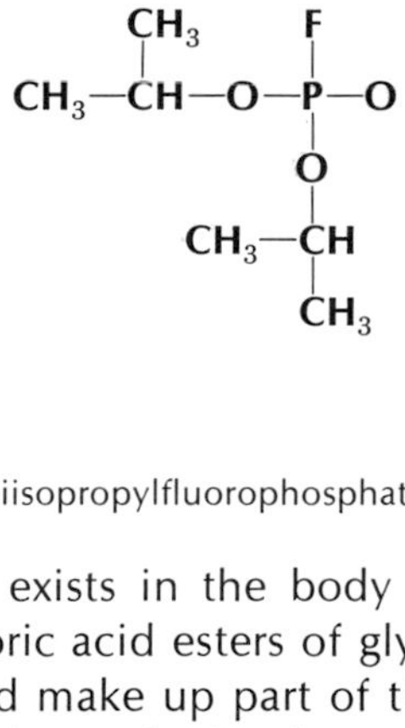

$$(CH_3)_2CH{-}O{-}\underset{O}{\overset{F}{P}}{-}O{-}CH(CH_3)_2$$

diisopropylfluorophosphate

$$\begin{array}{l} CH_2{-}O{-}\overset{O}{\overset{\|}{C}}{-}R \\ | \\ CH{-}O{-}\overset{O}{\overset{\|}{C}}{-}R \\ | \\ CH_2{-}O{-}\underset{O}{\overset{O}{P^-}}{-}O{-}CH_2CH_2\overset{+}{N}(CH_3)_3 \end{array}$$

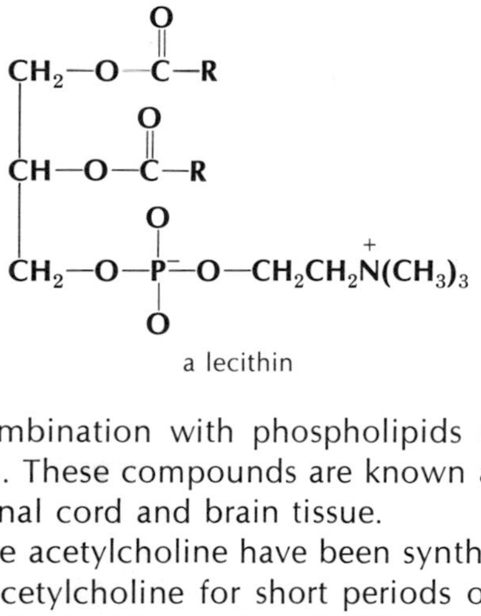

a lecithin

Choline exists in the body in combination with phospholipids (partial phosphoric acid esters of glycerol). These compounds are known as lecithins and make up part of the spinal cord and brain tissue.

Local anesthetics that resemble acetylcholine have been synthesized. They mitigate pain by replacing acetylcholine for short periods of time. An example of a local anesthetic is procaine hydrochloride, commonly known as Novocaine:

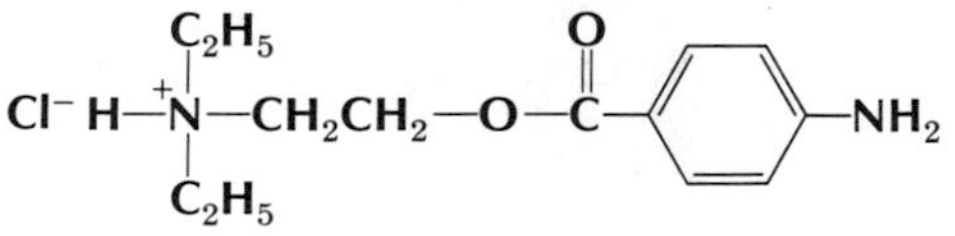

$$Cl^-\ H{-}\underset{C_2H_5}{\overset{C_2H_5}{\overset{+}{N}}}{-}CH_2CH_2{-}O{-}\overset{O}{\overset{\|}{C}}{-}C_6H_4{-}NH_2$$

Amino alcohols that have an effect on the sympathetic nervous system are known as *sympathomimetic* agents. Two of these are ephedrine and epinephrine (Adrenalin). Structurally these compounds resemble benzedrine, which is also a β-arylethylamine. Epinephrine is used to increase blood pressure and stimulate heart muscle, ephedrine relieves bronchial spasms, and benzedrine is used to reduce nasal congestion.

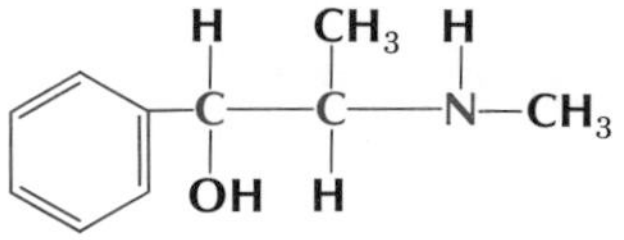

ephedrine

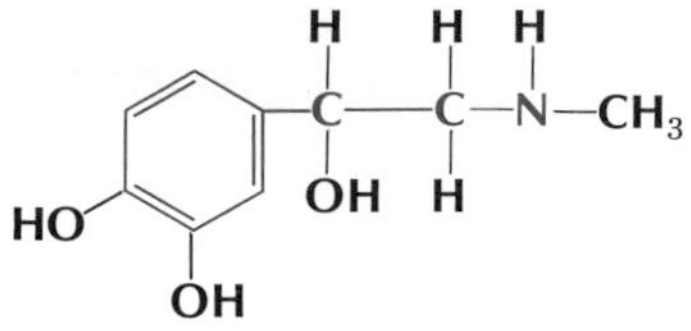

epinephrine

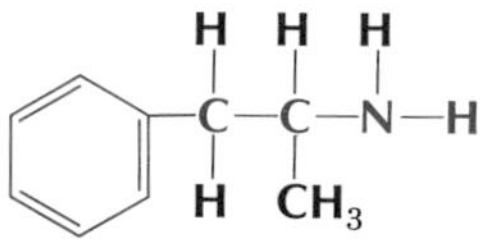

benzedrine

22.5 PHYSIOLOGICALLY ACTIVE AMIDES

Probably the most important amides are the high molecular mass proteins. Muscle, blood, and enzymes all contain these vital substances, which are discussed separately in Chapter 24.

Acetanilide, also known as antifebrin, is used as an *antipyretic* to relieve fever and as an *analgesic* to relieve pain in the treatment of neuralgia, headaches, and fevers. Phenacetin is contained in APC tablets (in combination with aspirin and caffeine) and can cause liver and kidney damage if used excessively.

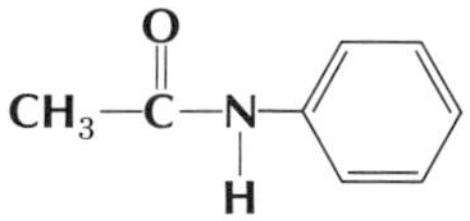

acetanilide

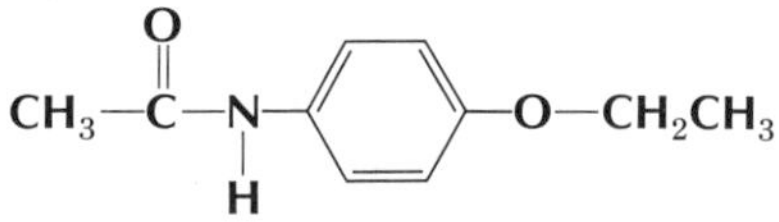

phenacetin

The general public has become aware of the term *amide* in an unfortunate way. Lysergic acid derived from ergot, a fungus disease of rye, when converted to its diethylamide is the hallucinogenic drug called LSD:

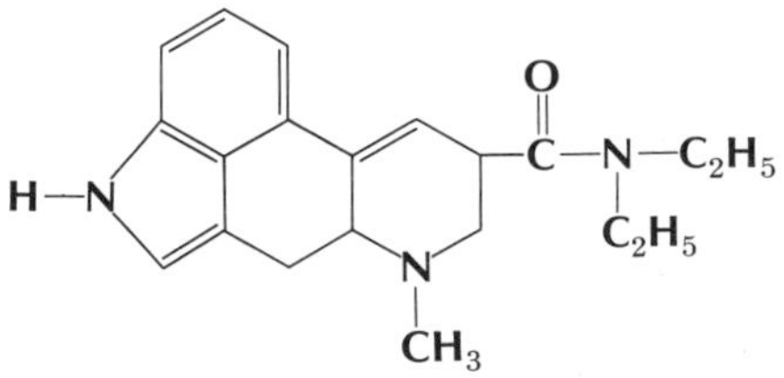

lysergic acid diethylamide

It is interesting to note that LSD contains components of nitrogen ring systems found in vitamins that are important and beneficial to man.

Lysergic acid diethylamide has been established to be more than an interesting compound that produces temporary hallucinations. Although its use has given rise to well-publicized tragedies involving persons under its influence, it now appears that its effects are longer lasting and dangerous in a more subtle way. Chromosome damage is being detected even in individuals who have had rather casual contact with LSD. The effects on their children and grandchildren may prove to be unfortunate and lasting.

Sulfanilic acid, which is a sulfuric acid derivative, forms amides with amines in the same way as carboxylic acids. The amide produced from ammonia is called sulfanilamide. It is effective against streptococcal and staphylococcal infections and was widely used during World War II to prevent the infection of wounds. The general class name for the sulfanilamide compounds that are active against bacterial infections is *sulfa drugs*. Two of the members of this class of compounds that contain nitrogen-based aromatic fragments are sulfapyridine and sulfadiazine.

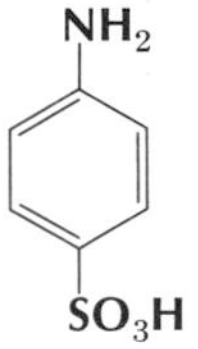

NH_2 / SO_3H — sulfanilic acid

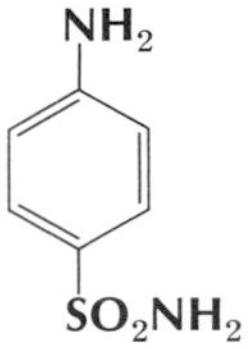

NH_2 / SO_2NH_2 — sulfanilamide

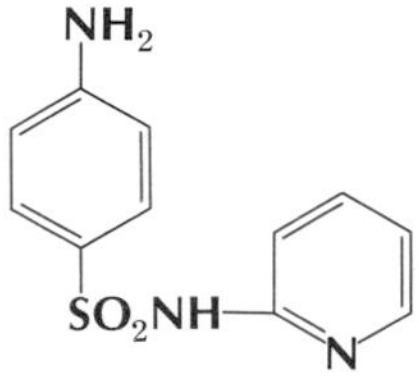

NH_2 / SO_2NH— N — sulfapyridine

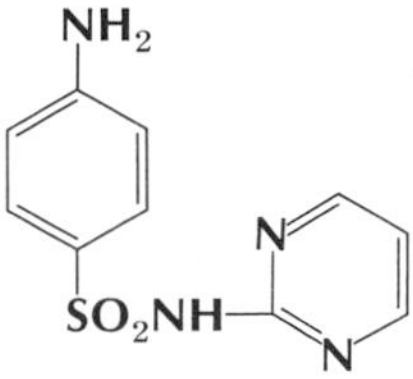

NH_2 / SO_2NH— N N — sulfadiazine

22.6 PYRROLE AND INDOLE COMPOUNDS

Natural products containing the pyrrole and indole ring systems are numerous. The structures of heme, chlorophyll, and vitamin B_{12} are illustrated in Section 14.8, and all contain pyrrole-derived rings.

Indole and its 3-methyl derivative are degradation products of the amino acid (Chapter 24) tryptophan and are responsible for the odor of feces.

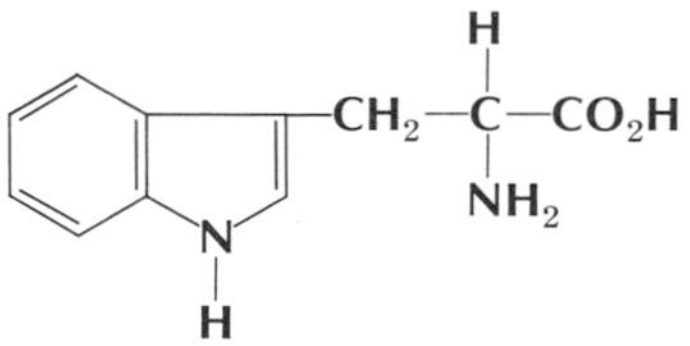

H, $-CH_2-C-CO_2H$, NH_2, N, H — trytophan

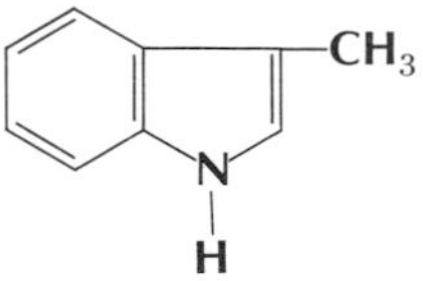

$-CH_3$, N, H — (Skatole) 3-methylindole

Reserpine, which contains the indole nucleus, is one of the alkaloids obtained from a species of Indian snake root. The crude extracts from the naturally occurring material had long been used on the Asian continent to treat dysentery, epilepsy, and insanity. The purified form is now one of a class of behavioral drugs called tranquilizers. The molecule was synthesized in 1956 by R. B. Woodward at Harvard, who received the Nobel prize in 1966 for his many synthetic accomplishments.

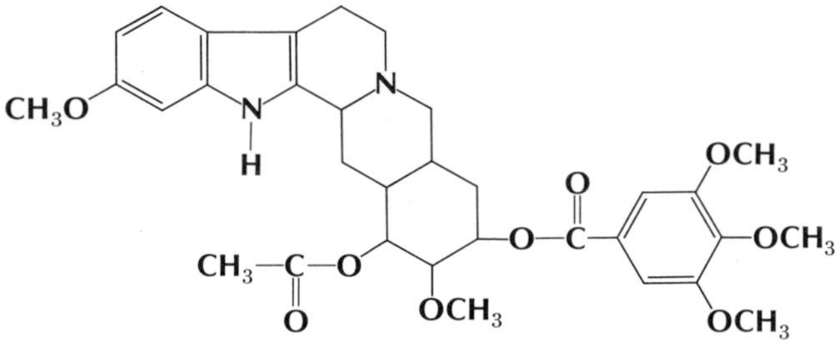

reserpine

EXAMPLE 22.3

Pyrrole is a planar molecule. What is the hybridization of the nitrogen atom of the ring?

The hybridization must be such that the nitrogen atom and the three groups attached to it are contained in one plane. The only hybridization that is consistent with this requirement is sp^2. The five electrons of the nitrogen atom must be distributed as follows:

1. one electron in each of the three sp^2 orbitals
2. two electrons in the p orbital perpendicular to the plane of the molecule

Note that the two electrons in a p orbital perpendicular can interact with the electrons in the p orbitals of carbon to form a delocalized π system similar to that of benzene. Pyrrole is aromatic and reacts in many ways similar to benzene.

22.7 PYRIDINE AND PIPERIDINE COMPOUNDS

The pyridine ring is contained in the deadly poison nicotine. However, the related oxidation product nicotinic acid (niacin) is one of the B vitamins essential to the human body. Niacin prevents pellagra, which is characterized by dermatitis, increased pigmentation, thickening of the skin, and soreness of the mouth and tongue. Niacin is found in all meats including liver. Whole cereal grains also contain niacin, but it is lost in the milling process.

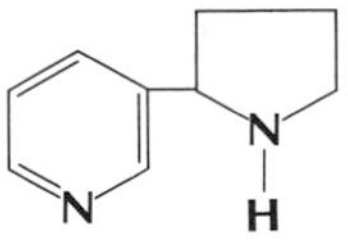

nicotine 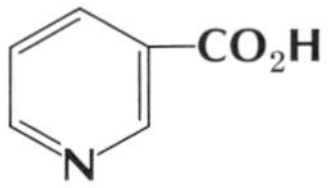 niacin

Vitamin B_6 (pyridoxine) is a pyridine-based compound that is contained in cereal grains, meat, fish, and egg yolk. The structurally related pyridoxal and pyridoxamine also exhibit vitamin B_6 activity.

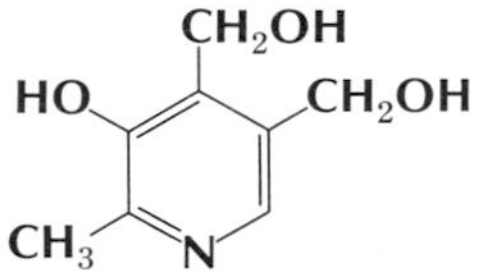

pyridoxine 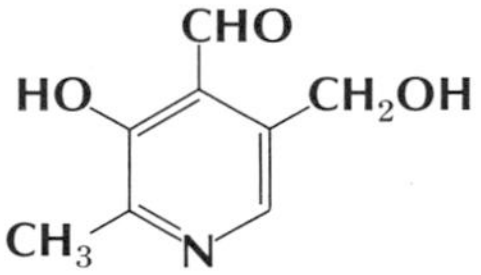 pyridoxal 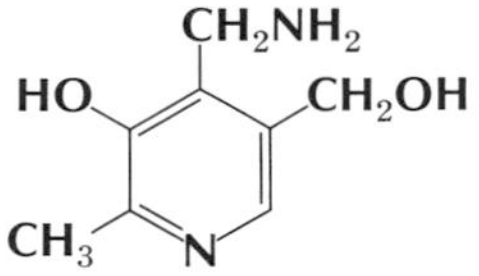 pyridoxamine

Reduction of pyridine yields piperidine. From a historical point of view the propyl derivative of piperidine called *coniine* is interesting because it is the poison of the hemlock plant that Socrates drank.

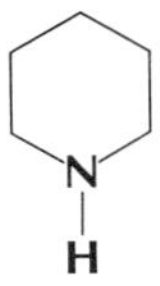

piperidine 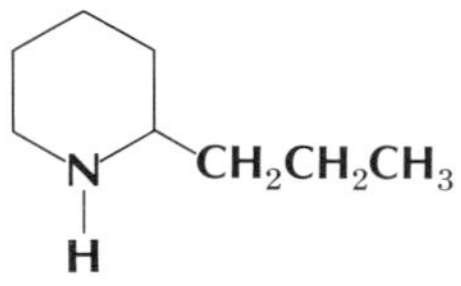 coniine

The piperidine ring system is found in a class of compounds called *tropane alkaloids,* which are bicyclic (contain two rings).

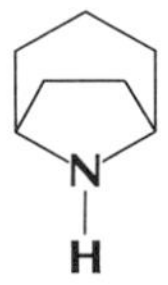

tropane 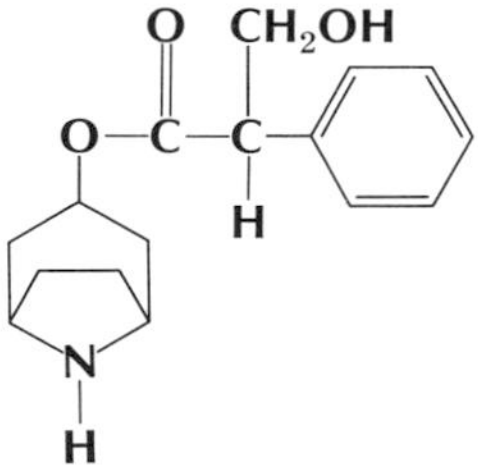 atropine 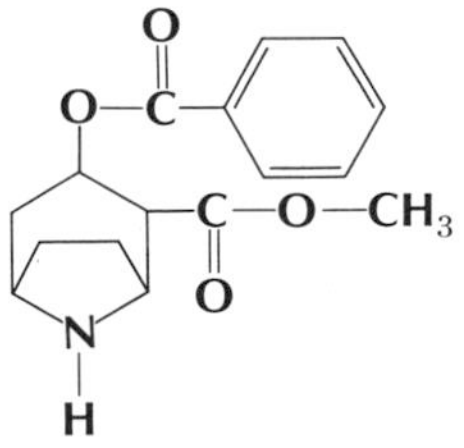 cocaine

Atropine occurs in the belladonna plant and is used as a *mydriatic* for the dilation of the pupil of the eye. It was used for cosmetic purposes by Europeans when enlarged pupils were considered beautiful. Cocaine was one of the earliest anesthetics, but it is habit forming and very toxic. The synthetic substitute, procaine (Section 22.4), has largely replaced cocaine because its side effects are less severe.

Quinine, which can be obtained from the bark of the tree *Cinchona officinalis,* contains a piperidine ring in a bicyclic skeleton attached to quinoline, a compound related to pyridine. Quinine is used to treat malaria.

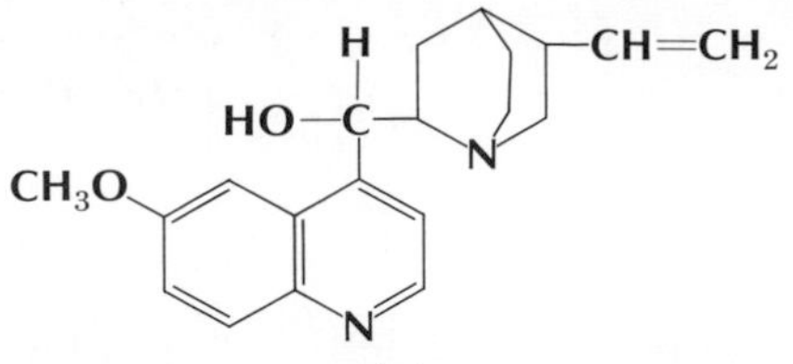

quinine

EXAMPLE 22.4

Pyridine is an aromatic planar compound. What is the hybridization of the nitrogen atom?

The nitrogen atom is contained in a structure in which it is bound by two σ bonds and a π bond. Each of these bonds requires the contribution of one electron from nitrogen. The electron involved in the π bond has to be located in a *p* orbital perpendicular to the plane of the molecule. The σ bonds are at a 120° angle because the molecule is shaped like benzene. The electrons of nitrogen forming these bonds must be in sp^2 hybrid orbitals. Therefore, the remaining two electrons of nitrogen have to be in the third sp^2 hybrid orbital, which is directed outward and in the plane of the pyridine molecule at a 120° angle with respect to the two σ bonds.

22.8 PYRIMIDINE COMPOUNDS

Thiamine (vitamin B_1) is necessary to prevent the disease beriberi, which was common in Asia where the diet consisted mainly of polished rice. The vital vitamin is present in the rice hulls, which are removed in the polishing process. Only 1 mg of thiamine a day is necessary for humans, and a deficiency causes nervous system disorders. Alcoholism and thiamine deficiencies apparently are related. The consumption of alcohol diminishes the desire for food, which in turn leads to thiamine deficiency and increases the desire for alcohol. The exact nature of the relationship is not known and may be a combination of physiological and psychological effects.

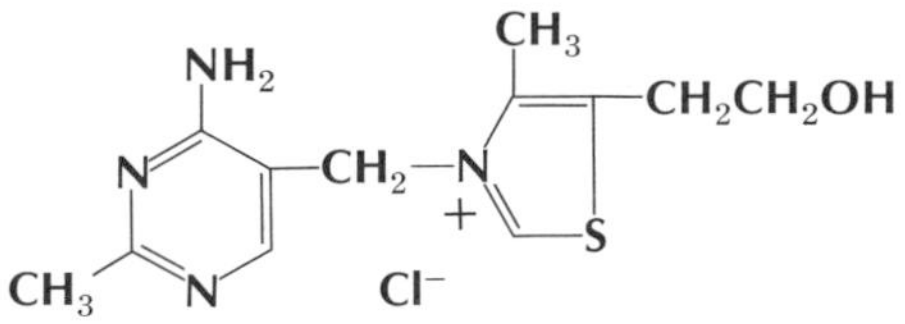

thiamine chloride

Riboflavin (vitamin B_2) contains a partially reduced pyrimidine ring. It is found in wheat germ, yeast, and liver. Approximately 1.5 mg/day are required in the human diet.

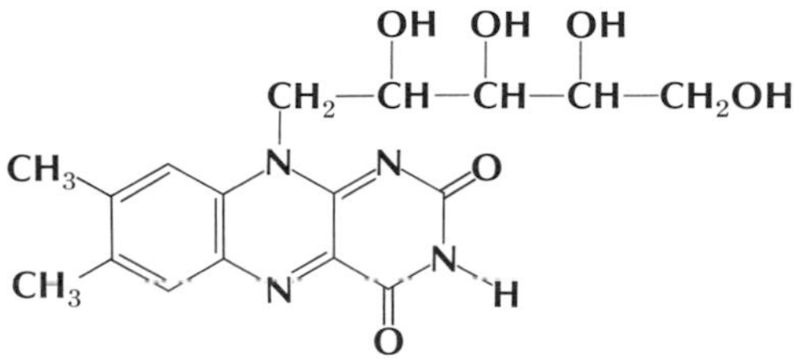

riboflavin

Folic acid is widespread in nature. In addition it can be synthesized by bacteria in the intestine. Folic acid deficiencies usually result in improper functioning of the digestive tract.

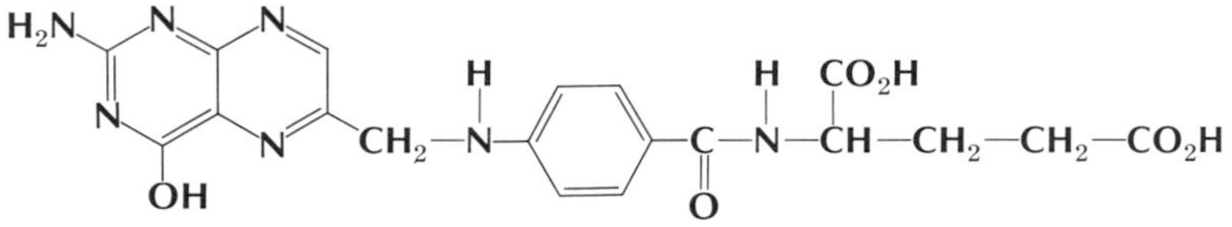

folic acid

The bases thymine, cytosine, and uracil are pyrimidine compounds that are partly responsible for maintaining the shape and structure of the nucleic acids DNA and RNA (Chapter 27). The nucleic acids play primary roles in the transmission of hereditary information.

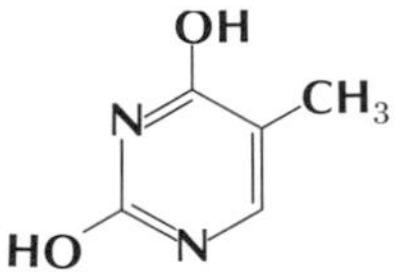

thymine

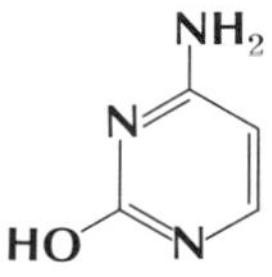

cytosine

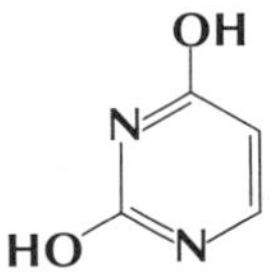

uracil

22.9 PURINE COMPOUNDS

The purine bases adenine and guanine are contained in nucleic acids.

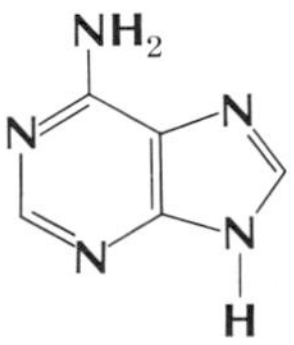

adenine

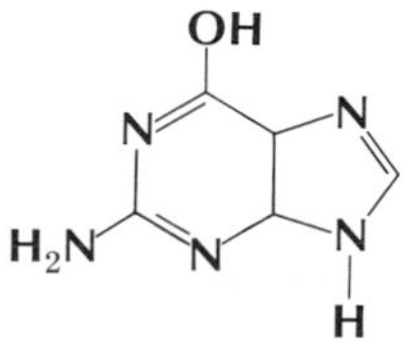

guanine

Caffeine, which is present in coffee, tea, and cola nuts (used for soft drinks), is a stimulant. The related theobromine is present in the cocoa bean. It is interesting to note that some individuals will not drink coffee on religious grounds because of the stimulant it contains but have no concern over cocoa.

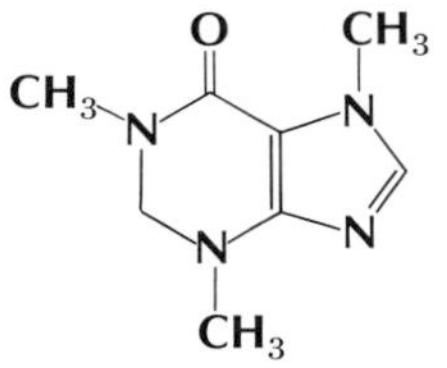

caffeine

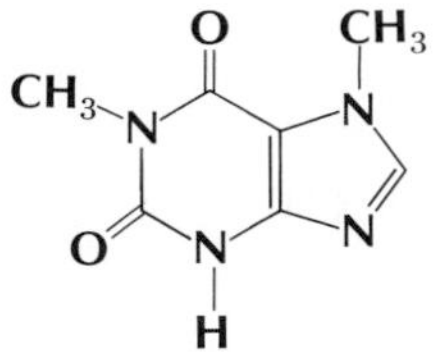

theobromine

22.10 OPIUM COMPOUNDS

The milky fluid of unripe oriental poppy seeds yields morphine and codeine. Morphine can be used to relieve severe pain but is highly addictive. Codeine, which differs from morphine by one methyl group, is present in small amounts in cough syrups. Heroin, which is not used for any medical purpose, is produced by esterifying the two hydroxyl groups of morphine.

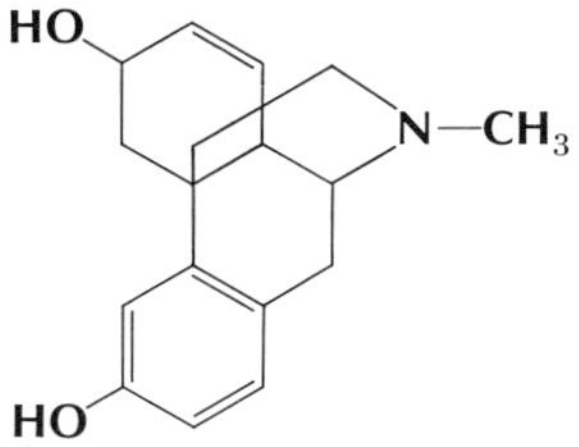

morphine

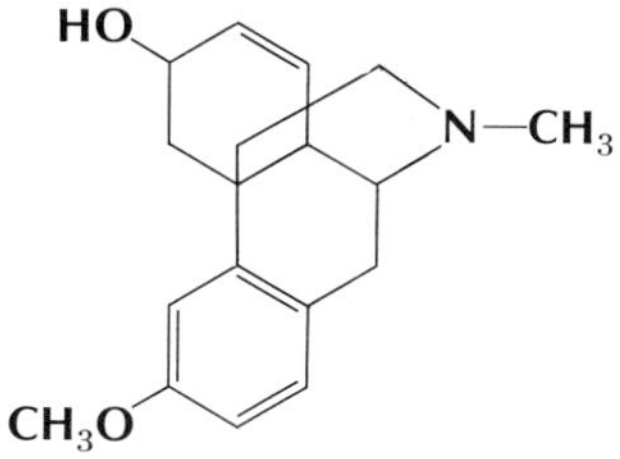

codeine

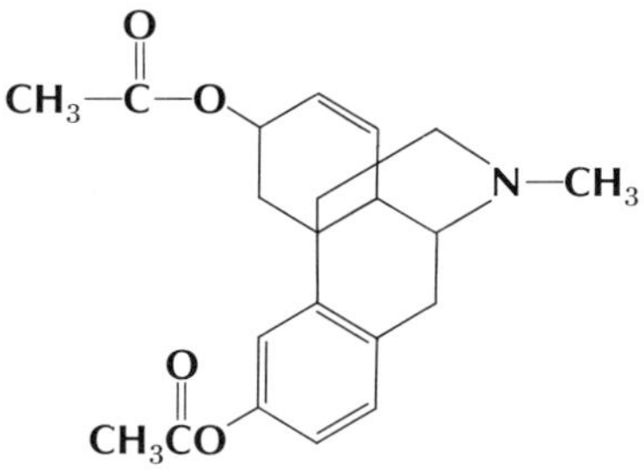

heroin

Suggested further readings

Amundsen, L. H., "Sulfanilamide and Related Chemotherapeutic Agents," *J. Chem. Educ.*, **19,** 167 (1942).

Fox, H. H., "The Chemotherapy of Tuberculosis," *J. Chem. Educ.*, **29,** 29 (1952).

Gates, M., "Analgesic Drugs," *Sci. Amer.*, offprint 304 (November 1966).

Mosher, H. S., Fuhrman, F. A., Buchwald, H. D., and Fischer, H. G., "Tarichatoxin–Tetrodotoxin: A Potent Neurotoxin," *Science,* **144,** 1100 (1964).

Robinson, T., "Alkaloids," *Sci. Amer.* (July 1959).

Terms and concepts

amides
amines
indole
invert soap
lysergic acid diethylamide
purine
pyridine
pyrimidine
pyrrole
quaternary ammonium salt
quinoline
sulfa drugs

Questions and problems

1. What is the general molecular formula for aliphatic amines containing one nitrogen atom? What is the general formula for compounds containing two nitrogen atoms?
2. How many amines are possible with the molecular formula C_3H_9N?
3. Write the structural formulas and names for all eight isomeric amines of the molecular formula $C_4H_{11}N$.
4. Why is the boiling point of propylamine higher than the boiling point of the isomeric trimethylamine (3.5°)?
5. The K_b values for aniline and methylaniline (methylphenylamine) are 4×10^{-10} and 7×10^{-10}, respectively. Explain why the two values differ.
6. The reduction of a nitrile, R—CN, by lithium aluminum hydride produces an amine, RCH_2NH_2. Suggest a way to produce hexamethylenediamine from 1,4-dichlorobutane using this reaction in one of the steps.
7. Amides are neutral compounds in spite of the fact that the nitrogen atoms in these molecules have an unshared pair of electrons. Suggest a reason for this fact.
8. The fishy odor of a solution of propylamine in water is eliminated when an equimolar amount of hydrochloric acid is added. Explain.
9. What structural features of Novacaine are similar to those of choline?
10. What are sulfa drugs?
11. Cite two cases in which a methyl group attenuates the physiological reactivity of compounds.
12. Some individuals claim that cola-based soft drinks are as addictive as coffee. Why?
13. Pyrrole is not very basic. Suggest a reason for this fact.
14. Pyridine is completely miscible with water, but benzene is insoluble. Why?

15. What substituted aniline would be expected to be a stronger base than aniline?

16. Will invert soaps be affected by the presence of Group II metal ions in water?

17. Compare the Williamson ether synthesis and the formation of a secondary amine from a primary amine and a haloalkane. Show how they are mechanistically related reactions.

23
STEREOISOMERISM

Three types of isomerism among organic compounds have been considered to this point in the text. Position isomerism was encountered in Chapter 16 when the isomers butane and isobutane were presented. From that point on the existence of position isomers in subsequent chapters became more commonplace. Functional isomerism was introduced in Chapter 17 when it was pointed out that alkenes and cycloalkanes have the same general molecular formula. Alcohols and ethers are functional isomers of each other, as are aldehydes and ketones. The third type of isomerism is *stereoisomerism,* in which atoms in a molecule can exist in more than one spatial arrangement. Examples of this type were considered under the terms *cis* and *trans* and were called *geometric isomers.* In this chapter a second type of stereoisomerism called *optical isomerism* is considered. The difference in the molecules of optical isomers lies in the asymmetry of the molecules, which in turn is a function of the spatial arrangement of the atoms. These compounds exhibit the interesting property of interacting with *plane polarized light* and rotating it in individually characteristic ways.

23.1 OPTICAL ACTIVITY

PLANE POLARIZED LIGHT

Ordinary white light consists of wave motion in which the waves have a variety of lengths (which correspond to different colors) and are vibrating in all possible planes perpendicular to the direction of propagation. Light can be monochromatic (one color) if filters are used to eliminate the other colors of white light or if it is generated by a special source. A sodium lamp, a monochromatic source of yellow light, is often used in the study of optical activity. Although the light consists of waves of only one wavelength, the individual light waves are vibrating in all possible planes (Figure 23.1).

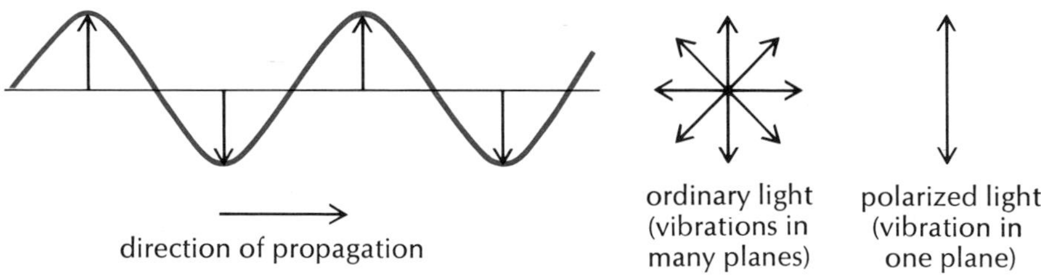

Figure 23.1
Representation of light as a wave motion.

A Nicol prism, which consists of a specially prepared Iceland spar crystal, has the property of serving as a screen with which to restrict the passage of light waves. Waves that are vibrating in one plane are transmitted while those in a perpendicular plane are rejected. Figure 23.2 gives a schematic representation of the action of Nicol prisms.

THE POLARIMETER

An instrument called a *polarimeter* is used to measure the optical activity of compounds. It has two Nicol prisms, one of which is a polarizer and the other an analyzer. Between these two portions of the apparatus is placed a polarimeter tube of known length that serves as the vessel for a solution of the compound to be examined. Many compounds do not affect the passage of the plane polarized light. The polarized light emerges from the sample tube without any change in the direction of polarization of the light. Such substances are said to be *optically inactive*. Optically

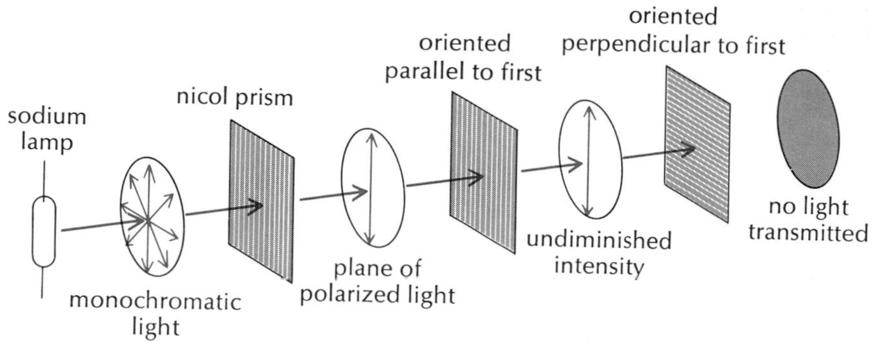

Figure 23.2
Effect of nicol prisms on light transmission.

active compounds alter the plane of polarized light, and it is necessary to rotate the analyzer in order to allow transmission of the light. The number of degrees and the direction of rotation of the analyzer correspond to the angle which the light is rotated by the compound (Figure 23.3).

SPECIFIC ROTATION

The rotation of plane polarized light by dissolved molecules must involve some interaction of the light with the molecules. Without specifying the nature of this phenomenon, the amount of the rotation is directly proportional to the number of molecules through which the light passes. Therefore, the observed rotation increases directly with the length of the polarimeter tube and with the concentration of the solution. In order to standardize and obtain meaningful reproducible results a value called the

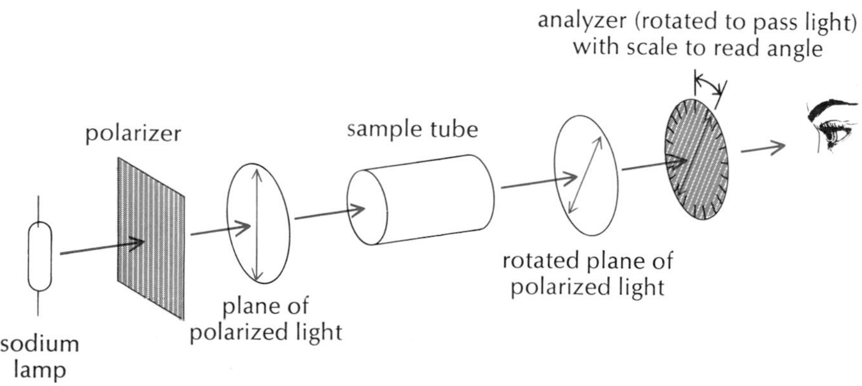

Figure 23.3
Schematic representation of a polarimeter.

specific rotation is calculated. It is symbolized by $[\alpha]$ which is the rotation of a solution at a concentration of 1 g/ml in a tube 1 decimeter (dm) long. The equation relating the observed rotation α, the concentration c, and the length of the tube l with the specific rotation is

$$[\alpha] = \frac{\alpha}{cl}$$

The specific rotation is a function of the temperature and the light source used. The specific rotation for experiments carried out at 25°C using the yellow D line of the spectrum of sodium is symbolized by $[\alpha]_{\mathrm{D}}^{25°}$.

EXAMPLE 23.1

A solution of a compound in a liquid when placed in a polarimeter does not exhibit any apparent rotation of the plane of polarized light. What does this observation suggest about the compound?

One possible reason for the lack of any apparent rotation is that the compound is optically inactive. A second possibility is that the compound indeed rotates the plane of polarized light, but it is rotated exactly 360° or any multiple of that quantity.

EXAMPLE 23.2

How could one distinguish between a solution of an optically inactive compound and one whose rotation is 360°?

The sample could be diluted by a factor of two. If it is inactive, the rotation of the plane polarized light will still be 0°. If it is optically active, the rotation will be diminished by one-half and will be 180°.

23.2 MOLECULAR ASYMMETRY

HISTORICAL OBSERVATIONS

The optical activity of a limited number of compounds was known in the middle of the nineteenth century. Louis Pasteur made the observation that the optically active salts of tartaric acid obtained from the residue in wine kegs could be crystallized in two forms that resembled each other in the same way that his left hand resembled his right hand. The two types of crystals were mirror images. The crystals of one form rotated the plane of polarized light in the opposite direction from that of the second form. He suggested that the atoms of the molecule in one optically active form are arranged asymmetrically and that the asymmetrical array of the atoms of the second form is in an opposite order. However, he did not arrive at a model for the origin of the asymmetry.

In 1874 the Dutch chemist Jacobus van't Hoff and the Frenchman Joseph Le Bel, who were both in their midtwenties, boldly proposed a possible model for the asymmetry of molecules. They realized that carbon usually was associated with four atoms or groups of atoms and pointed out that if these four groups are different and are arranged in a tetrahedral configuration about carbon, then two possible arrangements are possible. The arrangements bear the same relationship to each other as the right hand does to the left. The mirror image of one is equivalent to the second. Regardless of how the entire molecule of one arrangement is turned in space it can never have the same configuration as the second or be entirely superimposable on it. Two different molecules that are related by the equivalence of one isomer with the mirror image of the second isomer are called *enantiomers* (Figure 23.4).

At the time of the development of the theory of van't Hoff and Le Bel, electrons, the nature of bonding, and atomic structure were unknown quantities. Van't Hoff and Le Bel's simple picture of a molecule in three dimensions was viewed as unrealistic imagery. Some scientists of the day ridiculed the suggestion that man would ever arrive at an understanding of the arrangement of atoms in space. When the theory was proposed approximately 1 dozen optically active compounds were known. Most of

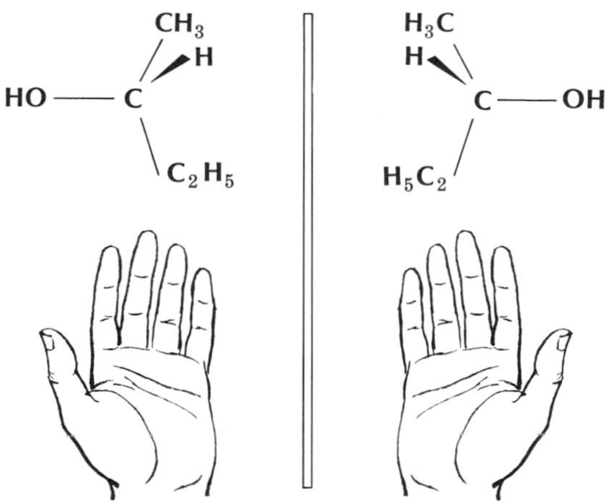

Figure 23.4
Mirror image relationships.

the compounds contained at least one carbon atom that was associated with four different groups. Such a carbon atom has come to be called an *asymmetric carbon atom,* although it is not asymmetric but rather is bonded to four different groups, giving rise to an asymmetric arrangement or asymmetric molecule. Some of the compounds that were reported to be optically active did not have four different groups about carbon. Van't Hoff suggested that these substances were impure and contained optically active contaminants. His suggestion proved correct.

PROJECTION FORMULAS

Lactic acid ($CH_3CH(OH)CO_2H$) contains a carbon atom with four different groups about it. Therefore, it can exist as two enantiomeric forms. These are illustrated in Figure 23.5, in which their mirror image relationship is shown. In order to write two-dimensional formulas that adequately represent these three-dimensional molecules, a specific convention must be followed (Figure 23.6). The carboxyl group, the methyl group, and the asymmetric carbon atom are arranged vertically on the printed page, but only the carbon atom is in the plane of the paper. Both the carboxyl group and the methyl group are directed behind the plane of the page, which then means that the hydrogen and hydroxyl group attached to the carbon center are above the plane. The projection of this arrangement on a plane is the standard representation of lactic acid.

In the early chapters of the organic chemistry section it was emphasized that a compound such as 2-bromopropane could be written in several two-dimensional ways, all of which represent the molecule:

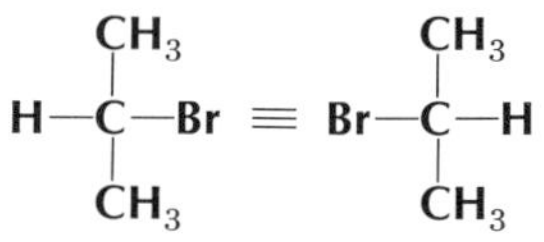

$$\begin{array}{c} CH_3 \\ | \\ H-C-Br \\ | \\ CH_3 \end{array} \equiv \begin{array}{c} CH_3 \\ | \\ Br-C-H \\ | \\ CH_3 \end{array}$$

There is no right-handedness or left-handedness of molecules, providing there is no asymmetry. Therefore, the two representations of 2-bromopropane are equivalent. One could be rotated in the plane of the paper to produce the second. Similarly, the three-dimensional ball and stick models of the two planar representations can be superimposed on each other.

The two enantiomeric lactic acid molecules present a representa-

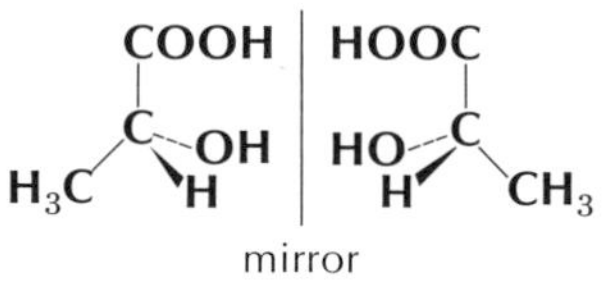

COOH | HOOC
C—OH | HO—C
H_3C H | H CH_3
mirror

Figure 23.5
Enantiomers of lactic acid.

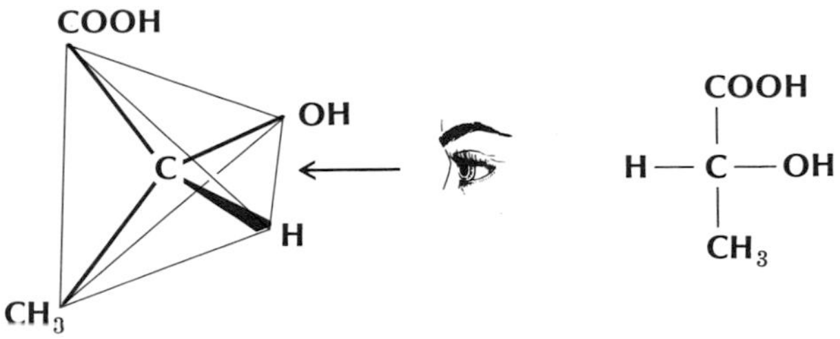

COOH
OH
C
H
CH_3

$$\begin{array}{c} COOH \\ | \\ H-C-OH \\ | \\ CH_3 \end{array}$$

Figure 23.6
Projection of lactic acid.

tional problem. In projecting the atoms into a two-dimensional representation the essence of the asymmetry, the three dimensionality, has been lost. Of itself, a two-dimensional structure can never be asymmetric.

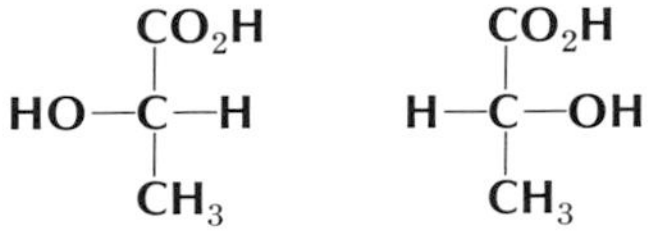

$$\begin{array}{c} CO_2H \\ | \\ HO-C-H \\ | \\ CH_3 \\ \text{(A)} \end{array} \qquad \begin{array}{c} CO_2H \\ | \\ H-C-OH \\ | \\ CH_3 \\ \text{(B)} \end{array}$$

The two molecular perspective formulas can never be made equivalent by rotation in the plane of the paper; whenever two groups occupy the same relative positions the other two groups are not in identical positions but rather are opposite each other across the carbon center. However, if one representation is lifted out of the plane and flipped over, it becomes identical to the second. Obviously this is an inherent error in the two-dimensional representations of molecules that results from ignoring the three dimensionality of molecules. If A is flipped over the carboxyl group and methyl group, which originally were behind the plane of the paper, are now in front of the plane. These groups do not occupy identical positions with respect to the carboxyl group and methyl group of B, which are behind the plane of the paper. In order to avoid the error of apparently achieving a two-dimensional equivalence of nonequivalent three-dimensional molecules it is important not to lift molecular representations out of the plane of the paper without simultaneously recalling the three dimensionality of the structure.

EXAMPLE 23.3

What is the lowest molecular mass bromoalkane that can be optically active?

A primary bromoalkane cannot be optically active because the primary carbon atom contains two identical groups, that is, two hydrogen atoms. A secondary bromoalkane could be optically active if the two alkyl groups attached to the secondary carbon atom are nonequivalent.

$$\begin{array}{c} \mathbf{R} \\ | \\ \mathbf{H{-}C{-}Br} \\ | \\ \mathbf{R'} \end{array}$$

The lowest molecular mass combination of alkyl groups is a methyl and an ethyl group.

$$\begin{array}{c} \mathbf{CH_3} \\ | \\ \mathbf{H{-}C{-}Br} \\ | \\ \mathbf{C_2H_5} \end{array}$$

23.3
RELATIVE AND ABSOLUTE CONFIGURATION

Two structures can be drawn to represent the two enantiomeric lactic acids. Early chemists were unable to ascertain which structure each of the two compounds corresponded to because they were unable to see the arrangement of the atoms about carbon in space. Therefore, it was necessary to assign arbitrarily a configuration to one of the members of the enantiomeric pair. The probability of being correct obviously was 50 percent. However, it was realized that if this procedure were adopted for all enantiomeric pairs of compounds utter chaos would result and progress in the chemistry of optically active compounds would be retarded. In order to circumvent this problem it was decided to define the configuration of one standard material and then chemically relate the configurations of all other compounds to it. If the wrong configuration was initially chosen then all other configurations would be wrong, but at least the system would be internally consistent. If this were the case then all configurations could be reversed if a means of determining absolutely the configuration of any substance were devised in the future.

As a standard the form of glyceraldehyde that rotates light in a clockwise direction was called *dextrorotatory* and was assigned the configuration with the hydroxyl group on the right side in the projection formula.

$$\begin{array}{c} \mathbf{CHO} \\ | \\ \mathbf{H{-}C{-}OH} \\ | \\ \mathbf{CH_2OH} \end{array}$$

D(+)-glyceraldehyde

The prefix D indicates the configuration at the asymmetric carbon; the symbol (+) indicates that the compound is dextrorotatory. The mirror image compound is L(−)-glyceraldehyde and corresponds to the structure in which the hydroxyl group is on the left in the projection formula.

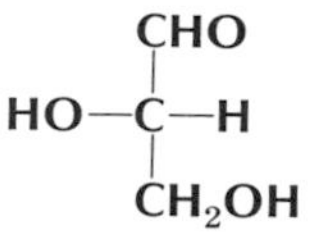

$$\begin{array}{c} CHO \\ | \\ HO{-}C{-}H \\ | \\ CH_2OH \end{array}$$

L(−)-glyceraldehyde

It rotates plane polarized light in a counterclockwise direction and is called a *levorotatory* compound. The designations D and L refer to configuration only and are not to be interpreted as dextrorotatory and levorotatory. The direction of rotation of light is symbolized by the plus and minus signs.

The D or L configuration of all other asymmetric compounds is defined according to whether the compound is derived from D- or L-glyceraldehyde. If D(+)-glyceraldehyde is converted into lactic acid by converting the CH_2OH group into CH_3 and the CHO group into CO_2H without any carbon–carbon bond cleavage, then the resulting lactic acid must have the same relative configuration as D(+)-glyceraldehyde. The lactic acid that results is levorotatory and is designated D(−)-lactic acid. This one transformation serves to emphasize that the D and L notations do not refer to the direction of rotation.

$$\begin{array}{c} CHO \\ | \\ H{-}C{-}OH \\ | \\ CH_2OH \end{array} \xrightarrow{\text{several steps}} \begin{array}{c} CO_2H \\ | \\ H{-}C{-}OH \\ | \\ CH_3 \end{array}$$

D(+)-glyceraldehyde — D(−)-lactic acid

A similar conversion of L(−)-glyceraldehyde yields L(+)-lactic acid.

In 1950 the absolute configuration of an optically active substance was finally determined by X-ray analysis. The absolute arrangement of the atoms in space corresponded to those assigned by relation to the arbitrary configuration of glyceraldehyde. Therefore, the original guess was correct, and no reversal of the configurations of the optically active compounds is necessary.

23.4 RACEMIC MIXTURES

When lactic acid is synthesized from compounds that are not asymmetric the product is optically inactive. Two routes that produce inactive material are illustrated.

$$CH_3CH_2CO_2H \xrightarrow[Br_2]{P} CH_3CHBrCO_2H \xrightarrow{AgOH} CH_3CH(OH)CO_2H$$

$$CH_3CHO + HCN \longrightarrow CH_3CH(OH)CN \xrightarrow[H^+]{H_2O} CH_3CH(OH)CO_2H$$

In the α bromination of propionic acid the two α hydrogen atoms are equivalent, so that the probability of replacing them with bromine is the same for each one. The result is that an equimolar mixture of D- and L-α-bromopropionic acids is obtained that upon hydrolysis yields an equimolar mixture of D- and L- lactic acids.

CO_2H, C, CH_3, H, H $\longrightarrow$ CO_2H, C, CH_3, Br, H $+$ CO_2H, C, CH_3, H, Br

L-α-bromopropionic acid   D-α-bromopropionic acid

In the second synthetic route the addition of HCN to the planar carbonyl group of acetaldehyde can proceed with equal facility from either side of the plane. An equimolar mixture of D and L isomers results, which upon hydrolysis yields an equimolar mixture of D- and L- lactic acids.

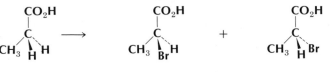

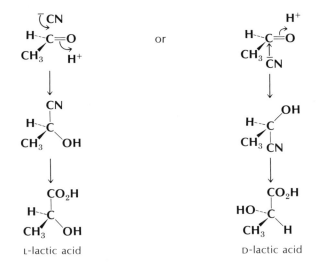

$^-$CN, H–C=O, CH_3, H^+ → CN, H–C, CH_3, OH → CO_2H, H–C, CH_3, OH

L-lactic acid

or

H^+, H–C=O, CH_3, $^-$CN → OH, H–C, CH_3, CN → CO_2H, HO–C, CH_3, H

D-lactic acid

Any laboratory synthesis starting from compounds that are not asymmetric always yields an equimolar mixture of D and L forms of an asymmetric compound if a center of asymmetry is generated in the reaction. A mixture of enantiomers is called a *racemic mixture*. The individual molecules of the mixture are optically active, but any rotation of plane polarized light by the D form is canceled by the opposite rotation of the L form.

Chemical reactions carried out in living systems almost always lead to the formation of optically active substances. The compounds of the living system include enzymes, hormones, carbohydrates, lipids, proteins, and nucleic acids, which are all optically active. When a compound containing an asymmetric center is produced in an asymmetric environment, formation of only one enantiomer is favored.

23.5 COMPOUNDS WITH TWO NONEQUIVALENT ASYMMETRIC CENTERS

The number of optically active isomers for a molecule containing n nonequivalent asymmetric carbon atoms is 2^n. This can be demonstrated for compounds of the formula $CH_3CH(OH)CH(OH)CO_2H$. The two internal carbon atoms, both of which are centers of asymmetry, contribute to the total rotatory property of the molecule. Since the carbon centers are nonequivalent their contributions to the total rotation of plane polarized light by the molecule is not likely to be equal. If the contributions to the rotation for the 2-carbon is either $+\alpha$ or $-\alpha$, depending on its configuration, and if that for the 3-carbon center is $+\beta$ or $-\beta$, then four possible total rotations can be written: $+\alpha + \beta$ and $+\alpha - \beta$; $-\alpha + \beta$ and $-\alpha - \beta$. Figure 23.7 shows a two-dimensional representation of the four isomers.

Structures II and III are mirror images, are not superimposable on each other, and are classified as enantiomers. This can be verified in two dimensions by imagining a mirror placed between II and III. Neglecting the fact that reversal of the printed letter occurs and remembering that the letters actually represent atoms, then the mirror image of any portion of molecule II is equivalent to the same region in molecule III. The sum of the rotatory contributions of II $(+\alpha - \beta)$ is equal but opposite in sign to that of III $(-\alpha + \beta)$.

Structures I and IV are mirror images and are nonsuperimposable. These enantiomers will rotate light in opposite directions but by the same absolute value.

Structures I and II are related to each other but are not enantiomers. They are called *diastereomers,* which is a term indicating pairs of isomers that are optically active but are not enantiomers. The pairs II and IV, I and III, and III and IV are all diastereomeric pairs. The physical properties of enantiomers are the same, but because diastereomers are molecules with different internal relative configurations they have different physical properties.

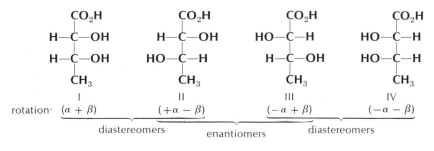

	I	II	III	IV
	CO_2H	CO_2H	CO_2H	CO_2H
	H—C—OH	H—C—OH	HO—C—H	HO—C—H
	H—C—OH	HO—C—H	H—C—OH	HO—C—H
	CH_3	CH_3	CH_3	CH_3
rotation:	$(\alpha + \beta)$	$(+\alpha - \beta)$	$(-\alpha + \beta)$	$(-\alpha - \beta)$

diastereomers (I and II) — enantiomers (II and III) — diastereomers (III and IV)

Figure 23.7
Enantiomers and diastereomers.

How many optically active compounds of the structure $CH_3CH(OH)CH(OH)CH(OH)CHO$ are there?

The compound contains three asymmetric carbon atoms. Therefore, the number of possible optically active compounds is $2^3 = 8$. Four compounds that are diastereomeric are:

```
   CHO            CHO              CHO            CHO
   |              |                |              |
H—C—OH        HO—C—H           H—C—OH        HO—C—H
   |              |                |              |
H—C—OH         H—C—OH         HO—C—H         HO—C—H
   |              |                |              |
H—C—OH         H—C—OH          H—C—OH         H—C—OH
   |              |                |              |
   CH3            CH3              CH3            CH3
```

The enantiomer of each of the above structures is the mirror image and accounts for the remaining four of the eight predicted structures.

23.6
COMPOUNDS WITH TWO EQUIVALENT ASYMMETRIC CENTERS

Historically, tartaric acid was the first compound examined that has two equivalent asymmetric centers. The rule 2^n for predicting the number of isomers does not apply to compounds in which the asymmetric centers are equivalent. The combination of rotatory contributions that can be written are $+\alpha + \alpha$, $+\alpha - \alpha$, $-\alpha + \alpha$, and $-\alpha - \alpha$. The structures corresponding to these individual contributions are given in Figure 23.8.

Because the terms $+\alpha - \alpha$ and $-\alpha + \alpha$ are equal to zero, the structures to which they correspond cannot be optically active. In fact, II and III are merely different ways of drawing the same compound. If III is rotated 180° in the plane of the paper, II results. The upper half of II is the mirror image

```
    CO2H             CO2H             CO2H              CO2H
    |                |                |                 |
 H—C—OH           H—C—OH         HO—C—H            HO—C—H
    |                |                |                 |
HO—C—H            H—C—OH         HO—C—H             H—C—OH
    |                |                |                 |
    CO2H             CO2H             CO2H              CO2H
     I                II               III               IV
 (+α + α)         (+α − α)         (−α + α)          (−α − α)
L-tartaric acid       meso-tartaric acid             D-tartaric acid
```

Figure 23.8
Tartaric acids.

of the bottom half, and any rotation of plane polarized light by one half is exactly balanced by the other half. This structure represents an optically inactive compound called *meso*-tartaric acid.

Structures I and IV are mirror images of each other and represent an enantiomeric pair of compounds. Therefore, tartaric acid exists in D, L, and *meso* forms.

23.7
RESOLUTION OF OPTICAL ISOMERS

The process of separating racemic mixtures into the active D and L components is called *resolution*. In 1848 Louis Pasteur accomplished the resolution of a racemic mixture of tartaric acid by separating the crystals of a salt according to their "right-handed" and "left-handed" forms. By carefully selecting the two possible crystal forms by means of a pair of tweezers and a magnifying glass he was able to resolve the compound. The rotation values of solutions of each of these compounds are of equal magnitude but of opposite direction. This process is possible only in rare instances in which the asymmetry of the molecule is reflected in the crystal form.

Occasionally microorganisms can be used to obtain one enantiomer from a racemic mixture. They metabolize one enantiomer that is compatible with the asymmetry of their particular biochemical systems. For example, the mold Penicillium glaucum will metabolize D-tartaric acid. A racemic mixture of tartaric acid will become enriched with L-tartaric acid as a result of this selective destruction process. This process is obviously limited by the availability of the proper organism to accomplish the resolution. One of the enantiomers is lost in the process and makes the process less desirable than others in which both isomers can be obtained.

The most general way of separating enantiomers involves reacting the racemic mixture with an optically active compound to produce a pair of diastereomers. The resultant diastereomers can be separated because they have different physical properties. Crystallization of the diastereomers is the most useful separation technique since they usually have different solubility properties. As an example of this process consider a mixture of acids reacting with a base of D configuration, as outlined in Figure 23.9.

racemic pair $\left\{\begin{array}{l}\text{D-acid}\\ \text{L-acid}\end{array}\right.$ + D-base ⟶ D-acid—D-base / L-acid—D-base

enantiomers — diastereomers (separable)

D-acid—D-base + H^+ ⟶ D-acid + D-base-H^+
L-acid—D-base + H^+ ⟶ L-acid + D-base-H^+

Figure 23.9
Resolution of enantiomers.

The diastereomeric product pair is separable, and each component of the original racemic mixture is then obtained by regeneration of the acids.

23.8 BIOLOGICAL ACTIVITY OF ENANTIOMERS

As illustrated in the following chapters, the biological activity of compounds such as carbohydrates, lipids, proteins, and nucleic acids in living systems is dependent on their optical activity. At this point only one example of the difference in the reaction of enantiomers is considered. Natural adrenaline, which rotates light counterclockwise, is about 20 times as biologically active as its enantiomer. In the human system only the proper form is generated. If a racemic mixture of adrenaline is administered to a patient, his body will utilize only the proper form and neglect the mirror image. There is no need, then, to resolve a racemic mixture of synthetic adrenaline before administering it to a patient. This is not the case for some lower organisms to which the improper enantiomer may be toxic.

Suggested further readings

Bent, R. L., "Aspects of Isomerism and Mesomerism. III. Stereoisomerism," *J. Chem. Educ.,* **30,** 328 (1953).

Figueras, J., Jr., "Stereochemistry of Simple Ring Systems," *J. Chem. Educ.,* **28,** 134 (1951).

Garvin, J. E., "Inexpensive Polarimeter for Demonstrations and Student Use," *J. Chem. Educ.,* **37,** 515 (1960).

Gill, E. J., "A Demonstration of Optical Activity with an Eskimo Yo-Yo," *J. Chem. Educ.,* **38,** 263 (1961).

Holleman, A. F., "My Reminiscences of van't Hoff," *J. Chem. Educ.,* **29,** 379 (1952).

Moseley, H. W., "Pasteur: The Chemist," *J. Chem. Educ.,* **5,** 50 (1928).

Noyce, W. K., "Stereoisomerism of Carbon Compounds," *J. Chem. Educ.,* **38,** 23 (1961).

Thompson, H. B., "The Criterion for Optical Isomerism," *J. Chem. Educ.,* **37,** 530 (1960).

Williams, F. T., "Resolution by the Method of Racemic Modification," *J. Chem. Educ.,* **39,** 211 (1962).

Terms and concepts

absolute configuration
asymmetry
dextrorotatory
diastereomers
enantiomers
glyceraldehyde

Le Bel
levorotatory
meso form
monochromatic light
Pasteur
plane polarized light
polarimeter
projection formula
racemic mixture
relative configuration
resolution
specific rotation
van't Hoff

Questions and problems

1. Define each of the following terms:

- **a.** monochromatic light
- **b.** plane polarized light
- **c.** optically active substance
- **d.** asymmetric carbon atom
- **e.** specific rotation
- **f.** enantiomers
- **g.** racemic mixture
- **h.** resolution
- **i.** diastereomers
- **j.** *meso* compound
- **k.** dextrorotatory
- **l.** levorotatory

2. What do the symbols L and (+) mean in L(+)-lactic acid?

3. Which of the following structures can exist in optically active forms?

- **a.** $CH_3CH_2CHBrCH_3$
- **b.** $CH_3CH_2CH(CH_3)_2$
- **c.** $CH_3CH_2CH(CH_3)CH_2CH_2CH_3$
- **d.** 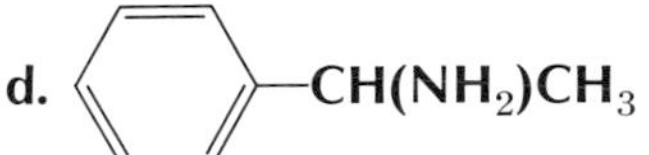

- **e.** $CH_3COCH_2CH_3$
- **f.** $CH_3COCHBrCH_3$
- **g.** $CH_3CH_2CO_2H$
- **h.** $CH_3CH(Br)CO_2CH_3$

4. How many optically active isomers of $CH_3CH(OH)CH(OH)CH(OH)CO_2H$ are there? Draw planar representations of each. Which structures are enantiomeric pairs?

5. How many asymmetric centers are contained in $CH_3CH(OH)CH(OH)CH(OH)CH_3$? Draw all the possible stereoisomers of this structure. Which are optically active?

6. Can a quaternary ammonium salt of the general structure $R_1R_2R_3R_4N^+Cl^-$ exist as enantiomers?

7. Which of the following structures are optically active?

- **a.** methylcyclobutane
- **b.** 1,1-dimethylcyclobutane
- **c.** *cis*-1,2-dimethylcyclobutane
- **d.** *trans*-1,2-dimethylcyclobutane
- **e.** *cis*-1,3-dimethylcyclobutane
- **f.** *trans*-1,3-dimethylcyclobutane

8. Can $CH_3CHBrCD_3$ exhibit optical activity?

9. What product would result from the reaction of D-2-bromobutane with hydroxide ion in water? (See Section 19.8.)

10. A solution of D-2-iodobutane gradually loses its optical activity when placed in contact with a solution of iodide ions. Why?

11. Suggest a means of resolving a racemic mixture of 2-aminobutane.

12. The specific rotation of the D form of a compound is $-20°$; that is, light is rotated to the left, and the compound is levorotatory. What is the specific rotation of the L form of the compound? What will be the apparent specific rotation for a mixture consisting of 25 percent of the D form and 75 percent of the L form.

13. What ways can be used to separate mixtures of enantiomers? What ways can be used to separate mixtures of diastereomers?

PART FIVE

BIOCHEMISTRY

The beginning of biochemistry, its scope, and its definition are difficult to summarize. Perhaps the simplest definition is that biochemistry is the chemistry of living animals and plants. Therefore, the composition, physical properties, and chemical reactions of the matter in living organisms are of interest to the biochemist.

The field of biochemistry is growing at such a rapid rate and in so many directions that molecular biologists, cytochemists, and biophysicists all contribute to the science. To understand fully the chemistry of life one must view it from the perspective of the life scientist, the physicist, and the chemist. The following chapters outline biochemistry from the viewpoint of the organic chemist.

24

AMINO ACIDS AND PROTEINS

The agents for the control, maintenance, growth, and reproduction of life are formed by approximately 20 compounds that contain an amino group and a carboxyl group and accordingly are called *amino acids.* These amino acids combine with each other to form proteins, which perhaps are the most complex organic molecules in nature. Proteins are major constituents of the skin, blood, muscles, hair, and vital tissues of the body. They also serve as one of the components in enzymes that catalyze biochemical reactions. Many hormones, which regulate metabolic processes, and antibodies, which resist the effects of toxic substances, are proteins.

24.1 STRUCTURE OF AMINO ACIDS

The amino acids contained in plants and animals in the form of proteins have an amino group attached to the carbon atom adjacent to the carboxylic acid group. This position is called the α-carbon, and such amino acids are α-amino acids. The general formula for an α-amino acid is:

$$\begin{array}{c} NH_2 \\ | \\ R—C—CO_2H \\ | \\ H \end{array}$$

The R of this general formula may be hydrogen, an alkyl group, an aromatic ring, or part of a heterocyclic ring. Amino acids that contain only one amino group and one carboxyl group are *neutral amino acids.* Some amino acids have additional carboxyl or amino groups and are classified as *acidic* or *basic amino acids,* depending on which functional group predominates. Except for glycine, in which R═H, all naturally occurring amino acids contain at least one center of asymmetry, are optically active, and are of L configuration.

$$\begin{array}{c} CO_2H \\ | \\ H_2N—C—H \\ | \\ R \end{array}$$

an L-amino acid

The human body, which consists of proteins composed of L-amino acids, either synthesizes the L-amino acids it needs or utilizes only the L-amino acids it obtains from nutritional sources.

All of the amino acids contained in proteins are known by their common name. Table 24.1 lists the most frequently encountered amino acids and names them in both a systematic and a conventional manner. In addition, the abbreviations of the common names, which are useful in describing proteins, are given. The table is divided according to the neutral, acidic, or basic properties of the compounds.

Certain amino acids that cannot be synthesized by the body and are necessary for proper metabolism and growth are said to be *essential amino acids.* They must be obtained directly from food sources. These amino acids are marked with an asterisk in Table 24.1.

24.2 SYNTHESIS OF AMINO ACIDS

The need for pure amino acids in nutritional studies with animals makes it desirable to develop general synthetic methods for these compounds. Some of them can be isolated from protein sources quite readily and are, of course, optically active. Laboratory synthetic methods lead to racemic mixtures (Section 23.4), which then must be resolved.

α-Aminonitriles can be produced by the addition of ammonia and hydrogen cyanide to aldehydes. The hydroxynitrile initially formed reacts with ammonia to produce the aminonitrile. Hydrolysis of the nitrile group yields the amino acid. This process is given in the following equations, which illustrate the conversion of acetaldehyde into alanine.

$$\underset{\text{aldehyde}}{CH_3{-}\overset{\displaystyle O}{\overset{/\!/}{C}}{-}H} + \underset{\text{hydrogen cyanide}}{HCN} \longrightarrow \underset{\text{hydroxynitrile}}{CH_3{-}\overset{\displaystyle OH}{\underset{\displaystyle H}{\overset{|}{\underset{|}{C}}}}{-}CN}$$

$$CH_3{-}\overset{\displaystyle OH}{\underset{\displaystyle H}{\overset{|}{\underset{|}{C}}}}{-}CN + NH_3 \longrightarrow \underset{\text{aminonitrile}}{CH_3{-}\overset{\displaystyle NH_2}{\underset{\displaystyle H}{\overset{|}{\underset{|}{C}}}}{-}CN} + H_2O$$

$$CH_3{-}\overset{\displaystyle NH_2}{\underset{\displaystyle H}{\overset{|}{\underset{|}{C}}}}{-}CN + 2H_2O \xrightarrow{H^+} \underset{\text{amino acid}}{CH_3{-}\overset{\displaystyle NH_2}{\underset{\displaystyle H}{\overset{|}{\underset{|}{C}}}}{-}CO_2H} + NH_3$$

The procedure, called the Strecker synthesis, is useful in the preparation of the simpler neutral amino acids.

Treatment of α-halogen substituted acids with ammonia replaces the halogen atom with an amino group.

$$CH_3{-}\overset{\displaystyle H}{\underset{\displaystyle Br}{\overset{|}{\underset{|}{C}}}}{-}CO_2H + 2NH_3 \longrightarrow CH_3{-}\overset{\displaystyle NH_2}{\underset{\displaystyle H}{\overset{|}{\underset{|}{C}}}}{-}CO_2H + NH_4Br$$

Note that an S_N2 displacement (Section 19.8) of bromide by NH_3 in the D-α-bromopropionic acid would yield L-alanine.

24.3 REACTIONS OF AMINO ACIDS

In general the reactions of amino acids are controlled by the two individual functional groups. The carboxyl group forms esters with alcohols.

$$R\underset{\displaystyle NH_2}{\underset{|}{C}}HCO_2H + CH_3CH_2OH \xrightarrow{HCl} R{-}\underset{\displaystyle NH_2}{\underset{|}{C}}HCO_2CH_2CH_3 + H_2O$$

Conversion of the carboxyl group into an acyl halide yields a derivative of an amino acid that will react with amines to form amides. This reaction is illustrated again in Section 24.5 in which the synthesis of peptides is examined.

$$R\underset{\displaystyle NH_2}{\underset{|}{C}}HCO_2H \xrightarrow{SOCl_2} R\underset{\displaystyle NH_2}{\underset{|}{C}}H\overset{\displaystyle O}{\overset{\|}{C}}{-}Cl \xrightarrow{RNH_2} R\underset{\displaystyle NH_2}{\underset{|}{C}}H\overset{\displaystyle O}{\overset{\|}{C}}{-}\overset{\displaystyle H}{\overset{|}{N}}{-}R$$

Table 24.1
Amino acids

name	*abbreviation*	*formula*
neutral amino acids		
glycine (aminoacetic acid)	Gly	$H_2NCH_2CO_2H$
alanine (α-aminopropionic acid)	Ala	$CH_3-C(NH_2)(H)-CO_2H$
serine (α-amino-β-hydroxypropionic acid)	Ser	$HOCH_2-C(NH_2)(H)-CO_2H$
threonine* (α-amino-β-hydroxybutyric acid)	Thre	$CH_3-C(OH)(H)-C(NH_2)(H)-CO_2H$
valine* (α-aminoisovaleric acid)	Val	$(CH_3)_2CH-C(NH_2)(H)-CO_2H$
leucine* (α-aminoisocaproic acid)	Leu	$(CH_3)_2CHCH_2-C(NH_2)(H)-CO_2H$

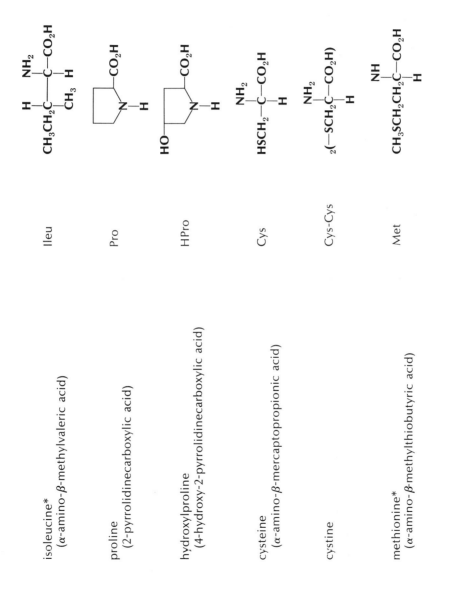

isoleucine* (α-amino-β-methylvaleric acid)	Ileu	$CH_3CH_2CH(CH_3)-CH(NH_2)-CO_2H$
proline (2-pyrrolidinecarboxylic acid)	Pro	pyrrolidine ring (N–H) with $-CO_2H$ at C-2
hydroxylproline (4-hydroxy-2-pyrrolidinecarboxylic acid)	HPro	HO– pyrrolidine ring (N–H) with $-CO_2H$ at C-2
cysteine (α-amino-β-mercaptopropionic acid)	Cys	$HSCH_2-CH(NH_2)-CO_2H$
cystine	Cys-Cys	$_2(-SCH_2CH(NH_2)-CO_2H)$
methionine* (α-amino-β-methylthiobutyric acid)	Met	$CH_3SCH_2CH_2CH(NH)-CO_2H$

Table 24.1 (cont.)
Amino acids

name	*abbreviation*	*formula*
phenylalanine* (α-amino-β-phenylpropionic acid)	Phe	$C_6H_5-CH_2-CH(NH_2)-CO_2H$
tyrosine [α-amino-β-(*p*-hydroxyphenyl)propionic acid]	Tyr	$HO-C_6H_4-CH_2-CH(NH_2)-CO_2H$
tryptophan* [α-amino-β-(3-indolyl)propionic acid]	Try	$C_8H_5N(H)-CH_2-CH(NH_2)-CO_2H$ (3-indolyl, N–H)
thyroxine	Thy	$HO-C_6H_2I_2-O-C_6H_2I_2-CH_2-CH(NH_2)-CO_2H$

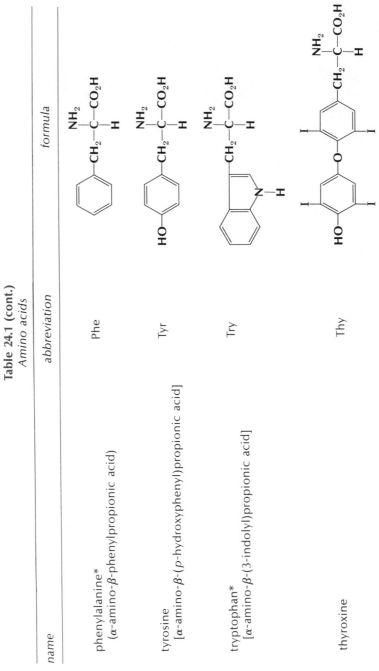

acidic amino acids		
aspartic acid (aminosuccinic acid)	Asp	$HO_2CCH_2-C(NH_2)(H)-CO_2H$
glutamic acid (α-aminoglutaric acid)	Glu	$HO_2CCH_2CH_2-C(NH_2)(H)-CO_2H$
basic amino acids		
arginine (α-amino-δ-guanidinovaleric acid)	Arg	$H_2N-C(=NH)-NH-CH_2CH_2CH_2-C(NH_2)(H)-CO_2H$
lysine (α,ϵ-diaminocaproic acid)	Lys	$H_2NCH_2CH_2CH_2CH_2-C(NH_2)(H)-CO_2H$
histidine [α-amino-β-(4-imidazolyl)propionic acid]	His	(4-imidazolyl)$-CH_2-C(NH_2)(H)-CO_2H$

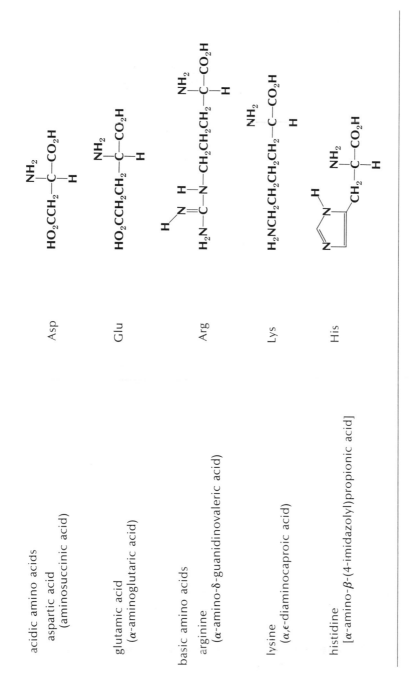

The amino group will react with acyl chlorides to form amides.

$$NH_2\underset{\displaystyle R}{\underset{|}{C}}HCO_2H + R{-}\overset{\displaystyle O}{\overset{\|}{C}}{-}Cl \longrightarrow R{-}\overset{\displaystyle O}{\overset{\|}{C}}{-}\overset{\displaystyle H}{\overset{|}{N}}{-}\underset{\displaystyle R}{\underset{|}{C}}HCO_2H + HCl$$

This type of reaction is used to render the amino group of amino acids unreactive in peptide synthesis (Section 24.5).

The primary amino group of amino acids reacts with nitrous acid to liberate nitrogen gas (Section 22.2):

$$H_2NCH_2CO_2H + HONO \longrightarrow HOCH_2CO_2H + N_2\uparrow + H_2O$$

This reaction, when used to determine the "free" (uncombined) amino groups in proteins is called the Van Slyke method, after D. D. Van Slyke, an American biochemist who first used this method in which the volume of nitrogen liberated is measured.

EXAMPLE 24.1

What volume of nitrogen will be liberated at STP in the Van Slyke analysis of 89 mg of alanine?

The molecular mass of alanine is 89 g/mole. Therefore, 89 mg of alanine corresponds to 0.001 mole.

$$\frac{0.089\text{ g}}{89\text{ g/mole}} = 0.001\text{ mole}$$

The quantity of nitrogen liberated will be 0.001 times that of the molar volume of nitrogen at STP.

$$(22.4\text{ liters/mole})0.001\text{ mole} = 0.0224\text{ liter}$$

The volume liberated is expected to be 22.4 ml.

Ninhydrin reacts with amino acids to produce a blue-colored compound. The amino acid becomes oxidized to yield ammonia, carbon dioxide, and an aldehyde, while ninhydrin is reduced to hydrindantin. The blue-colored compound is produced from the reaction of ninhydrin, hydrindantin, and ammonia. Approximately 10^{-6} mole of an amino acid will produce sufficient color in its reaction with ninhydrin to allow visual confirmation of the presence of an amino acid in a sample.

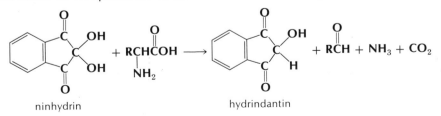

$$\text{ninhydrin} + R\underset{\displaystyle NH_2}{\underset{|}{C}}H\overset{\displaystyle O}{\overset{\|}{C}}OH \longrightarrow \text{hydrindantin} + R\overset{\displaystyle O}{\overset{\|}{C}}H + NH_3 + CO_2$$

ninhydrin

hydrindantin

ninhydrin + hydrindantin + NH_3 ⟶ [structure: two indanone rings joined by C=N—C, with C=O groups and one C—OH]

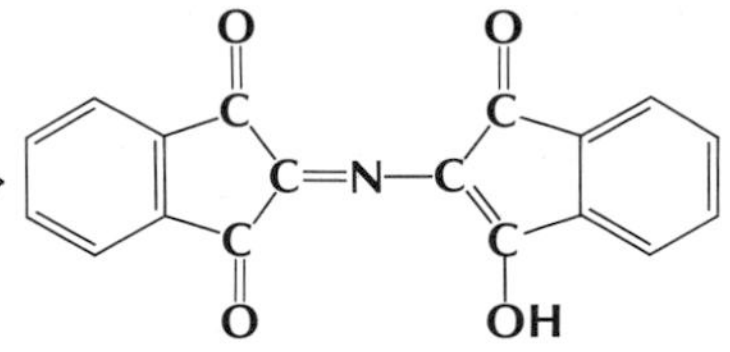

blue

EXAMPLE 24.2

What is the minimum quantity of alanine that can be detected by the ninhydrin test?

The test is sensitive enough to detect approximately 10^{-6} mole of an amino acid. The molecular mass of alanine is 89 g/mole. Therefore, the quantity of alanine that can be detected is

$$(89 \text{ g/mole})(10^{-6} \text{ mole}) = 0.000089 \text{ g}$$

or 0.089 mg.

24.4 PEPTIDES

Proteins, the high molecular mass substances, consist of numerous amino acids bound to each other by *peptide bonds.*

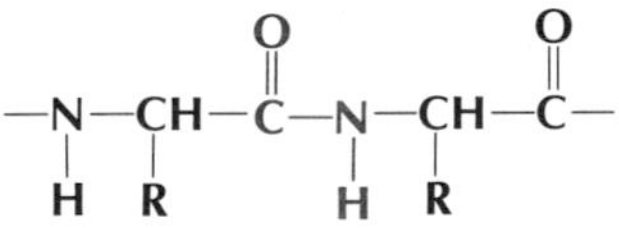

$$\text{—N—CH—}\overset{\text{O}}{\overset{\|}{\text{C}}}\text{—N—CH—}\overset{\text{O}}{\overset{\|}{\text{C}}}\text{—}$$
(with H on each N and R on each CH)

peptide bond

The peptide bond can be hydrolyzed like any amide bonds (Section 22.2) to yield the component acid and amine fragments. However, partial hydrolysis of proteins produces proteinlike substances called *peptides,* which consist of a relatively small number of amino acids (fewer than 50) joined by amide bonds. A peptide containing two amino acid units is called a *dipeptide,* one containing three amino acids, a *tripeptide;* and so on. Peptides are named by combining the names or abbreviations of the individual amino acids. The name starts with the amino acid whose amino group is free (the *N-terminal* amino acid) and ends with that whose carboxyl group is free (the *C-terminal* amino acid). Two examples of this system of nomenclature are:

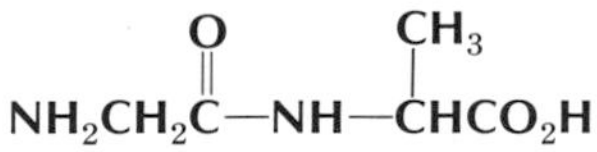

$$NH_2CH_2\overset{\text{O}}{\overset{\|}{\text{C}}}\text{—NH—}\overset{CH_3}{\overset{|}{\text{C}}}HCO_2H$$

glycyl-alanine (gly-ala)

$$NH_2\underset{CH_3}{\underset{|}{\text{C}}}H\text{—}\overset{\text{O}}{\overset{\|}{\text{C}}}\text{—NHCH}_2CO_2H$$

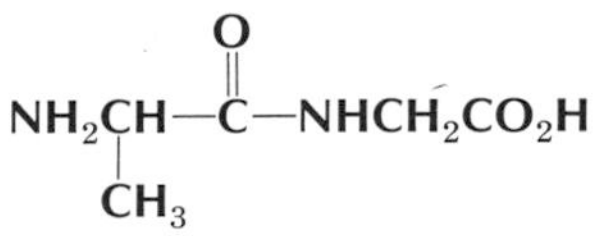

alanyl-glycine (ala-gly)

The complexity of peptide structures can be recognized by application of the mathematical concepts of permutations and combinations. While there are only two dipeptides possible from a given pair of amino acids (see above), the number of possible isomers increases rapidly with the number of amino acids that are to be combined. With three different amino acids in a tripeptide there are six possible isomers. The amino acids glycine, alanine, and valine can be combined to give gly-ala-val, gly-val-ala, val-gly-ala, val-ala-gly, ala-gly-val, and ala-val-gly. For 10 different amino acids combined in a decapeptide there are 3,628,800 possible isomers. The large number of isomers results not only from the number of amino acids present but also from the sequences of bonding that are possible.

EXAMPLE 24.3

How many pentapeptides containing five different amino acids are there?

There are five possible positions in a pentapeptide in which the five amino acids may be located. Any of the five could be in the N-terminal position. For each of those possible five different N-terminal peptides there could be any of the remaining four amino acids at the adjacent position. Therefore, the number of combinations for these two positions is 5×4. In a similar manner of reasoning, for each peptide having a specific combination of amino acids at the terminal positions there are three possibilities for different amino acids at the next position, two at the next, and finally once the preceeding four positions are fixed only one possibility for the C-terminal position. The total number of isomers is:

$$5 \times 4 \times 3 \times 2 \times 1 = 120$$

24.5 SYNTHESIS OF PEPTIDES

From the previous section it can be concluded that the synthesis of a specific peptide cannot be achieved readily by simply reacting amino acids with each other in a single step. Heating glycine and alanine together produces four dipeptides—gly-gly, gly-ala, ala-gly, and ala-ala—in addition to a large number of tri, tetra, etc. peptides. In order to achieve a specific synthesis the ways in which two amino acids can react must be restricted. Usually the amino group of the amino acid or N-terminal group of a peptide is blocked or rendered unreactive in some way before the reaction with another amino acid.

$$C_6H_5CH_2O\overset{O}{\overset{\|}{C}}Cl + H_2N\overset{CH_3}{\overset{|}{C}}HCO_2H \longrightarrow C_6H_5CH_2O\overset{O}{\overset{\|}{C}}-NH\overset{CH_3}{\overset{|}{C}}HCO_2H$$

blocking group

The carboxyl group of the amino acid or peptide derivative is then converted to the reactive acyl chloride.

$$C_6H_5CH_2O\overset{O}{\overset{\|}{C}}-NH\overset{CH_3}{\overset{|}{C}}HCO_2H \xrightarrow{SOCl_2} C_6H_5CH_2O\overset{O}{\overset{\|}{C}}\ NH\overset{CH_3}{\overset{|}{C}}HCOCl$$

The acyl chloride will react readily with the amino group of another amine or the N-terminal position of a peptide.

$$C_6H_5CH_2O\overset{O}{\overset{\|}{C}}-NH\overset{CH_3}{\overset{|}{C}}HCOCl + H_2NCH_2CO_2H$$

$$\longrightarrow C_6H_5CH_2O\overset{O}{\overset{\|}{C}}-NH\underset{CH_3}{\underset{|}{C}}H-\overset{O}{\overset{\|}{C}}-NHCH_2CO_2H$$

Finally, the original blocking group can be removed and the specific peptide desired can be synthesized.

$$C_6H_5CH_2O\overset{O}{\overset{\|}{C}}-NH\underset{CH_3}{\underset{|}{C}}H-\overset{O}{\overset{\|}{C}}-NHCH_2CO_2H \xrightarrow[Pd]{H_2}$$

$$C_6H_5CH_3 + CO_2 + H_2N\underset{CH_3}{\underset{|}{C}}H-\overset{O}{\overset{\|}{C}}-NHCH_2CO_2H$$

24.6
STRUCTURE DETERMINATION OF PEPTIDES

The component amino acids of a peptide can be ascertained by complete hydrolysis of the peptide in hydrochloric acid. A tripeptide consisting of glycine, alanine, and valine would yield these amino acids. However, as previously noted, there are six possible isomeric tripeptides that can be produced from three different amino acids. In order to ascertain the amino acid sequence it is necessary to determine somehow which amino acid is N-terminal and which is C-terminal. This is often accomplished by labeling the peptide at either or both the N-terminal amino acid and C-terminal amino acid prior to hydrolysis. The identification of the labeled amino acid after hydrolysis determines its position in the original peptide.

The N-terminal amino acid can be labeled by a reaction with 2,4-dinitrofluorobenzene.

$$\text{2,4-}(NO_2)_2C_6H_3F + NH_2CH(R)C(=O)\text{—(amino acids)} \longrightarrow \text{2,4-}(NO_2)_2C_6H_3\text{—NHCH(R)C(=O)—(amino acids)}$$

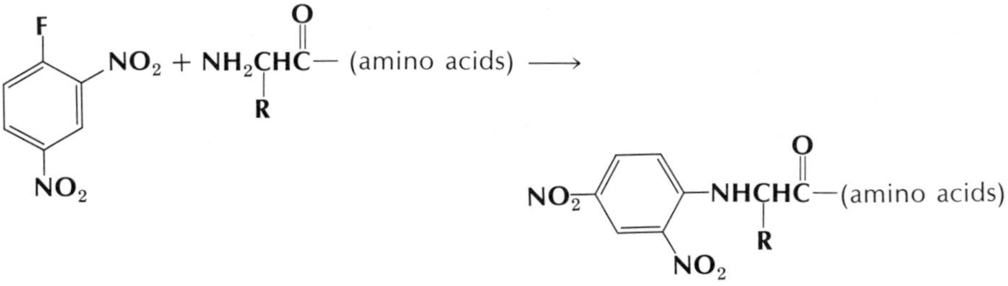

Upon hydrolysis the 2,4-dinitrophenyl derivative of the N-terminal amino acid is obtained along with free amino acids from the remainder of the peptide.

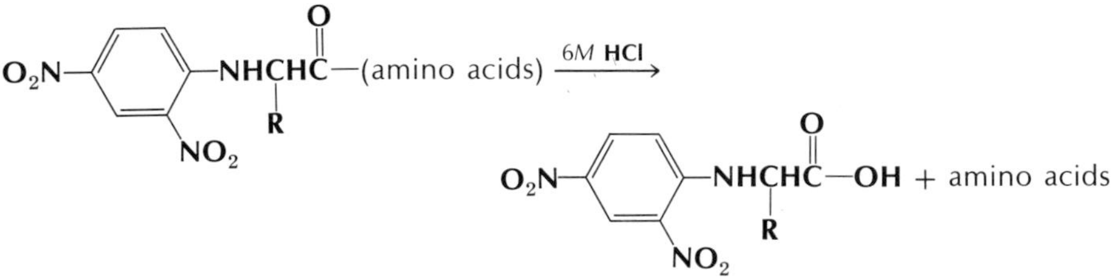

$$\text{2,4-}(O_2N)_2C_6H_3\text{—NHCH(R)C(=O)—(amino acids)} \xrightarrow{6M\ HCl} \text{2,4-}(O_2N)_2C_6H_3\text{—NHCH(R)C(=O)—OH} + \text{amino acids}$$

An example of the use of this reaction in structure determination is shown in Figure 24.1.

The C-terminal amino acid can be identified by heating the peptide with hydrazine at 100°. Under these conditions the carbonyl groups of the amide bonds are converted to hydrazides. The only amino acid left is the

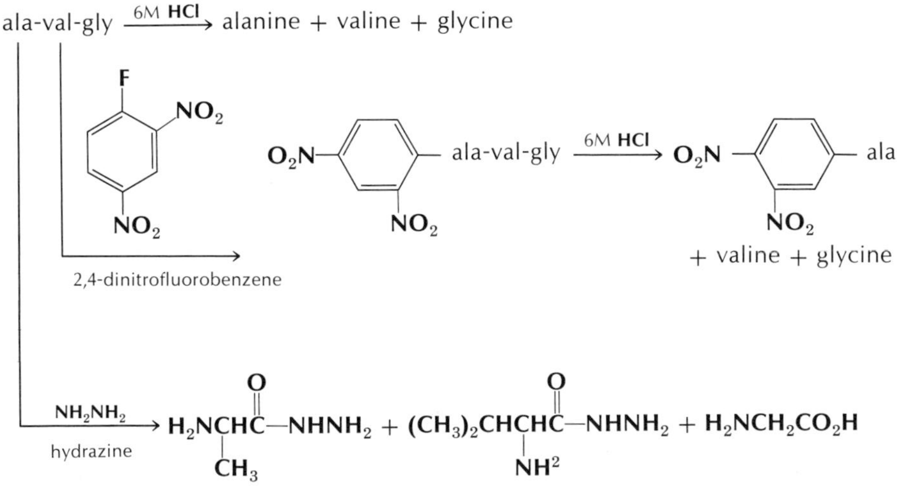

$$\text{ala-val-gly} \xrightarrow{6M\ HCl} \text{alanine + valine + glycine}$$

$$\text{ala-val-gly} \xrightarrow{\text{2,4-dinitrofluorobenzene}} \text{2,4-}(O_2N)_2C_6H_3\text{— ala-val-gly} \xrightarrow{6M\ HCl} \text{2,4-}(O_2N)_2C_6H_3\text{— ala} + \text{valine + glycine}$$

$$\text{ala-val-gly} \xrightarrow[\text{hydrazine}]{NH_2NH_2} H_2NCH(CH_3)C(=O)\text{—}NHNH_2 + (CH_3)_2CHCH(NH^2)C(=O)\text{—}NHNH_2 + H_2NCH_2CO_2H$$

Figure 24.1
Determination of the structure of a tripeptide.

C-terminal one. This reaction is illustrated with glycylalanine:

$$NH_2CH_2\overset{\overset{\displaystyle O}{\|}}{C}{-}NH\underset{\underset{\displaystyle CH_3}{|}}{C}HCO_2H + NH_2NH_2 \longrightarrow NH_2CH_2\overset{\overset{\displaystyle O}{\|}}{C}{-}NHNH_2 + NH_2\underset{\underset{\displaystyle CH_3}{|}}{C}HCO_2H$$

An example of the use of this reaction in structure determination is given in Figure 24.1.

For the tripeptide illustrated in Figure 24.1 the determination of the N-terminal and C-terminal amino acids automatically fixes the amino acid sequence in the tripeptide. However, for tetra-, penta-, etc. peptides the structure cannot be determined by analysis of the terminal groups alone; partial hydrolysis methods are employed to break the original peptide into smaller fragments. The structure of the lower molecular mass peptide fragments of the parent peptide then are determined. The structure of the parent peptide can be deduced from the structure of the hydrolyzed fragments. A pentapeptide consisting of glycine, alanine, valine, leucine, and isoleucine can yield tetra-, tri-, and dipeptides upon partial hydrolysis. There are two possible tetrapeptides, three possible tripeptides, and four possible dipeptides that can be produced. If the structures of enough of these can be determined, the structure of the pentapeptide can be deduced. For example, if the tripeptides obtained are ala-leu-val, gly-ala-leu, and leu-val-ileu, then the pentapeptide must be gly-ala-leu-val-ileu.

EXAMPLE 24.4

A tripeptide consisting of alanine, glycine, and leucine yields the dipeptide alanylglycine upon partial hydrolysis. What information is needed to ascertain the structure of the tripeptide?

There are only two possible isomeric compounds that could give the indicated dipeptide. These are ala-gly-leu and leu-ala-gly. Therefore, one end group determination experiment will indicate which of the possible structures is correct. The researcher could determine either the N-terminal amino acid or the C-terminal amino acid.

EXAMPLE 24.5

A hexapeptide yields the following tripeptides upon partial hydrolysis: leu-ala-val, ala-val-ileu, and ser-gly-leu. What is the structure of the hexapeptide?

By arranging the tripeptides in a vertical array such that the identical amino acids are in the same column the answer can be determined.

```
      leu-ala-val
          ala-val-ileu
ser-gly-leu
--------------------------
ser-gly-leu-ala-val-ileu
```

24.7
STRUCTURE OF SOME PEPTIDES

The determination of the amino acid sequence in peptides is a complex task, as illustrated in the previous section with low molecular mass peptides. At this point the structure of two important nonapeptides will be given without structural proof. They have been synthesized by Vincent Du Vigneaud of Cornell University, who received the Nobel prize in 1955 for his research.

Oxytocin is formed by the pituitary gland and causes the contraction of smooth muscle, such as that of the uterus. It is used to induce delivery or to make uterine contractions more effective in certain cases. The structure of oxytocin is shown in Figure 24.2. The arrows indicate the direction of attachment from a carboxyl group to an amine group. Note the presence of aspartic acid and glutamic acid, both of which contain two carboxyl groups. One carboxyl group is used in the peptide chain in each case, while the second is bound in the form of an amide with ammonia. In addition, the terminal glycine is bound with ammonia to yield a terminal amide.

Vasopressin is also produced in the pituitary gland. It serves to regulate the excretion of water by the kidneys and thus affects blood pressure. Vasopressin deficiency causes the excessive excretion of urine and leads to the disease diabetes insipidus. The structure of vasopressin differs from that of oxytocin by only two amino acids. It is remarkable how these differences account for the difference in physiological action. By contrast,

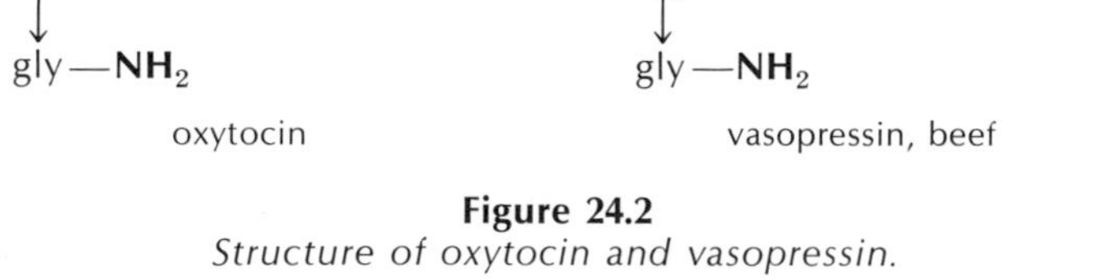

```
           ↗tyr↘                            ↗tyr↘
       cys       ileu                   cys       phe
        |          ↓                     |          ↓
       cys        glu—NH2               cys        glu—NH2
     ↙    ↖      ↙                    ↙    ↖      ↙
  pro       asp                    pro       asp
   ↓         |                      ↓         |
  leu       NH2                    arg       NH2
   ↓                                ↓
  gly—NH2                          gly—NH2
        oxytocin                     vasopressin, beef
```

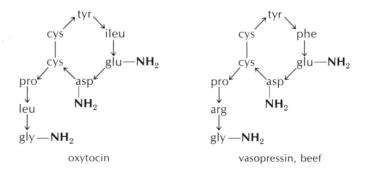

Figure 24.2
Structure of oxytocin and vasopressin.

the vasopressin of swine, which differs in structure from that given in Figure 24.2 by a molecule of lysine in place of arginine, has unchanged physiological action and can be safely administered to patients with diabetes insipidus.

24.8 COMPOSITION OF PROTEINS

PRIMARY STRUCTURE

In principle the determination of amino acid sequences in proteins is merely an extension of the principles outlined for peptides. The application of these principles to proteins involves arduous years of work because there are many structural variations possible for a large collection of constituent amino acids. There are approximately 10^{55} isomeric proteins consisting of the same 50 constituent amino acids. Obviously the problem of structure determination is immense.

Insulin, which is an antidiabetic hormone, consists of two peptide chains: one of 30 amino acids and the other of 21 (Figure 24.3). The two chains are linked by two *disulfide bonds* (—S—S—) between cysteine molecules.

```
   |                          |
 H—N                          N—H
   |                          |
 H—C—CH2—S—S—CH2—C—H
   |    disulfide             |
 O=C       bond               C=O
   |                          |
```

In addition another disulfide bond causes one chain to contain a cyclic structure. The Nobel prize was awarded to the English biochemist F. Sanger in 1958 for his determination of the amino acid sequence in insulin.

The insulin from different animals has slightly differing sequences. The differences are illustrated in Figure 24.3. Because the sequence within the cyclic portion of the shorter chain does not affect the physiological function of the insulin, diabetics who become allergic to one type of insulin can use insulin from another animal source.

The enzyme ribonuclease (Figure 24.4) contains 114 amino acids in a single chain. The four disulfide bonds form several rings in the protein.

Human hemoglobin contains two pairs of identical proteins that are called α and β. The number of amino acids in each chain are 141 and 146, respectively. The amino acid sequence of the β chain is given in Figure 24.5. In some people the sixth amino acid from the N-terminal position of their hemoglobin β chain is valine rather than glutamic acid. Such people suffer from the disease sickle cell anemia (Chapter 27). The difference of only one amino acid out of 146 has a profound physiological effect in this case.

The complete three-dimensional structure of proteins involves determining more than a sequence of amino acids. The amino acid sequence

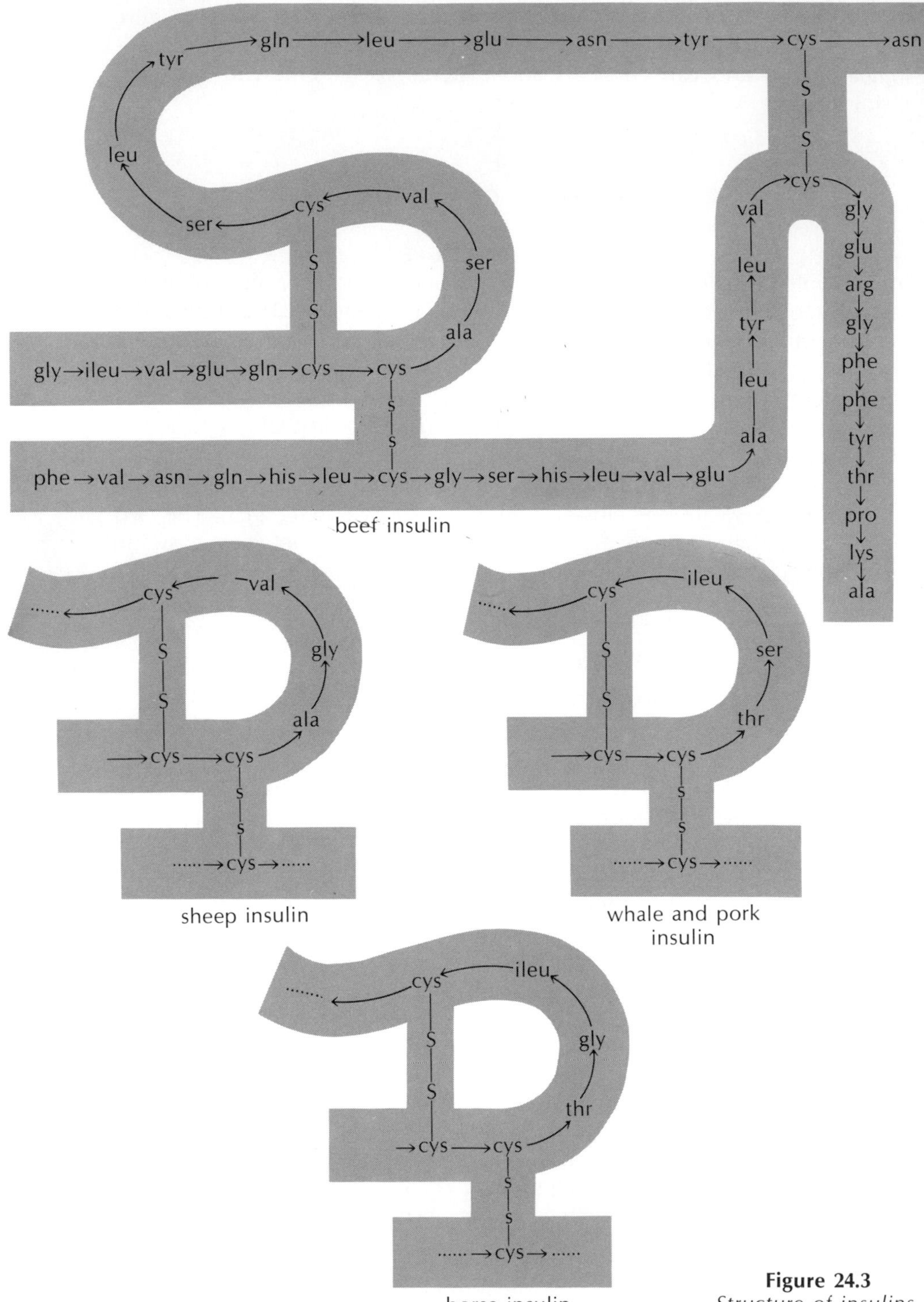

Figure 24.3
Structure of insulins.

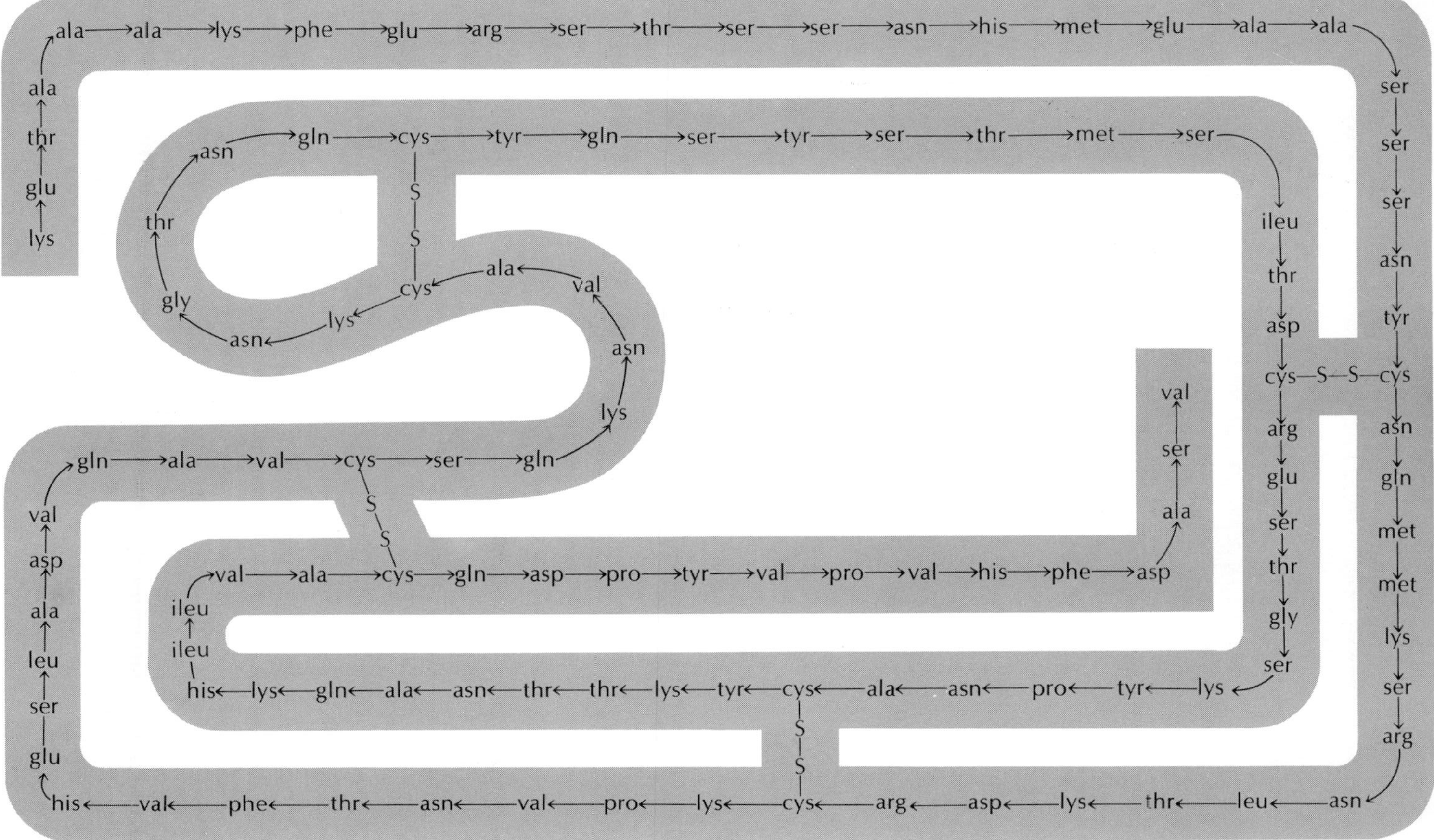

Figure 24.4
Structure of ribonuclease.

val → his → leu → thr → pro → glu → glu → lys → ser → ala → val → thr →
ala → leu → try → gly → lys → val → asp → val → asp → glu → val → gly → gly →
glu → ala → leu → gly → arg → leu → leu → val → val → tyr → pro → try → thr →
glu → arg → phe → phe → glu → ser → phe → gly → asp → leu → ser → thr → pro →
asp → ala → val → met → gly → asp → pro → lys → val → lys → ala → his → gly →
lys → lys → val → leu → gly → ala → phe → ser → asp → gly → leu → ala → his →
leu → asp → asp → leu → lys → gly → thr → phe → ala → thr → leu → ser → glu →
leu → his → cys → asp → lys → leu → his → val → asp → pro → glu → asp → phe →
arg → leu → leu → gly → asp → val → leu → val → cys → val → leu → ala → his →
his → phe → gly → lys → glu → phe → thr → pro → pro → val → glu → ala → ala →
tyr → glu → lys → val → val → ala → gly → val → ala → asp → ala → leu → ala →
his → lys → tyr → his

Figure 24.5
Amino acid sequence in β chain of hemoglobin.

constitutes the *primary structure*. There are secondary, tertiary, and quaternary structural features and perhaps more that are important. That these features exist and are important is well known from the *denaturation* phenomenon. When a protein is heated, irradiated, or treated with some solvents, a disruption of its three-dimensional structure occurs. The frying of an egg and the curdling of milk casein are common examples of denaturation. The primary structural sequence of amino acids is not disturbed, but there is a disruption of other bonds that is irreversible.

SECONDARY STRUCTURE

Many proteins consist of chains coiled into a spiral known as a helix. Such a helix may be either right- or left-handed, as in the case of screws, and constitutes the *secondary structure* of many proteins (Figure 24.6). For proteins consisting of L-amino acids the right-handed (or α) helix is more stable than the left-handed helix. The spiral is held together by hydrogen bonds between the proton of the N—H group of one amino acid and the oxygen of the C═O group of another amino acid in the next turn of the helix. There are approximately 3.6 amino acid units for each turn in the helix. After five turns or eighteen amino acids, the helix positions repeat themselves (Figure 24.7).

TERTIARY STRUCTURE

The presence of disulfide bonds in a protein tends to prevent the formation of a regular helix with a constant pitch. The molecule has to assume a shape consistent with the structural restrictions of both the disulfide bonds and the hydrogen bonds. The resultant course of the chain is roughly helical but with a bent structure that gives the molecules a spherical shape. Features of the molecule that contribute to the gross three-dimensional shape excluding those contributing to the formation of the α helix are contributors to the so-called *tertiary structure* of the protein.

The oxygen-carrying protein myoglobin, which colors the muscle tissue

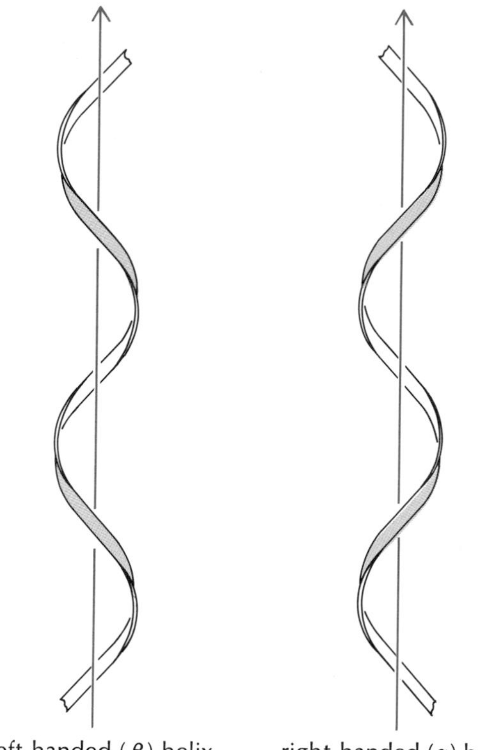

Figure 24.6
α and β helices.

of whales red, consists of a single chain of 153 amino acids. Its structure was determined by X-ray analysis by J. C. Kendrew and M. F. Perutz of Cambridge University. The Nobel Prize of 1963 was awarded for this work. The roughly helical chains are folded on each other in a complicated, unsymmetrical way. The molecule does not contain cysteine units and thus must be held together by hydrogen bonds. Therefore, the shape of the molecule must be controlled by features even more subtle than the commonly encountered types of bonds.

One of the reasons for the twisting of the protein chain of myoglobin appears to be associated with the presence of the amino acid proline.

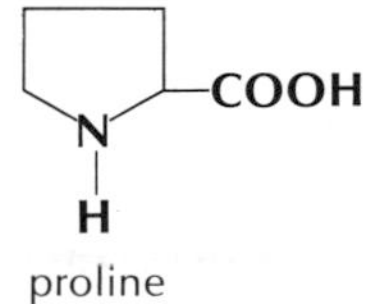

proline

It contains a secondary amino group which, when combined in a protein, does not have any N—H bond so that the protein structure cannot hydrogen bond at this point. The helix bends at each point where a proline

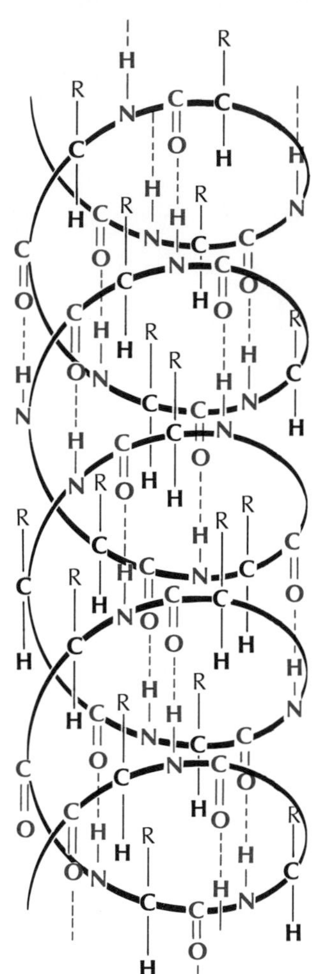

Figure 24.7
Hydrogen bonding in an α helix.

molecule occurs. There are four proline molecules in myoglobin, and each is located approximately in a corner of the shape defined by the structure. The exact reason for the contribution of proline to the tertiary structure of myoglobin is unknown. There probably are many other subtle contributors to tertiary structure that will be discovered in time.

QUATERNARY STRUCTURE

The *quaternary structure* of proteins corresponds to the manner in which several protein subunits are associated in larger molecules. Hemoglobin consists of two pairs of different proteins, each protein containing a molecule of heme (Chapter 14). The two identical α chains and two identical

β chains are arranged tetrahedrally in a three-dimensional structure. Each chain enfolds a heme group. All of these units are held together by ionic or polar forces but not by formal covalent bonds.

Suggested further readings

Asimov, I., "Potentialities of Protein Isomerism," *J. Chem. Educ.*, **31,** 125 (1954).
Doty, P. "Proteins," *Sci. Amer.*, offprint 7 (September 1957).
Fruton, J. S., "Proteins," *Sci. Amer.*, offprint 10 (June 1950).
Kendrew, J. C., "The Three Dimensional Structure of a Protein Molecule," *Sci. Amer.*, offprint 121 (December 1961).
Lande, S., "Conformation of Peptides," *J. Chem. Educ.*, **45,** 587 (1968).
Pauling, L., Corey, R. B., and Hayward, R., "The Chemical Structure of Proteins," *Sci. Amer.* offprint 31 (July 1954).
Perutz, M. F., "The Hemoglobin Molecule," *Sci. Amer.*, offprint 196 (November 1964).
Phillips, D. C., "The Three Dimensional Structure of An Enzyme Molecule," *Sci. Amer.* (November 1966).
Stein, W. H., and Moore, S., "The Chemical Structure of Proteins," *Sci. Amer.*, offprint 80 (February 1961).
Thompson, E. O. P. "The Insulin Molecule," *Sci. Amer.*, offprint 42 (May 1955).
Wheeler, T. A., "Models of the α-Helix Configuration in Polypeptides," *J. Chem. Educ.*, **34,** 136 (1957).

Terms and concepts

α-helix
amino acid
chymotrypsin
C-terminal amino acid
denaturation
disulfide bond
insulin
ninhydrin
N-terminal amino acid
oxytocin
peptide
primary structure
protein
quaternary structure
secondary structure
tertiary structure
Van Slyke test
vasopressin

Questions and problems

1. Write the structural formula for each of the following peptides:
a. glycylglycine
b. glycylglycylglycine
c. glycylalanine
d. glycylalanylleucine
e. alanylglycylleucine
f. glycylalanylleucylserine

2. Which amino acids can form peptides with free amino groups at internal positions in the peptide chain?

3. What volume of nitrogen at STP will be liberated in the Van Slyke reaction from **a.** 75 mg of glycine and **b.** 146 mg of lysine?

4. Write the equations for the steps necessary in order to synthesize glycylalanylglycine.

5. Write the condensed formulas for the 24 isomeric tetrapeptides containing glycine, alanine, leucine, and isoleucine.

6. How many possible dipeptides are there of the 21 amino acids listed in Table 24.1?

7. How many dipeptide structures would have to be established in order to ascertain the structure of a pentapeptide that gives the dipeptides upon hydrolysis?

8. A pentapeptide consists of leucine, valine, alanine, and two molecules of glycine. After the peptide is heated with hydrazine, the only free amino acid obtained is valine. When the pentapeptide is treated with 2,4-dinitrofluorobenzene followed by hydrolysis, glycine, alanine, leucine, and valine are obtained. Partial hydrolysis of the pentapeptide yields gly-leu-val. What is the structure of the pentapeptide?

9. Hypertensin is a decapeptide formed in the kidneys; it causes an increase in blood pressure. The N-terminal and C-terminal amino acids are aspartic acid and leucine, respectively. Upon partial hydrolysis the tripeptides formed are val-tyr-val, asp-arg-val, phe-his-leu, and his-pro-phe. What is the structure of hypertensin?

10. A hexapeptide (A) consists of glycine, leucine, isoleucine, phenylalanine, and two molecules of alanine. When treated with 2,4-dinitrofluorobenzene followed by hydrolysis only glycine, leucine, isoleucine, and alanine are obtained as free amino acids. When heated with hydrazine the free amino acid obtained is alanine. Partial hydrolysis of A yields two tripeptides, B and C. Complete hydrolysis of B produces alanine, isoleucine, and phenylalanine. Upon heating B with hydrazine, isoleucine is formed. The tripeptide C upon complete hydrolysis yields isoleucine, alanine, and glycine. What is the structure of A?

11. What amino acids would have no hydrogen attached to the amide nitrogen if they are components of a peptide?

12. Would a stable α-helix be formed from a mixture of L and D amino acids randomly distributed in a protein chain? Why?

13. How would the stability of the α-helix be affected by the presence of large numbers of proline molecules?

25

CARBOHYDRATES

The name *carbohydrate* originated from the French term *hydrate de carbon,* which indicated these substances could be represented by the general formula $C_n(H_2O)_n$. For example, the sugar glucose has a molecular formula $C_6H_{12}O_6$ or $C_6(H_2O)_6$. Although carbohydrates do not contain hydrated carbon atoms and there are many substances classified as carbohydrates that do not have the ratio of carbon, hydrogen, and oxygen indicated by the general formula, the name still remains.

The carbohydrates are polyhydroxy aldehydes and ketones or substances that when hydrolyzed give these compounds. Carbohydrates include sugars, starches, cellulose, and other substances found in the roots, stems, and leaves of all plants. These compounds contained in food are an important source of energy for animals. Among the many industrial carbohydrate-based materials are paper, cellophane, cotton, rayon, and confections.

25.1 CLASSIFICATION OF CARBOHYDRATES

The carbohydrates are subdivided into three classes: *monosaccharides, oligosaccharides* (from the Greek *oligos,* a few), and *polysaccharides.* The class of monosaccharides includes all sugars that contain a single carbohydrate unit; monosaccharides cannot yield simpler carbohydrates upon hydrolysis. Although monosaccharides can contain as few as three carbon atoms, the five and six carbon structures, called *pentoses* and *hexoses,* respectively, are the most numerous and important. If the sugar is an aldehyde either free or combined as an hemiacetal or acetals it is called an *aldose,* whereas it is called a *ketose* if it is a ketone.

Oligosaccharides contain a relatively small number of monosaccharide units joined by acetal bonds (Section 20.5), which are called *glycosidic* linkages in carbohydrate chemistry. These bonds are between the aldehyde or ketone site of one monosaccharide and the hydroxyl group of another. Hydrolysis of this glycosidic linkage produces the component monosaccharides. Two sugar units combine to form a disaccharide, three form a trisaccharide, and so on.

Polysaccharides contain hundreds to thousands of monosaccharide units joined by a series of glycosidic linkages. These molecules are the second example presented of naturally occurring polymers; the first was proteins.

25.2 MONOSACCHARIDES

ALDOTRIOSES

The simplest aldose is the *triose* (three-carbon sugar) glyceraldehyde, which can exist in two optically active forms:

$$\begin{array}{ccc} & \text{CHO} & \\ & | & \\ \text{H}- & \text{C} & -\text{OH} \\ & | & \\ & \text{CH}_2\text{OH} & \end{array} \qquad \begin{array}{ccc} & \text{CHO} & \\ & | & \\ \text{HO}- & \text{C} & -\text{H} \\ & | & \\ & \text{CH}_2\text{OH} & \end{array}$$

D-glyceraldehyde L-glyceraldehyde

Glyceraldehyde is important because the other sugars may be regarded as being derived from it by lengthening the carbon chain at the carbonyl carbon atom. The sugars then are designated as D or L, depending upon the configuration of the asymmetric carbon atom farthest from the aldehyde group, that is, the original carbon bonded to the secondary hydroxyl group in glyceraldehyde. All naturally occurring monosaccharides are of D configuration. Chapter 23 pointed out that the symbol D does not correspond to the direction of the rotation of the plane polarized light. Thus both D(+) and D(−) sugars are known. For the most part in this chapter the signs of rotation will not be given.

ALDOTETROSES

An *aldotetrose* (four-carbon aldehyde sugar) contains two nonequivalent asymmetric centers, making possible the four isomers illustrated in Figure 25.1. The student should convince himself that there are only four possible

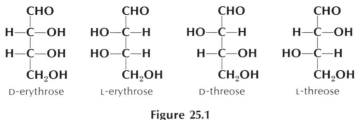

D-erythrose	L-erythrose	D-threose	L-threose
CHO	CHO	CHO	CHO
H—C—OH	HO—C—H	HO—C—H	H—C—OH
H—C—OH	HO—C—H	H—C—OH	HO—C—H
CH_2OH	CH_2OH	CH_2OH	CH_2OH

Figure 25.1
The aldotetroses.

right-left combinations for the location of the hydroxyl groups (or hydrogens) in the standard projection formula. It also should be noted that the bottom half of the projection formulas of D-erythrose and D-threose are identical to the last two carbon atoms with their attached functional groups in D-glyceraldehyde. Both D-erythrose and D-threose can be thought of as derived from D-glyceraldehyde by the insertion of a H—C—OH group between the number 1 and 2 carbon atoms. The H—C—OH group can be inserted in either of two ways, yielding the tetroses D-erythrose and D-threose. In a similar manner the isomers L-erythrose and L-threose can be mentally derived from L-glyceraldehyde.

The molecules D-erythrose and D-threose are designated D because of the location of the hydroxyl group on the right side of the asymmetric center farthest away from the aldehyde. D-Erythrose and L-erythrose are enantiomers (Section 23.2) as are D-threose and L-threose. Other combinations such as D-erythrose and D-threose or D-erythrose and L-threose are diastereomers (Section 23.5). Recall that enantiomers possess the same physical properties; diastereomers do not.

ALDOPENTOSES

There are eight isomeric *aldopentoses* since such five-carbon sugars contain three nonequivalent asymmetric centers (Figure 25.2). The eight isomers can be considered as four pairs of enantiomers. In other words, there are four aldopentoses of the D configuration and four of the L configuration. The four compounds of the D configuration are diastereomers, as are the four compounds of the L configuration.

The eight aldopentoses can be derived mentally from the aldotetroses by the insertion of an H—C—OH group between the aldehyde group and the remainder of the molecule. The H—C—OH group can be inserted in two possible ways in each of the four aldotetroses, and the eight aldopentoses result.

CHO	CHO	CHO	CHO
H—C—OH	HO—C—H	H—C—OH	HO—C—H
H—C—OH	H—C—OH	HO—C—H	HO—C—H
H—C—OH	H—C—OH	H—C—OH	H—C—OH
CH_2OH	CH_2OH	CH_2OH	CH_2OH
D-ribose	D-arabinose	D-xylose	D-lyxose

CHO	CHO	CHO	CHO
HO—C—H	H—C—OH	HO—C—H	H—C—OH
HO—C—H	HO—C—H	H—C—OH	H—C—OH
HO—C—H	HO—C—H	HO—C—H	HO—C—H
CH_2OH	CH_2OH	CH_2OH	CH_2OH
L-ribose	L-arabinose	L-xylose	L-lyxose

Figure 25.2
The aldopentoses.

EXAMPLE 25.1

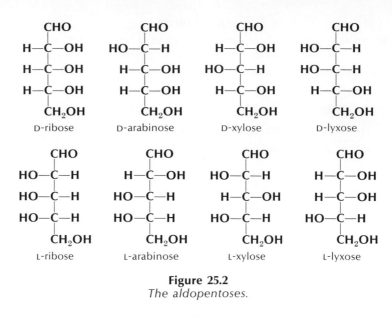

Would you expect ribose to be soluble in water?

The compound 1-propanol is completely miscible with water and contains one hydroxyl group bound to three carbon atoms (see Table 19.1). The solubility of 1-pentanol is 2.7 g/100 g of water. Therefore, D-ribose, which contains four hydroxyl groups attached to a five-carbon structure, should be very soluble in water.

ALDOHEXOSES

There are 16 isomeric *aldohexoses.* The eight diastereomeric compounds of D configuration are given in Figure 25.3. Of course there are eight mirror image compounds or enantiomers of the L configuration, one for each compound shown in Figure 25.3. The specific rotations (Section 23.1) are listed in order to emphasize the differences for diastereomers, all of the D configuration. The specific rotation of the enantiomer of each compound is of the same magnitude but of opposite sign.

EXAMPLE 25.2

What is the structure of L-glucose? What is its specific rotation?

L-Glucose is the enantiomer of D-glucose and is the mirror image of the latter compound. Therefore, the isomer can be

CHO	CHO	CHO	CHO
H—C—OH	HO—C—H	H—C—OH	H—C—OH
H—C—OH	H—C—OH	HO—C—H	H—C—OH
H—C—OH	H—C—OH	H—C—OH	HO—C—H
H—C—OH	H—C—OH	H—C—OH	H—C—OH
CH_2OH	CH_2OH	CH_2OH	CH_2OH
D-allose(+14°)	D-altrose(+33°)	D-glucose(+53°)	D-gulose(−20°)
CHO	CHO	CHO	CHO
HO—C—H	HO—C—H	H—C—OH	HO—C—H
HO—C—H	H—C—OH	HO—C—H	HO—C—H
H—C—OH	HO—C—H	HO—C—H	HO—C—H
H—C—OH	H—C—OH	H—C—OH	H—C—OH
CH_2OH	CH_2OH	CH_2OH	CH_2OH
D-mannose(+14°)	D-idose(+15°)	D-galactose(+80°)	D-talose(+21°)

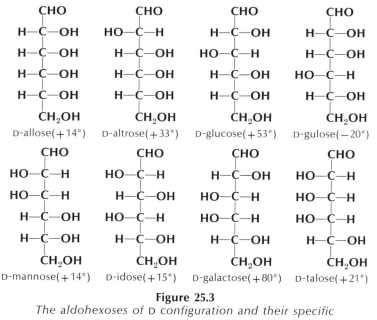

Figure 25.3
The aldohexoses of D configuration and their specific rotations.

drawn by reflecting the planar projection formula in an imagined mirror perpendicular to the plane of the page:

CHO	CHO
H—C—OH	HO—C—H
HO—C—H	H—C—OH
H—C—OH	HO—C—H
H—C—OH	HO—C—H
CH_2OH	CH_2OH
D-glucose	L-glucose

The specific rotation of L-glucose must be equal in magnitude but opposite in sign to that of D-glucose. The predicted specific rotation of L-glucose is −53°.

EXAMPLE 25.3

Describe the isomeric aldoheptoses.

These compounds have five asymmetric centers and there should be $2^5 = 32$ isomers. Each isomer will have a mirror image isomer and, therefore, there will be 16 pairs of enantiomers. Each member of the D series of compounds will be diastereomeric with the other members of the series. All members of the D series will contain a hydroxyl group on the asymmetric carbon atom farthest removed from the aldehyde group on the right in the projection formula.

EXAMPLE 25.4

What relationship exists between each possible pair of the three compounds given below?

$$
\begin{array}{ccc}
\begin{array}{rcl} & CHO & \\ H- & C & -OH \\ HO- & C & -H \\ HO- & C & -H \\ H- & C & -OH \\ H- & C & -OH \\ & CH_2OH & \end{array}
&
\begin{array}{rcl} & CHO & \\ H- & C & -OH \\ HO- & C & -H \\ H- & C & -OH \\ H- & C & -OH \\ H- & C & -OH \\ & CH_2OH & \end{array}
&
\begin{array}{rcl} & CHO & \\ HO- & C & -H \\ H- & C & -OH \\ H- & C & -OH \\ HO- & C & -H \\ HO- & C & -H \\ & CH_2OH & \end{array}
\\
\text{(A)} & \text{(B)} & \text{(C)}
\end{array}
$$

A and B represent compounds of the D series of aldoheptoses. They are diastereomers because they cannot be enantiomers. Enantiomers are always of opposite configuration and have to be in different series, that is, D and L. The structure C is of the L series and is the mirror image of A and therefore is the enantiomer of A. The structures B and C are of different series, but since they cannot be mirror images (only one mirror image is possible per compound and that has been indicated as A with C) they must be diastereomers.

KETOSES

The simplest ketose is dihydroxyacetone, which does not contain any asymmetric centers. Of the various possible isomeric ketotetroses, ketopentoses, and ketohexoses, the most important naturally occuring one is the ketohexose D-fructose.

$$
\begin{array}{cc}
\begin{array}{c} CH_2OH \\ C{=}O \\ CH_2OH \end{array}
&
\begin{array}{rcl} & CH_2OH & \\ & C{=}O & \\ HO- & C & -H \\ H- & C & -OH \\ H- & C & -OH \\ & CH_2OH & \end{array}
\\
\text{dihydroxyacetone} & \text{D-fructose}
\end{array}
$$

D-Fructose occurs as the free sugar in fruit and honey as well as in a disaccharide called *sucrose* in which it is combined with glucose (Section 25.8).

What aldohexoses correspond to D*-fructose in configuration at the asymmetric centers on carbons 3 through 5?*

A search of Figure 25.3, which contains representations of the D aldohexoses, reveals that only two compounds have identical configurations with D-fructose at asymmetric centers on carbons 3 through 5. This should have been expected because the aldohexoses have one more asymmetric center than fructose. There can be only two possible different configurations at that one center.

```
      CH2OH               CHO                 CHO
        |                  |                   |
        C=O             H—C—OH             HO—C—H
        |                  |                   |
    HO—C—H             HO—C—H              HO—C—H
        |                  |                   |
     H—C—OH             H—C—OH              H—C—OH
        |                  |                   |
     H—C—OH             H—C—OH              H—C—OH
        |                  |                   |
      CH2OH              CH2OH               CH2OH
```

D-fructose D-glucose D-mannose

25.3 MUTAROTATION OF MONOSACCHARIDES

EXPERIMENTAL OBSERVATIONS

The structures given in the previous section for monosaccharides are called *open chain* formulas and were proposed by the German chemist Emil Fischer (1852–1919). Fischer received the Nobel prize in 1902 for his pioneering work in the chemistry of carbohydrates. However, extensive studies of the properties of sugars have led to the conclusion that they exist in cyclic forms both in the solid state and in solution. As an example of the type of experimental evidence available, consider the behavior of D-glucose. Crystallization of D-glucose from methanol yields material that has a specific rotation of 113° when dissolved in water. Upon standing the specific rotation of the solutions slowly drops to 52.5°. Crystallization of D-glucose from acetic acid yields material that has a specific rotation of 19° when dissolved in water. However, the specific rotation of this solution gradually increases to 52.5°. A gradual change in rotation to an equilibrium point is known as *mutarotation.* All aldopentoses and aldohexoses undergo mutarotation.

THE CONCEPT OF CYCLIC ACETALS

Mutarotation can be explained by the existence of two nonequivalent cyclic structures for each monosaccharide. The carbonyl group and one of the hydroxyl groups in the chain can react to yield a cyclic hemiacetal. When this occurs the carbonyl carbon becomes an asymmetric center, and two isomers can be generated.

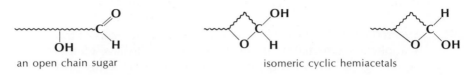

an open chain sugar

isomeric cyclic hemiacetals

Each of the isomers can be isolated under appropriate conditions. However, in water an equilibrium is set up with the open chain form since hemiacetal formation is reversible (Section 20.5). A mixture of both isomers with an intermediate specific rotation then results.

The names of the cyclic hemiacetals are derived from the names of the open chain sugar. Stable cyclic hemiacetals can be obtained with five- or six-membered rings. Rings containing six atoms (five carbons and one oxygen) are named as derivatives of pyran, and those containing five atoms are named as derivatives of furan.

pyran furan

Glucose in a six-membered ring form is called glucopyranose and, in a five membered ring form, glucofuranose. The configuration at the new asymmetric center is designated α or β, depending on whether the OH group of the hemiacetal is on the right or left in the Fischer projection formula. The two isomeric D-glucopyranoses are shown below:

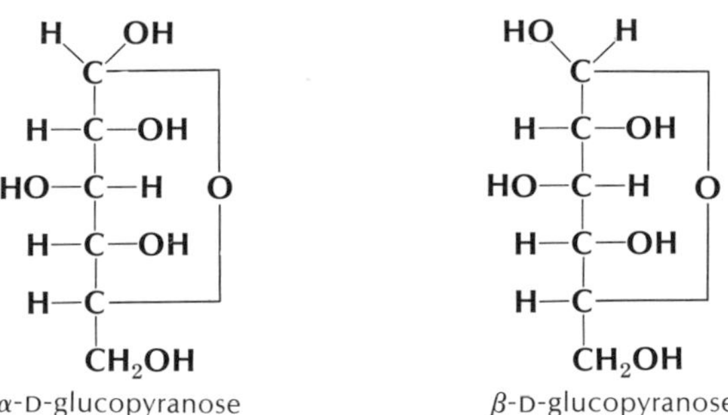

α-D-glucopyranose β-D-glucopyranose

The long right-angle bonds of course are not contained in the actual molecule. They are used only to indicate the positions bonded together in the hemiacetal.

HAWORTH PROJECTION FORMULAS

The Haworth projection structures for cyclic carbohydrates are a somewhat better way of depicting the stereochemistry (Figure 25.4). In these structures

Figure 25.4
Haworth projection formulas.

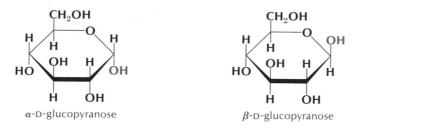

the ring is projected in a plane perpendicular to the plane of the page. The heavy lines indicate that portion of the ring coming out of the page toward the reader. The *cis* and *trans* relationships of the functional groups become evident in these structures. Note that the α configuration corresponds to the structure in which the hydroxyl group is directed below the plane of the ring in the Haworth projection of D sugars.

EXAMPLE 25.6

What is the Haworth projection of β-D-mannopyranose?

The only difference between the structure of β-D-mannopyranose and β-D-glucopyranose is in the configuration at the number 2 carbon atom of the open chain representation. Therefore, a simple reversal of the positions of the hydrogen and hydroxyl group at the asymmetric center closest to the hemiacetal carbon yields the desired structure:

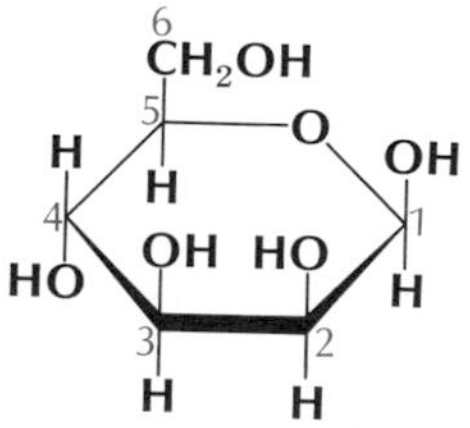

The α and β forms of a sugar such as glucose that may be regarded as geometrical isomers or diastereomers can be isolated in pure forms by recrystallization. However, in solution the hemiacetal is in equilibrium with a small amount (less than 0.1 percent) of the open chain aldehyde. Reclosure of the ring can form either isomer so that eventually equilibrium is achieved in solution. The specific rotation equilibrium value of 52.5° for glucose corresponds to 37 percent of the α isomer (specific rotation = 113°) and 63 percent of the β isomer (specific rotation = 19°).

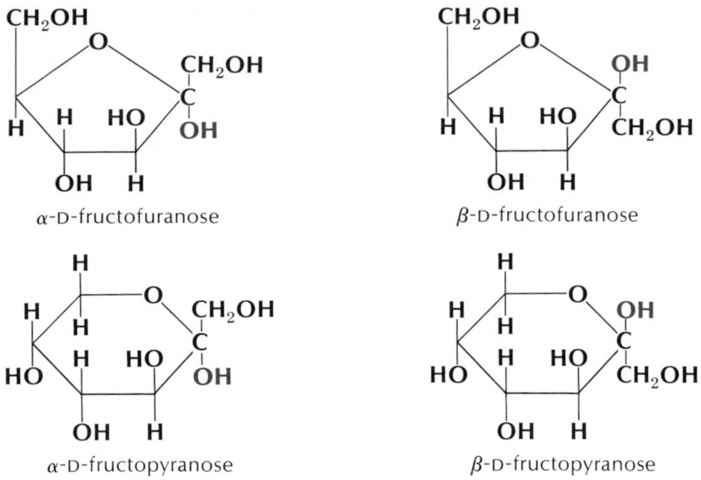

α-D-fructofuranose

β-D-fructofuranose

α-D-fructopyranose

β-D-fructopyranose

Figure 25.5
Isomeric D-fructoses.

Fructose can exist in the α and β forms of both pyranose and furanose rings (Figure 25.5).

25.4 SYNTHESIS OF MONOSACCHARIDES

GENERAL CONSIDERATIONS

Monosaccharides can be synthesized from other monosaccharides by either a chain-lengthening or degradation (shortening) technique. The series of reactions that extends a sugar chain is known as the *Kiliani synthesis,* whereas the degradation of the chain is known as the *Ruff degradation.* Both methods involve transformation at the carbonyl center.

KILIANI SYNTHESIS

The Kiliani synthesis involves addition of hydrogen cyanide to the carbonyl group (Section 23.4) followed by hydrolysis of the cyano group to an acid subsequently reduced to an aldehyde. The method is illustrated in Figure 25.6 for the conversion of D-glyceraldehyde into a mixture of diastereomers, D-erythrose and D-threose.

A mixture of two diastereomers always results in the Kiliani synthesis because an asymmetric center is generated in the addition reaction of HCN to the carbonyl center. The other asymmetric centers in molecules undergoing the Kiliani synthesis are not altered. The entire sequence of reactions can be carried out on the mixture of diastereomers, and the products can be separated by means of their differing physical properties at the end of the synthesis.

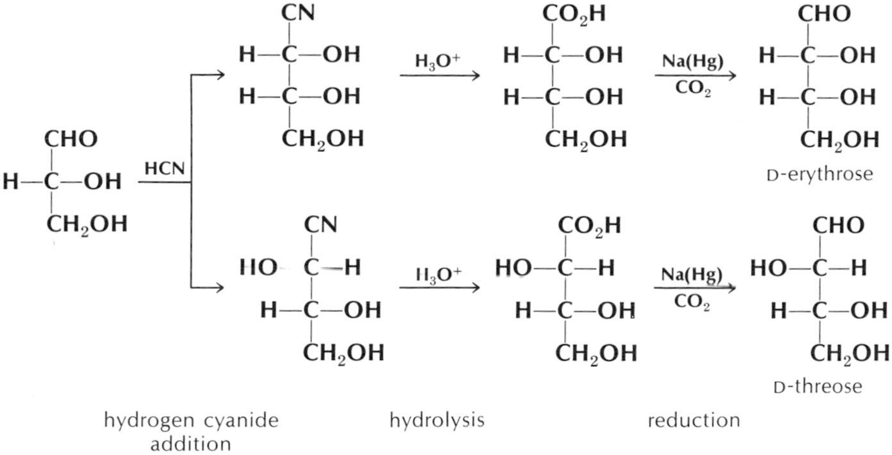

CHO
H—C—OH
CH_2OH

HCN

CN
H—C—OH
H—C—OH
CH_2OH

H_3O^+

CO_2H
H—C—OH
H—C—OH
CH_2OH

Na(Hg)
CO_2

CHO
H—C—OH
H—C—OH
CH_2OH
D-erythrose

CN
HO—C—H
H—C—OH
CH_2OH

H_3O^+

CO_2H
HO—C—H
H—C—OH
CH_2OH

Na(Hg)
CO_2

CHO
HO—C—H
H—C—OH
CH_2OH
D-threose

hydrogen cyanide addition   hydrolysis   reduction

Figure 25.6
The Kiliani synthesis of D-erythrose and D-threose.

EXAMPLE 25.7

Which aldotetrose is needed to synthesize D-ribose by the Kiliani method? What other product is formed?

The aldotetrose that must be used has to have the same configuration as D-ribose at the carbon atoms 3 and 4. The configuration of the carbon atom 2 of D-ribose cannot be controlled in the Kiliani method. Therefore, a compound differing from D-ribose at the carbon atom 2 will be generated in the chain-lenthening process. The compound that is needed is D-erythrose. D-Arabinose will be the second product formed in the reaction.

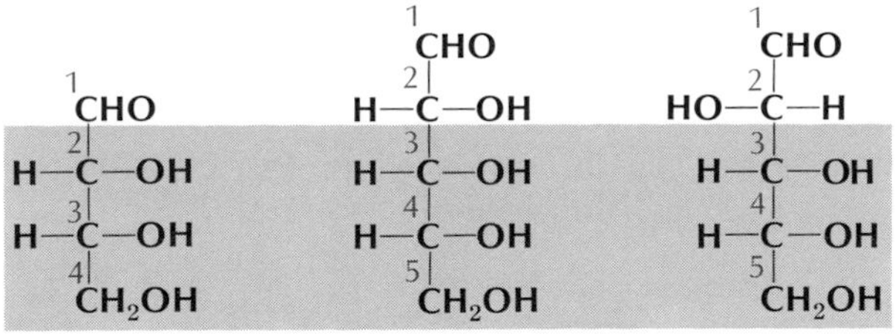

D-erythrose	D-ribose	D-arabinose
	1 CHO	1 CHO
1 CHO	2 H—C—OH	2 HO—C—H
2 H—C—OH	3 H—C—OH	3 H—C—OH
3 H—C—OH	4 H—C—OH	4 H—C—OH
4 CH_2OH	5 CH_2OH	5 CH_2OH

THE RUFF DEGRADATION

The Ruff degradation involves oxidation of the carbonyl group, which is then converted to a calcium salt and further oxidized (Figure 25.7). The net result of the total process is the loss of the original carbonyl group

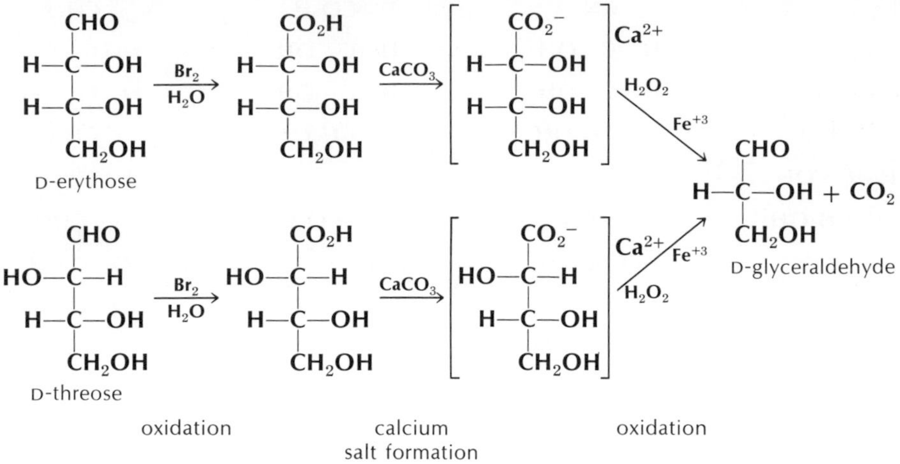

$$\begin{array}{c}CHO\\ H-C-OH\\ H-C-OH\\ CH_2OH\\ \text{D-erythose}\end{array} \xrightarrow[H_2O]{Br_2} \begin{array}{c}CO_2H\\ H-C-OH\\ H-C-OH\\ CH_2OH\end{array} \xrightarrow{CaCO_3} \left[\begin{array}{c}CO_2^-\\ H-C-OH\\ H-C-OH\\ CH_2OH\end{array}\right] Ca^{2+} \xrightarrow[Fe^{+3}]{H_2O_2}$$

$$\begin{array}{c}CHO\\ HO-C-H\\ H-C-OH\\ CH_2OH\\ \text{D-threose}\end{array} \xrightarrow[H_2O]{Br_2} \begin{array}{c}CO_2H\\ HO-C-H\\ H-C-OH\\ CH_2OH\end{array} \xrightarrow{CaCO_3} \left[\begin{array}{c}CO_2^-\\ HO-C-H\\ H-C-OH\\ CH_2OH\end{array}\right] Ca^{2+} \xrightarrow[H_2O_2]{Fe^{+3}}$$

$$\begin{array}{c}CHO\\ H-C-OH\\ CH_2OH\\ \text{D-glyceraldehyde}\end{array} + CO_2$$

oxidation — calcium salt formation — oxidation

Figure 25.7
The Ruff degradation.

and the formation of a new carbonyl group at a carbon that was originally an asymmetric center. The remaining asymmetric centers are not changed and, therefore, only one compound is obtained from the degradative procedure. Either of two diastereomeric compounds of n carbon atoms can be used to prepare a desired compound of $n - 1$ carbon atoms. This fact is illustrated in Figure 25.7 for D-erythrose and D-threose.

EXAMPLE 25.8

What aldotetrose will be formed by the Ruff degradation of D-arabinose?

The aldotetrose that will be formed will have the same configuration as the carbon atoms 3 and 4 of D-arabinose. The aldotetrose that will be formed is D-erythrose.

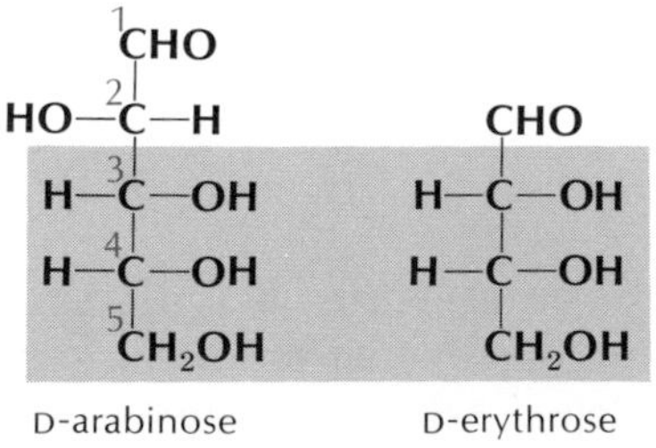

$$\begin{array}{c}{}^{1}CHO\\ HO-{}^{2}C-H\\ H-{}^{3}C-OH\\ H-{}^{4}C-OH\\ {}^{5}CH_2OH\\ \text{D-arabinose}\end{array} \qquad \begin{array}{c}CHO\\ H-C-OH\\ H-C-OH\\ CH_2OH\\ \text{D-erythrose}\end{array}$$

25.5 OXIDATION OF MONOSACCHARIDES

It might be expected that aldoses could be oxidized by Tollens' reagent (Chapter 20), whereas ketoses could not be oxidized. While ketones in general are not oxidized by Tollens' reagent, α-hydroxyketones are and, therefore, ketoses such as fructose are oxidized. A distinction between aldoses and ketoses is not possible with this reagent.

Bromine in water oxidizes aldoses to *glyconic acids;* ketoses are not affected. This reaction is the first step in the Ruff degradation.

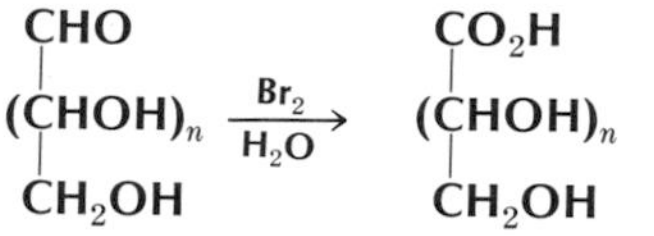

$$\underset{\text{an aldose}}{\begin{array}{c} CHO \\ | \\ (CHOH)_n \\ | \\ CH_2OH \end{array}} \xrightarrow[H_2O]{Br_2} \underset{\text{a glyconic acid}}{\begin{array}{c} CO_2H \\ | \\ (CHOH)_n \\ | \\ CH_2OH \end{array}}$$

The more powerful oxidizing agent nitric acid brings about oxidation not only of the aldehyde group of aldoses but also of the primary hydroxyl group. The dicarboxylic acid produced is called a *glycaric acid.*

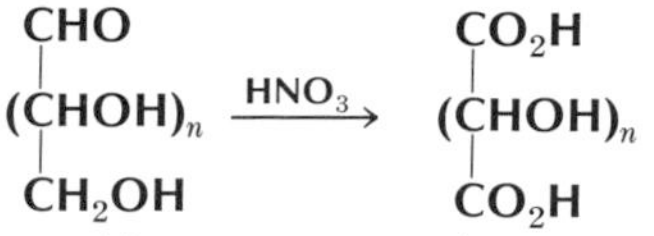

$$\underset{\text{an aldose}}{\begin{array}{c} CHO \\ | \\ (CHOH)_n \\ | \\ CH_2OH \end{array}} \xrightarrow{HNO_3} \underset{\text{a glycaric acid}}{\begin{array}{c} CO_2H \\ | \\ (CHOH)_n \\ | \\ CO_2H \end{array}}$$

This conversion of an aldose is a useful way of ascertaining whether the hydroxyl groups are oriented in an internally symmetric or asymmetric manner. The oxidation of D-threose produces an optically active glycaric acid because the rotatory powers of the two asymmetric centers do not cancel (Section 23.6). The molecule is asymmetric.

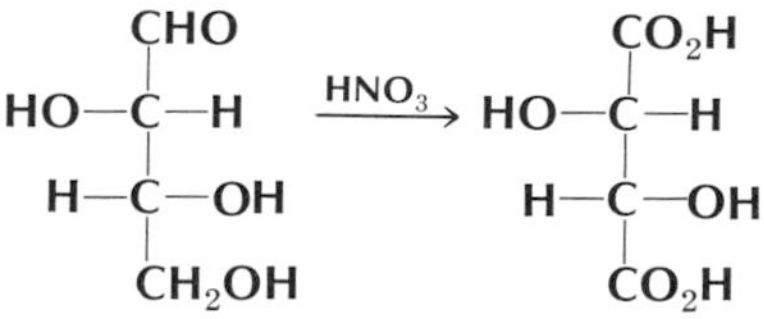

$$\underset{\text{D-threose}}{\begin{array}{c} CHO \\ | \\ HO-C-H \\ | \\ H-C-OH \\ | \\ CH_2OH \end{array}} \xrightarrow{HNO_3} \underset{\substack{\text{D-tartaric acid} \\ \text{(optically active)}}}{\begin{array}{c} CO_2H \\ | \\ HO-C-H \\ | \\ H-C-OH \\ | \\ CO_2H \end{array}}$$

Oxidation of D-erythrose yields an optically inactive glycaric acid since the rotatory powers of the two equivalent asymmetric centers cancel each other (Section 23.6). The molecule is symmetric and cannot be optically active.

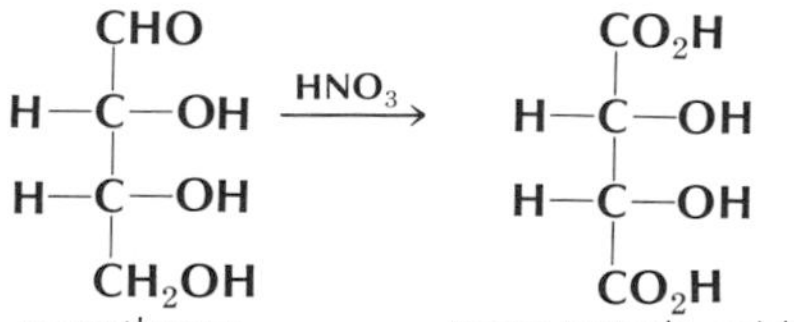

$$\underset{\text{D-erythrose}}{\begin{array}{c} CHO \\ | \\ H-C-OH \\ | \\ H-C-OH \\ | \\ CH_2OH \end{array}} \xrightarrow{HNO_3} \underset{\substack{\textit{meso}\text{-tartaric acid} \\ \text{(optically inactive)}}}{\begin{array}{c} CO_2H \\ | \\ H-C-OH \\ | \\ H-C-OH \\ | \\ CO_2H \end{array}}$$

Will the glycaric acid derived from D*-xylose be optically active?*

D-xylose	the glycaric acid
CHO	CO_2H
H—C—OH	H—C—OH
HO—C—H	HO—C—H
H—C—OH	H—C—OH
CH_2OH	CO_2H

No. The glycaric acid is symmetrical. The mirror image is superimposable upon the original by simply rotating the mirror image in the plane of the paper.

mirror image of the glycaric acid	mirror image of the glycaric acid rotated 180° in the plane of the page
CO_2H	CO_2H
HO—C—H	H—C—OH
H—C—OH	HO—C—H
HO—C—H	H—C—OH
CO_2H	CO_2H

25.6 OSAZONE FORMATION

Phenylhydrazine reacts with α-hydroxyketones to yield products in which 2 moles of the hydrazine are incorporated. One mole reacts with the carbonyl group and one with the α-hydroxy group. A third mole of the hydrazine is converted into aniline and ammonia. Aldoses and ketoses, like fructose, react with phenylhydrazine to yield products known as *osazones*.

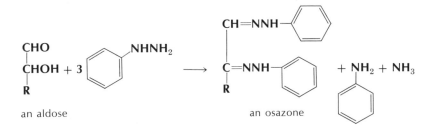

$$\underset{\text{an aldose}}{\text{CHO—CHOH—R}} + 3\ C_6H_5NHNH_2 \longrightarrow \underset{\text{an osazone}}{\text{CH=NNH}C_6H_5\text{—C(=NNH}C_6H_5\text{)—R}} + C_6H_5NH_2 + NH_3$$

Osazone formation destroys the asymmetric center adjacent to the carbonyl group. The configuration of the remaining portion of the molecule is unaffected.

A pair of diastereomeric aldoses which differ only in configuration at the center adjacent to the carbonyl group are called *epimers.* It follows that glucose and mannose are epimeric sugars. Epimeric sugars will yield the same osazone since the sole difference between the two is eliminated in the osazone reaction. This fact is illustrated for the reaction of D-glucose and D-mannose:

```
     CHO                                HC=NNH—C6H5                      CHO
     |                                  |                                |
  H—C—OH                                C=NNH—C6H5                   HO—C—H
     |                                  |                                |
 HO—C—H     --C6H5NHNH2-->          HO—C—H      <--C6H5NHNH2--     HO—C—H
     |                                  |                                |
  H—C—OH                             H—C—OH                          H—C—OH
     |                                  |                                |
  H—C—OH                             H—C—OH                          H—C—OH
     |                                  |                                |
     CH2OH                              CH2OH                            CH2OH
  D-glucose                                                          D-mannose
```

EXAMPLE 25.10

Which aldohexose will yield the same osazone as D-talose?

The structure of D-talose is given in Figure 25.3. A search of this collection of structures reveals that the compound that contains the same configuration at all carbon centers except for the one adjacent to the carbonyl group is D-galactose.

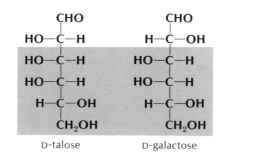

```
      CHO                 CHO
      |                   |
  HO—C—H               H—C—OH
      |                   |
  HO—C—H              HO—C—H
      |                   |
  HO—C—H              HO—C—H
      |                   |
   H—C—OH              H—C—OH
      |                   |
      CH2OH               CH2OH
   D-talose           D-galactose
```

25.7
ESTABLISHMENT OF CONFIGURATION

Up to this point the structures of the sugars presented have been accepted on faith. However, it was necessary to establish them rigorously at some time in the past. The exact methods employed involve logical reasoning and putting together of experimental facts. A few examples of the interrelationships used in such reasoning should suffice to illustrate the general principles employed.

The four D-aldopentoses can be oxidized separately with nitric acid to yield the corresponding glycaric acids. D-Arabinose and D-lyxose give optically active substances, whereas D-ribose and D-xylose yield optically inactive (*meso*) compounds. Therefore, the hydroxyl groups of D-ribose and D-xylose must be symmetrically placed. The two possible arrangements within the D-glycaric acids are:

$$\begin{array}{c} CO_2H \\ H-C-OH \\ H-C-OH \\ H-C-OH \\ CO_2H \end{array} \qquad \begin{array}{c} CO_2H \\ H-C-OH \\ HO-C-H \\ H-C-OH \\ CO_2H \end{array}$$

Therefore, D-ribose and D-xylose must have one of the following structures:

$$\begin{array}{c} CHO \\ H-C-OH \\ H-C-OH \\ H-C-OH \\ CH_2OH \end{array} \qquad \begin{array}{c} CHO \\ H-C-OH \\ HO-C-H \\ H-C-OH \\ CH_2OH \end{array}$$

These two possibilities must be distinguished in some way.

Ruff degradation of D-ribose yields an aldotetrose that upon oxidation yields an optically inactive glycaric acid. Since the glycaric acid must be *meso*-tartaric acid the two hydroxyl groups in the aldotetrose must both be on the right-hand side in the projection formula. Therefore, the hydroxyl groups on the carbon atoms 3 and 4 of the D-ribose molecule also must be on the right-hand side in the projection formula. Only one of the two previously suggested structures for D-ribose corresponds to this arrangement.

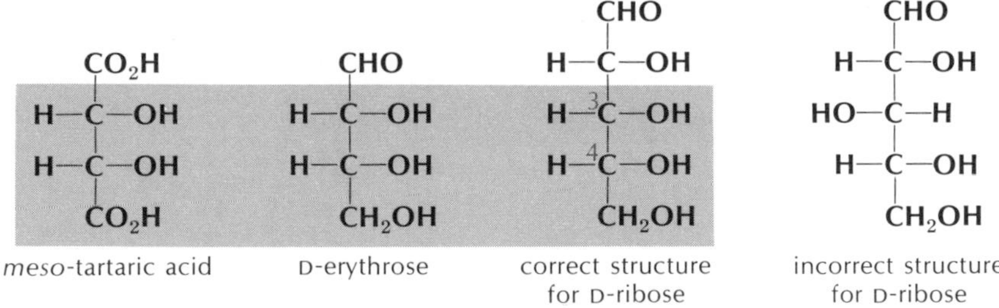

meso-tartaric acid — D-erythrose — correct structure for D-ribose — incorrect structure for D-ribose

Since it has been determined that D-ribose or D-xylose must have one or the other of two structures, the assignment of structure to D-ribose automatically leads to the assignment of the structure of D-xylose.

The structures of the two D-aldopentoses that yield optically active glycaric acids must be one of the following:

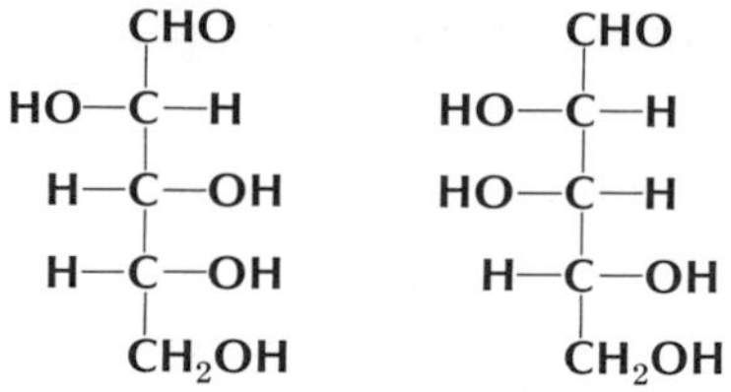

```
      CHO                CHO
       |                  |
  HO—C—H           HO—C—H
       |                  |
   H—C—OH          HO—C—H
       |                  |
   H—C—OH           H—C—OH
       |                  |
      CH2OH              CH2OH
```

The structure of D-arabinose can be assigned by means of the osazone reaction. The osazone of D-arabinose is identical to the osazone of D-ribose, whose structure is known. Therefore, the arrangement of the hydroxyl groups in D-arabinose and D-ribose must be identical except for the one adjacent to the carbonyl group. The only choice for the structure of D-arabinose is the one in which the configuration at the number 2 carbon atom is opposite that of D-ribose.

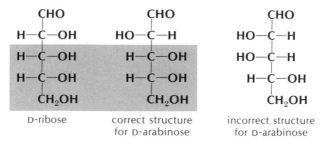

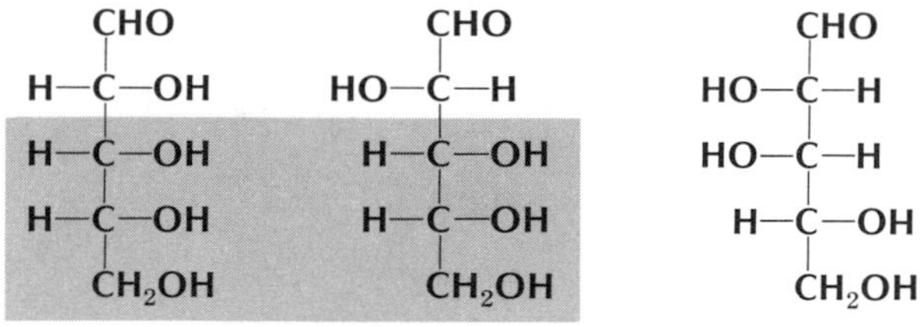

```
     CHO              CHO                 CHO
      |                |                   |
  H—C—OH          HO—C—H             HO—C—H
      |                |                   |
  H—C—OH           H—C—OH            HO—C—H
      |                |                   |
  H—C—OH           H—C—OH             H—C—OH
      |                |                   |
     CH2OH            CH2OH               CH2OH
```

D-ribose | correct structure for D-arabinose | incorrect structure for D-arabinose

The structure of D-lyxose follows from the assignment of structure of D-arabinose because the two compounds had to have one or the other of two structures suggested on the basis of the oxidation to an optically active glycaric acid. The structure of D-lyxose can be confirmed as correct by the fact that it yields the same osazone as D-xylose.

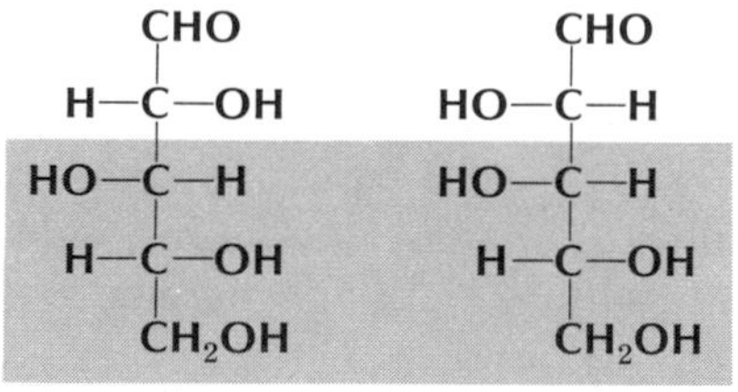

```
      CHO              CHO
       |                |
   H—C—OH          HO—C—H
       |                |
  HO—C—H           HO—C—H
       |                |
   H—C—OH           H—C—OH
       |                |
      CH2OH            CH2OH
```

D-xylose | correct structure for D-lyxose

25.8 GLYCOSIDES

STRUCTURE

The cyclic hemiacetals of sugars can be converted into acetals (Section 20.5) specifically called *glycosides* by treatment with alcohol and acid. This reaction is shown below for α-D-glucopyranose and methyl alcohol to yield an α-D-methyl glycoside. The β isomer reacts in a similar manner to yield a β-D-methyl glycoside.

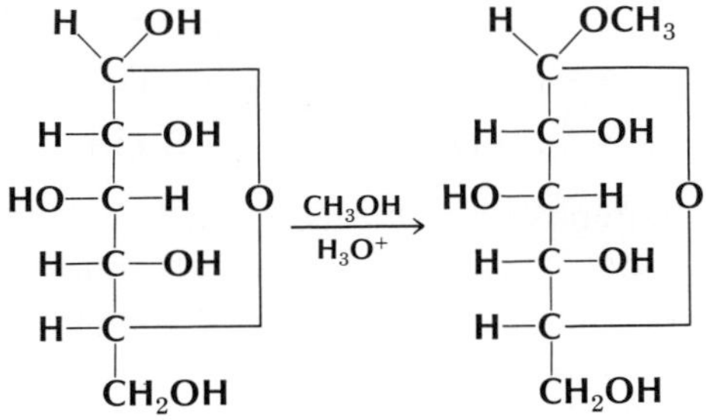

H OH
C
H—C—OH
HO—C—H O
H—C—OH
H—C
CH_2OH
α-D-glucopyranose

$\xrightarrow[H_3O^+]{CH_3OH}$

H OCH_3
C
H—C—OH
HO—C—H O
H—C—OH
H—C
CH_2OH
α-D-methyl glycoside

Some sugars such as glucose and fructose occur free in nature. The majority of sugars are found bound as glycosides to a nonsugar fragment such as a steroid. Nucleic acids (Chapter 27) are glycosides of the sugars ribose and deoxyribose.

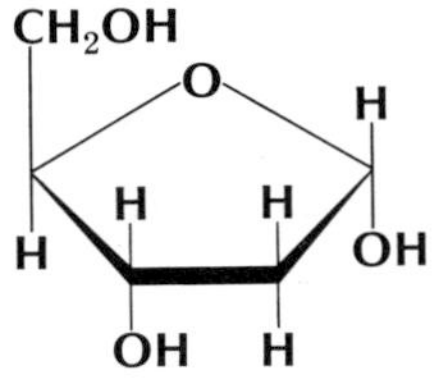

CH_2OH, O, H, H, H, H, OH, OH, H

2-deoxyribose

Glycosides in which the fragment alcohol is another sugar are disaccharides and higher saccharides.

DISACCHARIDES

Disaccharides are glycosides formed from two monosaccharides, one acting as the hemiacetal and the other as the alcohol. Hydrolysis of a disaccharide in the presence of an acid or the proper enzyme catalyst yields the component monosaccharides. The most common bonding in disaccharides is from the carbon atom number 1 of the hemiacetal to the carbon atom number 4 of the monosaccharide serving as the alcohol. Other types of linkages also are possible.

Maltose consists of α-D-glucose serving as the hemiacetal bound to another molecule of D-glucose whose hemiacetal center may be α or β:

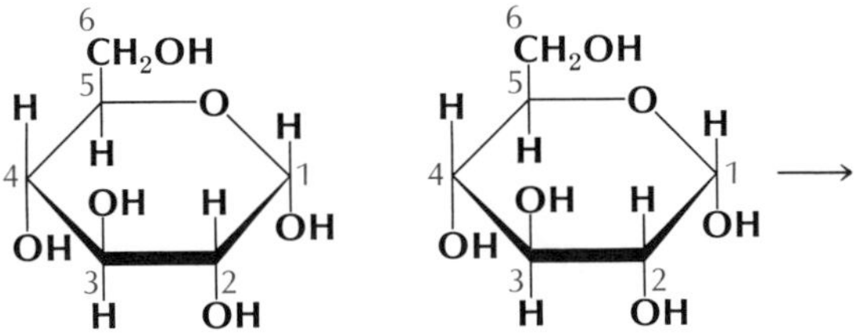

6 CH_2OH, 5, O, H, H, H, 4, 1, OH, H, OH, OH, 3, 2, H, OH

α-D-glucose

6 CH_2OH, 5, O, H, H, H, 4, 1, OH, H, OH, OH, 3, 2, H, OH

α-D-glucose →

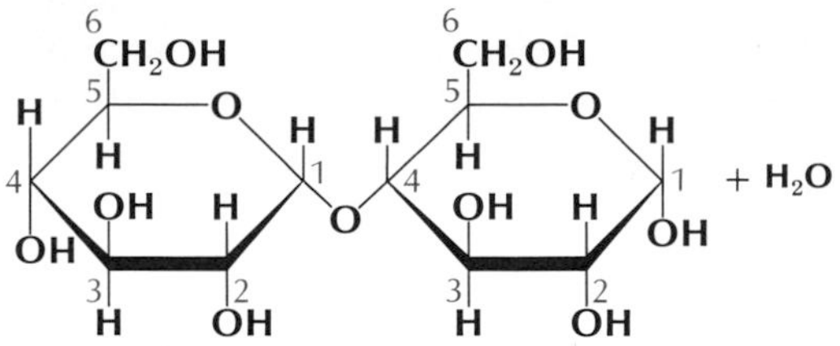

α-maltose

Mutarotation of α-maltose by reversal of configuration of the hemiacetal center would yield a mixture of this compound and the β-maltose. Both compounds are called maltose because they both contain glucose joined in a characteristic way at the acetal carbon. Maltose is said to contain an α linkage because of the α configuration at the acetal carbon of the glycoside. Maltose is produced from the polysaccharide starch by the action of the enzyme diastase of malt. It is then converted to glucose by yeast and eventually fermented to ethyl alcohol. When the β form of glucose forms a disaccharide with another mole of glucose, cellobiose is formed. The second mole of glucose may be either α or β at the hemiacetal center. The structure of β-cellobiose, which contains a β linkage at the acetal carbon and has the β configuration at the hemiacetal carbon, is shown below:

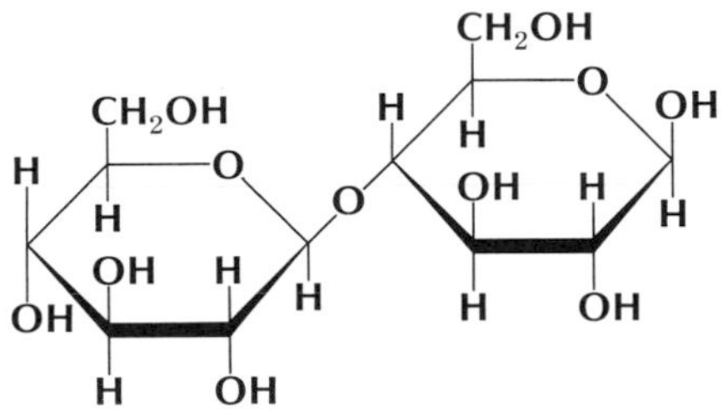

β-cellobiose

The isomeric α-cellobiose differs in configuration only at the hemiacetal carbon. Both isomeric forms are called cellobiose because they consist of glucose units joined by a β linkage. Cellobiose can be obtained by the hydrolysis of cellulose.

The sugar lactose is present to the extent of 5–8 percent in human milk and 4–6 percent in cow's milk. It is a disaccharide of β-D-galactose and a glucose unit that may be either α or β.

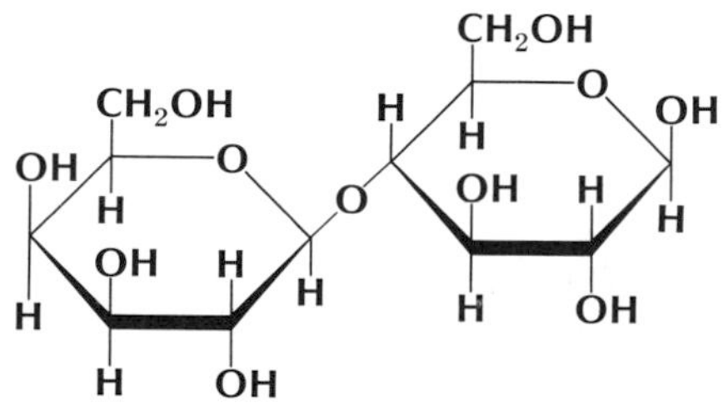

β-lactose

Sucrose is contained in sugar cane and sugar beet, each of which furnishes approximately one-half of the world's sugar supply. Its principal use is for food.

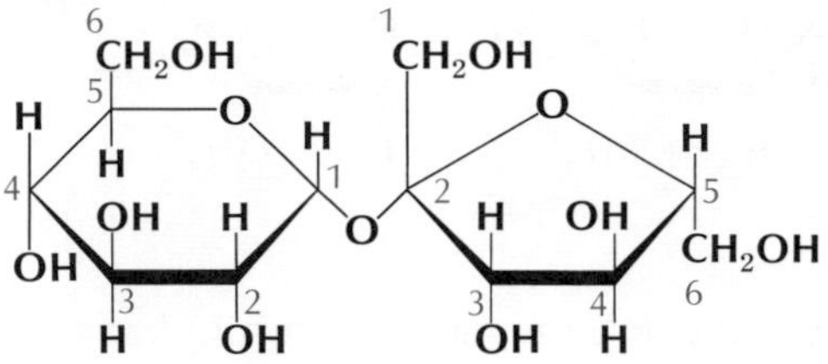

sucrose

Sucrose consists of glucose and fructose. It does not reduce Tollens' reagent, nor does it form an osazone. All of the other dissaccharides listed above do reduce Tollens' reagent and give osazones because of the potential aldehyde functional group that is in equilibrium with the hemiacetal. The units in sucrose are joined in such a way as to eliminate the possibility of the reagents reacting with a free aldehyde or α-hydroxyketone group. This linkage exists between the carbon atom number 1 of glucose and the carbon atom number 2 of fructose. Therefore, the molecule contains two acetal centers and cannot give the test that a hemiacetal does by simple reversal to yield an aldehyde.

POLYSACCHARIDES

The exact number of saccharide units in naturally occurring polysaccharides from sources such as plants is difficult to determine. The very process of separation may partially hydrolyze them to lower molecular mass polysaccharides or cause further polymerization to yield higher molecular mass material. In spite of the experimental difficulties pure samples of polysaccharides with molecular masses of several thousand to approximately 1 million have been obtained from plant sources. The molecular mass of the polysaccharides is probably a specific characteristic of the source from which they are obtained.

There is a fundamental difference in the reactivity of polysaccharides, depending on whether they are α- or β-linked. The α-linked polymers can be more easily digested by most animals than the β-linked polymers. *Starches* are α-linked polymers of glucose (Figure 25.8) and serve as the major source of food for some animals. *Celluloses* (Figure 25.8), which are β-linked polymers of glucose, can be utilized by cattle and other herbivores for food. However, this can be accomplished only because of the presence of microorganisms in their digestive tracts that can hydrolyze the cellulose to glucose. This difference in reactivity illustrates the specificity of enzymes (Section 9.2).

Starch is available in potatoes, rice, wheat, and cereal grains. The exact molecular mass and structure of the starch is a function of its source. Two fractions can be isolated from starch, amylose and amylopectin. Amylose is a linear polymer, as depicted in Figure 25.8. Amylopectin contains

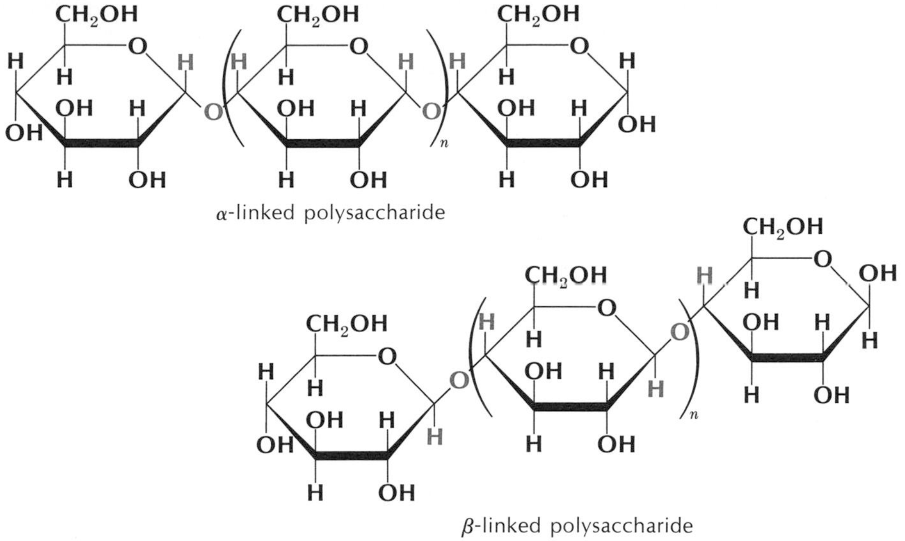

Figure 25.8
Structures of starch and cellulose.

branches of glucose units to form a series of connected chains. The branched chain is formed when the hydroxyl group on carbon atom number 6 (the primary hydroxyl group) of one linear polymer bonds in acetal formation with the hemiacetal end of another linear polymer.

The most abundant organic molecule is cellulose, which accounts for approximately half of all of the carbon atoms contained in the plant kingdom. Unlike starch, which is soft and has no structural stability, cellulose is tough and forms the support material for plants.

Suggested further readings

Darmstaedter, L., and Oesper, R. E., "Emil Fischer," *J. Chem. Educ.,* **5,** 37 (1928).

Frohwein, Y. Z., "A Simplified Proof of the Constitution and the Configuration of D-Glucose," *J. Chem. Educ.,* **46,** 55 (1969).

Heuser, E., "The High Points in the Development of Cellulose Chemistry Since 1876," *J. Chem. Educ.* **29,** 449 (1952).

Horecker, B. L., "Pathways of Carbohydrate Metabolism and Their Physiological Significance," *J. Chem. Educ.,* **42,** 244 (1965).

Hudson, C. S., "The Basic Work of Fischer and van't Hoff in Carbohydrate Chemistry," *J. Chem. Educ.,* **30,** 120 (1953).

Hudson, C. S., "Emil Fischer's Discovery of the Configuration of Glucose," *J. Chem. Educ.* **18,** 353 (1941).

Hurd, C. D., "Hemiacetals, Aldals and Hemialdals," *J. Chem. Educ.,* **43,** 527 (1966).

Scloch, T. J. "Cellulose, Glycogen and Starch," *J. Chem. Educ.,* **18,** 353 (1941).

aldose
carbohydrate
cellobiose
cellulose
deoxyribose
epimers
fructose
glucose
glycaric acid
glyconic acid
glycoside
Haworth projection
hexose
ketose
Kiliani synthesis
lactose
maltose
monosaccharide
mutarotation
oligosaccharide
osazone
pentose
polysaccharide
ribose
Ruff degradation
starch
sucrose

Questions and problems

1. What is meant by D(−)-fructose?
2. Draw all the isomeric ketotetroses and ketopentoses in open chain form. Which compounds are optically active?
3. Draw L-glucose and L-mannose in open chain form.
4. Which of the D-aldohexoses will yield optically inactive glycaric acids when oxidized?
5. Which aldopentose is needed for the Kiliani synthesis of D-glucose? Write the steps for this transformation.
6. Which aldohexoses will be produced by the extension of the carbon chain of D-xylose? Write the steps for this transformation.
7. What aldopentose will be produced by the Ruff degradation of D-galactose? Write the steps for the transformation.
8. Which aldohexose will yield the same osazone as that derived from D-gulose?
9. Which aldohexoses upon Ruff degradation followed by oxidation with nitric acid will yield optically inactive glycaric acids?
10. Why do both maltose and lactose react with Tollens' reagent whereas sucrose does not?
11. Can methyl-α-D-glucopyranoside reduce Tollens' reagent? Explain.
12. Draw a trisaccharide containing α-D-glucose units.
13. There are four D-aldoheptoses which yield optically inactive glycaric acids when oxidized. Draw them in open chain form.
14. Which D-aldoheptoses will yield optically inactive glycaric acids when they are first degraded by one carbon atom followed by oxidation with nitric acid? Draw the structures.

26
LIPIDS

In the previous chapters two of the major components of living systems, the amino acids and carbohydrates, were discussed. The last major components of living organisms to be presented are the *lipids*. Lipids cannot be represented as a class of compounds containing similar functional groups. They are a conglomerate of relatively nonpolar compounds whose solubilities differ from those of the polar amino acids and carbohydrates and which are relatively low molecular mass materials.

Since lipids are actually several different types of compounds they are usually considered by subclassifications in which common functional groups are found. These subclassifications are triglycerides, phosphatides, waxes, steroids, and terpenes.

26.1 TRIGLYCERIDES

STRUCTURE, PROPERTIES, AND SOURCES

Triglycerides are esters of glycerol and high molecular mass carboxylic acids. In Chapter 21 such compounds were described as fats and oils that differ only in the degree of unsaturation in the carboxylic acid. The approximate compositions of some common fats and oils are given in Table 26.1.

Animals accumulate fat when their intake of food exceeds their energy output requirements. Vital organs such as kidneys are enclosed in a thick layer of fat, which helps to prevent damage when they are subjected to a blow. A subcutaneous layer of fat helps insulate the animal against heat loss. Although plants do not generally store fats and oils for energy requirements, some (such as peanuts and olives) produce triglycerides in abundance.

Triglycerides can be separated from natural sources by simple physical methods. Oils from vegetables such as olives, corn, or soybeans can be obtained by pressing them. Animal fats can be separated from other tissues by melting the fat and removing the liquid in a process much like cooking bacon.

The exact composition of triglycerides varies within certain limits, as indicated in Table 26.1. The conditions in which vegetables are grown affect the degree of unsaturation: for example, linseed oil obtained from flaxseed grown in warm climates may contain up to twice the degree of unsaturation found in oil obtained from seed grown in cold climates. Similarly, the composition of lard from hogs depends on their diet; the fat of corn-fed hogs has less unsaturation than that of peanut-fed hogs.

Purified triglycerides are tasteless, colorless, and odorless. The characteristic tastes of various vegetable oils are a result of impurities. When triglycerides are left at warm temperatures in the presence of air they develop odors and become rancid. Hydrolysis to yield the free acids is one of the processes that accounts for the odor. Oxidation of the double bonds

Table 26.1

Acid components of triglycerides

	component acids (percent)[a]					
triglyceride	*myristic*	*palmitic*	*stearic*	*oleic*	*linoleic*	*linolenic*
fats						
butter	7-10	24-26	10-13	30-40	4-5	
lard	1-2	28-30	12-18	40-50	6-7	
tallow	3-6	24-32	20-25	37-43	2-3	
edible oils						
corn oil	1-2	8-12	2-5	19-49	34-62	
cottonseed oil	0-2	20-25	1-2	23-35	40-50	
olive oil		9-10	2-3	83-84	3-5	
peanut oil		8-9	2-3	50-65	20-30	
safflower oil		6-7	2-3	12-14	75-80	0.5-0.15
soybean oil		6-10	2-5	20-30	50-60	5-11

[a] *Totals less than 100 percent indicate the presence of lower or higher acids in small amounts.*

by oxygen also occurs to yield aldehydic fragments, which in turn are oxidized to the odoriferous low molecular mass acids.

HYDROGENATION

Vegetable oils are liquids at room temperature, whereas animal fats are solids. When the double bonds in molecules of vegetable oils are hydrogenated the resultant product is a solid at room temperature. Completely hydrogenated vegetable oil is very brittle, like tallow. Commercial producers of Crisco, Spry, and Fluffo control the extent of the hydrogenation such that some unsaturation remains. The product is a mixture of saturated and unsaturated materials whose iodine value (Section 21.9) is approximately 50.

Oleomargarine is obtained by the partial hydrogenation of oils from corn, cottonseed, peanut, and soybean. The final product is an emulsion containing approximately 15 percent by weight of milk. A yellow vegetable dye and vitamins A and D also are added. The characteristic flavor of butter is due in part to acetoin and diacetyl and, therefore, these substances are added to many products.

$$CH_3-\underset{\underset{H}{|}}{\overset{\overset{OH}{|}}{C}}-\overset{\overset{O}{\|}}{C}-CH_3 \qquad CH_3-\overset{\overset{O}{\|}}{C}-\overset{\overset{O}{\|}}{C}-CH_3$$

acetoin　　　diacetyl

The relationship between human consumption of saturated fats and arterial disease has been the object of extensive medical research. Unsaturated fats are believed to be beneficial in preventing arterial deposits. Safflower oil, because of its high content of unsaturated material, is now a popular product.

RANCIDITY

Butter in warm, moist surroundings turns rancid; that is, it acquires a disagreeable odor. The odor results from products formed by hydrolysis and oxidation of triglycerides. In addition to the high molecular mass acids listed in Table 26.1, triglycerides contain lower molecular mass acids such as caproic, caprylic, and capric. Hydrolysis of the triglycerides then releases these relatively volatile and odorous compounds.

Oxidation of the centers of unsaturation by oxygen of the air is another cause of rancidity. The oxidation process leads to the cleavage of the unsaturated compounds into short chain aldehydes and carboxylic acids, both classes having unpleasant odors.

HARDENING OF OILS

Some oils such as linseed oil react with oxygen of the air to form hard, tough films and are called *drying oils.* They find use in the liquid medium of paints and varnishes. Linoleum and oilcloth contain drying oils in addition to some fillers, which when dried make durable surface coverings.

The drying or hardening process is the result of the formation of *hydroperoxides* (—O—O—H) at carbon atoms flanked on both sides by double bonds, as illustrated for linoleic acid:

$$CH_3-(CH_2)_4-CH=CH-CH_2-CH=CH-(CH_2)_7-CO_2H$$

linoleic acid

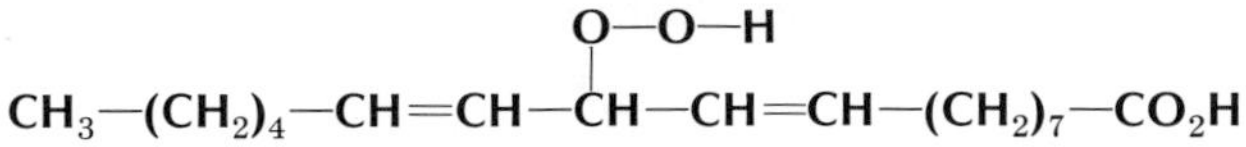

$$CH_3-(CH_2)_4-CH=CH-\underset{}{\overset{O-O-H}{\overset{|}{C}H}}-CH=CH-(CH_2)_7-CO_2H$$

hydroperoxide of linoleic acid

In subsequent steps the hydroperoxide reacts with neighboring molecules of triglycerides containing linoleic acid. As a result of this reaction a *peroxide* (—O—O—) bridge is formed:

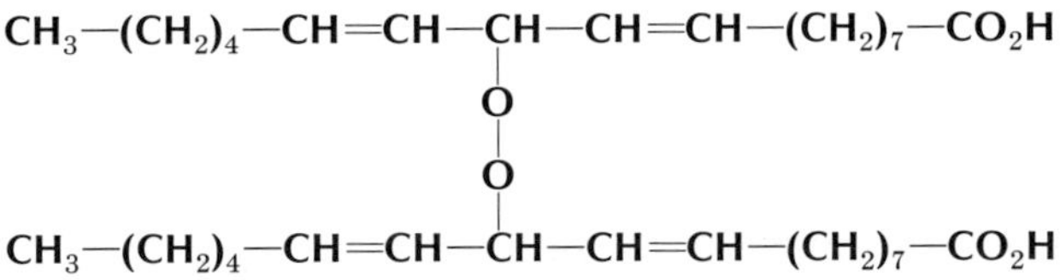

$$\begin{array}{c} CH_3-(CH_2)_4-CH=CH-CH-CH=CH-(CH_2)_7-CO_2H \\ | \\ O \\ | \\ O \\ | \\ CH_3-(CH_2)_4-CH=CH-CH-CH=CH-(CH_2)_7-CO_2H \end{array}$$

peroxide bridge between linoleic acid molecules

Continued reaction of any of the three sites within the triglyceride-containing linoleic acid leads to a network of peroxide bridges in a polymeric substance.

26.2 PHOSPHATIDES

Phosphatides are either derivatives of esters of glycerol and phosphoric acid or sphingosine, which is a compound combining the amino acid serine and the aldehyde corresponding to palmitic acid.

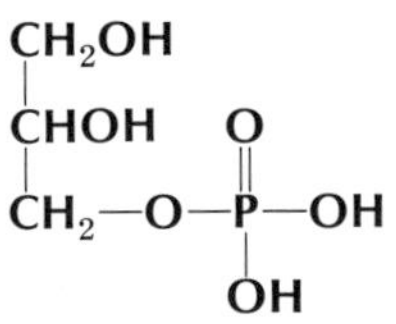

$$\begin{array}{l} CH_2OH \\ | \\ CHOH \quad\quad O \\ | \quad\quad\quad\quad \| \\ CH_2-O-P-OH \\ \quad\quad\quad\quad\;\; | \\ \quad\quad\quad\quad OH \end{array}$$

glycerol phosphate

$$\begin{array}{l} \quad\quad\;\; H \\ \quad\quad\;\; | \\ HO-C-CH=CH(CH_2)_{13}CH_3 \\ \quad\quad\;\; | \\ NH_2-CH \\ \quad\quad\;\; | \\ \quad\quad CH_2OH \end{array}$$

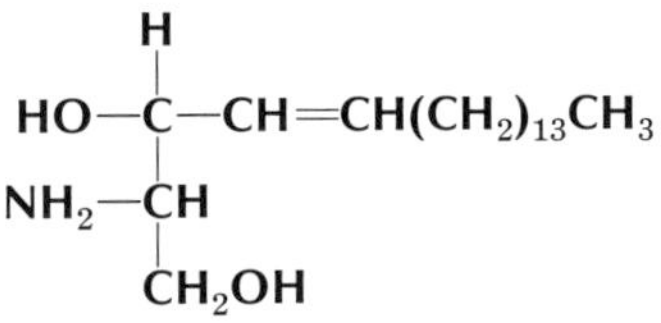

sphingosine

Lecithins, which are contained in egg yolks and liver, are derivatives of glycerol phosphate and the quaternary methyl ammonium derivative of β-hydroxyethylamine (Section 22.4).

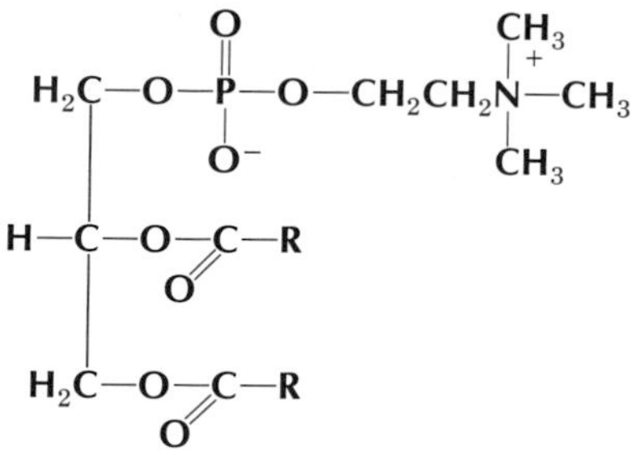

$$
\begin{array}{l}
\qquad\qquad\quad O \qquad\qquad\qquad\quad CH_3 \\
\qquad\qquad\quad \| \qquad\qquad\qquad\quad\ |\,+ \\
H_2C-O-P-O-CH_2CH_2N-CH_3 \\
\ \ |\qquad\qquad\ | \qquad\qquad\qquad\quad\ | \\
\ \ |\qquad\qquad O^- \qquad\qquad\quad\ \ CH_3 \\
H-C-O-C(=O)-R \\
\ \ | \\
H_2C-O-C(=O)-R
\end{array}
$$

a lecithin

The choline from lecithins provides the mechanism for transmission of nerve impulses. Acetylcholine is formed at a nerve cell that reacts with a receptor protein in the next cell. In this manner the nerve impulse is sent along. However, in order for the cell to be prepared for the next message the acetylcholine must be hydrolyzed. Cholinesterase catalyzes this reaction, which can occur in millionths of a second. The nerve gases act by deactivating cholinesterase and thus preventing the cells from returning to a state of preparedness for the next nerve impulse.

Cephalins, which are derivatives of glycerol phosphate, are part of the brain and spinal tissue. In cephalins the nitrogen is not quaternized as it is in lecithin. In living systems at a pH near 7 the cephalins are protonated.

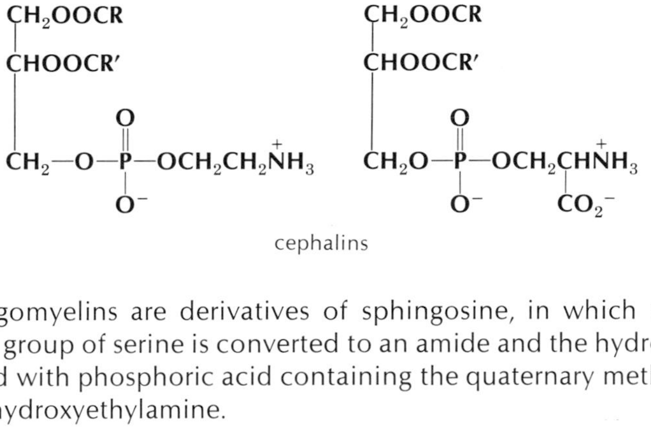

$$
\begin{array}{ll}
CH_2OOCR & CH_2OOCR \\
| & | \\
CHOOCR' & CHOOCR' \\
| & | \\
|\qquad\quad O & |\qquad\quad O \\
|\qquad\quad \| & |\qquad\quad \| \\
CH_2-O-P-OCH_2CH_2\overset{+}{N}H_3 & CH_2O-P-OCH_2CH\overset{+}{N}H_3 \\
\qquad\quad\ \ | & \qquad\quad\ \ |\qquad\qquad\ | \\
\qquad\quad\ O^- & \qquad\quad\ O^- \qquad\quad CO_2^-
\end{array}
$$

cephalins

Sphingomyelins are derivatives of sphingosine, in which the amine functional group of serine is converted to an amide and the hydroxyl group is esterified with phosphoric acid containing the quaternary methyl derivative of β-hydroxyethylamine.

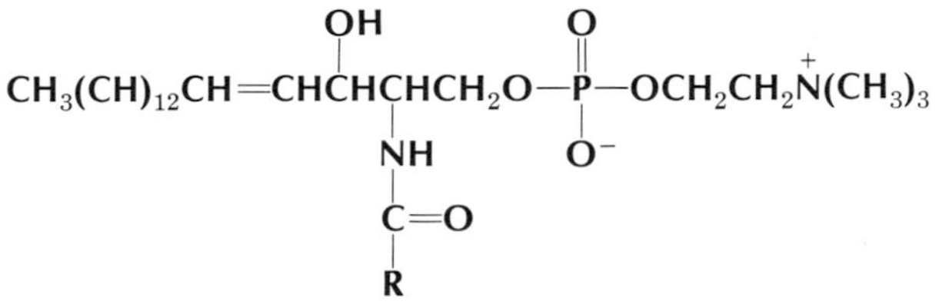

$$
\begin{array}{l}
\qquad\qquad\qquad\qquad\quad OH \qquad\qquad\quad O \\
\qquad\qquad\qquad\qquad\quad\ | \qquad\qquad\quad\ \ \| \\
CH_3(CH)_{12}CH{=}CHCHCHCH_2O-P-OCH_2CH_2\overset{+}{N}(CH_3)_3 \\
\qquad\qquad\qquad\qquad\qquad\ | \qquad\qquad\ \ | \\
\qquad\qquad\qquad\qquad\qquad NH \qquad\quad\ \ O^- \\
\qquad\qquad\qquad\qquad\qquad\ | \\
\qquad\qquad\qquad\qquad\qquad C{=}O \\
\qquad\qquad\qquad\qquad\qquad\ | \\
\qquad\qquad\qquad\qquad\qquad R
\end{array}
$$

sphingomyelin

26.3 STEROIDS

STRUCTURE

The ring system of steroids was illustrated in Chapter 16. The molecules of this class of compounds are roughly flat, and the ring substituents may be on either side of the rings.

There are profound differences in the physiological effects of steroids in the human body. Some function as vitamins and some as sex hormones. Steroids are responsible for both the stimulation of the heart and hardening of the arteries. Most steroids are known by their common name, which either reflects their action or the source of the material.

Cholesterol, which is very abundant in nerve and brain tissue, was first isolated in 1775 from gallstones. The prefix *chole-* is the Greek word for bile. The cholesterol contained in consumed food is absorbed through the intestines and forms part of the body's supply. However, cholesterol is also produced in the body from unsaturated material such as fats via the intermediacy of squalene (Section 17.8).

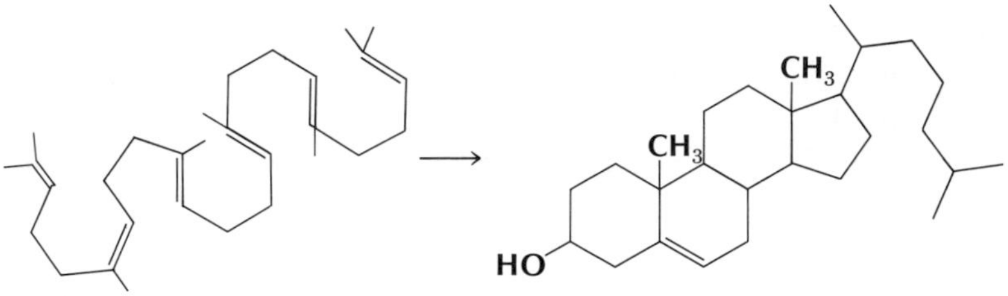

CH_3 CH_3 HO

squalene cholesterol

If the body's supply of cholesterol becomes too high, it precipitates in the gallbladder as gallstones and in the arteries, leading to restricted blood flow and high blood pressure. The German chemist Adolf Windaus devoted most of his research to the field of steroids. In 1928 he was awarded the Nobel prize for his many contributions, among which was a proposed structure for cholesterol. Ironically, the structure was incorrect and the correct one was proposed in 1932.

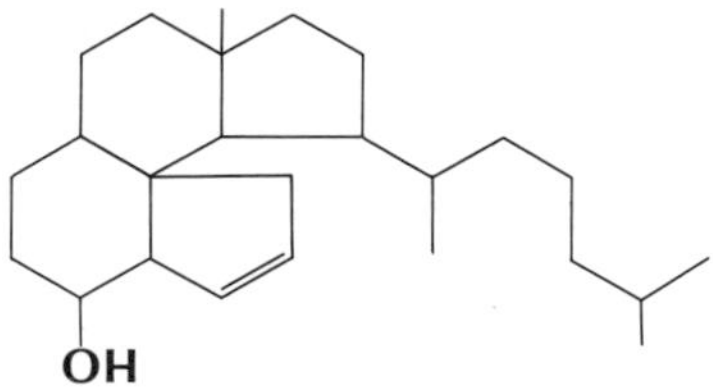

OH

Windaus structure for cholesterol

SEX HORMONES

The male sex hormone testosterone causes the development of secondary male sexual characteristics such as facial hair and deep voice. Estradiol, the female sex hormone, performs similar but characteristically different functions with the female. The oxidized form of estradiol, estrone, is responsible for the changes of estrus (cycle of ovulation).

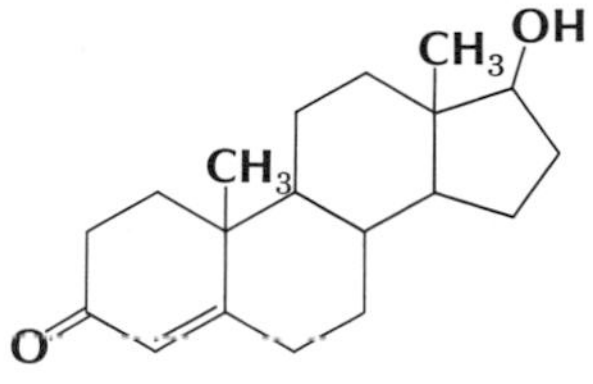

testosterone

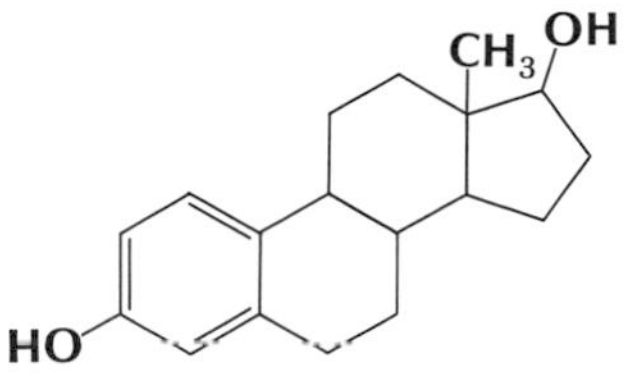

estradiol

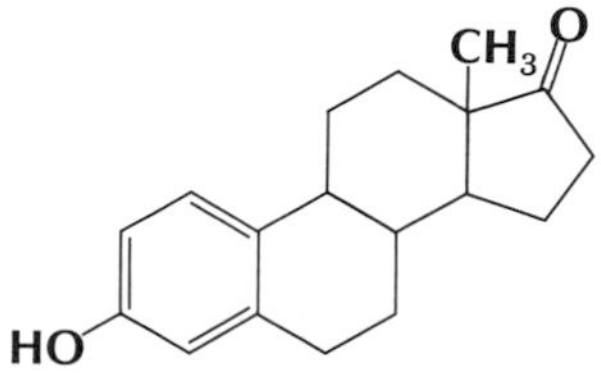

estrone

Progesterone, which aids the maintenance of pregnancy in the female and is known to inhibit the release of ova from the ovaries, is remarkably similar to the structure of testosterone.

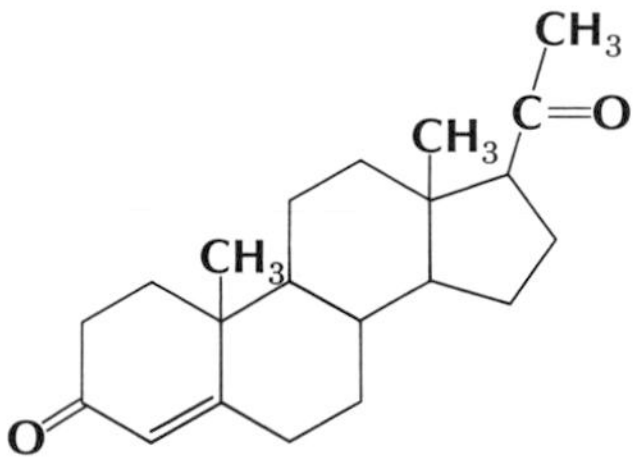

progesterone

The biological activity of sex hormones is very high, and relatively small quantities are produced in animals. To isolate these substances from animals required extensive purification. Estradiol as the diester of α-naphthoic acid was first obtained in a 12-mg quantity from the processing of 1.5 tons of hog ovaries. The related estrone initially was produced by extraction of the urine of pregnant mares in which the concentration is approximately 0.3 mg/liter. Early analytical methods for determining the biological activity of crude extracts of female sex hormones included the use of spayed female mice. The dosage necessary to reestablish the normal sexual cycle of these mice was defined as one *international mouse unit.* This unit corresponds to 0.001 mg of estrone.

The first male sex hormone isolated was androsterone, which was obtained from the extensive processing of thousands of liters of urine from steers. Testosterone was first isolated from the testis tissue of cattle. The biological activity of male sex hormones initially was determined by their effect on castrated cocks, capons. The comb and wattle of capons atrophy and almost disappear. Administration of androsterone or testosterone causes a regeneration of these secondary sex characteristics. The amount of material that when injected into each of three capons on two successive days produces an average increase of 20 percent in the area of the comb on the fourth day is defined as a *capon unit*. The capon unit corresponds to approximately 0.1 mg.

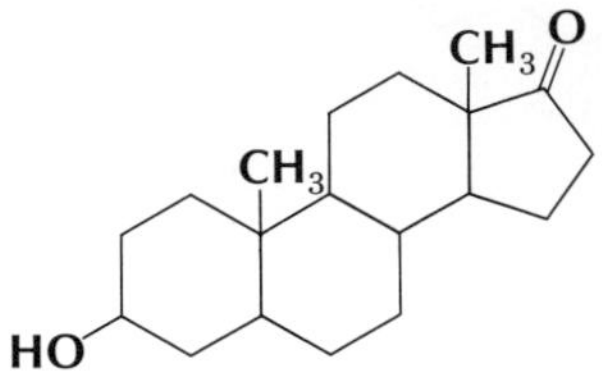

CH_3 CH_3 O HO

androsterone

ORAL CONTRACEPTIVES

The active compounds of oral contraceptives are compounds related to progesterone and estradiol. Three of the compounds now employed in pill form are norethindrone, mestranol, and norethynodrel:

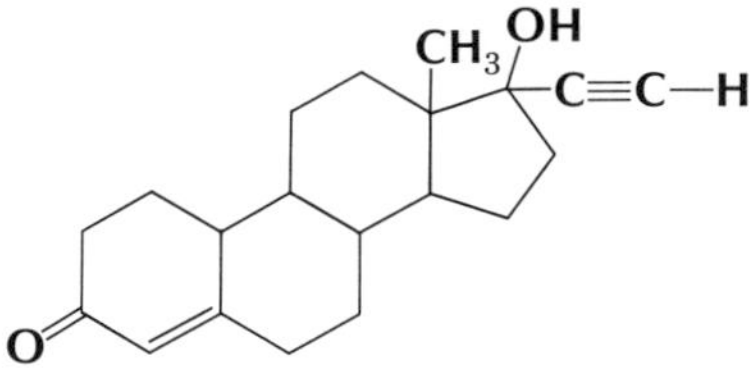

CH_3 OH —C≡C—H O

norethindrone

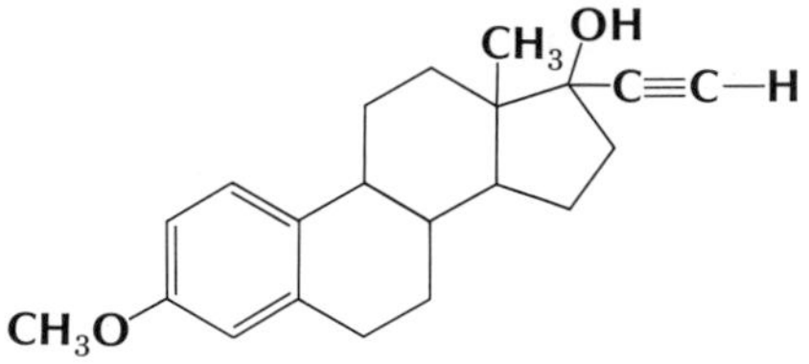

CH_3 OH —C≡C—H CH_3O

mestranol

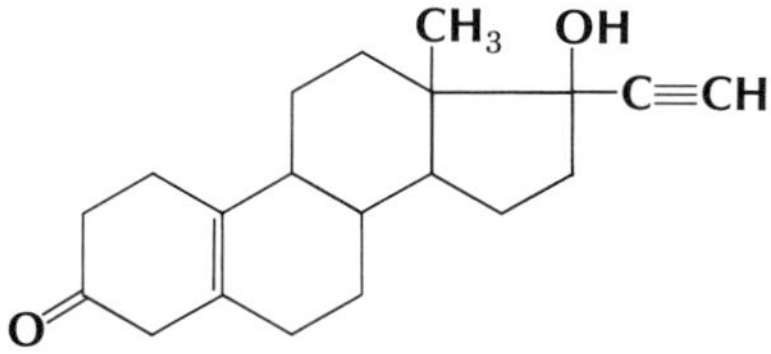

CH_3 OH —C≡CH O

norethynodrel

The compounds allow a regular menstrual cycle, but no ova are released. Estradiol and progesterone would serve the same function, but they are less effective than the synthetic compounds and also must be given by injection for proper control.

CORTICOSTEROIDS

The adrenals are two glands located in the body, one above each of the kidneys. The outer layer (adrenal cortex) produces a number of important steroid hormones called *corticosteroids*. The function of these compounds is to regulate physiological processes such as kidney function, electrolyte balance, growth, metabolism, and resistance to disease. Approximately 40 steroids have been isolated from the adrenal cortex. *Cortisone,* which is now available by synthesis, is effective in the treatment of rheumatic fever, rheumatoid arthritis, and some skin disorders.

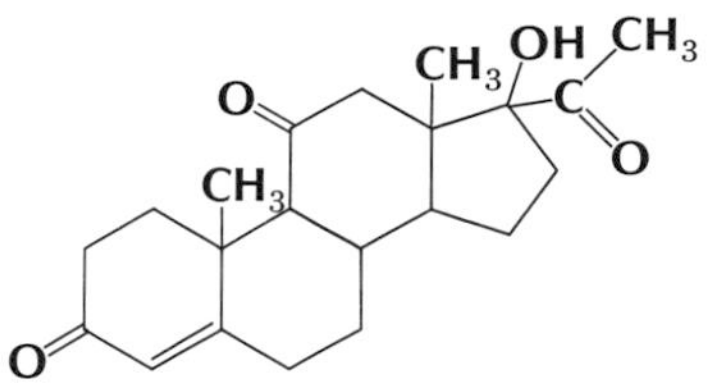

cortisone

Aldosterone is a corticosteroid that is used in the treatment of Addison's disease, which is caused by adrenocortical atrophy.

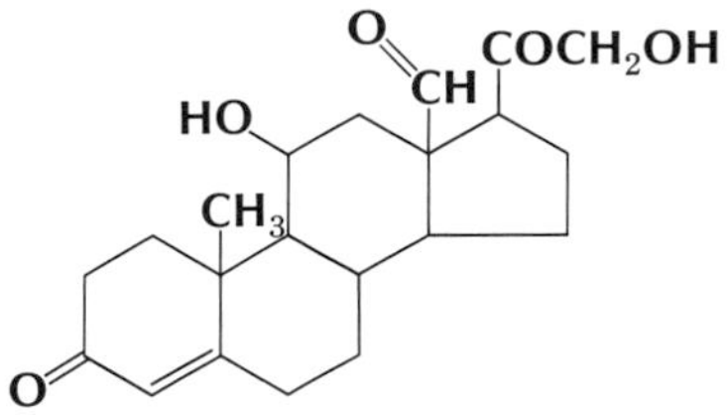

aldosterone

BILE ACIDS

The steroid bile acid, cholic acid, combines with amino acids to form substances necessary for the digestion of lipids.

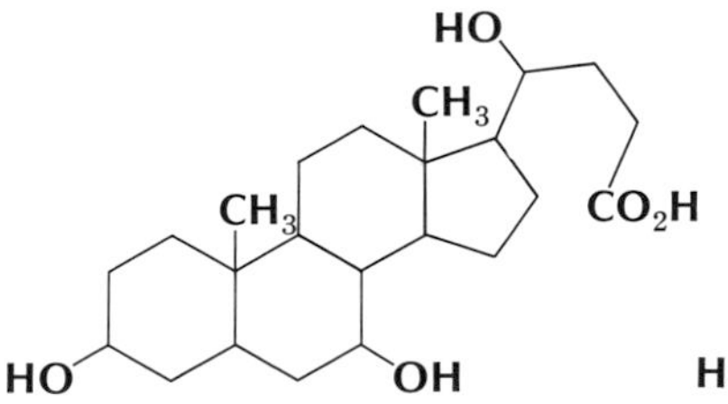

cholic acid

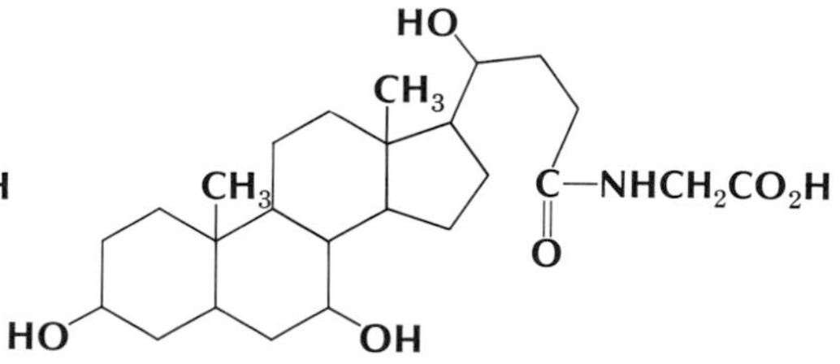

glycocholic acid

As salts the bile acids emulsify ingested fats (Chapter 28) and aid in their hydrolysis and absorption through the intestine.

STEROID SAPOGENINS

Steroid sapogenins are the alcohol portion of glycosides (Section 25.8) called *steroid saponins.* The sugar component of the saponin is usually glucose, galactose, or xylose. Hydrolysis of the saponin yields the steroid sapogenin, two examples of which are sarsasapogin and diosgenin. The sarsaparilla root is the source of sarsasapogin, and diosgenin is obtained from yams. Both plant substances are cultivated to provide starting material that can be converted into progesterone and other important steroids for medical use.

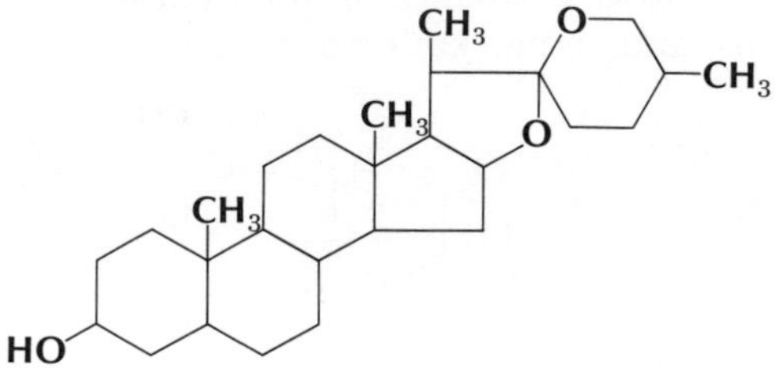

sarsasapogin

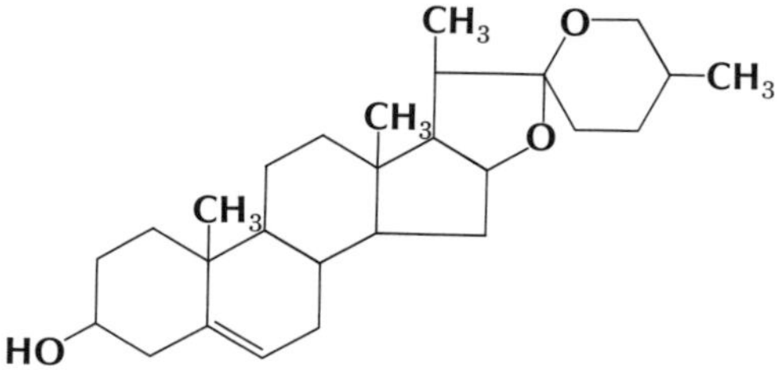

diosgenin

CARDIAC GLYCOSIDES

Cardiac glycosides have a powerful effect on the heart muscle. Digitalis, which is obtained from the seeds of purple foxglove, is used in the treatment of heart disease.

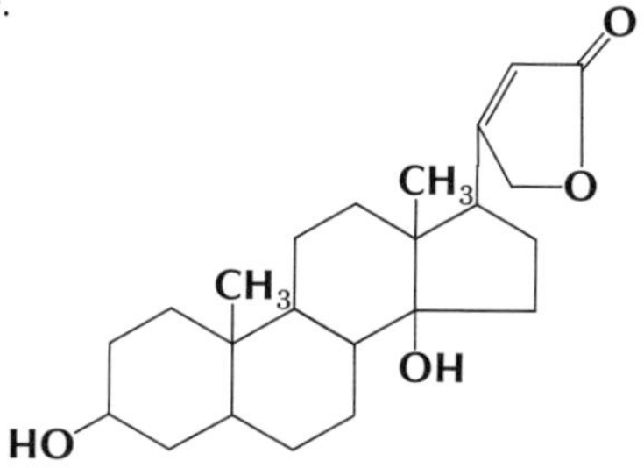

digitalis

In contrast to digitalis, strophanthin, one of the toxic components in African arrow poison, can stop the heart of a 20-g mouse within minutes at dosages of 0.00006 g.

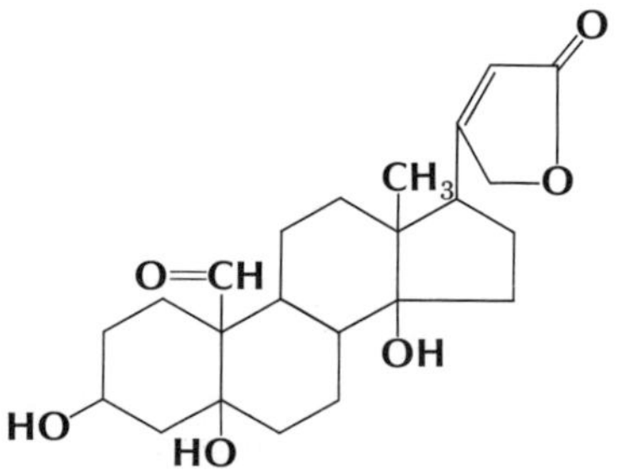

strophanthin

The red squill (sea onion) contains scilliroside, a highly toxic, specific steroid poison for rats that was used as early as the thirteenth century.

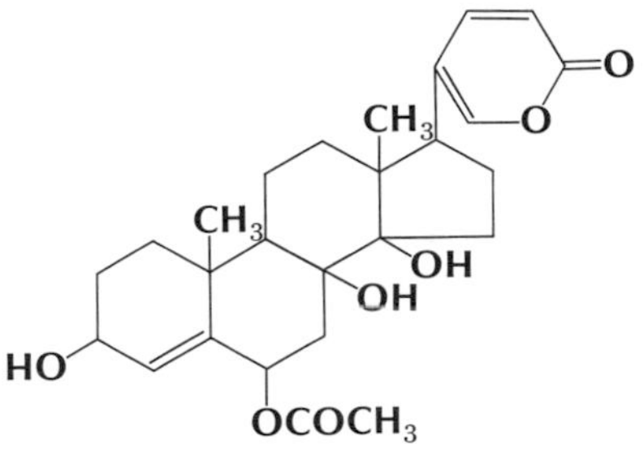

scilliroside

26.4 TERPENES

STRUCTURE

The terpenes were introduced in Chapter 17. In general they are of significance predominately in plants and are of relatively minor biochemical significance in humans. The terpenes have in common the monomeric structural unit isoprene (Section 17.8). Each terpene can be mentally divided into from two to eight isoprene units. The divisibility of terpene structures is known as the isoprene rule and was first suggested by O. Wallack in 1887 while he was at the University of Göttingen. Wallack received the Nobel prize in chemistry in 1910.

The terpene molecules in nature are classified according to the number of isoprene units they contain (Table 26.2). Classes containing five and seven isoprene units are not found in nature.

MONOTERPENES

The monoterpene hydrocarbon has the molecular formula $C_{10}H_{16}$ and is deficient six hydrogen atoms relative to the saturated hydrocarbon $C_{10}H_{22}$. This deficiency can be structurally accommodated by four structural types: (1) an open chain with three double bonds, (2) a monocyclic structure

Table 26.2
Classes of terpenes

class	*number of carbon atoms*	*number of isoprene units*
monoterpenes	10	2
sesquiterpenes	15	3
diterpenes	20	4
triterpenes	30	6
tetraterpenes	40	8

with two double bonds, (3) a bicyclic structure with one double bond, and (4) a tricyclic structure. Only the first three classes occur in nature. Examples of derivatives of each of these classes are shown below:

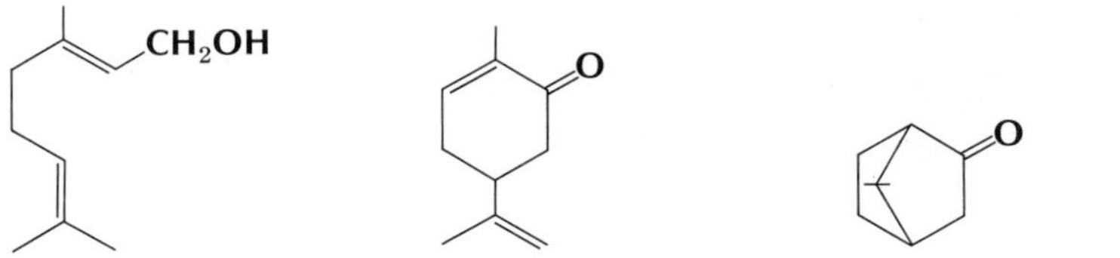

CH_2OH O O

geraniol
(Turkish geranium oil)

carvone
(oil of caraway)

camphor
(formosan camphor tree)

SESQUITERPENES

Sesquiterpenes derived from the general formula $C_{15}H_{26}$ are found in nature predominately as open chain, monocyclic, and bicyclic structures. Several of these structures are given below:

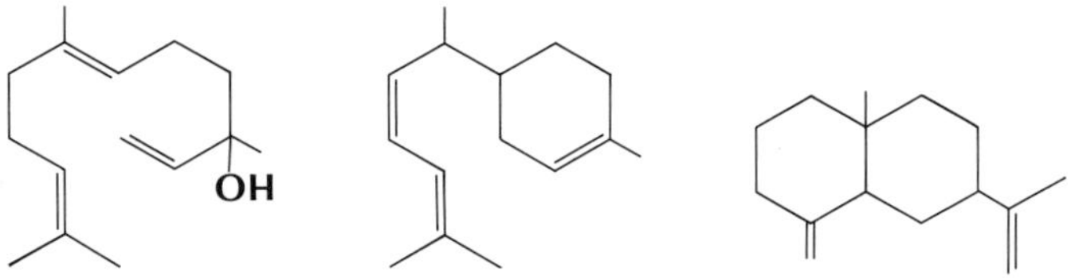

OH

nerolidol
(oil of neroli)

zingiberone
(oil of ginger)

β-selinene
(oil of celery)

DITERPENES

The diterpenes occur in nature as open chain, mono-, bi-, and tricyclic structures. The most important open chain compound is the alcohol phytol $C_{20}H_{39}OH$, which occurs as an ester in chlorophyll (Section 14.8).

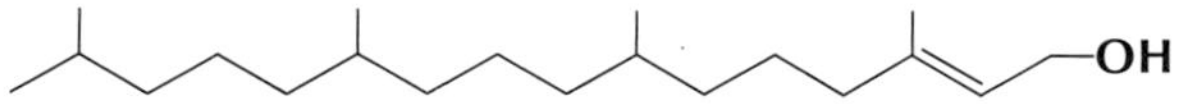

—OH

phytol

Vitamin A is a monocyclic diterpene (Section 17.8).

TRITERPENES

The open chain triterpene squalene was illustrated in Section 17.8. Its structure was confirmed by P. Karrer of the University of Zürich in 1931; he received the Nobel prize in 1937.

Lanosterol is a tetracyclic triterpene found in wool fat (Section 17.8). It can be derived by cyclization of squalene.

The reddish yellow pigments that occur in plants are tetraterpenes called *carotenes*. Lycopene is an open chain tetraterpene that occurs in ripe tomatoes and watermelon.

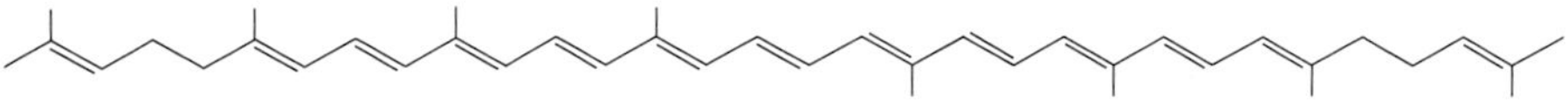

lycopene

Many oxygenated derivatives of monocyclic and bicyclic tetraterpenes called *xanthins* are known. They make up the red and yellow pigments of plants and some animals.

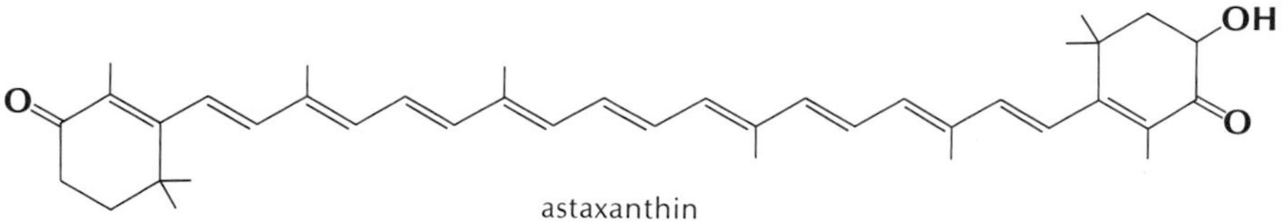

OH

O

O

astaxanthin
(lobster shells)

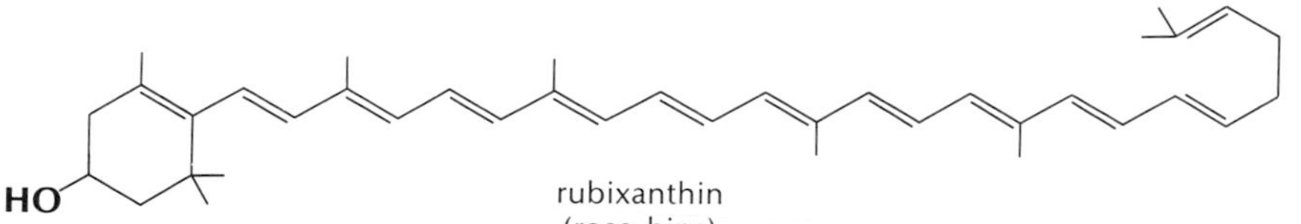

HO

rubixanthin
(rose hips)

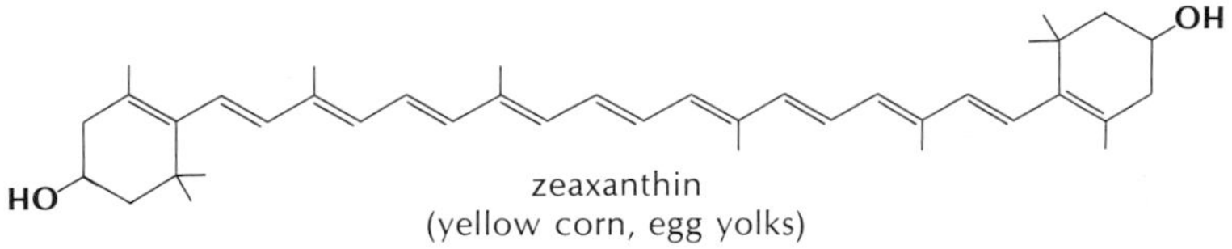

OH

HO

zeaxanthin
(yellow corn, egg yolks)

Suggested further readings

Beyler, R. E., "Some Recent Advances in the Field of Steroids," *J. Chem. Educ.* **37,** 491 (1960).

Fieser, L. F., "Steroids," *Sci. Amer.* (March 1950).

Gray, G. W., "Cortison and ACTH," *Sci. Amer.* (March 1950).

Partridge, W. S., and Schierz, E. R., "Otto Wallach: The First Organizer of the Terpenes," *J. Chem. Educ.,* **24,** 106 (1947).

Sterrett, F. S., "Nature of Essential Oils. I. Production," *J. Chem. Educ.,* **39,** 203 (1962).

Sterrett, F. S., "Nature of Essential Oils. II. Chemical Constituents, Analysis," *J. Chem. Educ.,* **39,** 246 (1962).

Terms and concepts

androsterone	lecithin
bile acid	lipid
cephalins	oil
cholesterol	oral contraceptive
cholic acid	peroxide
digitalis	phosphatide
estradiol	progesterone
estrone	steroid
fat	testosterone
hydroperoxide	triglyceride

Questions and problems

1. What products are produced from the hydrolysis of the following?
- **a.** triglycerides
- **b.** phosphatides
- **c.** waxes
- **d.** steroids
- **e.** terpenes

2. Are oils or fats more likely to become rancid?

3. How could propionic acid arise from linolenic acid?

4. How many moles of hydrogen are required to completely hydrogenate 1 mole of lycopene?

5. List the chemical reactions that testosterone and estradiol will undergo. Suggest one chemical reaction that each one will undergo and the other will not. What spectroscopic difference might be used to distinguish one sample from the other?

27

NUCLEIC ACIDS

Nucleic acids constitute less than 1 percent of the organic compounds contained in living organisms. However, these relatively minor constituents of the cell are responsible for the reproduction and the transmission of hereditary information. Nucleic acids are found both in the nuclei of cells and in the surrounding cytoplasm, with *deoxyribonucleic acid* (DNA) being found almost totally in the nucleus and *ribonucleic acid* (RNA) in both the nucleus and cytoplasm. DNA is responsible for the transfer of genetic information while RNA directs the synthesis of proteins.

27.1 PRIMARY STRUCTURE OF NUCLEIC ACIDS

Nucleic acids are high molecular mass molecules consisting of phosphoric acid, a pentose that may be either ribose or 2-deoxyribose and a variety of bases.

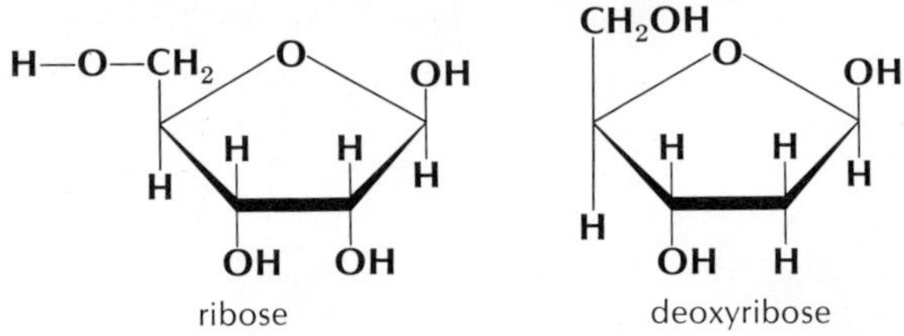

ribose deoxyribose

The most commonly encountered bases are adenine and guanine, which are purine bases, and thymine, cytosine, and uracil, which are pyrimidine bases (Chapter 22). There is always one pentose and one phosphate group for every base. The amount of adenine in DNA on a mole basis is equal to that of thymine, and the amount of guanine is equal to that of cytosine.

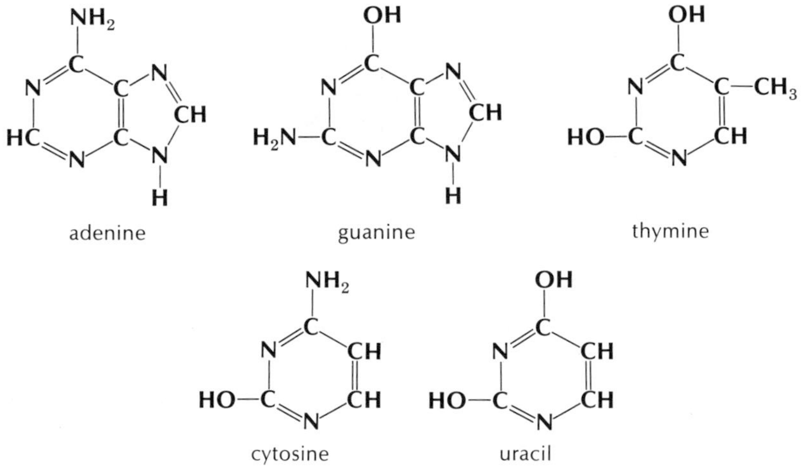

adenine guanine thymine

cytosine uracil

Nucleic acids are made up of structural units called *nucleotides,* which in turn contain a simpler unit called a *nucleoside. Nucleosides* consist of a pentose and a base condensed together with the elimination of a molecule of water. This condensation is illustrated for 2-deoxyribose and adenine in the following equation:

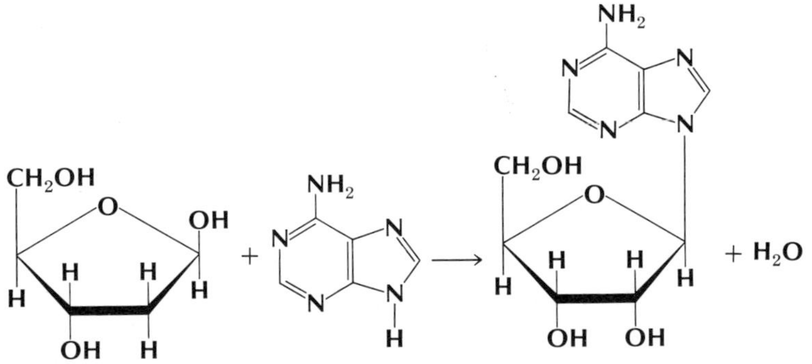

$$+ H_2O$$

Thus nucleoside formation is analogous to glycoside formation, that is, the conversion of a hemiacetal to an acetal; the difference is the substitution of a nitrogen compound for an oxygen compound as the nucleophile. Combination of a molecule of phosphoric acid at the number 5 carbon atom of either a ribose or deoxyribose nucleoside produces a *nucleotide,* which constitutes the monomeric unit of nucleic acids. The reaction between the alcohol function and phosphoric acid is classed as an *esterification reaction.* The nucleotide of deoxyribose, adenine, and phosphoric acid is given below:

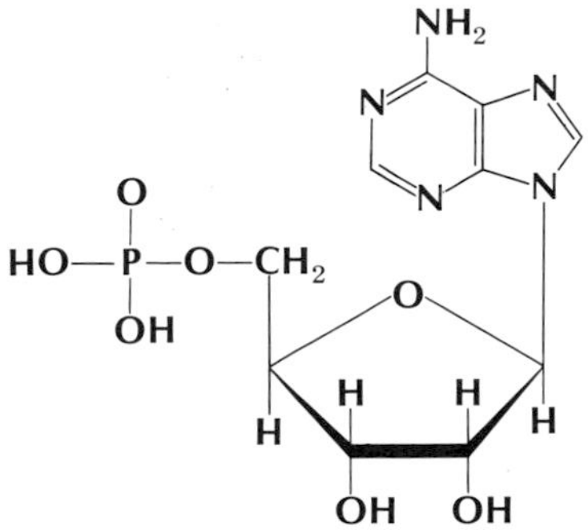

deoxyladenylic acid

The names of the nucleosides and nucleotides are given in Table 27.1.

Nucleic acids are polymers of nucleotides arranged with alternating phosphate and pentose units forming a "backbone" produced by condensation (ester formation) of the phosphoric acid of one nucleotide with the hydroxyl group on carbon number 3 of the pentose of another nucleotide. The base portion of each nucleotide is attached to the backbone. A representation of a portion of deoxyribonucleic acid structure is shown in Figure 27.1 for deoxyadenosine, deoxycytidine, and deoxythymidine. A partial structure of a ribonucleic acid containing adenosine, uridine, and cytidine is given in Figure 27.2. The accompanying block diagrams serve to simplify the complex structures and emphasize the structural relationship between the chain and the attached bases.

Table 27.1
Names of nucleosides and nucleotides

base	*pentose*	*nucleoside*	*nucleotide (acid)*
cytosine	ribose	cytidine	cytidylic
cytosine	deoxyribose	deoxycytidine	deoxycytidylic
thymine	deoxyribose	thymidine	thymidylic
uracil	ribose	uridine	uridylic
uracil	deoxyribose	deoxyuridine	deoxyuridylic
adenine	ribose	adenosine	adenylic
adenine	deoxyribose	deoxyadenosine	deoxyladenylic
guanine	ribose	guanosine	guanylic
guanine	deoxyribose	deoxyguanosine	deoxyguanylic

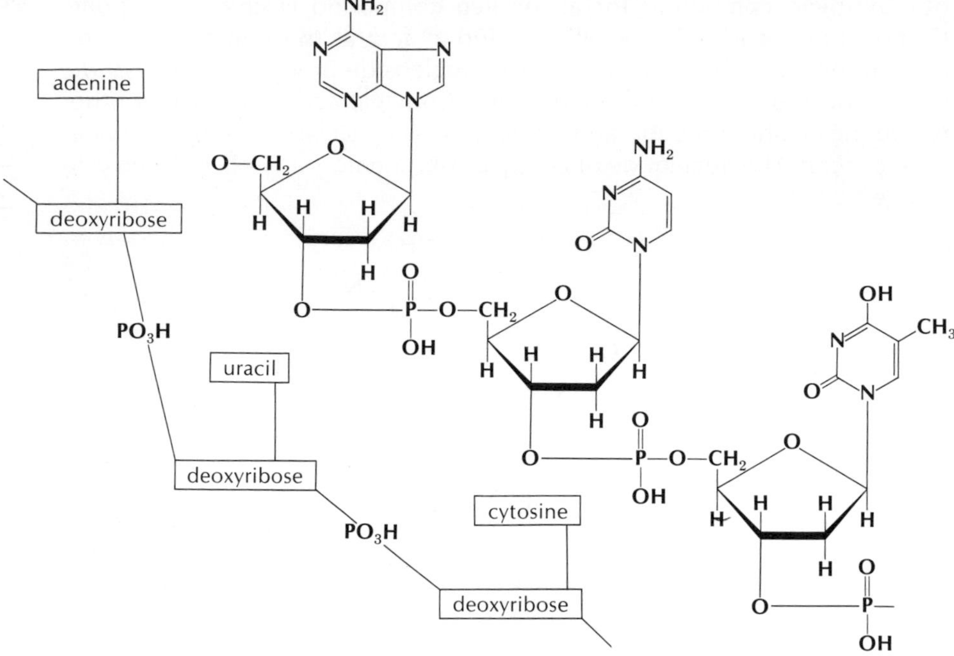

Figure 27.1
A portion of a deoxyribonucleic acid.

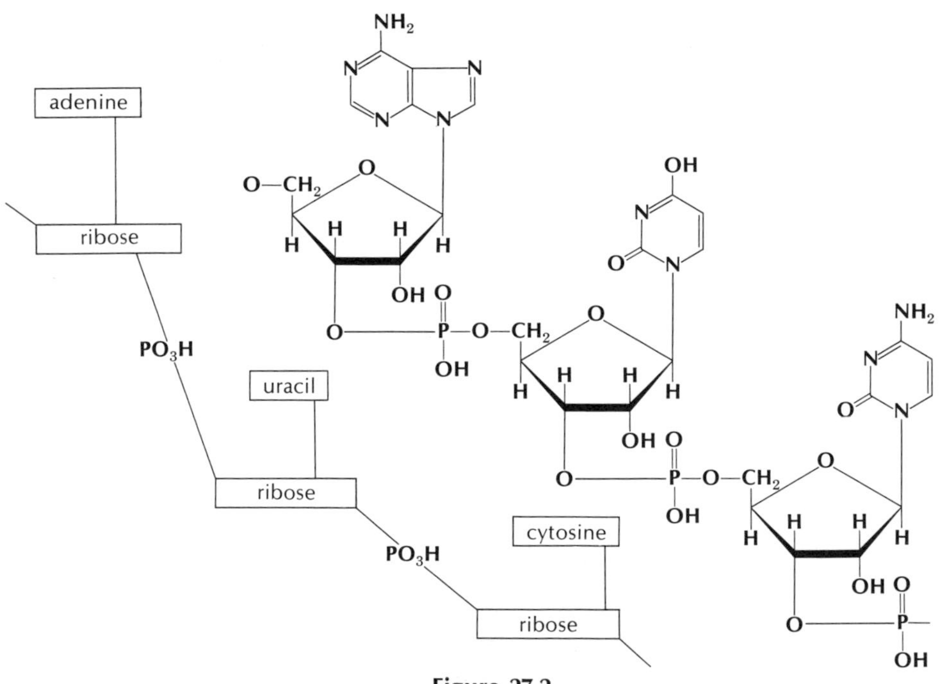

Figure 27.2
A portion of a ribonucleic acid.

27.2 HYDROGEN BONDS IN NUCLEIC ACIDS

The DNA molecule is a double-stranded polymer of total molecular mass more than 100 million. The two intertwined strands are held together by extensive hydrogen bonding in a fashion first correctly described by F. H. C. Crick and J. D. Watson in 1953. Knowing that every molecule of DNA contains equimolar quantities of adenine and thymine and of cytosine and guanine, they predicted that these two base pairs form sets of hydrogen bonds that determine the secondary structure of the DNA molecule. Two hydrogen bonds can be formed between the adenine and thymine units, and three hydrogen bonds can be formed between the cytosine and guanine units (Figure 27.3). Each member of the base pair is said to complement the other.

The two strands of DNA are coiled in an α helix and are intertwined in such a way that a thymine unit of one chain is always opposite an adenine unit of the other chain. Similarly, the guanine of one chain is paired with the cytosine of the second chain. A schematic representation of the double strand of DNA with its hydrogen bonds is shown in Figure 27.4. The horizontal bars between the helices represent hydrogen bonds in the appropriate quantity between complementary base pairs. The DNA molecule is approximately 20Å in diameter.

In later discussions it will be convenient to use the symbol G for guanine, C for cytosine, A for adenine, and T for thymine. The complementary base pairs in DNA are A-T and G-C. Note that because of the unfavorable relationship of the number of hydrogen bonds formed the base pairs A-G and C-T do not occur.

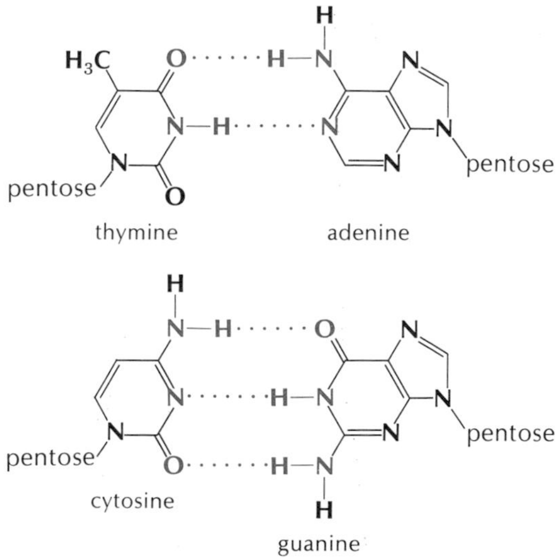

Figure 27.3
Hydrogen bonding of base pairs in DNA.

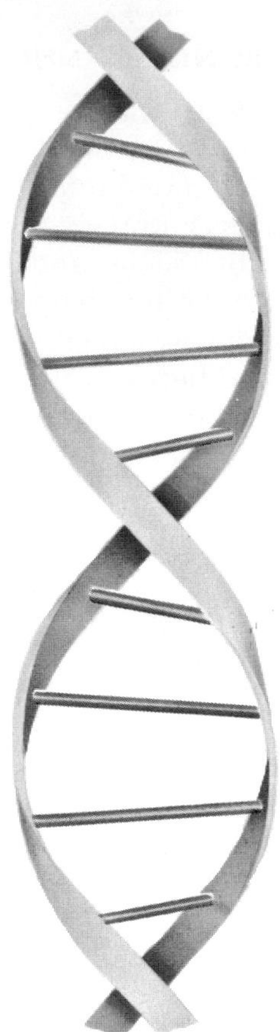

Figure 27.4
DNA as a double-stranded helix.

RNA contains uracil instead of thymine and is of lower molecular mass than DNA. Since there is no uniform ratio of base pairs in the molecule it is postulated that RNA is single stranded although it is helical at least in part. There are actually at least two types of RNA—messenger RNA and transfer (soluble) RNA—that play different roles in biochemical processes.

27.3 REPLICATION OF DNA

Cellular division necessary in a growing organism requires the replication of DNA for new cells. Each individual has his own specific sequence of nucleotide units that must be reproduced for the maintenance of his chemical identity. This continuous replication of a specific arrangement for molecules with molecular masses in the millions requires a very special chemical process. It is postulated that a temporary uncoiling of the two strands of DNA occurs. Each of these strands then serves as a mold for the formation of the complementary strand from a pool of available nucleotide units. The nucleotides do not randomly join but rather are guided by the structural requirements of the bases of the "parent" DNA strands. Two new "daughter" strands are produced that are exact replicas of the parents. This process is illustrated in Figure 27.5, in which geometric shapes

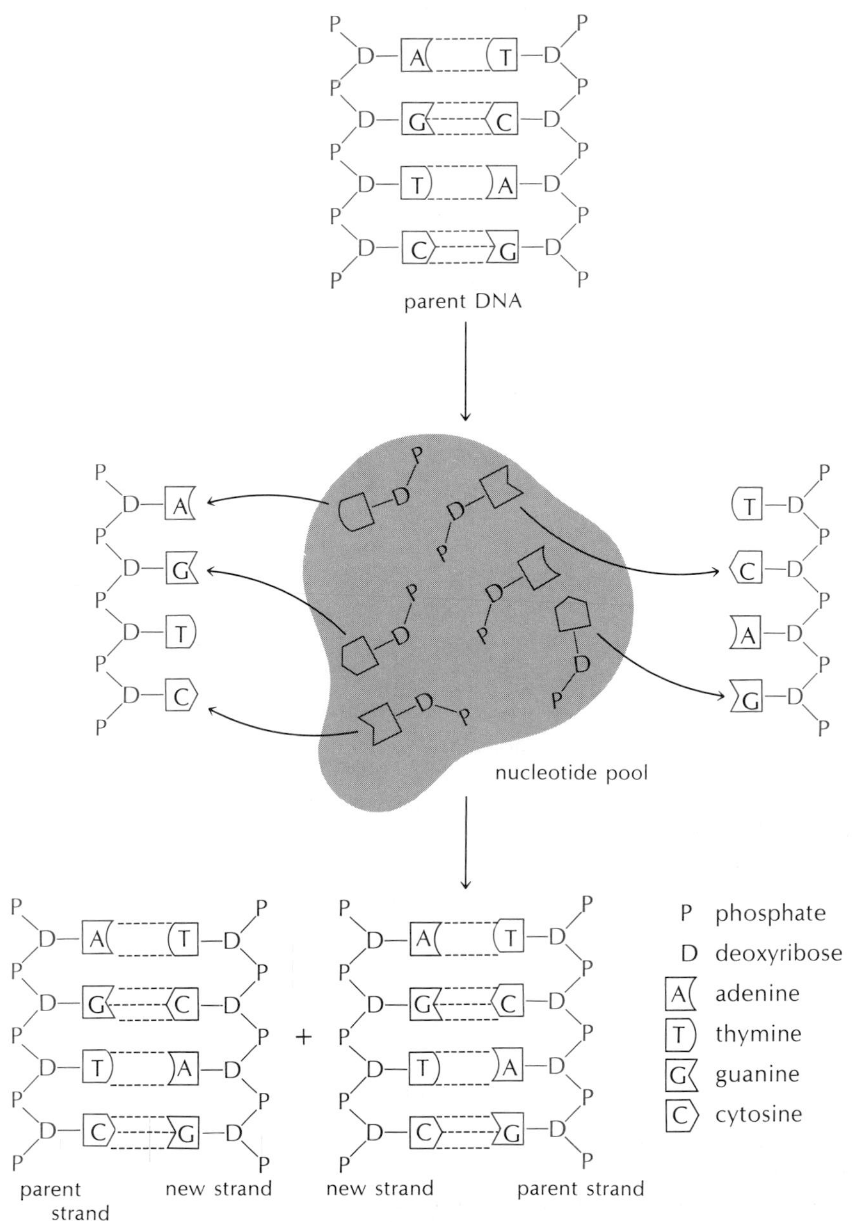

Figure 27.5
Replication of DNA.

represent the bases and the letters contained within are the first letters of the bases. For the sake of simplicity separate short chains are illustrated. Actually it is unlikely that the entire coil of DNA would unravel and create two other coils directly. Rather it seems more likely that the process involves a gradual, partial uncoiling with the subsequent formation of the daughter chains at each of the two original components of the coil. The unraveling of a double-stranded rope with simultaneous recoiling of each strand with newly developing strands in the continuous process of generation is a close macroscopic analogy.

27.4 SYNTHESIS OF PROTEINS BY DNA AND RNA

The destruction and reproduction of proteins in living systems is incredibly fast considering the complexity of the molecules. The cell structure that produces hemoglobin must synthesize one new protein every 90 sec. Since there are about 150 amino acids in the protein chain, it must grow at the rate of approximately 2 amino acids per second. Moreover, this growth rate must be controlled by DNA and the two types of RNA. DNA serves as the director of chemical processes, while the RNA molecules work under its control. While the structure of DNA is a function of the individual, the RNA molecules of different individuals appear to be similar.

The first step in the formation of proteins involves the association of an amino acid with the relatively low molecular mass soluble RNA. There is at least one type of soluble RNA for each amino acid. Evidence indicates that the specificity of a particular soluble RNA molecule for an amino acid is due to the base sequence in the RNA molecule. The sequence can be regarded as a sentence written in a four-letter alphabet composed of the RNA bases. The sequences of bases are called *codons,* and they are read three at a time by the amino acid. For example, the base sequence U-U-U specifies phenylalanine. Soluble RNA with only uridylic acid units therefore can produce only polyphenylalanine. There are actually several codons for each amino acid (Table 27.2). For example, G-C-A, G-C-C, G-C-G, and G-C-U all specify alanine. The grouping of four bases in trios gives rise to $4^3 = 64$ different codons, more than enough to specify all the common amino acids.

The complex of soluble RNA and the amino acid proceeds to the site of protein formation, the *ribosome.* After the amino acid is added to the growing protein chain the soluble RNA returns to join with another molecule of its specific kind of amino acid. The order in which the amino acids are added to the protein is not controlled by soluble RNA but rather by messenger RNA. Messenger RNA is apparently synthesized by DNA, as are the other kinds of RNA. The guanine in a strand of DNA serves as a mold for placement of cytosine in RNA molecules, but since RNA contains uracil instead of thymine, it is likely that adenine in DNA directs the placement of uracil in RNA. The RNA molecules are thus complements of DNA strands.

Table 27.2
Amino acids and codons

amino acid	*codons*	*amino acid*	*codons*
alanine	GCA, GCC, GCG, GCU	leucine	CUA, CUC, CUG, CUU, UUA, UUG
arginine	CGA, CGC, CGG, CGU, AGA, AGG	lysine	AAA, AAG
asparagine	AAC, AAU	methionine	AUG
aspartic acid	GAC, GAU	phenylalanine	UUC, UUU
cystine	UGC, UGU	proline	CCA, CCC, CCG, CCU
glutamic acid	GAA, GAG	serine	UCA, UCC, UCG, UCU, AGC, AGU
glutamine	CAA, CAG	threonine	ACA, ACC, ACG, ACU
glycine	GGA, GGC, GGG, GGU	tryptophan	UGA, UGG
histidine	CAC, CAU	tyrosine	UAC, UAU
isoleucine	AUA, AUC, AUU	valine	GUA, GUC, GUG, GUU

A partial DNA strand and the related RNA chain whose formation it directs are shown below, using letter symbols for the bases. In this symbolism the backbone of the sugar and phosphate groups are not shown.

```
DNA chain:  —A—G—G—C—T—A—
             |  |  |  |  |  |
RNA chain:  —U—C—C—G—A—U—
```

Messenger RNA at the ribosome site must carry a code to pick out the proper soluble RNA molecule with its attached amino acid. This probably occurs by a matching of base pairs between the two RNA molecules:

```
messenger RNA:  —U—C—C—G—A—U—
                 |  |  |  |  |  |
  soluble RNA:  —A—G—G—C—U—A—
```

Messenger RNA is of higher molecular mass than soluble RNA. Its complexity can be illustrated by considering the formation of the amino acid sequence in the protein of the chain of hemoglobin that contains 146 amino acids (Section 24.8). In order to direct proper placement of the amino acids so that protein formation can occur, the messenger RNA molecule must contain a specific sequence of $146 \times 3 = 438$ bases along its chain.

27.5
DNA AND GENETICS

GENERAL CONSIDERATIONS

Chromosomes are complex substances of DNA and protein that are the chemical carriers of hereditary information. The sperm cell of the male contains DNA surrounded by protein. Egg cells contain DNA, proteins, and lipids. When the sperm fertilizes the egg a pairing of chromosomes occurs. Both parents contribute chromosomes that contain all the necessary hereditary information that the growing fertilized egg will need. The specifications for the synthesis of hemoglobin, muscle, hormones, and enzymes for a lifetime are packed in a small but extremely complicated unit. The exact manner of operation and of development of all the features of life processes is a research area of intense interest.

Occasionally intervening forces such as radiation produce changes in DNA or the replication of the molecule is imperfect. A mutant organism that has different hereditary characteristics from the parents results. Such mutations probably occur only once in a million times, and most mutants are less able to survive than the normal members of the species and therefore do not reproduce. Thus, species are preserved. However, there are environmental conditions, such as in the case of sickle cell anemia, which tend to make some mutants more fit for survival than the parent. In other cases, proper diet and medical care preserve victims of genetic defects who otherwise would die. Such is the case with patients having phenylketonuria and galactosemia.

SICKLE CELL ANEMIA

The red blood cells of normal humans are roughly disk shaped; those mutants with sickle cell anemia have poorly shaped sickled cells. Sickled cells cannot provide the proper amounts of oxygen, and the individual with this inborn disease must either produce many more red blood cells than a normal human or suffer the effects of oxygen deficiency.

The sickling of the cells results from the presence of a glutamic acid residue at the point in the hemoglobin protein where a valine is located in a normal person (Section 24.8). This minor difference of one amino acid in a chain of 146 amino acids causes a change in the tertiary and quaternary structure of the protein, which in turn leads to the sickle shape. The medical result is a weak person who is frequently ill and is unable to fight infections that normal persons can handle easily.

The abnormal hemoglobin protein chain is a product of a hereditary trait and cannot be cured by medicine. The only hope of preventing the birth of children with sickle cell anemia is in genetic counseling. It might be expected that the debilitating features of sickle cell anemia would make the person with this disease a rarity. Individuals who receive the sickle cell trait from both parents do die while young. Those who receive the sickle cell trait from one parent also should be less able to survive than those with normal parents. However, the sickle cells have one advantage over normal cells: The malaria-producing parasite that lives in red blood cells cannot survive in sickle cells, making sufferers of sickle cell anemia immune to malaria. Therefore, certain African tribes have as many as 50 percent of their members with sickle cell anemia. These individuals are better able to stand their environmental conditions than normal people and have continued to reproduce others with the same genetic disease. Approximately 8 percent of Afro-Americans have the abnormal hemoglobin. The absence of malaria in the United States causes a change in the natural selection principles.

PHENYLKETONURIA

In individuals suffering from the disease called *phenylketonuria* a defect in the DNA apparently makes it unable to direct the production of the enzyme that converts phenylalanine to the amino acid tyrosine. In such individuals phenylalanine in excess of the amount used for the formation of protein or hormones is converted to phenylpyruvic acid:

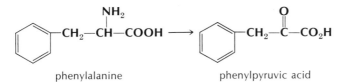

$$C_6H_5\text{—}CH_2\text{—}\underset{\displaystyle}{\overset{\displaystyle NH_2}{\overset{|}{C}H}}\text{—}COOH \longrightarrow C_6H_5\text{—}CH_2\text{—}\overset{\displaystyle O}{\overset{\|}{C}}\text{—}CO_2H$$

phenylalanine — phenylpyruvic acid

Apparently a high concentration of phenylpyruvic acid in an infant causes mental damage because many mentally retarded children are found to excrete it in their urine.

Tests have been developed to detect the presence of the chromosomes of carriers of the phenylketonuria disease. The probability of producing children with this genetic defect can be calculated and is an aid to the parents in deciding whether to have children.

Restriction of the amount of phenylalanine in food consumed by the young child allows normal development. After the age of six the restricted diet can be discontinued without ill effects. Some states require the testing of the urine and blood of newborn babies in order to detect this genetic defect at the earliest possible time.

GALACTOSEMIA

Some infants have a high level of galactose in the blood. In addition the monosaccharide is also excreted in the urine. Such children vomit their food, do not gain weight, and eventually may die. Children that survive are either dwarfed or mentally retarded and may develop cataracts.

This disease, called galactosemia, results from the inability to metabolize galactose. The absence of the necessary enzyme in turn reflects a deficiency in the genetic information of DNA. There is no cure for galactosemia, and its detrimental effects can be avoided only by the elimination of galactose from the diet. This means that lactose in human milk and cow's milk and all related dairy products must be restricted. Fortunately, with care, the child with galactosemia can grow up without any noticeable effects. In some cases the body eventually develops alternate metabolic pathways, and galactose then can be safely consumed.

Suggested further readings

Clark, B. F. C., and Marcker, K. A., "How Proteins Start," *Sci. Amer.* (January 1968).

Crick, F. H. C., "The Genetic Code," *Sci. Amer.,* offprint 123 (October 1962).

Crick, F. H. C., "The Genetic Code III," *Sci. Amer.* (October 1966).
Crick, F. H. C. "Nucleic Acids," *Sci. Amer.*, offprint 54 (September 1957).
Dobzhansky, T., "The Genetic Basis of Evolution," *Sci. Amer.*, offprint 6 (January 1954).
Hanawalt, P. C., and Hayes R. H., "The Repair of DNA," *Sci. Amer.* (February 1967).
Hoagland, M. B., "Nucleic Acids and Proteins," *Sci. Amer.*, offprint 68 (December 1959).
Holley, R. W., "The Nucleotide Sequence of a Nucleic Acid," *Sci. Amer.* (February 1966).
Horowitz, N. H., "The Gene," *Sci. Amer.*, offprint 17 (October 1956).
Muller, H. J., "Radiation and Human Mutation," *Sci. Amer.*, offprint 29 (November 1955).
Nirenberg, M., "The Genetic Code II," *Sci. Amer.*, offprint 153 (March 1963).
Puck, T. T., "Radiation and the Human Cell," *Sci. Amer.*, offprint 71 (April 1960).
Zuckerkandle, E., "The Evolution of Hemoglobin," *Sci. Amer.*, offprint 1012 (May 1965).

Terms and concepts

adenine
chromosome
codon
cytosine
deoxyribose
DNA
galactosemia
guanine
messenger RNA
nucleoside
nucleotide
phenylketonuria
replication
ribose
RNA
sickle cell anemia
thymine
transfer RNA
uracil

Questions and problems

1. What are the chemical differences between DNA and RNA?
2. What are the functional differences between DNA and RNA?
3. What hydrolysis products will be produced from adenylic acid?
4. Draw the trinucleotide specified by the symbol G-C-A.
5. Consider the structural difference between glutamic acid and valine and suggest a reason for the change in secondary and tertiary structure that occurs in hemoglobin when one is substituted for the other.
6. Compare the codons for glutamic acid and valine. Indicate the differences in RNA that could lead to the improper placement of valine for glutamic acid in the formation of sickle cell hemoglobin.

28

METABOLISM

Most of the chemical reactions discussed in the organic chemistry section of this text can occur in living systems. Aldehydes can be either oxidized or reduced; esters can be formed or hydrolyzed. However, unlike laboratory processes, the life processes must occur in a nearly neutral aqueous solution at a relatively low temperature. Moreover, the reactions must occur rapidly in order to provide the responses necessary in a living system. In this chapter the essential functions of enzymes, coenzymes, and hormones in allowing rapid reactions to occur will be presented. The role of these agents in the metabolism of amino acids, carbohydrates, and lipids will be discussed. Photosynthesis, the most important reaction of plants, will be examined briefly.

28.1 BIOCHEMICAL REACTIONS

ENZYMES

Biochemical reactions follow the same laws of conservation of mass and energy as do ordinary chemical reactions, but they must have an energy of activation (Section 9.3) low enough to allow reactions to occur rapidly at body temperature. A catalyst makes this possible by providing a mechanistic pathway with a low energy of activation. In biological systems these catalysts are protein molecules called *enzymes*. More than a thousand different enzymes are known, and many more undoubtedly will be discovered. Enzymes are classified according to the functional transformations they catalyze. Table 28.1 gives the current classification of enzymes.

The rate of enzyme-catalyzed reactions has to be very rapid for the normal functioning of living systems, and the catalyst must continue to function efficiently without being consumed. The enzymes do not promote energetically unfavorable reactions but rather only accelerate energetically favorable reactions. For example, α-chymotrypsin hydrolyzes proteins at body temperature in a neutral media. In the absence of α-chymotrypsin the very same process requires hot concentrated hydrochloric acid for a comparable rate of hydrolysis.

Table 28.1
Types of enzymes

enzyme	*type of reaction*	*example of reaction*
oxireductases	oxidation-reduction	$CH_3CHO \rightleftharpoons CH_3CH_2OH$
transferases	transfer of a functional group between molecules	$CH_2OHC(=O)CO_2H + CH_3CH(NH_2)CO_2H \rightleftharpoons CH_2OHCH(NH_2)CO_2H + CH_3C(=O)CO_2H$
hydrolases	hydrolysis	$R{-}C(=O){-}NHR \longrightarrow RCO_2H + RNH_2$
lyases	removal or addition of groups	$HO_2CCH_2CH(OH)CO_2H \rightleftharpoons HO_2CCH{=}CHCO_2H$
isomerases	isomerization	$CH_3{-}C(NH_2)(H){-}CO_2H \rightleftharpoons CH_3{-}C(H)(NH_2){-}CO_2H$
ligases	bond formation between molecules	$NH_2CHCO_2H + CH_3CH(NH_2)CO_2H \longrightarrow NH_2CHC(=O){-}N(H){-}CH(CH_3)CO_2H$

Enzymes are highly specific in their catalytic activity. A given hydrolase cannot hydrolyze all peptide linkages. For example, α-chymotrypsin catalyzes the hydrolysis of amide linkages containing L-phenylalanine, L-methionine, L-tryptophan, and L-tyrosine, whereas the hydrolase trypsin catalyzes the hydrolysis of amide linkages containing either L-arginine or L-lysine. The specificities of other enzymes are indicated in this chapter.

According to current theory of the action of enzymes, the reacting substance, the *substrate,* becomes bound in some specific manner to the surface of the enzyme, permitting easy access to the reaction site by the second reagent. The specificity of the enzyme is accounted for by its particular composition and its secondary and tertiary structure. This theory is called the *lock and key model;* the enzyme is pictured as the key that can function by opening only certain locks or substrates.

An important goal of enzyme chemistry at present is to elucidate the specific sites in the protein chain that contribute the catalytic activity for reactions of one special type. In addition, once the sites are located it is of interest to ascertain what physical and chemical processes occur at the active site to promote a rapid reaction. Part of the activity undoubtedly is the result of specific alignment of reactants for a reaction. It is postulated that binding to an enzyme surface is accompanied by stretching of potentially reactive bonds. This allows for a more facile attack by another molecule such as water. By contrast the reaction in the absence of an enzyme requires chance alignment upon collision; the proper energy for achieving a transition state must be contained in the reactant partners to cause the rupture and formation of bonds.

COENZYMES

Although enzymes are generally proteins, they are not always entirely so. Many enzymes contain a nonprotein group. In these complex enzyme molecules neither the protein nor nonprotein portion alone is active; both portions must be combined for enzyme activity. The protein portion is called the *apoenzyme,* and the nonprotein portion is called the *coenzyme.* Coenzymes are usually of lower molecular mass than apoenzymes.

In some cases the coenzymes appear to join with the apoenzyme to provide the special molecular configuration required for the specific reactions that the enzyme catalyzes. The coenzyme itself may contain the active site but may need to be attached to the larger apoenzyme in order to be in the proper environment for its reactions to occur.

Coenzymes often serve as counterfoils for the reaction that the enzyme catalyzes. In other words, if an enzyme can oxidize a substrate it may be because the coenzyme can be reduced. If the substrate releases a phosphate unit the coenzyme may react with it.

Many coenzymes incorporate vitamins in their structure. In addition coenzymes are usually esters of phosphoric acid, as illustrated by the structures of several coenzymes listed in Table 28.2. Note that adenosine

Table 28.2
Structures of some coenzymes

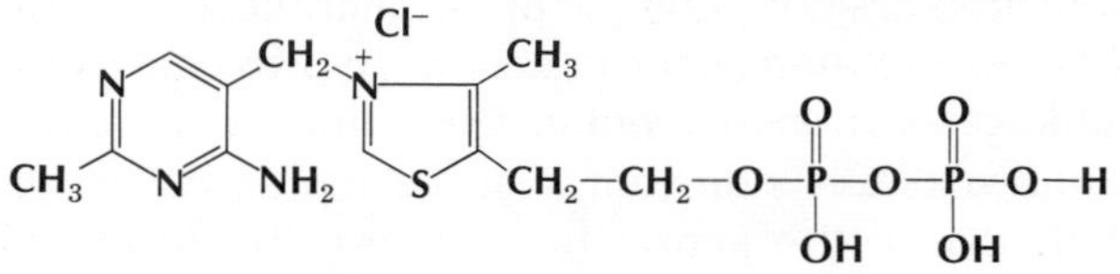

thiamine pyrophosphate

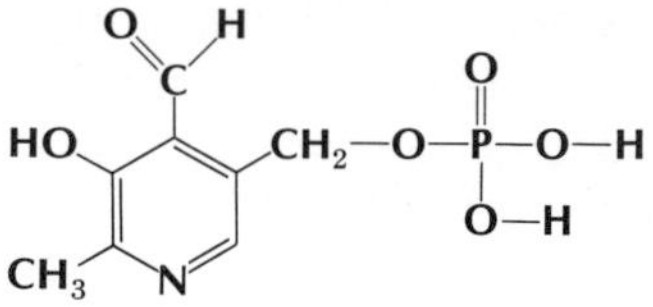

pyridoxal phosphate

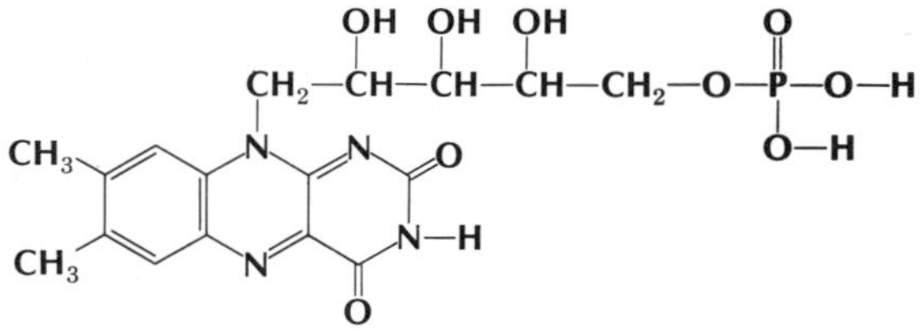

riboflavin phosphate

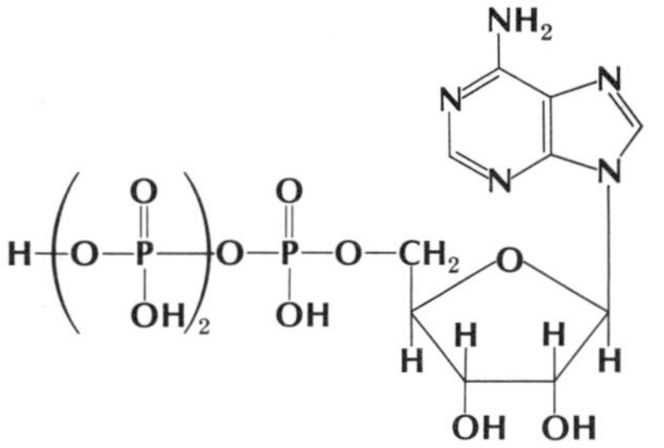

adenosine triphosphate

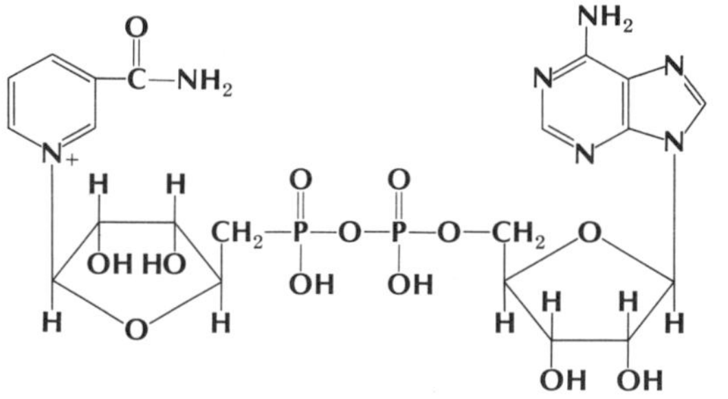

nicotiamide adenine dinucleotide

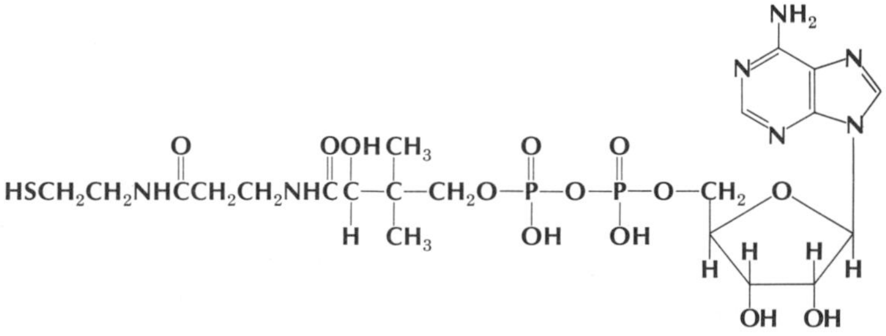

coenzyme A

triphosphate (ATP), nicotinamide adenine dinucleotide (NAD), and coenzyme A (CoA) all contain the nucleoside adenosine. In the previous chapter it was pointed out that RNA is responsible for the synthesis of proteins. The special relationship between proteins and nucleic acids is again shown here in the case of enzyme activity. The coenzyme, which is a derivative of a nucleoside, must fit into the apoenzyme surface in a special way in order to function properly.

ADENOSINE TRIPHOSPHATE AND ENERGY STORAGE

Chemical reactions in biochemical systems conform to the same energy requirements as any other reaction. A spontaneous reaction is accompanied by a negative free energy change (Chapter 9). When this energy is released in a biochemical system during a metabolic process the organism makes as efficient use of it as possible. Usually metabolic processes involve a series of steps in which relatively minor chemical changes occur so that the total energy change is released gradually. The energy then can be stored efficiently in a number of moles of adenosine triphosphate (ATP), which is the product of the endothermic reaction of phosphoric acid and adenosine diphosphate (ADP):

$$\mathbf{ADP} + \mathbf{H_3PO_4} + 12\ \text{kcal/mole} \longrightarrow \mathbf{ATP} + \mathbf{H_2O}$$

Upon demand of the organism the adenosine triphosphate can be hydrolyzed to adenosine diphosphate with the release of the same energy:

$$\mathbf{ATP} + \mathbf{H_2O} \longrightarrow \mathbf{ADP} + \mathbf{H_3PO_4} + 12\ \text{kcal/mole}$$

OXIDATION REDUCTION IN BIOCHEMICAL SYSTEMS

Oxidation in living systems releases energy. The changes in electron distribution in molecules are accompanied by conversion of the coenzymes nicotinamide adenine dinucleotide (NAD) and flavine adenine dinucleotide (FAD) into the reduced molecules $NADH_2$ and $FADH_2$.

$$\mathbf{NAD} + \mathbf{2H^+} + 2e^- \longrightarrow \mathbf{NADH_2}$$
$$\mathbf{FAD} + \mathbf{2H^+} + 2e^- \longrightarrow \mathbf{FADH_2}$$

In effect electrons are removed from substances that are being metabolized and transferred to the coenzyme, which at some later time serves as a reducing agent in another reaction.

ATP is synthesized from ADP whenever NAD and FAD are reduced. The exact course of the necessary reactions is not understood, but reduction of NAD to $NADH_2$ is accompanied by formation of 3 moles of ATP whereas 2 moles of ATP are generated when FAD is reduced. The ultimate result of the oxidation of a substance is the storage of energy in ATP via the intermediacy of FAD and NAD.

In the functioning of complex living organisms the various organs must act in concert and therefore must have a communications network to coordinate their activities. The nervous system provides part of this communications network. The humoral system, which consists of the fluids that circulate throughout the organism, provides a second means of communication. The *endocrine glands,* glands of internal secretion, synthesize and send forth hormones that act as chemical messengers through the humoral system.

The exact manner in which hormones act in a living organism is not well understood, but they apparently control the availability of reactants to cells. Hormones may work outside the cell wall to promote or retard admittance of reactants through the cell wall. The hormone insulin (Section 24.8) acts by increasing the permeability of cell walls to glucose.

28.2 PROTEINS

DIGESTION OF PROTEINS

Proteins must be digested in order to be converted into amino acids that can be absorbed through the intestinal wall. The amino acids then undergo further metabolic processes in which necessary proteins are produced or degradation occurs to produce energy.

The various enzymes that catalyze the digestion of proteins act by hydrolyzing amide linkages and are known as *peptidases.* Enzymes that act on the terminal positions of proteins are *exopeptidases;* enzymes that act in the interior positions of the proteins are *endopeptidases.* Exopeptidases may be either *aminopeptidases* or *carboxypeptidases,* depending on which end of the protein they hydrolyze.

Pepsinogen is an inactive protein of molecular mass 42,600, which when partially hydrolyzed yields pepsin, a protein-digesting enzyme of molecular mass 34,500. The nature of this partial hydrolysis is unknown, but the reason for its occurrence is obvious. A gland that manufactures pepsinogen would be digested by the very enzyme it produced if the enzyme were prepared in active form.

In the stomach the combination of HCl and pepsin affects the digestion of protein-containing food. Approximately 10 percent of the amide linkages are hydrolyzed, and peptides of molecular masses in the range of 500 to several thousand are produced.

In the small intestine the peptides encounter the endopeptidases trypsin and chymotrypsin. Like pepsin these enzymes are produced in the form of the inactive proteins of trypsinogen and chymotrypsinogen by the pancreas. If the active enzymes were produced in the pancreas the organ itself would be hydrolyzed. Trypsinogen and chymotrypsinogen contain 229 and 246 amino acid units, respectively. Approximately 40 percent of the amino acids are located in analogous positions within the structures. There are

10 cysteine molecules in the coiled chymotrypsinogen that are joined by 5 disulfide bonds. The catalytic activity of chymotrypsin is due to a serine molecule that occupies position number 195 and to histidine molecules at positions 40 and 57 of the chain. Trypsin is quite similar to chymotrypsin in the location of its active sites.

METABOLISM OF AMINO ACIDS

After hydrolysis catalyzed by trypsin and chymotrypsin the relatively small peptides are further hydrolyzed by aminopeptidases and carboxypeptidases.

$$\text{peptides} \xrightarrow[\text{carboxypeptidase}]{\text{aminopeptidase}} \text{amino acids}$$

The free amino acids formed pass through the intestinal wall and into the bloodstream to be transported to tissues where they can be used in life-maintaining processes.

Proteins are synthesized in the body under the direction of nucleic acids (Chapter 27). The eight essential amino acids lysine, leucine, isoleucine, valine, methionine, tryptophan, phenylalanine, and threonine that cannot be synthesized by the body from carbohydrates, lipids, and inorganic nitrogen compounds must be obtained from digested proteins.

Each amino acid undergoes its own characteristic metabolic reactions. The large number of these reactions is not unexpected, considering the diversity of amino acid structures. A description of each of these processes is beyond the scope of this text. Alanine is converted into pyruvic acid, which is one of the intermediates in the Krebs cycle that is presented later in this chapter. Aspartic acid and glutamic acid are converted into oxaloacetic acid and α-ketoglutaric acid, which also are part of the Krebs cycle.

$$\underset{\text{pyruvic acid}}{CH_3-\overset{\overset{\displaystyle O}{\|}}{C}-CO_2H} \qquad \underset{\text{oxaloacetic acid}}{HO_2CCH_2-\overset{\overset{\displaystyle O}{\|}}{C}-CO_2H} \qquad \underset{\alpha\text{-ketoglutaric acid}}{HO_2CCH_2CH_2-\overset{\overset{\displaystyle O}{\|}}{C}-CO_2H}$$

These keto acids and their related amino acids can be interconverted by the action of the transferase transaminase.

One metabolic product of tryptophane is nicotinic acid (Chapter 22). Phenylalanine and tyrosine are metabolized into thyroxine and adrenaline (Chapter 22). Histidine yields histamine; tyrosine yields melanin, a pigment of skin and hair; cysteine yields taurine, which combines with cholic acid (Section 26.3) to give bile salts.

28.3
CARBOHYDRATES

DIGESTION OF CARBOHYDRATES

Starch-containing foods are the major source of carbohydrates in humans and many other animals. Digestion of starch commences in the mouth, catalyzed by amylases contained in saliva. The starches are broken down to the disaccharide maltose and some higher molecular mass polysaccharides called dextrins. In the stomach the hydronium ions are primarily responsible for the further hydrolysis of dextrins and maltose. A mixture of glucose, maltose, and higher saccharides then proceeds to the small intestine, where the remaining higher molecular mass materials are converted to maltose by the action of pancreatic amylase. The enzyme maltase completes digestion by converting maltose to glucose. Glucose is absorbed through the walls of the intestine into the bloodstream, where it is distributed to the vital organs. In the liver the glucose can be stored in the form of *glycogen* (partially polymerized glucose) until it is needed. The pancreatic hormone insulin is responsible for the regulation of the proper blood sugar level and chemically signals for the depolymerization of glycogen when the level drops to a low value.

METABOLISM OF GLUCOSE

Glucose is ultimately oxidized to carbon dioxide and water with the release of 690 kcal of energy per mole. This energy is stored as efficiently as possible in ATP.

$$\text{glucose} + \text{oxygen} \longrightarrow \text{carbon dioxide} + \text{water} + 690 \text{ kcal}$$

The enzyme hexokinase catalyzes the conversion of glucose into glucose-6-phosphate. The energy input required for the reaction is supplied by the conversion of 1 mole of ATP into ADP.

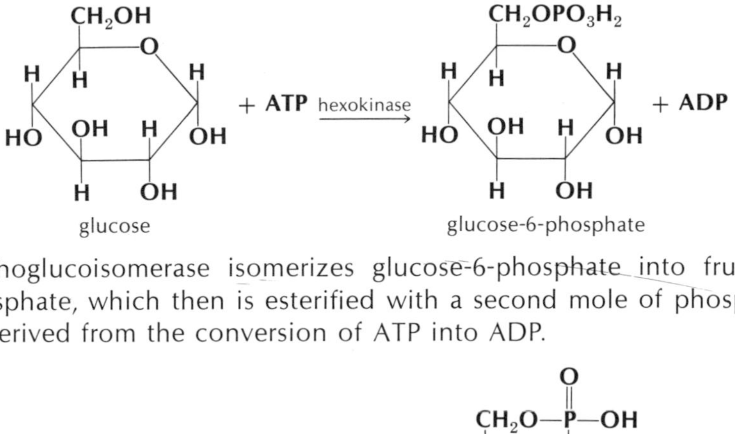

glucose + **ATP** $\xrightarrow{\text{hexokinase}}$ glucose-6-phosphate + **ADP**

Phosphoglucoisomerase isomerizes glucose-6-phosphate into fructose-6-phosphate, which then is esterified with a second mole of phosphoric acid derived from the conversion of ATP into ADP.

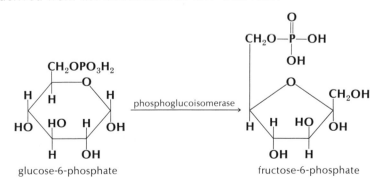

glucose-6-phosphate $\xrightarrow{\text{phosphoglucoisomerase}}$ fructose-6-phosphate

$$\text{fructose-6-phosphate} + ATP \xrightarrow{\text{hexokinase}} ADP + \begin{array}{c} CH_2OPO_3H_2 \\ | \\ C{=}O \\ | \\ HO{-}C{-}H \\ | \\ H{-}C{-}OH \\ | \\ H{-}C{-}OH \\ | \\ CH_2OPO_3H_2 \end{array}$$

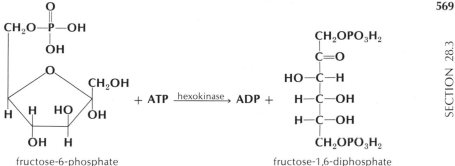

fructose-6-phosphate fructose-1,6-diphosphate

The enzyme aldolase effects the cleavage of fructose-1,6-diphosphate into glyceraldehyde-3-phosphate and dihydroxyacetone phosphate.

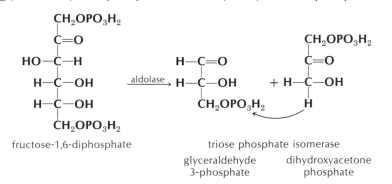

$$\begin{array}{c} CH_2OPO_3H_2 \\ | \\ C{=}O \\ | \\ HO{-}C{-}H \\ | \\ H{-}C{-}OH \\ | \\ H{-}C{-}OH \\ | \\ CH_2OPO_3H_2 \end{array} \xrightarrow{\text{aldolase}} \begin{array}{c} H{-}C{=}O \\ | \\ H{-}C{-}OH \\ | \\ CH_2OPO_3H_2 \end{array} + \begin{array}{c} CH_2OPO_3H_2 \\ | \\ C{=}O \\ | \\ H{-}C{-}OH \\ | \\ H \end{array}$$

fructose-1,6-diphosphate

triose phosphate isomerase

glyceraldehyde 3-phosphate dihydroxyacetone phosphate

Dihydroxyacetone would be a dead end in metabolic pathways and would result in the loss of half of the potential energy derived from glucose if it were not for triose phosphate isomerase, which catalyzes the conversion of dihydroxyacetone phosphate into glyceraldehyde-3-phosphate. Up to this point 2 moles of ATP per mole of glucose have been used and none of the energy derivable from glucose has been stored.

Glyceraldehyde-3-phosphate can be oxidized to glyceric acid 1,3-diphosphate with the simultaneous reduction of NAD. Note that 3 moles of ATP are produced per mole of NAD and, therefore, 6 moles of ATP now are generated from the original mole of glucose since 1 mole of glucose yields 2 moles of glyceraldehyde-3-phosphate.

$$\begin{array}{c} H{-}C{=}O \\ | \\ H{-}C{-}OH \\ | \\ CH_2OPO_3H_2 \end{array} + NAD \longrightarrow \begin{array}{c} O \\ \| \\ C{-}OPO_3H_2 \\ | \\ H{-}C{-}OH \\ | \\ CH_2OPO_3H_2 \end{array} + NADH_2$$

glyceraldehyde-3-phosphate glyceric acid 1,3-diphosphate

The enzyme phosphoglyceryl kinase catalyzes the conversion of glyceric acid 1,3-diphosphate into glyceric acid 3-phosphate with the simultaneous conversion of ADP into ATP. The glyceric acid 3-phosphate then is converted in a series of steps into pyruvic acid and another mole of ADP is converted into ATP.

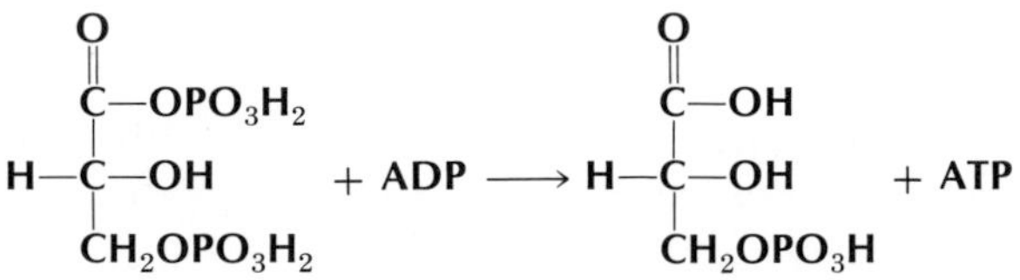

$$\begin{array}{c} O \\ \| \\ C\text{—}OPO_3H_2 \\ | \\ H\text{—}C\text{—}OH \\ | \\ CH_2OPO_3H_2 \end{array} + ADP \longrightarrow \begin{array}{c} O \\ \| \\ C\text{—}OH \\ | \\ H\text{—}C\text{—}OH \\ | \\ CH_2OPO_3H \end{array} + ATP$$

glyceric acid 1,3-diphosphate glyceric acid 3-phosphate

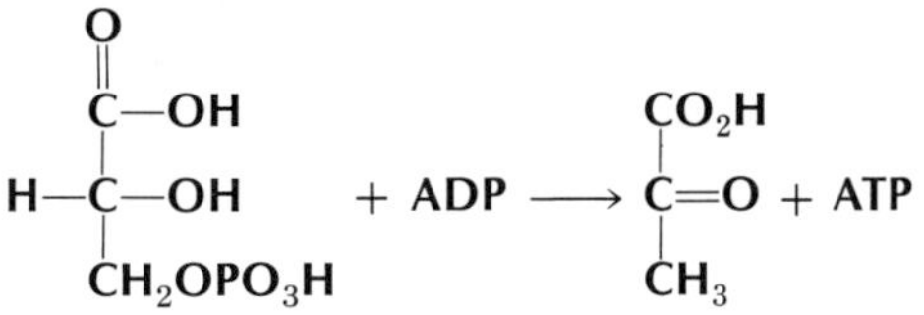

$$\begin{array}{c} O \\ \| \\ C\text{—}OH \\ | \\ H\text{—}C\text{—}OH \\ | \\ CH_2OPO_3H \end{array} + ADP \longrightarrow \begin{array}{c} CO_2H \\ | \\ C{=}O \\ | \\ CH_3 \end{array} + ATP$$

glyceric acid 3-phosphate pyruvic acid

The fate of pyruvic acid depends on the presence or absence of oxygen in the biochemical system. In the absence of oxygen the metabolism occurs *anaerobically;* in the presence of oxygen the metabolism occurs *aerobically.*

ANAEROBIC METABOLISM OF PYRUVIC ACID

In the absence of oxygen $NADH_2$ reduces pyruvic acid to lactic acid. Simultaneously 3 moles of ATP are converted into ADP. Since 1 mole of glucose produces 2 moles of pyruvic acid, 6 moles of ATP are consumed. This eliminates the 6 moles originally produced in the oxidation of glyceraldehyde-3-phosphate.

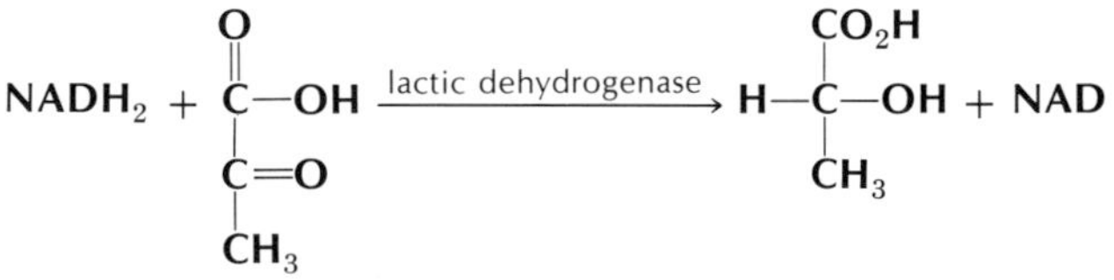

$$NADH_2 + \begin{array}{c} O \\ \| \\ C\text{—}OH \\ | \\ C{=}O \\ | \\ CH_3 \end{array} \xrightarrow{\text{lactic dehydrogenase}} \begin{array}{c} CO_2H \\ | \\ H\text{—}C\text{—}OH \\ | \\ CH_3 \end{array} + NAD$$

The anaerobic metabolism of 1 mole of glucose into 2 moles of lactic acid results in the overall formation of 2 moles of ATP. Two moles of ATP are required in the initial steps. The oxidation by NAD and the reduction by $NADH_2$ balance ATP requirements, and the steps converting glyceric acid 1,3-diphosphate and glyceric acid 3-phosphate each generate 1 mole of ATP. Since 1 mole of glucose yields 2 moles of each of these latter intermediates 4 moles of ATP are produced. Subtracting the 2 moles of ATP required in the initial steps results in the net production of 2 moles of ATP.

The energy content of glucose and lactic acid relative to carbon dioxide and water are 690 and 315 kcal/mole, respectively.

$$\text{glucose} + \text{oxygen} \longrightarrow \text{carbon dioxide} + \text{water} + 690 \text{ kcal/mole}$$
$$\text{lactic acid} + \text{oxygen} \longrightarrow \text{carbon dioxide} + \text{water} + 315 \text{ kcal/mole}$$

Therefore, the anaerobic metabolism of glucose to lactic acid liberates $690 - 2(315) = 60$ kcal/mole of glucose. This means that only $60/690 \times 100 = 9$ percent of the total energy of glucose is liberated.

$$\text{glucose} \longrightarrow \text{lactic acid} + 60 \text{ kcal/mole}$$

The energy stored in the 2 moles of ATP produced is 24 kcal and, therefore, only $24/60 \times 100 = 40$ percent of the energy liberated in the conversion of glucose into pyruvic acid is trapped.

AEROBIC METABOLISM OF PYRUVIC ACID

The aerobic metabolism of pyruvic acid involves a much more efficient storage of released energy. If oxygen is present the reduction of pyruvic acid by $NADH_2$, which consumes some of the energy stored in ATP, need not occur. The pyruvic acid then can be metabolized in a sequence of reactions known as the *Krebs cycle.*

The energy liberated by the conversion of 1 mole of glucose into 2 moles of lactic acid is less than for the conversion into 2 moles of pyruvic acid. The energy content of pyruvic acid relative to carbon dioxide and water is 275 kcal/mole.

$$\text{pyruvic acid} + \text{oxygen} \longrightarrow \text{carbon dioxide} + \text{water} + 275 \text{ kcal/mole}$$

Therefore, conversion of glucose to pyruvic acid releases $690 - 2(275) = 140$ kcal/mole.

$$\text{glucose} \longrightarrow \text{pyruvic acid} + 140 \text{ kcal/mole}$$

In the course of the reactions 2 moles of ATP are used in initial steps, 6 are produced in the oxidation step by NAD, and 4 are produced in subsequent steps in which glyceric acid 3-phosphate reacts. (Remember that 2 moles of each of these substances are produced per mole of glucose.) The net result is the formation of 8 moles of ATP and the storage of 96 kcal of the energy liberated, giving $96/140 \times 100 = 68$ percent efficient reaction sequence. The most efficient release of energy from glucose occurs in the Krebs cycle. The compounds and reactions of the Krebs cycle are shown in Figure 28.1.

Loss of carbon dioxide from pyruvic acid yields an acetyl group ($CH_3—C(=O)—$) which, combined with coenzyme A, produces acetyl coenzyme A, which plays the central role in the metabolism of carbohydrates. As will be shown later coenzyme A is important in the metabolism of fats and amino acids as well. The net result of the Krebs cycle is to convert the two-carbon acetyl unit into two molecules of carbon dioxide and four of water. In this sequence the 3 moles of NAD that are converted generate 9 moles of ATP. Oxidation by FAD produces 2 moles of ATP. In addition the conversion of α-ketoglutaric acid to succinic acid by the agency of coenzyme A is associated with the production of another mole of ATP. A turn about the Krebs cycle causes the formation of 12 moles of ATP.

The conversion of pyruvic acid into acetyl coenzyme A generates 3

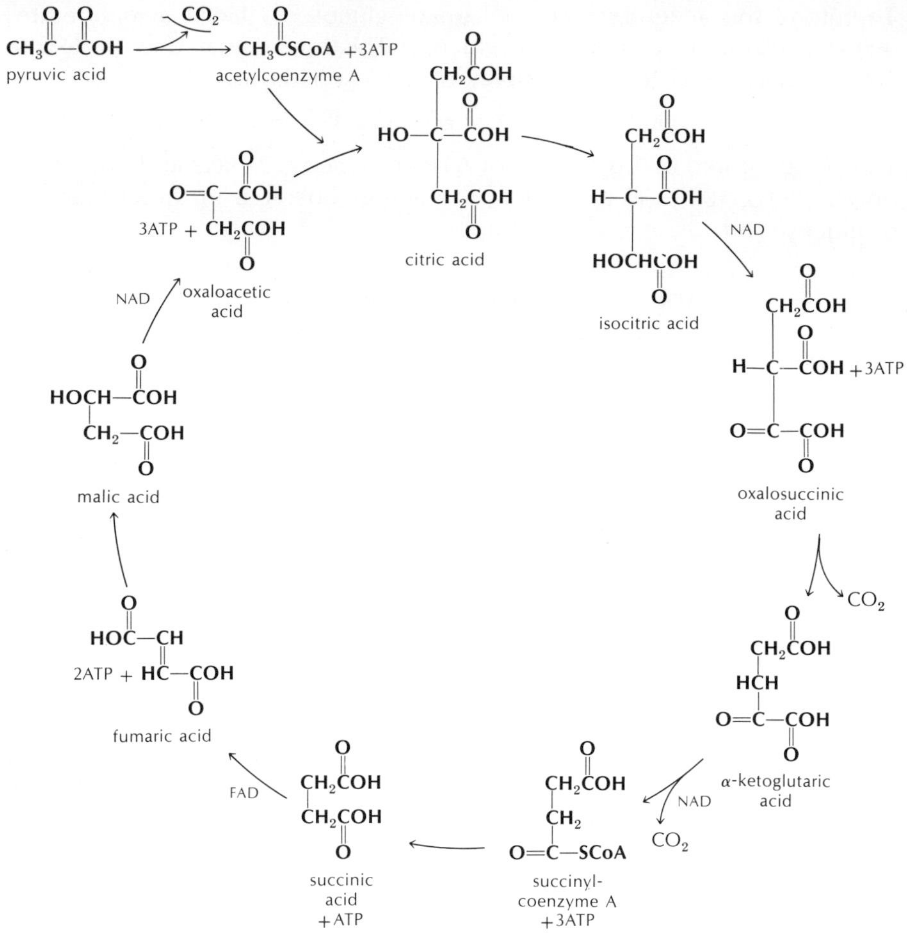

Figure 28.1
Conversion of pyruvic acid into carbon dioxide and water in the Krebs cycle.

moles of ATP. Therefore, the conversion of the 2 moles of pyruvic acid produced from 1 mole of glucose yields 2(3) + 2(12) = 30 moles of ATP. The conversion of glucose into pyruvic acid produces 8 more moles and, therefore, the total metabolism of 1 mole of glucose into carbon dioxide and water yields 38 moles of ATP. Only 456 kcal of the 690 kcal that glucose liberates is stored; the process is 66 percent efficient.

28.4 LIPIDS

DIGESTION OF LIPIDS

The digestion of the triglycerides, which are the principal food source of the lipid class, occurs predominately in the small intestine. A combination of lipases from the pancreas and intestine catalyzes the hydrolysis of the ester bonds of the triglycerides. Glycerol and a mixture of mono- and diesters of glycerol and acids then are absorbed through the wall of the intestine and enter the blood and lymph systems. Glycerol is metabolized by the addition of phosphate from ATP to yield glyceraldehyde-3-phosphate, which is an intermediate in the sequence previously outlined in the metabolism of glucose. It is metabolized in the same sequence of reactions.

METABOLISM OF ACIDS

The acids and glycerol recombine after passage through the intestinal wall and are deposited at storage sites as "depot fat," which can be used after hydrolysis to form energy if the living system requires it. In the liver the triglycerides are rehydrolyzed and metabolized, causing the release of large quantities of energy.

The acids of triglycerides contain even numbers of carbon atoms and are metabolized in two atom chuncks. The process resembles the set of reactions of the Krebs cycle except that in each cycle the substance being metabolized becomes degraded by an additional two carbon atoms so that the reaction system may be more properly described as a *spiral reaction system*. The acids react according to the sequence outlined in Figure 28.2.

In the sequence of reactions coenzyme A is converted into acetyl coenzyme A with the formation of adenosine monophosphate (AMP), and the acid is degraded by two carbon atoms. The acetyl coenzyme A then enters the Krebs cycle and is degraded to carbon dioxide and water. Each time two carbon atoms are removed from the acid one FAD and one NAD react and, therefore, 5 moles of ATP are produced. The oxidation of 1 mole of stearic acid ($C_{17}H_{35}CO_2H$) occurs via the spiral eight times to yield 9 moles of acetyl coenzyme A and $8 \times 5 = 40$ moles of ATP. Oxidation of 1 mole of acetyl CoA in the Krebs cycle yields 12 moles of ATP and, therefore, $12 \times 9 = 108$ moles of ATP are generated in the oxidation of the acetyl coenzyme A derived from stearic acid. One mole of ATP was required to start the spiral sequence of reactions. The total number of moles of ATP generated is $40 + 108 - 1 = 144$. The energy stored in the process is 1764 kcal. Complete oxidation of stearic acid liberates 2700 kcal/mole.

$$\text{Stearic acid} + \text{oxygen} \longrightarrow \text{carbon dioxide} + \text{water} + 2700 \text{ kcal}$$

The energy storage is $1764/2700 \times 100 = 65$ percent efficient.

28.5
PLANT BIOCHEMISTRY

In a biochemical sense plants and animals are quite similar. Amino acids, carbohydrates, and lipids are contained in both types of living systems. While plants contain starch, which is α linked like the glycogen stored in animals, they also contain the β-linked cellulose (Chapter 25) for which

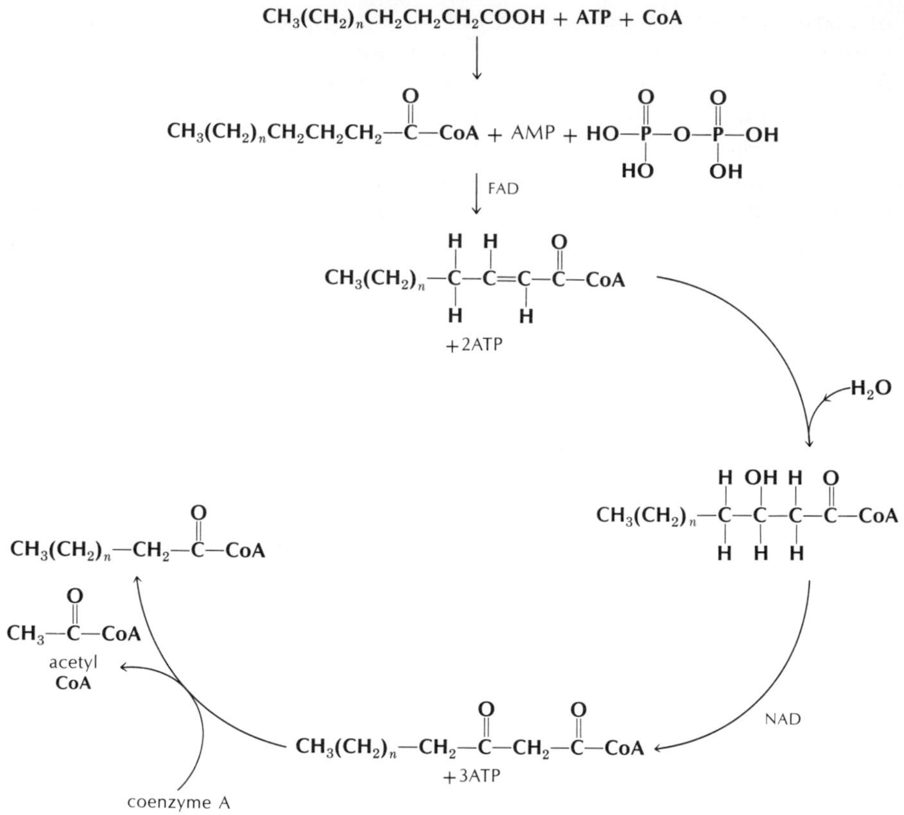

$$CH_3(CH_2)_nCH_2CH_2CH_2COOH + ATP + CoA$$

$$\downarrow$$

$$CH_3(CH_2)_nCH_2CH_2CH_2\text{—}\overset{O}{\overset{\|}{C}}\text{—}CoA + AMP + HO\text{—}\underset{HO}{\overset{O}{P}}\text{—}O\text{—}\underset{OH}{\overset{O}{P}}\text{—}OH$$

$$\downarrow \text{FAD}$$

$$CH_3(CH_2)_n\text{—}\underset{H}{\overset{H}{C}}\text{—}\overset{H}{C}\text{=}\underset{H}{C}\text{—}\overset{O}{\overset{\|}{C}}\text{—}CoA$$

+2ATP

H_2O

$$CH_3(CH_2)_n\text{—}\underset{H}{\overset{H}{C}}\text{—}\underset{H}{\overset{OH}{C}}\text{—}\underset{H}{\overset{H}{C}}\text{—}\overset{O}{\overset{\|}{C}}\text{—}CoA$$

NAD

$$CH_3(CH_2)_n\text{—}CH_2\text{—}\overset{O}{\overset{\|}{C}}\text{—}CH_2\text{—}\overset{O}{\overset{\|}{C}}\text{—}CoA$$

+3ATP

coenzyme A

$$CH_3\text{—}\overset{O}{\overset{\|}{C}}\text{—}CoA$$

acetyl
CoA

$$CH_3(CH_2)_n\text{—}CH_2\text{—}\overset{O}{\overset{\|}{C}}\text{—}CoA$$

Figure 28.2
Metabolism of acids.

there is no counterpart in animals. The triglycerides of plants are on the average more unsaturated than those of animals (Chapters 21 and 26).

The Krebs cycle is common to plants as well as to animals. However, in plants that contain chlorophyll there is another cycle that produces larger molecules from carbon dioxide and water in the presence of sunlight. This process, called *photosynthesis,* is one of the most important reaction sequences that occur on earth.

PHOTOSYNTHESIS

The general equation for photosynthesis is:

$$6CO_2 + 6H_2O \xrightarrow{\text{light}} C_6H_{12}O_6 + 6O_2$$

In addition to carbohydrates, plants can synthesize lipids and, given a source of nitrogen, amino acids as well as a wide variety of other nitrogen compounds (Chapter 22).

Photosynthesis occurs by a series of reactions that yield a variety of compounds in very short periods of time. In fact, although photosynthesis has been known for a long time, it was not until the development of methods for using radioisotopes in biochemistry two decades ago that the details of the reaction could be elucidated. The radioisotope carbon-14 in the form of radioactive carbon dioxide was used by M. Calvin and J. A. Bassham to follow the chemistry of photosynthesis.

The process of photosynthesis may be divided into parts. One of these is the so-called *light reaction,* which requires sunlight to release free oxygen. The other, the *dark reaction,* involving the conversion of carbon dioxide to carbohydrates, includes reactions that can occur in the absence of light. Of course, the starting materials for the dark reaction must be produced by the light reaction. In addition, the dark reaction can proceed equally as well in the light.

The chlorophyll molecule is a large organometallic complex containing a magnesium atom and a conjugated organic portion:

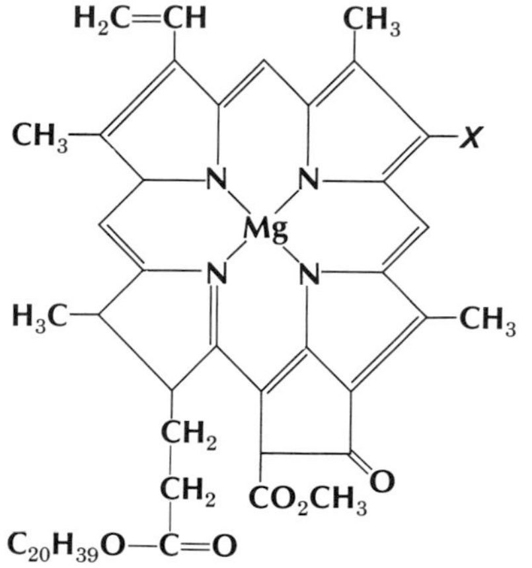

$H_2C{=}CH$ CH_3
CH_3 X
N N Mg N N
H_3C CH_3
CH_2
CH_2 CO_2CH_3 O
$C_{20}H_{39}O{-}C{=}O$

$X = -C_2H_5$ or $-CHO$

chlorophyll

When light energy is absorbed by chlorophyll, the electrons of the conjugated π system that extends over the molecule are affected. The energy is used to form chlorophyll in an excited electronic state with properties and chemistry different from those of the molecule in the ground electronic state. Some of the energy of the excited chlorophyll is transferred to compounds such as ATP and the reduced form of nicotinamide adenine dinucleotide phosphate ($NADPH_2$). These two compounds are used by the plant in the dark reaction. Part of the energy of the excited chlorophyll is used to photolyze water, that is to break up water into oxygen, electrons, and protons. It is the formation of oxygen that is so important in serving as a balance to animals, which consume oxygen in the atmosphere and release carbon dioxide. The electrons released in the photolysis of water are transferred to $NADPH_2$.

The actual absorption of light by chlorophyll requires approximately 10^{-15} to 10^{-9} sec. Obviously the study of such a process is one of considerable difficulty. Nevertheless, biochemists are probing the nature of the light absorption, the subsequent formation of ATP and $NADPH_2$, and the photolysis reaction.

The dark reaction is a well-understood process that involves the conversion of carbon dioxide into carbohydrates. A series of rapid, enzyme-catalyzed reactions occurs. The processes have been studied using radioactive carbon dioxide. As the dark reaction proceeds, compounds containing radioactive carbon are synthesized in the plant. The order of appearance of radioactive compounds provides information about the sequence of chemical reactions. In only 5 sec five radioactive compounds—pyruvic acid, malic acid, aspartic acid, glyceric acid 3-phosphate, and glyceric acid 2-phosphate—are produced. Approximately 60 percent of the radioactivity is contained in the latter two compounds. Therefore, they are the first produced in the photosynthesis reaction. After 90 sec approximately 14 radioactive compounds are present in the plant.

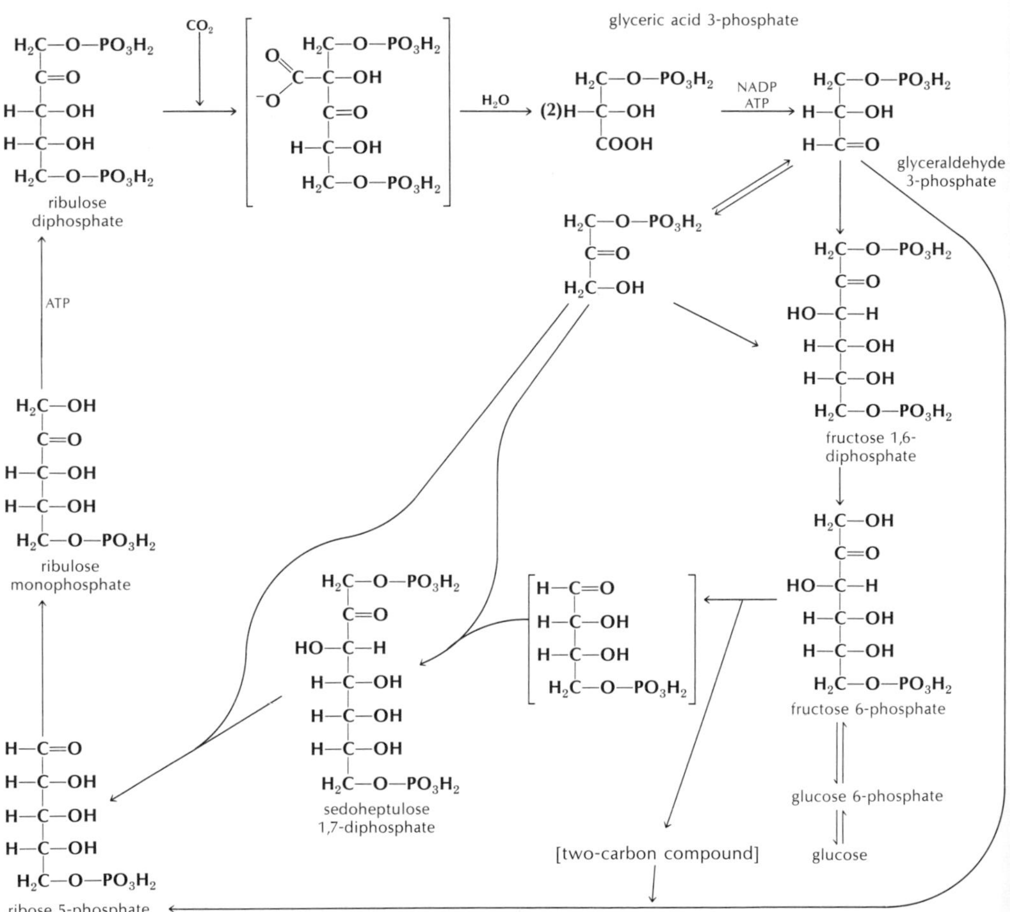

Figure 28.3
The dark reaction.

It was thought initially that a two-carbon compound combined with carbon dioxide to form the related compounds glyceric acid 3-phosphate and glyceric acid 2-phosphate. However, such a compound never has been detected. Instead, a five-carbon compound, ribulose diphosphate, first combines with carbon dioxide and then rapidly cleaves to form two molecules of glyceric acid 3-phosphate. The formation of glucose from water and carbon dioxide is shown in Figure 28.3.

The synthesis of pyruvic acid, malic acid, and aspartic acid, which are the other three compounds detected in the 5 sec after exposure to radioactive carbon dioxide, is outlined in Figure 28.4.

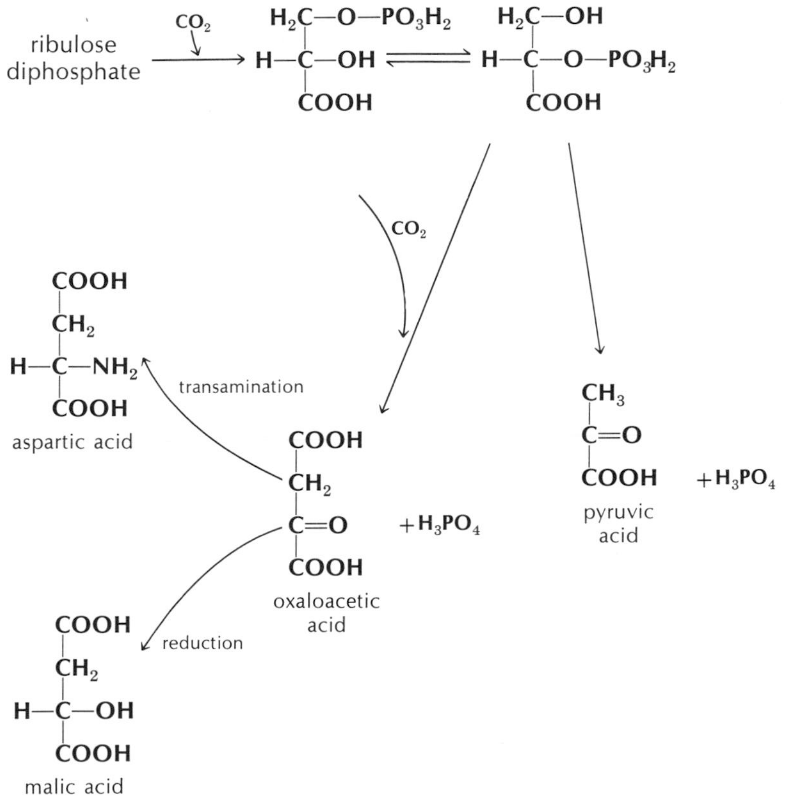

Figure 28.4
Synthesis of pyruvic acid, malic acid, and aspartic acid.

NITROGEN IN PLANTS

Most plants cannot use the free nitrogen available in the atmosphere and therefore must obtain it in combination with other elements. The nitrogen in fertilizer is in the form of nitrates, ammonium ion, and urea.

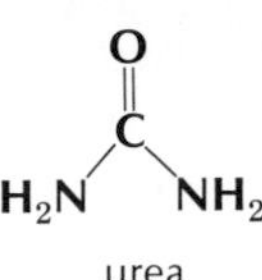

urea

The organic fertilizer, manure, contains nitrogen products excreted by animals.

Certain plants called *legumes,* such as clover, peas, and beans, are able to convert atmospheric nitrogen into nitrates and nitrites. A reaction caused by bacteria occurs in the nodules of the roots. For this reason legumes are useful in enriching soil with nitrogen if they are not harvested but rather are plowed under.

Plants are able to synthesize benzene rings and the closely related nitrogen-containing aromatic rings (Chapter 22). Phenylalanine and tryptophan are two of the essential amino acids produced by plants. Both contain aromatic rings necessary for proper nutritional balance in humans.

Suggested further readings

Arnon, D. I., "The Role of Light in Photosynthesis," *Sci. Amer.,* offprint 75 (November 1960).

Bassham, J. A., "The Path of Carbon in Photosynthesis," *Sci. Amer.,* offprint 122 (June 1962).

Dawkings, M. J. R., and Hull, D., "The Production of Heat by Fat," *Sci. Amer.,* offprint 1018 (August 1965).

Green, D. E., "The Metabolism of Fats," *Sci. Amer.,* offprint 16 (January 1954).

Green, D. E., "The Synthesis of Fats," *Sci. Amer.,* offprint 67 (February 1960).

Levin, R., and Goldstein, M. S., "The Action of Insulin," *Sci. Amer.* (May 1958).

Neurath, H., "Protein Digesting Enzymes," *Sci. Amer.,* offprint 198 (December 1964).

Rabinowitch, E. I., and Govindjee, "The Role of Chlorophyll in Photosynthesis," *Sci. Amer.,* offprint 1016 (July 1965).

Sprain, D. M., "Atherosclerosis," *Sci. Amer.* (August 1966).

Terms and concepts

adenosine triphosphate
aerobic
anaerobic
apoenzyme
chymotrypsin
coenzyme
depot fat
digestion
endocrine glands
enzyme

flavine adenine dinucleotide
glycogen
hormone
humoral system
Krebs cycle
nicotinamide adenine dinucleotide
pepsin
peptidase
photosynthesis
substrate
trypsin

Questions and problems

1. What reaction of pyruvic acid and aspartic acid will occur by the action of the transferase transaminase? Write the reaction and name the products. Which of these products is a component of the Krebs cycle?
2. Which component of the Krebs cycle can be produced from glutaric acid?
3. Which of the following peptides hydrolyze in the presence of chymotrypsin? of trypsin?

a. val-thr-met
b. gly-ala-lys
c. ala-tyr-phe
d. try-leu-arg

4. Sucrose is a disaccharide of fructose and glucose. Indicate how fructose can be metabolized.
5. Lactose is a disaccharide of galactose and glucose. Galactose can enter the metabolic cycle by first being isomerized to glucose. The process involves the conversion of galactose-1-phosphate into glucose-1-phosphate. Which asymmetric center changes configuration in this reaction?
6. In what form are carbohydrates stored in the body?
7. When bread is chewed for a prolonged period without swallowing, the bread tastes sweet. Why?
8. Does sucrose undergo any change in the mouth?
9. Calculate the energy stored in the ATP during the metabolism of palmitic acid.
10. When 9-phenylnonanoic acid [$C_6H_5(CH_2)_8CO_2H$] is fed to dogs, benzoic acid is excreted. Why? What product would be expected if 10-phenyldecanoic acid [$C_6H_5(CH_2)_9CO_2H$] were fed to a dog?
11. In what form are triglycerides stored in the body?
12. Calculate and compare the amount of energy that can be obtained from the complete oxidation of 1 g of stearic acid and 1 g of glucose.
13. What would occur if all animal life were removed from the earth? What would occur if all plant life were removed from earth? What type of balance is required for the survival of plants and animals?
14. Which compound is produced first from radioactive carbon dioxide in a green plant? Which carbon atom in the compound should contain the carbon-14?

APPENDIX 1

MATHEMATICAL REVIEW

1
MULTIPLICATION

The process of adding a given number a certain number of times is called *multiplication*. Therefore, 3 times 4 means 3 added four times or 4 added three times to give the *product* 12.

The various ways of expressing multiplication of 4 and 3 are:

$$4 \times 3 \qquad 4 \cdot 3 \qquad 4(3) \qquad (4)(3)$$

When a complex quantity contained within parentheses is multiplied, each term must be multiplied.

$$\tfrac{1}{2}(4 + 8) = \tfrac{1}{2}(4) + \tfrac{1}{2}(8) = 2 + 4 = 6$$
$$\tfrac{1}{2}(2x + y) = \tfrac{1}{2}(2x) + \tfrac{1}{2}(y) = x + y/2$$

2
DIVISION

The process of finding out how many times a number is contained in another number is called *division*. The ways of expressing division are:

$$8 \div 4 \qquad \frac{8}{4} \qquad 8/4$$

The number to the left of the divide symbol, above the horizontal line or to the left of the slant line, is called the *numerator;* the other number is the *denominator.* The *quotient* is the number obtained by dividing one number into another.

3
FRACTIONS

The division of one number by another can be expressed as a fraction. *Proper fractions* are those in which the numerator is smaller than the denominator; others are *improper fractions.*

In order to add or subtract two fractions, their denominators must be changed to a common number. The numerators then can be added or subtracted and the resultant placed over the *common denominator.*

$$\frac{4}{7} + \frac{2}{5} = \frac{4(5)}{7(5)} + \frac{2(7)}{5(7)} = \frac{20}{35} + \frac{14}{35} = \frac{34}{35}$$

Note that conversion of fractions to a common denominator requires that both the numerator and denominator be multiplied by the same number. Of course, identical operations of either multiplication or division on both the numerator and denominator of a fraction leave the fraction mathematically unchanged.

Fractions are multiplied by multiplying both the numerators and the denominators. The product of the numerators is placed over the product of the denominators. The fraction is then reduced to lowest terms by division of numerator and denominator by identical quantitities.

$$\frac{4}{7} \times \frac{2}{5} = \frac{4 \times 2}{7 \times 5} = \frac{8}{35}$$

$$\frac{2}{5} \times \frac{3}{8} = \frac{2 \times 3}{5 \times 8} = \frac{6}{40} = \frac{6/2}{40/2} = \frac{3}{20}$$

Division of fractions can be regarded as a multiplication of the numerator by the inverse of the fraction in the denominator.

$$\frac{2/5}{3/7} = \frac{2}{5} \times \frac{7}{3} = \frac{14}{15}$$

The equivalence is achieved by multiplying both numerator and denominator by the inverse of the fraction in the denominator. This changes the denominator to 1.

$$\frac{\frac{2}{5}}{\frac{3}{7}} = \frac{\frac{2}{5} \times \frac{7}{3}}{\frac{3}{7} \times \frac{7}{3}} = \frac{\frac{14}{15}}{\frac{21}{21}} = \frac{\frac{14}{15}}{1} = \frac{14}{15}$$

4
DECIMALS

A decimal can be regarded as a fraction in which the denominator is an unexpressed power of 10. Addition and subtraction of decimals are accomplished in the same manner as similar operations using whole numbers, but the decimals must be placed in the proper column.

add: 5.46 + 130.21

$$\begin{array}{r} 5.46 \\ \underline{130.21} \\ 135.67 \end{array}$$

subtract: 130.21 − 5.46

$$\begin{array}{r} 130.21 \\ \underline{-5.46} \\ 124.75 \end{array}$$

Decimals are multiplied as if they were whole numbers. The location of the decimal point in the product is determined by adding the number of digits to the right of the decimal point in all of the numbers multiplied together. The product contains the same total number of digits to the right of the decimal point.

$$\begin{array}{r} 12.041 \\ \underline{\times 0.15} \\ 60205 \\ \underline{12041} \\ 1.80615 \end{array}$$

Division of decimals involves relocation of the decimal points in both the numerator and denominator prior to actually carrying out the division. The relocation is accomplished by multiplying by a power of 10 such that the denominator becomes a whole number. Then the division is carried out and the decimal point is located immediately above its position in the dividend.

$$\frac{84.42}{2.1} = \frac{84.42 \times 10}{2.1 \times 10} = \frac{844.2}{21}$$

$$\begin{array}{r} 40.2 \\ 21\overline{)844.2} \\ \underline{84} \\ 42 \\ \underline{42} \end{array}$$

5
EXPONENTS

Many numbers encountered in science are best expressed in terms of powers of 10. An exponent is a number that is a superscript following another number and indicates how many times the latter number must be multiplied by itself.

$$2^4 = 2 \cdot 2 \cdot 2 \cdot 2 = 16$$
$$10^3 = 10 \cdot 10 \cdot 10 = 1000$$

For both large and small numbers exponents of the base 10 are employed in order to make the number more compact and easier to handle. A number multiplied by 10^2 is equivalent to another number in which the decimal point is moved two places to the right.

$$3 \times 10^2 = 300$$

If a number is multiplied by 10^{-3} the decimal is moved three places to the left.

$$3467 \times 10^{-3} = 3.467$$

In expressing a number in exponential form the decimal point is moved to a new position so that the number value is between 1 and 10. This new number is then multiplied by the proper power of 10 to maintain its original value. The exponent is determined by counting the number of places that the decimal point is moved. If the decimal point is moved to the left the exponent is a positive number; if moved to the right the exponent is a negative number:

$$5243 = 5.243 \times 10^3$$
$$0.0467 = 4.67 \times 10^{-2}$$

The movement of the decimal point and the introduction of a power of 10 actually involves simultaneous multiplication and division by the same power of 10.

$$5243 = 5243 \times \frac{10^3}{10^3} = \frac{5243}{10^3} \times 10^3 = 5.243 \times 10^3$$

$$0.0467 = 0.0467 \times \frac{10^{-2}}{10^{-2}} = \frac{0.0467}{10^{-2}} \times 10^{-2} = 4.67 \times 10^{-2}$$

In multiplying exponential numbers first multiply the numerical portion of the number and then algebraically add the exponents of the powers of 10.

$$(4 \times 10^2)\ (2 \times 10^3) = (4 \times 2) \times (10^2 \times 10^3) = 8 \times 10^5$$

Division is carried out in the usual manner with the numerical portion of the number. The powers of 10 then are calculated.

$$\frac{4 \times 10^2}{2 \times 10^3} = \frac{4}{2} \times \frac{10^2}{10^3} = 2 \times \frac{10^2}{10^3} = 2 \times 10^{-1}$$

6
PROPORTIONALITY

It is said that x is *proportional* to y if x is related to or depends on y. If a certain percentage change in x produces an equal percentage change in y then x is *directly proportional* to y. For example, mass and volume are directly proportional to each other because the mass of a substance depends on the volume of the sample under consideration. Mathematically the direct proportionality between mass m and volume V is expressed as follows:

$$m \propto V \qquad m = kV$$

The symbol $\propto$ means "is directly proportional to" and is the *proportionality sign.* In the second equation the k is a *proportionality constant* and allows the replacement of the proportionality sign by an equal sign. In the case of mass and volume the k is equal to the density of the substance m/V.

The proportionality constant k in a direct proportion indicates the ratio of the two quantities. If one quantity is changed by a factor of 2, then the other quantity also must change by a factor of 2 in order to maintain the equality. For example, the mass and volume of a given substance always must be related such that if a larger mass is considered, the volume of the sample also must be larger. Thus for a given substance

$$\frac{m_1}{V_1} = \frac{m_2}{V_2} = \frac{m_3}{V_3} = k = \text{density}$$

where the subscripts refer to different quantities of the material.

An *inverse proportionality* indicates that as one variable x is increased by a certain percentage change the related variable y is decreased by the same percentage change. The inverse proportionality may be indicated as:

$$x \propto \frac{1}{y} \qquad x \propto y^{-1}$$

If a proportionality constant k is introduced the following equations result:

$$x = k\left(\frac{1}{y}\right) \qquad x = kx^{-1}$$

An alternate way of expressing the inverse relationship between the two variable results from rearranging the above equations:

$$xy = k$$

Such an expression indicates that whatever value of x is chosen, y must be such that the product of the two equal a constant k.

At constant temperature the pressure P of a gas and its volume V are inversely related. Therefore, it follows that the equalities listed below are correct. The subscripts refer to different experimental conditions for the same sample of a gas.

$$P_1V_1 = P_2V_2 = P_3V_3 = k$$

7
GRAPHS

The relationship between two properties can be conveniently represented by a graph in which the values of one property are plotted along the horizontal axis (*abscissa*) and the other along the vertical axis (*ordinate*).

In Figure 1 a graph of a direct proportion $y = kx$ is illustrated. The points define a straight line that goes through the *origin,* the point at which both y and x are zero. The slope of the line is equal to k and represents the ratio of the change in the quantity y per unit change in the quantity x.

In Figure 2 the straight line does not pass through the origin. The equation of the line is $y = kx + c$. The value of c is a constant that is equal to the value of y when x is zero, that is, the y intercept or the point on the ordinate when x is zero. The value k is still the slope of the line. The equation is said to indicate a linear relationship between x and y, but the two quantities are not directly proportional.

In Figure 3 the inverse proportionality $xy = k$ is graphically illustrated.

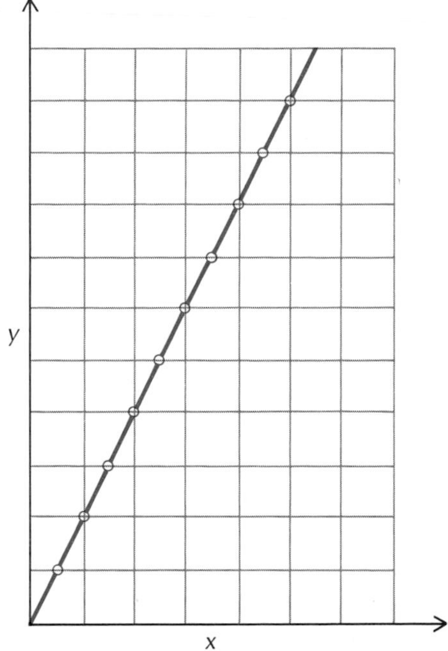

Figure 1

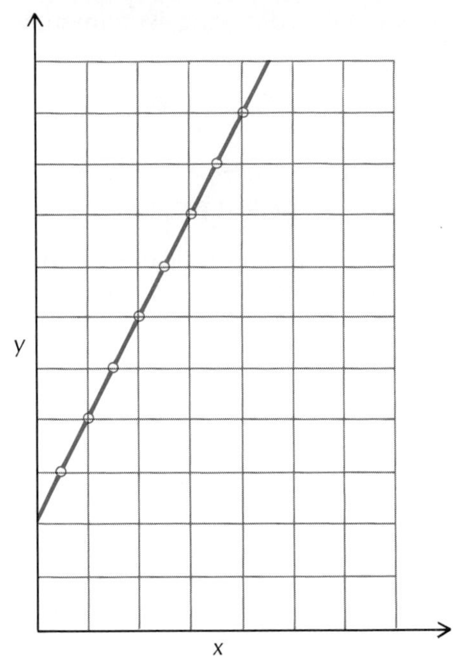

Figure 2

Note that at no point does the curve intersect the x or y axis. As either quantity approaches zero the other quantity approaches infinity. An inverse proportionality can be graphed to yield a straight line if one quantity is plotted vs the reciprocal of the other quantity, as illustrated in Figure 4. The labels of the points in Figures 3 and 4 are for identical values of x and y.

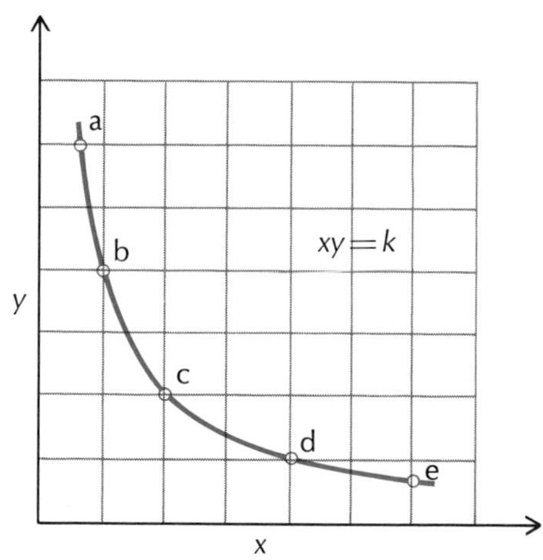

Figure 3

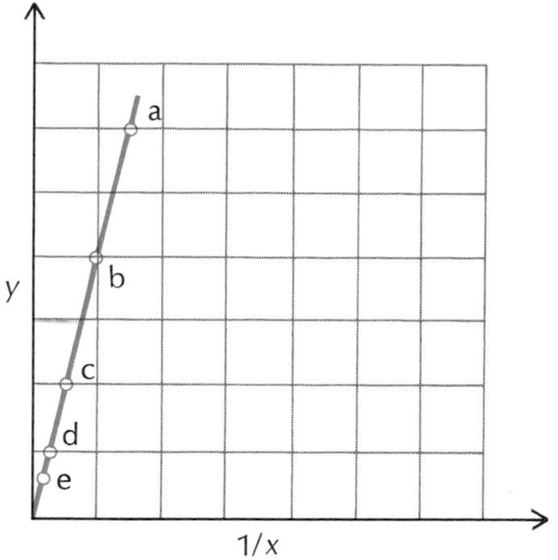

Figure 4

8
LOGARITHMS

The power to which the number 10 must be raised to equal a desired number is its *logarithm*. The log of 100 is 2, because $10^2 = 100$. Similarly, the log of 0.0001 is -4. Since most numbers are not integral powers of 10 the logarithms may be nonintegers. Table 1 gives logarithms of numbers between 1 and 10 to the number of decimal places needed in this text. In order to obtain the logarithm of 6.3, look for 6 in the first vertical column and then across to the column 0.3 to obtain the logarithm 0.799. For numbers greater than 10 or less than 1 write the number in exponential form so it is a number between 1 and 10 multiplied by a power of 10. The number 6300 is 6.3×10^3. Since the logarithm of a product $m \times n$ is

Table 1
Table of logarithms

	0.0	0.1	0.2	0.3	0.4	0.5	0.6	0.7	0.8	0.9
1	000	041	079	114	146	176	204	230	255	279
2	301	322	342	362	380	398	415	431	447	462
3	477	491	505	519	532	544	556	568	580	591
4	602	613	623	634	644	653	663	672	681	690
5	699	708	716	724	732	740	748	756	763	771
6	778	785	792	799	806	813	820	826	833	839
7	845	851	857	863	869	875	881	887	892	898
8	903	909	914	919	924	929	935	940	945	949
9	954	959	964	969	973	978	982	987	991	996

equal to the log of m plus the log of n, then

$$\begin{aligned}\log(6.3 \times 10^3) &= \log 6.3 + \log 10^3 \\ &= 0.799 + 3 \\ &= 3.799\end{aligned}$$

The only use of logarithms in this text is in the calculation of pH, which is the negative logarithm of the hydronium ion concentration. For a hydronium ion concentration of 0.00063 the pH is calculated as follows:

$$\begin{aligned}\text{pH} &= -\log(0.00063) \\ &= -\log(6.3 \times 10^{-4}) \\ &= -(\log 6.3 + \log 10^{-4}) \\ &= -(0.799 - 4) \\ &= -(-3.201) \\ &= +3.201\end{aligned}$$

9
QUADRATIC EQUATIONS

An equation of the general form

$$ax^2 + bx + c = 0$$

is called a quadratic equation. The solution of the equation is given by

$$x = \frac{-b \pm \sqrt{b^2 - 4ac}}{2a}$$

The $\pm$ sign indicates that two roots or values of x are possible. In chemical problems usually one of the roots is clearly inadmissible as a solution. One such case is a negative value when only a positive value is physically possible. The other case commonly encountered is a value that is larger than is possible, considering the quantities involved in the chemistry. In either case common sense allows one to choose the correct root.

APPENDIX 2

NOMENCLATURE OF INORGANIC COMPOUNDS

1 BINARY COMPOUNDS

By the demands of popular usage certain compounds are known by their *common* or *trivial* name. Such names do not convey the composition of the compound and in general systematic nomenclature is preferred. Examples of common names are water for H_2O and ammonia for NH_3.

The names of ionic *binary* (containing two elements) compounds are derived from the name of the metal followed by a stem of the nonmetal with the addition of -*ide* as the ending:

$\mathbf{CaCl_2}$	calcium chloride
$\mathbf{Na_2O}$	sodium oxide
ZnS	zinc sulfide

Identical rules are used for binary covalent compounds such as hydrogen chloride (HCl). The -*ide* ending is placed on the more electronegative element. These simple considerations do not apply when more than one possible compound can be formed between two elements. In the examples chosen there is no ambiguity in the name.

Many metals can exist in several oxidation states and combine with nonmetals to give a number of compounds. There are, for example, two chlorides of iron, $FeCl_2$ and $FeCl_3$. In order to distinguish the two the ending *-ous* is used for the compound in which the metal is in the lower oxidation state. The ending *-ic* is used for the metal in the higher oxidation state. Therefore, $FeCl_2$ is ferrous chloride and $FeCl_3$ is ferric chloride. This system of nomenclature is limited because there are some pairs of elements that form more than two compounds. Furthermore, the system requires that the individual know that ferrous refers specifically to iron in the 2+ oxidation state. In the case of copper the 2+ oxidation state is in the higher oxidation state and is designated *cupric*.

The use of Roman numerals after the English name of the metal to indicate the oxidation state is now recommended. By this system $FeCl_2$ becomes iron(II) chloride and $FeCl_3$ is iron(III) chloride.

2
TERNARY COMPOUNDS

In many ternary compounds two of the elements are covalently bonded to yield a polyatomic anion or cation, as in the case of the nitrate ion NO_3^- and the ammonium ion NH_4^+. Names are specifically given to these species because they behave as integral units in some chemical reactions. The names of some compounds containing the polyatomic ions are given below:

$\mathbf{NH_4Cl}$	ammonium chloride
$\mathbf{NaNO_3}$	sodium nitrate
$\mathbf{NH_4NO_3}$	ammonium nitrate

A list of the names of common polyatomic anions is given in Table 1.

Table 1
Polyatomic anions

$\mathbf{C_2H_3O_2^-}$	acetate	$\mathbf{ClO^-}$	hypochlorite
$\mathbf{BO_3^{3-}}$	borate	$\mathbf{OH^-}$	hydroxide
$\mathbf{CO_3^{2-}}$	carbonate	$\mathbf{NO_3^-}$	nitrate
$\mathbf{ClO_3^-}$	chlorate	$\mathbf{NO_2^-}$	nitrite
$\mathbf{ClO_2^-}$	chlorite	$\mathbf{ClO_4^-}$	perchlorate
$\mathbf{CrO_4^-}$	chromate	$\mathbf{MnO_4^-}$	permanganate
$\mathbf{CN^-}$	cyanide	$\mathbf{PO_4^{3-}}$	phosphate
$\mathbf{Cr_2O_7^{2-}}$	dichromate	$\mathbf{SO_4^{2-}}$	sulfate
		$\mathbf{SO_3^{2-}}$	sulfite

3
OXYACIDS

When there are two common oxyacids, the one containing the nonmetal in the lower oxidation state is named with the *-ous* ending and the other is named with the *-ic* ending. Thus H_2SO_4 is sulfuric acid and H_2SO_3 is sulfurous acid. When there are more than two oxyacids the prefix *hypo-* and *per-* are used. The prefix hypo- refers to the compound that contains the nonmetal in an oxidation state lower than that of the *-ous* acid; the prefix *per-* is for the compound in the oxidation state higher than that of the *-ic* acid. The four oxyacids of chlorine are named

$HClO$ hypochlorous acid
$HClO_2$ chlorous acid
$HClO_3$ chloric acid
$HClO_4$ perchloric acid

APPENDIX

3

ANSWERS TO SELECTED PROBLEMS

Chapter 1

4. **a.** 2.5 cm **b.** 10^{-4} kg **c.** 0.25 l **d.** 2×10^6 mg **e.** 5×10^5 cm **f.** 3×10^3 ml
5. 1 liter or 10^{-3} m^3
6. 0.5 furlongs per fortnight
7. 75°C
8. 0°U
9. 3 g/cm^3
11. **a.** 65 cal **b.** 25 cal **c.** 625 cal
12. 5×10^3 cal
13. 1761.8°F
14. −178.6°F
16. $112,000

Chapter 2

1. 160 lb/in.^{2}
4. 516.8 cm
5. 1250 ml
6. 1 liter
7. 2 atm
8. 0.8 liter
9. a. 0.1 mole **b.** 0.075 mole **c.** 1 mole
10. 12.04×10^{23}
11. 44 amu
12. 64 g
13. the gas is of higher molecular mass (34 g/mole)
14. 0.4 atm, P of H_2 is unaffected, 3.4 atm
16. a. H_2 **b.** H_2 **c.** H_2 **d.** same for each
17. a. H_2 **b.** O_2 **c.** same for each

Chapter 3

3. 2500 cal
7. B
8. 100°C, 120°C
9. equal to
12. 730 cal
16. more than
17. 225 g/mole
18. 8400 cal/mole
20. 92 g/mole
23. 5.3 cal/mole-deg

Chapter 4

1. 28 amu
2. Fe_2O_3
3. HgO
4. AsH_3
5. As_4, AsH_3
6. the ratio of the atomic mass of *A* to *Y* is 9 to 1

Chapter 5

2. 1.16×10^{-23} g, 1.38×10^4 coulombs/g
4. less than by a factor of $\frac{1}{2}$
6. a. 4 **b.** 2 **c.** 8 **d.** 2 **e.** 6
11. O, Se, Mg, Xe, Ne, P
23. by its emission spectrum
26. 9 orbitals, 18 electrons, 50 electrons

Chapter 6

2. a. ionic **b.** ionic **c.** covalent **d.** covalent **e.** covalent
4. BrF

5. a. sp^2 trigonal planar **b.** sp^3 pyramidal **c.** sp^3 tetrahedral **d.** sp linear **e.** p linear **f.** sp^3 angular
7. NF_3 contains one more electron pair
8. NH_4^+ will be tetrahedral and NH_2^- will be angular
10. CCl_4 is nonpolar and NF_3 is polar

Chapter 7

1. Kr, HF, and SO_2
3. smaller than
5. a. yes **b.** yes **c.** no
7. Kr
9. 21.1 cal/mole-deg
10. 8400 cal/mole
12. the larger value for alcohol indicates larger intermolecular forces

Chapter 8

1. a. 1 mole/liter **b.** 2 mole/liter
2. 0.22, 0.78
4. heat would have to be supplied to maintain the temperature
5. increase of pressure increases solubility
6. no, for flavoring
7. exothermic
9. $-4.65°$
10. 100.26°
11. a. same **b.** less than **c.** same as
13. 120 g/mole

Chapter 9

4. the rate of reaction is slow at normal temperatures
5. doubled, quadrupled
6. rate $= k[A][B]$, $A + B \longrightarrow M$ is rate determining followed by $M + C \longrightarrow X + Y$ in a fast reaction
7. catalysis by enzymes
8. rate $= k[A]^2[B]^2$, 400
12. $K = \dfrac{[SO_3]^2}{[SO_2]^2[O_2]}$
13. at low temperature, endothermic
14. the density of diamond is larger than that of graphite
16. 4 moles
17. 9

Chapter 10

4. H_2SO_4 is acid, $HC_2H_3O_2$ is base, HSO_4^- is conjugate base, $H_2C_2H_3O_2^+$ is conjugate acid
5. smaller

8. pH values are 3, 5, 7.4, and 8.7
12. 4.5×10^{-4}
13. $5.7 \times 10^{-5} M$
14. $1.3 \times 10^{-3} M$
15. $H_3O^+ = 10^{-1} M$, $HC_2H_3O_2 = 10^{-1} M$, $OH^- = 10^{-13} M$, $C_2H_3O_2^- = 1.8 \times 10^{-5} M$
16. $K_h = 5.5 \times 10^{-11}$
17. pH = 4
18. $K_h = 10^{-2}$

Chapter 11

2. **a.** N = 3 **b.** S = 4 **c.** I = 5 **d.** Cr = 4 **e.** P = −3 **f.** Fe = O **g.** Cl = 3 **h.** Sn = 4 **i.** Pb = 2 **j.** N = 5
4. oxidizing agents are $CoCl_2$, H_2, SO_2, Br_2, NO_3^-, H_2O_2
7. all potentials would be more negative by 0.77 V
12. Pb^{2+}
13. **a.** F_2 **b.** Na^+ **c.** Mg **d.** F^-
16. six

Chapter 12

4. lower ionization potential, larger atomic and ionic radius
6. a and c
7. Cs, Li
11. b

18.
```
     H    H
     |    |
  H—Ge—Ge—H
     |    |
     H    H
```

Chapter 13

1. $H^- + H_2O \longrightarrow H_2 + OH^-$
3. six
8. b and c
9. $HClO_4$, $HClO_3$, HIO_4
11. hydrogen bonding
12. −2 and +2
13. lower molecular mass
14. three equivalent resonance contributors can be written
15. Na_2SO_4
16. H_2SeO_4
19. HNO_3
20. tetrahedral

Chapter 14

2. 4.90, 0, 4.90, 3.87, 2.84
3. Sc^{3+}, Ag^+, Hg^{2+}

4. they each contain an available electron pair
8. *dsp*², O
9. $CdBr_4^{2-}$ sp^3 tetrahedral, $Au(CN)_2^-$ sp linear
10. $6s^25d^14f^1$, no
14. the *trans* isomer does not have a dipole moment
15. there are only the *cis* and *trans* isomers

Chapter 15

1. a. $^{21}_{9}F \longrightarrow {}^{0}_{-1}e \quad {}^{21}_{10}Ne$ **b.** $^{220}_{86}Rn \longrightarrow {}^{4}_{2}He + {}^{216}_{84}Po$ **c.** $^{56}_{28}Ni \longrightarrow {}^{56}_{27}Co$
2. 0.25 g
3. $^{10}_{5}B$
4. $^{247}_{99}Es$
5. $^{2}_{1}H$
7. 27,850 years

Chapter 16

2. a. 3-methylheptane **b.** 2-iodohexane **c.** 1,1,2-trichloroethane **d.** methylcyclopropane **e.** 3-methylpentane **f.** must indicate *cis* or *trans*
4. 1,1-dimethylcyclobutane, *cis-* and *trans*-1,2-dimethylcyclobutane and *cis-* and *trans*-1,3-dimethylcyclobutane
5. cyclopentane, methylcyclobutane, the three isomeric dimethylcyclopropanes, and ethylcyclopropane
7. 43.3 percent 2-chloro-2,3-dimethylbutane, 56.7 percent 1-chloro-2,3-dimethylbutane, 25 percent 1-chloropentane, 50 percent 2-chloropentane, and 25 percent 3-chloropentane
9. 26 kcal/mole
12. C_nH_{2n-2}
13. C_8H_8
15. 2-bromobutane and sodium

Chapter 17

2. a. 2-methyl-1-butene **b.** 3-methyl-1-butyne **c.** 1-butyne **d.** 3,3-dimethyl-1-pentene
3. a. cyclopropane and propene **b.** cyclopropene, propadiene, and propyne
4. a. KOH in alcohol **b.** a followed by addition of Br_2 **c.** b followed by KOH in alcohol **d.** c followed by addition of HCl **e.** Na
5. a. add Br_2 **b.** add Ag^+ **c.** add Br_2
6. cyclopentene
7. 1-pentyne and 3-methyl-1-butyne
8. cyclopentane
9. a. propene **b.** 2,3-dimethyl-2-butene **c.** 2-methyl-2-butene **d.** *cis-* or *trans*-2-butene
10. 1-bromopentane
11. 3-methylcyclopentene
12. a. 230 nm **b.** 245 nm **c.** 265 nm

Chapter 18

1. a. *m*-nitrotoluene **b.** 1,3,5-trichlorobenzene **c.** *p*-methylphenol **e.** isopropylbenzene

4. 2,3

5. 10

7. 1,2-diethylbenzene, 1-methyl-2-propylbenzene, and 1-methyl-2-isopropylbenzene

9. a. chlorinate followed by nitration **c.** Friedel-Crafts with CH_3Cl followed by $KMnO_4$ oxidation

11. a. add Br_2 **b.** test solubility in NaOH solution **c.** test solubility in HCl solution **d.** add Ag^+

Chapter 19

3. a. B_2H_6 followed by H_2O_2 and OH^- **b.** H_3O^+ **c.** OH^- in water **d.** $ZnCl_2$ and HCl

6. b. dehydration to 1-pentene, add Br_2, and then dehydrohalogenate with OH^- in alcohol

7. a. $C_nH_{2n}O$ **b.** $C_nH_{2n-2}O$ **c.** $C_nH_{2n-2}O$ **d.** $C_nH_{2n-2}O$ **e.** $C_nH_{2n}O$ **f.** $C_nH_{2n+2}O_2$

8. a. Na **b.** NaOH **c.** iodoform test **u.** iodoform test **e.** Lucas reagent

9. $CH_2{=}CH{-}O{-}CH_3$

10. 2-methyl-2-propanol

11. 2-pentanol and 3-methyl-2-pentanol

12. cyclopentanol

13. 1-pentanol

14. either $CH_3CH_2CH_2CH_2CH_2NH_2$ or 1-pentene will be formed

Chapter 20

3. a. $Na_2Cr_2O_7$ **b.** $LiAlH_4$ **c.** OH^- in water

4. a. Na **b.** iodoform test **c.** Tollens' reagent **d.** iodoform test **e.** iodoform test **f.** phenylhydrazine

5. 2-pentanol and 3-methyl-2-butanol

6. 3-pentanone

7. cyclopentanol

8. cyclopentanone

9. a. 2-methyl-2-butanol **b.** 1-propanol **c.** 2-butanol

12. $(CH_3)_2CDCHO$

13. exchange of oxygen occurs via reversible formation of the hydrate

Chapter 21

2. butanoic acid, methylpropanoic acid, propyl methanoate, isopropyl methanoate, ethyl ethanoate, and methyl propanoate

4. CF_3CO_2H due to stronger electron withdrawal by the more electronegative fluorine

7. form methyl propanoate

8. propyl propanoate
9. add methyl Grignard reagent to ethyl propanoate
12. two

13. $HO\!\left(\overset{O}{\overset{\|}{C}}CH_2CH_2\overset{O}{\overset{\|}{C}}-O-CH_2CH_2CH_2-O\right)_{\!n}H$

Chapter 22

1. $C_nH_{2n+3}N$, $C_nH_{2n+4}N_2$
2. four
3. hydrogen bonding
4. the methyl group is electron releasing and will increase the basicity
8. the ammonium salt that is less volatile is formed
13. the electron pair on nitrogen is part of a delocalized π system around the aromatic ring
14. electron pair of nitrogen can form hydrogen bonds with water
15. *p*-methylaniline

Chapter 23

3. a, c, d, f, and h
4. eight
5. two centers, three compounds, two are optically active
6. yes
7. only d
8. yes
9. L-2-butanol
10. multiple S_N2 displacements by I^- leading to loss of specific configuration
12. +20°, +10°.

Chapter 24

2. the basic amino acids
3. a. 22.4 ml **b.** 44.8 ml
6. 210
7. three
8. gly-ala-gly-leu-val
9. asp-arg-val-tyr-val-his-pro-phe-his-leu
11. proline and hydroxyproline

Chapter 25

4. allose and galactose
5. D-arabinose
6. D-gulose and D-idose
7. D-lyxose
8. D-idose
9. D-allose, D-altrose, D-gulose, and D-idose

Chapter 26

2. oils
3. oxidative cleavage of the double bond at the end of the molecule opposite the carboxyl group
4. 13

Chapter 27

1. RNA contains uracil instead of thymine
2. DNA is double stranded and directs RNA
5. glutamic acid is an acidic amino acid; the additional polar group should cause changes in the three-dimensional structure of neighboring portions of the molecule.
6. Compare GAA for glutamic acid vs GUA for valine, also GAA vs GUU and GAG vs. GUG

Chapter 28

1. alanine and ketosuccinic acid will be formed
2. α-ketoglutaric acid
4. formation of fructose-6-phosphate and then along glucose pathway
5. carbon atom number 4
7. hydrolysis to yield maltose occurs in the mouth
8. no
10. phenylacetic acid

INDEX

Designed by Rita Naughton
Set in Zenith
Composed by Graphic Services, Inc.
Printed and bound by Von Hoffman Press, Inc.
Illustrated by BMA Associates, Inc.
HARPER & ROW, PUBLISHERS, INC.

70 71 72 73 7 6 5 4 3 2

TABLE OF RELATIVE ATOMIC MASSES

element	*symbol*	*atomic number*	*atomic mass*
Actinium	**Ac**	89	. . . .
Aluminum	**Al**	13	26.9815
Americium	**Am**	95	. . . .
Antimony	**Sb**	51	121.75
Argon	**Ar**	18	39.948
Arsenic	**As**	33	74.9216
Astatine	**At**	85	. . . .
Barium	**Ba**	56	137.34
Berkelium	**Bk**	97	. . . .
Beryllium	**Be**	4	9.0122
Bismuth	**Bi**	83	208.980
Boron	**B**	5	10.811
Bromine	**Br**	35	79.904
Cadmium	**Cd**	48	112.40
Calcium	**Ca**	20	40.08
Californium	**Cf**	98	. . . .
Carbon	**C**	6	12.01115
Cerium	**Ce**	58	140.12
Cesium	**Cs**	55	132.905
Chlorine	**Cl**	17	35.453
Chromium	**Cr**	24	51.996
Cobalt	**Co**	27	58.9332
Copper	**Cu**	29	63.546
Curium	**Cm**	96	. . . .
Dysprosium	**Dy**	66	162.50
Einsteinium	**Es**	99	. . . .
Erbium	**Er**	68	167.26
Europium	**Eu**	63	151.96
Fermium	**Fm**	100	. . . .
Fluorine	**F**	9	18.9984
Francium	**Fr**	87	. . . .
Gadolinium	**Gd**	64	157.25
Gallium	**Ga**	31	69.72
Germanium	**Ge**	32	72.59
Gold	**Au**	79	196.967
Hafnium	**Hf**	72	178.49
Helium	**He**	2	4.0026
Holmium	**Ho**	67	164.930
Hydrogen	**H**	1	1.00797
Indium	**In**	49	114.82
Iodine	**I**	53	126.9044
Iridium	**Ir**	77	192.2
Iron	**Fe**	26	55.847
Krypton	**Kr**	36	83.80
Lanthanum	**La**	57	138.91
Lawrencium	**Lw**	103	. . . .
Lead	**Pb**	82	207.19
Lithium	**Li**	3	6.939
Lutetium	**Lu**	71	174.97
Magnesium	**Mg**	12	24.312
Manganese	**Mn**	25	54.9380
Mendelevium	**Md**	101	. . . .